ANNALS OF THE NEW YORK ACADEMY OF SCIENCES

Volume 871

OTOLITH FUNCTION IN SPATIAL ORIENTATION AND MOVEMENT

Edited by Bernard Cohen and Bernhard J. M. Hess

The New York Academy of Sciences
New York, New York
1999

Copyright © 1999 by the New York Academy of Sciences. All rights reserved. Under the provisions of the United States Copyright Act of 1976, individual readers of the Annals *are permitted to make fair use of the material in them for teaching or research. Permission is granted to quote from the* Annals *provided that the customary acknowledgment is made of the source. Material in the* Annals *may be republished only by permission of the Academy. Address inquiries to the Executive Editor at the New York Academy of Sciences.*

Copying fees: *For each copy of an article made beyond the free copying permitted under Section 107 or 108 of the 1976 Copyright Act, a fee should be paid through the Copyright Clearance Center, Inc., 222 Rosewood Drive, Danvers, MA 01923. The fee for copying an article is $3.00 for nonacademic use; for use in the classroom, it is $0.07 per page.*

∞*The paper used in this publication meets the minimum requirements of American National Standard for Information Sciences—Permanence of Paper for Printed Library Materials, ANSI Z39.48-1984.*

Cover art (*softcover only*): The cover shows gray squirrel otoconia. Scanning electron micrographs supplied by Anna Lysakowski.
We thank Dr. Hansjürg Scherberger for supplying the photograph of Volker Henn that appears with the Tribute.

Library of Congress Cataloging-in-Publication Data

Otolith function in spatial orientation and movement / edited by Bernard Cohen and Bernhard J. M. Hess.
    p. cm. — (Annals of the New York Academy of Sciences; v. 871)
    Includes bibliographical references.
    ISBN 1-57331-217-7 (cloth : alk. paper). — ISBN 1-57331-218-5 (paper : alk. paper)
    1. Otoliths Congresses. I. Cohen, Bernard, 1929– . II. Hess, Bernhard J. M. III. Series.
Q11.N5 vol. 871
[QP471]
500 s — dc21
[573.8'7]
                                                                      99-24766
                                                                       CIP

CDP/PCP

Printed in the United States of America
ISBN 1-57331-217-7 (cloth)
ISBN 1-57331-218-5 (paper)
ISSN 0077-8923

ANNALS OF THE NEW YORK ACADEMY OF SCIENCES

Volume 871

EDITORIAL STAFF

Managing Editor
JUSTINE CULLINAN

Associate Editor
TRUMBULL ROGERS

The New York Academy of Sciences
2 East 63rd Street
New York, New York 10021

THE NEW YORK ACADEMY OF SCIENCES
(Founded in 1817)

BOARD OF GOVERNORS, October 1998–September 1999

ELEANOR BAUM, *Chairman of the Board*
BILL GREEN, *Vice Chairman of the Board*
RODNEY W. NICHOLS, *President and CEO* [ex officio]

Honorary Life Governors
WILLIAM T. GOLDEN   JOSHUA LEDERBERG
JOHN T. MORGAN, *Treasurer*

Governors

D. ALLAN BROMLEY	LAWRENCE B. BUTTENWIESER	PRAVEEN CHAUDHARI
JOHN H. GIBBONS	RONALD L. GRAHAM	HENRY M. GREENBERG
ROBERT G. LAHITA	MARTIN L. LEIBOWITZ	JACQUELINE LEO
WILLIAM J. McDONOUGH	KATHLEEN P. MULLINIX	SANDRA PANEM
CHARLES RAMOND	SARA LEE SCHUPF	JAMES H. SIMONS
	TORSTEN N. WIESEL	

RICHARD A. RIFKIND, *Past Chairman of the Board*

HELENE L. KAPLAN, *Counsel* [ex officio] PETER KOHN, *Secretary* [ex officio]

OTOLITH FUNCTION
IN SPATIAL ORIENTATION AND
MOVEMENT

ANNALS OF THE NEW YORK ACADEMY OF SCIENCES

Volume 871
May 28, 1999

OTOLITH FUNCTION IN SPATIAL ORIENTATION AND MOVEMENT[a]

Editors
BERNARD COHEN and BERNHARD J. M. HESS

CONTENTS

Foreword. *By* BERNARD COHEN and BERNHARD J. M. HESS................ xi

Tribute to Volker Henn. *By* BERNARD COHEN xiii

Part 1: Basic Mechanisms of Otolith Processing

Otolith Ocular Reflex Function of the Tangential Nucleus in Teleost Fish.
  By HIROSHI SUWA, EDWIN GILLAND, and ROBERT BAKER 1

Stimulus Processing by Type II Hair Cells in the Mouse Utricle.
  By JEFFREY R. HOLT, MELISSA A. VOLLRATH, and RUTH ANNE EATOCK... 15

The Planes of the Utricular and Saccular Maculae of the Guinea Pig.
  By I. S. CURTHOYS, G. A. BETTS, A. M. BURGESS, H. G. MACDOUGALL,
  A. D. CARTWRIGHT, and G. M. HALMAGYI 27

Miniature EPSPs and Sensory Encoding in the Primary Afferents of the
  Vestibular Lagena of the Toadfish, *Opsanus tau*. *By* RACHEL LOCKE,
  JEAN VAUTRIN, and STEPHEN HIGHSTEIN 35

Part 2: Otolith Processing in the Vestibular Nuclei and Vestibulocerebellum

A Review of Otolith Pathways to Brainstem and Cerebellum.
  By J. A. BÜTTNER-ENNEVER 51

Signal Processing Related to the Vestibulo-Ocular Reflex during Combined
  Angular Rotation and Linear Translation of the Head.
  By ROBERT A. MCCREA and CHIJU CHEN-HUANG................... 65

[a]This volume contains the papers from a conference entitled **Otolith Function in Spatial Orientation and Movement: Symposium in Memory of Volker Henn**, which was held by the Zurich University Hospital and the Betty and David Koetser Foundation for Brain Research in Zurich, Switzerland, on May 18–19, 1998.

Otolith Processing in the Deep Cerebellar Nuclei. *By* U. BÜTTNER,
S. GLASAUER, L. GLONTI, J. F. KLEINE, and C. SIEBOLD 81

Control of Spatial Orientation of the Angular Vestibulo-Ocular Reflex by the
Nodulus and Uvula of the Vestibulocerebellum. *By* BORIS M. SHELIGA,
SERGEI B. YAKUSHIN, ADAM SILVERS, THEODORE RAPHAN, and
BERNARD COHEN 94

Part 3: The Linear Vestibulo-Ocular Reflex and Otolith/Canal Interactions

Characteristics of the VOR in Response to Linear Acceleration.
By GARY D. PAIGE and SCOTT H. SEIDMAN 123

Functional Organization of Primate Translational Vestibulo-Ocular Reflexes
and Effects of Unilateral Labyrinthectomy. *By* DORA E. ANGELAKI,
M. QUINN MCHENRY, SHAWN D. NEWLANDS, and
J. DAVID DICKMAN 136

Inertial Processing of Vestibulo-Ocular Signals. *By* BERNHARD J. M. HESS
and DORA E. ANGELAKI 148

Cross-Striolar and Commissural Inhibition in the Otolith System.
By Y. UCHINO, H. SATO, K. KUSHIRO, M. ZAKIR, M. IMAGAWA,
Y. OGAWA, M. KATSUTA, and N. ISU 162

Human Ocular Counterrolling During Roll-Tilt and Centrifugation.
By H. G. MACDOUGALL, I. S. CURTHOYS, G. A. BETTS, A. M. BURGESS,
and G. M. HALMAGYI 173

Canal and Otolith Afferent Activity Underlying Eye Velocity Responses to
Pitching While Rotating. *By* T. RAPHAN, M. DAI, J. MARUTA, W. WAESPE,
V. HENN, J.-I. SUZUKI, and B. COHEN 181

Part 4: Clinical Implications of Otolith Function

Clinical Testing of Otolith Function. *By* G. M. HALMAGYI and
I. S. CURTHOYS 195

Directional Abnormalities of Vestibular and Optokinetic Responses in
Cerebellar Disease. *By* MARK F. WALKER and DAVID S. ZEE 205

Assessing Otolith Function by the Subjective Visual Vertical.
By ANDREAS BÖHMER and FRED MAST 221

Horizontal Linear Vestibulo-Ocular Reflex Testing in Patients with
Peripheral Vestibular Disorders. *By* THOMAS LEMPERT,
MICHAEL A. GRESTY, and ADOLFO M. BRONSTEIN 232

Part 5: Cortical Processing of Otolith Information

Vestibular–Pursuit Interactions: Gaze-Velocity and Target-Velocity Signals in the Monkey Frontal Eye Fields. *By* Kikuro Fukushima, Junko Fukushima, and Toshikazu Sato 248

Short-Latency Visual Stabilization Mechanisms that Help to Compensate for Translational Disturbances of Gaze. *By* F. A. Miles 260

Linear Vestibular Self-Motion Signals in Monkey Medial Superior Temporal Area. *By* F. Bremmer, M. Kubischik, M. Pekel, M. Lappe, and K.-P. Hoffmann. 272

The Contributions of Vestibular Signals to the Representations of Space in the Posterior Parietal Cortex. *By* Richard A. Andersen, Krishna V. Shenoy, Lawrence H. Snyder, David C. Bradley, and James A. Crowell 282

The Vestibular Cortex: Its Locations, Functions, and Disorders. *By* Thomas Brandt and Marianne Dieterich 293

Part 6: Spatial Orientation

Cortical Areas Activated by Bilateral Galvanic Vestibular Stimulation. *By* Elie Lobel, Justus F. Kleine, Anne Leroy-Willig, Pierre-François Van de Moortele, Denis Le Bihan, Otto-Joachim Grüsser, and Alain Berthoz 313

The Interaction of Otolith and Proprioceptive Information in the Perception of Verticality: The Effects of Labyrinthine and CNS Disease. *By* Adolfo M. Bronstein 324

The Role of the Otoliths in Perception of the Vertical and in Path Integration. *By* Horst Mittelstaedt 334

Replication of Passive Whole-Body Linear Displacements from Inertial Cues: Facts and Mechanisms. *By* R. Grasso, S. Glasauer, P. Georges-François, and I. Israël 345

Part 7: The Otoliths and Space

Artificial Gravity Considerations for a Mars Exploration Mission. *By* Laurence R. Young. 367

Poster Papers

The Role of Somatosensory Input for the Perception of Verticality. *By* Dimitri Anastasopoulos, Adolfo Bronstein, Thomas Haslwanter, Michael Fetter, and Johannes Dichgans 379

Otolith Vestibular-Evoked Potentials in Humans: Intensity, Direction of Acceleration (Z+, Z–), and BESA Modeling of Generators. *By* P. M. BAUDONNIÈRE, S. BELKHENCHIR, J. C. LEPECQ, and S. MERTZ...384

Measuring the Otolith–Ocular Response by Means of Unilatral Radial Acceleration. *By* A. H. CLARKE, A. ENGELHORN, CH. HAMANN, and U. SCHÖNFELD 387

Saccular Dysfunction in Ménière's Patients: A Vestibular-Evoked Myogenic Potential Study. *By* CATHERINE DE WAELE, PATRICE TRAN BA HUY, JEAN-PIERRE DIARD, GEORGES FREYSS, and PIERRE-PAUL VIDAL........392

VOR Gain Modulation in the Monkey Due to Convergence of Otolith and Semicircular Canal Afferences During Eccentric Sinusoidal Rotation. *By* L. FUHRY, J. NEDVIDEK, C. HABURCAKOVA, S. GLASAUER, G. BROZEK, and U. BÜTTNER..398

An Alternative Approach to the Central Processing of Canal and Otolith Signals. *By* ANDREA M. GREEN and HENRIETTA L. GALIANA 402

Otolith Signal Processing and Motion Sickness. *By* ERIC GROEN, JELTE BOS, BERND DE GRAAF, and WILLEM BLES 406

Otolith–Canal Interaction During Pitch While Rotating. *By* T. HASLWANTER, R. JAEGER, and M. FETTER 410

Phase Adaptation of the Linear Vestibulo-Ocular Reflex. *By* S. HEGEMANN, M. J. SHELHAMER, and D. S. ZEE 414

Separation Between On- and Off-Center Passive Motion in Darkness. *By* I. ISRAËL and S. GLASAUER 417

Development of Synaptic Innervation in the Rodent Utricle. *By* ANNA LYSAKOWSKI 422

Human Gaze Stabilization for Voluntary Off-Centric Head Rotations. *By* W. P. MEDENDORP, B. J. BAKKER, J. A. M. VAN GISBERGEN, and C. C. A. M. GIELEN 426

A Simple Model of Vestibular Canal–Otolith Signal Fusion. *By* THOMAS MERGNER and STEFAN GLASAUER.....................430

Centrifugal Force Affects Perception but not Nystagmus in Passive Rotation. *By* MARIE-LUISE MITTELSTAEDT and WILLI JENSEN..................435

Oculomotor, Postural, and Perceptual Asymmetries Associated with a Common Cause: Craniofacial Asymmetries and Asymmetries in Vestibular Organ Anatomy. *By* D. ROUSIE, J. C. HACHE, P. PELLERIN, J. P. DEROUBAIX, P. VAN TICHELEN, and A. BERTHOZ................439

Role of Otoliths in Spatial Orientation During Passive Travel in a Curve.
By I. SIEGLER, G. REYMOND, and P. LEBOUCHER................... 447

Comparison of Tilt Estimates Based on Time Settings, Saccadic Pointing,
and Verbal Reports. *By* A. D. VAN BEUZEKOM and
J. A. M. VAN GISBERGEN....................................... 451

Vestibular Projections in the Human Cortex. *By* P. P. VIDAL, C. DE WAELE,
P. M. BAUDONNIÈRE, J. C. LEPECQ, and P. TRAN BA HUY 455

Spatial Properties of Otolith Units Recorded in the Vestibular Nuclei.
By SERGEI B. YAKUSHIN, THEODORE RAPHAN, and BERNARD COHEN. . . . 458

Index of Contributors 463

Financial assistance was received from:

- BETTY AND DAVID KOETSER FOUNDATION, ZURICH, SWITZERLAND
- DEPARTMENT OF NEUROLOGY, UNIVERSITY HOSPITAL ZURICH

The New York Academy of Sciences believes it has a responsibility to provide an open forum for discussion of scientific questions. The positions taken by the participants in the reported conferences are their own and not necessarily those of the Academy. The Academy has no intent to influence legislation by providing such forums.

Foreword

BERNARD COHEN[a] AND BERNHARD M. HESS[b]

[a]*Department of Neurology, Mount Sinai School of Medicine, New York, New York 10029-6574, USA*

[b]*Department of Neurology, University Hospital Zürich, CH-8091 Zürich, Switzerland*

The otolith organs, the saccule and utricle, sense the linear acceleration of gravity, translational movements, and the centripetal acceleration generated during turning. They initiate reflexes that stabilize upright posture and provide critical information about the orientation of the head and body in space. They also have a widespread influence on motor responses and perception of the external world. Powerful otolith–ocular and otolith–collic reflexes are induced if the head is moved at the right frequencies and with the appropriate visual set. Despite this, there is a relative dearth of knowledge about central otolith processing. This volume grew from a conference, held in Zürich, Switzerland, in May 1998, to fill this need. It summarizes knowledge of static and dynamic processing of otolith activity in the labyrinth and provides new information about otolith processing through the vestibular nuclei and vestibulocerebellum. It also addresses how otolith information is processed in the cortex and how it can be used to diagnose and understand vestibular disease. The focus was on new research in this important and emerging area.

The conference was initiated by the late Volker Henn, a leader in the field of oculomotor and vestibular research. His sudden death in December of 1997 terminated a bright career and cost us a treasured friend. We dedicate this volume to his memory.

We thank the Betty and David Koetser Foundation for Brain Research of Zürich, which sponsored the conference. We also thank Professor Klaus Hess and the Department of Neurology of the University of Zürich for providing support as well as the opportunity to present this material at Volker Henn's home university. We are grateful to Dora Angelaki, Alain Berthoz, Thomas Brandt, Stephen M. Highstein, and Theodore Raphan for giving expert advice about the material that should be presented at the symposium. Bill Boland, Trumbull Rogers, and Sheila Kane of the New York Academy of Sciences have given us strong editorial support and encouragement and made it possible to realize the book from this symposium with only a short lead time. Special thanks go to Rebecca Rindlisbacher, Sepp Müller, Janine Weilenmann, Hubert Misslisch, Hansjürg Scherberger, and Dominick Straumann, who helped organize and run the conference. Finally, we are indebted to Anna Lysakowski for her elegant scanning electron microphotograph of the otoconia of the gray squirrel, used on the cover of the paper-bound version.

VOLKER HENN (1943–1997)

Tribute to Volker Henn

Volker Henn was born in 1943 in Gotha in the former German Democratic Republic, and moved to West Germany with his family in 1951. Despite a strong family tradition in architecture, he was attracted to medicine. While in medical school, he studied with O.-J. Grüsser in Berlin, who shaped his interest in the physiology of the visual system At Grüsser's suggestion, he came to my laboratory in New York when he finished medical school. We were just starting studies of single-unit activity and eye movements in the alert monkey, combining them with vestibular stimulation and computer analysis. Here he first developed his interest in the oculomotor and vestibular systems. It was a critical and productive year, and it laid the foundation for much future work for both of us. Volker then moved to Zürich, where he studied neurology with Günter Baumgartner. He became an excellent clinical neurologist and teacher of neurology. Baumgartner also encouraged his research interests, and helped him found a vestibular and oculomotor laboratory. Combining both animal and human studies, it has become world famous.

Volker was made Professor of Neurology in 1983. When Baumgartner died in 1991, he became Acting Director of the Department of Neurology at Zürich, serving for four years. In 1977, he won the Goetz Prize of the University of Zürich, the Franceschetti-Liebrecht Prize of the German Ophthalmological Society in 1978, and the prestigious Hallpike-Nylen-Prize of the Bárány Society in 1990. He was an active member of editorial boards of numerous scientific journals, advisory member for many research foundations, and president of the Betty & David Koetser Foundation for Brain Research, which sponsored the meeting that led to this book.

Volker Henn made both fundamental and clinical contributions to our knowledge of the oculomotor system. He showed how neural activity in the pontine reticular formation is spatially organized to produce horizontal saccadic eye movements. He did seminal work with Jean and Ulrich Büttner on the nature of neural activity responsible for vertical saccadic eye movements in the rostral iMLF. Through his experiments with kainic acid, he gave us a better understanding of the neural organizations that produce horizontal, vertical, and torsional saccades. With Klaus Hepp, he showed how neural activity is organized in motoneurons, and determined the coordinate frame for these movements. From this came his understanding that Listing's Law is a fundamental phenomenon that underlies eye movement. His investigation of movements in Listing's Plane is one of his major contributions. He was fascinated by the meaning and importance of this phenomenon, and did both behavioral and single-unit experiments to understand it.

Volker Henn also made important contributions to our knowledge of the vestibular system. His early papers with Walter Waespe, based on work begun in collaboration with Laurence Young, showed that visually related processes could be detected at the very first central synapse in the vestibular system. This was a fundamental finding and it was largely unappreciated up to that time. It demonstrated the importance of vision in vestibular function, especially in primates. It also highlighted the importance of studying brain function in alert animals, and it has led to a large body of work on visual-vestibular interaction.

Volker was a humanist and an intellectual. He was a warm, caring, and sympathetic friend. He loved art and literature, and was an avid bibliophile, collecting early editions about the history of science. Ernst Mach was someone that he particularly admired, and most recently, he did an imaginative translation of Mach's "Grundlinien der Lehre von den Bewegungsempfindungen" with Laurence Young. He was a circus buff, and he loved zoos, particularly if they had monkeys. Thus, a trip to Japan was not complete without seeing the Monkey Island at Beppo. He had other wide-ranging interests that included cycling, scuba diving, and biology wherever he found it. The accompanying photograph, taken during a trip to Madagascar, where he measured the eye movements of chameleons, shows a ring-tailed lemur responding to this interest.

His sudden, unexpected death shocked all that knew him. The man who initiated and helped plan the conference did not live to see it realized. Fortunately, the Koetser Foundation and the Department of Neurology of the University of Zürich enabled us to hold the conference, which was superbly organized by Bernhard Hess and his colleagues from the Department of Neurology.

The book that arose from this conference is the first to my knowledge that is devoted solely to the otoliths and to otolith function. We have dedicated it to the memory of Volker Henn, my close friend, for both his scientific contributions and his humanity. Volker Henn's death has left a large void that will not be filled easily, but the body of work that he left behind and the ideas that it will generate are his legacy to the field.

—BERNARD COHEN

Otolith Ocular Reflex Function of the Tangential Nucleus in Teleost Fish

HIROSHI SUWA, EDWIN GILLAND AND ROBERT BAKER[a]

Department of Physiology and Neuroscience, New York University School of Medicine, 550 First Avenue, New York, New York 10016, USA

ABSTRACT: In teleost fish, the tangential nucleus can be identified as a compact, separate cell group lying ventral to the VIIIth nerve near the middle of the vestibular complex. Morphological analysis of larval and adult hindbrains utilizing biocytin and fluorescent tracers showed the tangential nucleus to be located entirely within rhombomeric segment 5 with all axons projecting into the contralateral MLF. Combined single-cell electrophysiology and morphology in alert goldfish found three classes of neurons whose physiological sensitivity could be readily correlated with rotational axes about either the anterior (45°), posterior (135°), or horizontal (vertical axis) semicircular canals. Tangential neurons could be distinguished from those in semicircular-canal specific subnuclei by an irregular, spontaneous background of 10–15 sp/s and sustained static sensitivity after ±4° head displacements. Each axis-specific tangential subtype terminated appropriately onto oculomotor subnuclei responsible for either vertical, torsional, or horizontal eye movements and, in a few cases, axon collaterals descended in the MLF toward the spinal cord. We hypothesize, therefore, that the tangential nucleus consists of 3 axis-specific phenotypes that process gravitoinertial signals largely responsible for controlling oculomotor function, but that also in part, maintain body posture.

INTRODUCTION

Static head tilt relative to gravity induces compensatory otolith-ocular reflexes in all vertebrates.[1] In land animals, these reflexes have been shown to be functionally quite diverse in context including counterrolling of the eyes during head tilts, low-frequency enhancement of the angular vestibular ocular reflex dynamics, as well as directional changes in dynamics of canal–ocular reflexes as a function of head orientation relative to gravity.[2,3] Otolith-ocular reflexes recorded during linear accelerations accompanying head translation also exhibit a dependence on viewing distance in binocular foveate mammals.[4] Similarly, aquatic environments that provide a natural three-dimensional buoyancy also rely heavily on information of head orientation relative to gravity in which otolith afferents can provide both gravito-inertial (head-tilt) and acceleration-dependent (translation) signals to the vestibular nuclei. For example, otolith-ocular reflexes in teleost fish, in particular torsion, can account for up to 100% of the angle of head tilts up to 90°, as compared to roughly 30–50% in lateral-eyed mammals, such as the rabbit, and possibly less than 10% in frontal-eyed mammals, such as primates.[5,6] However, because primary otolith afferents probably cannot distinguish between head tilt and translation in fish, other gravity-dependent ocular and postural behaviors along with spatial orientation must be sorted out in the vestibular nucleus as in other vertebrates.[2]

---

[a]To whom correspondence may be addressed. Phone: 212-263-5402; fax: 212-689-9060; e-mail: bakerr01@popmail.med.nyu.edu

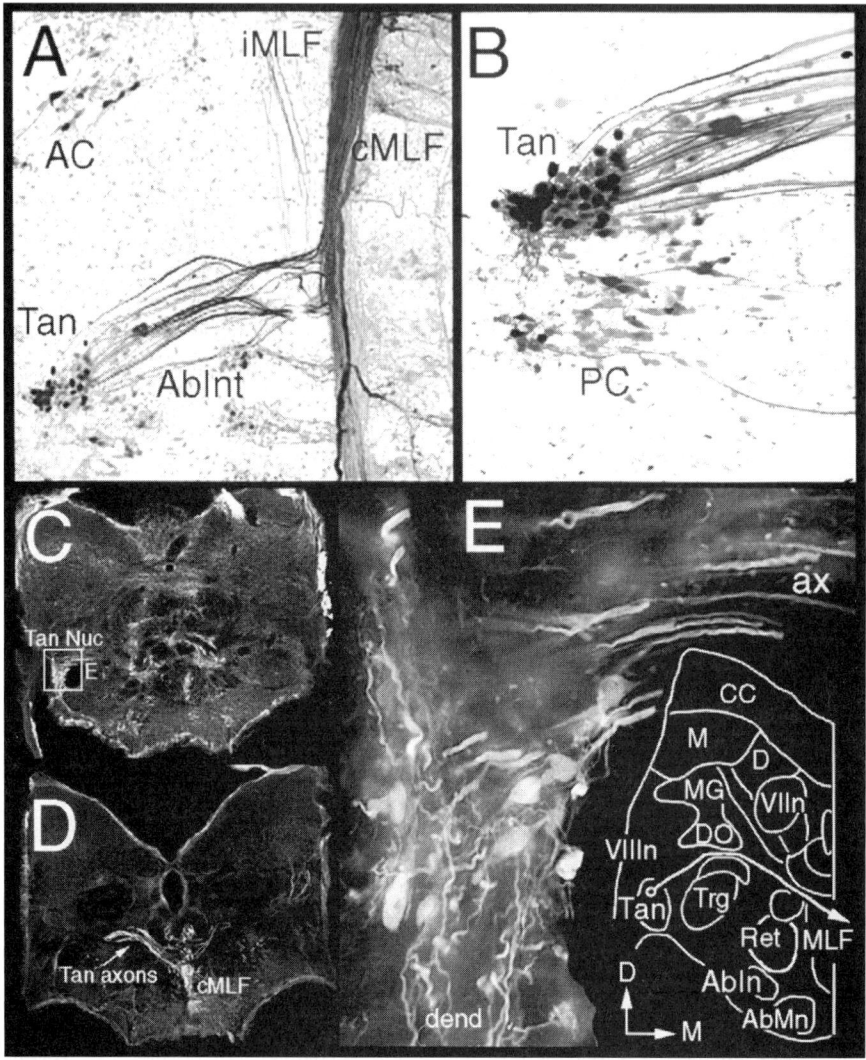

FIGURE 1. The tangential nucleus in juvenile and adult goldfish. (**A** and **B**) Low- and high-magnification confocal reconstruction of two juvenile (55 day) goldfish hindbrains 4 h after placement of rhodamine dextran in the right midbrain. The tangential (Tan) nucleus was located in rhombomere segment (rhs) 5. (**C–E**) Biocytin was placed in the right MLF just caudal to the oculomotor nucleus in adult goldfish. Axons of tangential neurons (C and E) crossed the midline 200 μm rostral in (D). The *schematic inset* in E shows the relationship of the tangential nucleus with other brainstem nuclei located in rhs 5. *Abbreviations*: AC and PC: anterior and posterior canal-related second-order subnuclei; CC: crista cerebelli; M, D, DO, Tan, and MG: medial, descending, descending octaval, tangential, and magnocellular octaval nuclei; VIIn and VIIIn: central sensory tracts; Trg: secondary gustatory tract; Ret: medullary reticular nuclei; Abd Int and Mn: rostral abducens internuclear and motoneuronal subnuclei; MLF: medial longitudinal fasciculus. A color version of this figure will be available on the New York Academy of Sciences Web Site (www.nyas.org).

Recording of single vestibular nuclei neurons in mammals during either head tilt or sinusoidal linear translation, mainly in the horizontal head plane, has shown an extensive convergence of inputs from the otolith organs and the semicircular canals.[7,8] Use of natural vestibular stimuli and electrical stimulation of separate sensory end organs suggested that vestibular neurons could also be broadly tuned, and that convergent otolith and canal driven responses are found throughout all subdivisions of the vestibular nucleus. In most instances, individual response profiles are consistent with vestibular neurons from all subnuclei playing an important role in one aspect or other of three-dimensional ocular responses.[9] Interestingly, and in contrast to the distributed otolithic organization in mammals, preliminary physiology of a second-order vestibular subnucleus in goldfish, the tangential, suggests the presence of a three-dimensional gravito-inertial input presumed to arise from the utricular end organ with overlapping convergence from each of the semicircular canals.[10]

The adult goldfish tangential nucleus consists of a tightly clustered group of neurons (FIG. 1), which may be largely responsible for producing both vertical and torsional eye motion (FIGS. 2–4), including a subtype responsive to horizontal angular, and probably linear, acceleration (FIGS. 5 and 6). Preliminary analysis shows that individual tangential neurons can be physiologically distinguished from other second-order anterior and posterior canal subnuclei neurons involved in angular acceleration ocular-specific reflex pathways (FIGS. 1A and 1B; FIG. 7). The single-cell morphological reconstruction of representative, individually identified, tangential neurons exhibited appropriate axonal trajectories and termination patterns to ocular motoneurons for each head-tilt reference frame (FIGS. 3 and 7). Hence, the tangential nucleus in teleost fish is hypothesized to act as a highly specific and unique source of gravito-inertial signals primarily responsible for otolith-induced oculomotor, and to some extent, other postural reflexes.

METHODS

Goldfish were operated on under 0.02% MS222 anesthesia, and all surgery was completed at least 3 days before experiments. Tapered screws were placed into the frontal and temporal bones to construct a dental acrylic pedestal for external support of the head during recording. The rostral valvula cerebelli or the facial lobe was removed by aspiration under visual guidance, and sharp glass electrodes were utilized to penetrate axons in the MLF at the level of either the trochlear nucleus or the tangential nucleus. Electrodes were filled with 3% neurobiotin (Vector Labs) dissolved in 0.5 M potassium acetate. After recording the electrophysiological signals, neurobiotin was injected at 4 Hz with 5–10 nA positive current pulses. Fish were transcardially perfused 10–20 h later with 4% paraformaldehyde and 1% glutaraldehyde. The brains were cut at 50–100 μm and the sections stained with a conventional DAB method and cresyl violet. Dextran-coupled fluorescent tracers (FIG. 1A and 1B) or biocytin crystals (FIG. 1C–1E) were placed in the oculomotor nucleus. Whole brains were cleared in glycerol and scanned on a Zeiss LSM-510 confocal microscope at 3–7 μm z-axis intervals.

Eye movements were recorded by use of the scleral search-coil technique after suturing a 2.8-mm-diameter coil to the center of each eye under 4% lidocaine anesthesia. Otolith and canal related axons were identified following natural angular rotation and head tilt. The fish tank, search-coil field, planetarium, and manipulators were situated on a vertical-axis (z) rotation table with ± 20° range. The entire experimental setup was placed on an elevated air table that could be displaced about the x and y axes at ± 4°. Static fore and aft displacements about axes perpendicular to either of the two anterior/posterior canal planes (i.e., the 45° and 135° axes) were used to characterize the intra-axonal recorded signals.[11] The behavioral profiles were then correlated with camera lucida morphological reconstructions. Analysis of electrophysiological data was accomplished with GW Instruments software. Detailed methods and materials can be found in references.[12–14]

RESULTS

Segmental Arrangement of Hindbrain Vestibular Neurons

To characterize the spatial segregation of neuronal groups within the postembryonic rhombomeric blueprint, the entire hindbrain vestibular and reticular scaffolds were reconstructed in juvenile goldfish with confocal microscopy after the placement of either biocytin or fluorescent dextran amine crystals in the MLF or spinal cord.[15–16] Since small whole hindbrains could be cleared in glycerol, and scanned with the confocal microscope at 3–5 μm z-axis intervals, each of the neuronal subgroups could be visualized, along with individual neurons and axon trajectories in whole-mount preparations. After placement of rhodamine dextran in the right midbrain, vestibular neurons with contralateral axons were located in rhombomere segments 1, 2, 5, and 6, as shown in pseudocolored depth profiles in FIG. 1A and 1B.

Biocytin crystals placed into the ascending MLF in adult goldfish were used to delineate the detailed hindbrain morphology of vestibuloocular projections.[15–17] Cytoarchitectural analysis of the octavolateralis area in teleosts has distinguished five first-order octaval nuclei consisting of the anterior, magnocellular (MG), descending (DO), tangential (Tan), and posterior octaval nuclei (FIG. 1 schematic). The tangential nucleus, in particular, lies at the base of the entering VIIIth nerve root as a cluster of about 50 neurons that could be reproducibly identified in all fish (FIGS. 1C, 1E; 5A, 5B). Axons of tangential neurons projected across the midline into the contralateral MLF about 200-μm rostral to the nucleus (FIG. 1D). The schematic inset in FIG. 1E shows the position of the tangential nucleus in respect to other brainstem nuclei located in rhombomeric segment 5.

Vestibular and Oculomotor Reference Frames

The semicircular canals on each side of the vertebrate head are aligned in an orientation that can be described by an axis perpendicular to that canal's plane.[18] Hence three principal axes (VA, 45°, and 135°) can represent measurement of head angular acceleration in three-dimensional space as well as provide convenient vestibular and oculomotor reference frames in three dimensions. Each horizontal semicircular canal lies in the horizontal plane and exhibits a vertical (VA) principal axis (i.e., rotation about z). The principal anterior and posterior axes, which include the anterior canal on one side and the posterior canal on the other side, lie in a horizontal plane, but at angles of 45° and 135° to the midline sagittal plane (see FIG. 7). The preferred rotational axes of all recorded vestibular neurons were examined in respect to these principal axes even though eye-rotation axes are not generally described in respect to either the 45° or 135° axis of vestibular stimulation.

Eye positions are generally presented as rotation vectors in a head-fixed coordinate system in which eye position is defined in coordinates of a rotation out of a particular reference position that is usually straight-ahead gaze.[3] This measurement system employs three rotation angles to define eye positions as horizontal (Hor), vertical (Ver), and torsional (Tor). For example, ocular torsion in the goldfish is rotation of the eye during naso–occipital motion (rotation about x) activating alternately, the bilateral anterior and posterior canals producing counterrotation of each eye.[5] Accordingly this type of coordinate frame, while not directly overlapping that of either the vestibular or visual 45° and 135° principal axes, can be conveniently used to describe neuronal responses in the fish because the sensitivity axis of each neuron can be estimated as the axis around which head rotation causes maximal modulation of firing rate (FIGS. 4 and 6). Notably, the structure and function of all 18 tangential neurons individually studied was consistent with a role in producing an appropriately directed three-dimensional eye motion behavior (FIGS. 2–7).

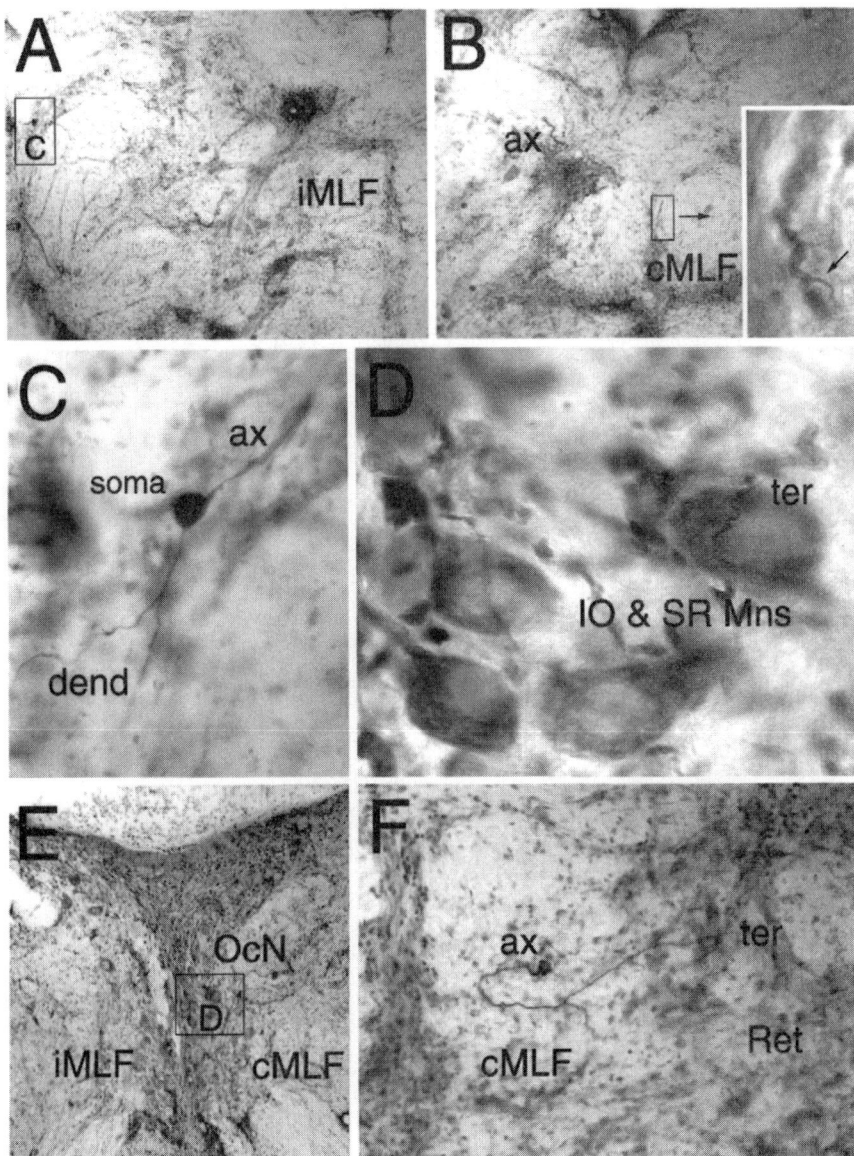

FIGURE 2. Morphology of a 45° axis-related tangential neuron. (**A**) The soma (*box*) was located in the same coronal plane as the rostral abducens nucleus. (**B**) Parent axon (ax) projecting through the ipsi-MLF toward the contra-MLF giving rise to a rostral and caudal axon collateral (*inset on right*). (**C**) High magnification of the neuron in part (A) showing the soma, initial segment (ax), and ventral dendrites (dend). (**D** and **E**) High and low magnification of axonal termination in the inferior oblique (IO) and superior rectus (SR) subdivisions of the oculomotor nucleus. (**F**) Axon collateral and terminals in reticular nuclei adjacent to the MLF at the level of rhs 7. A color version of this figure will be available on the New York Academy of Sciences Web Site (www.nyas.org).

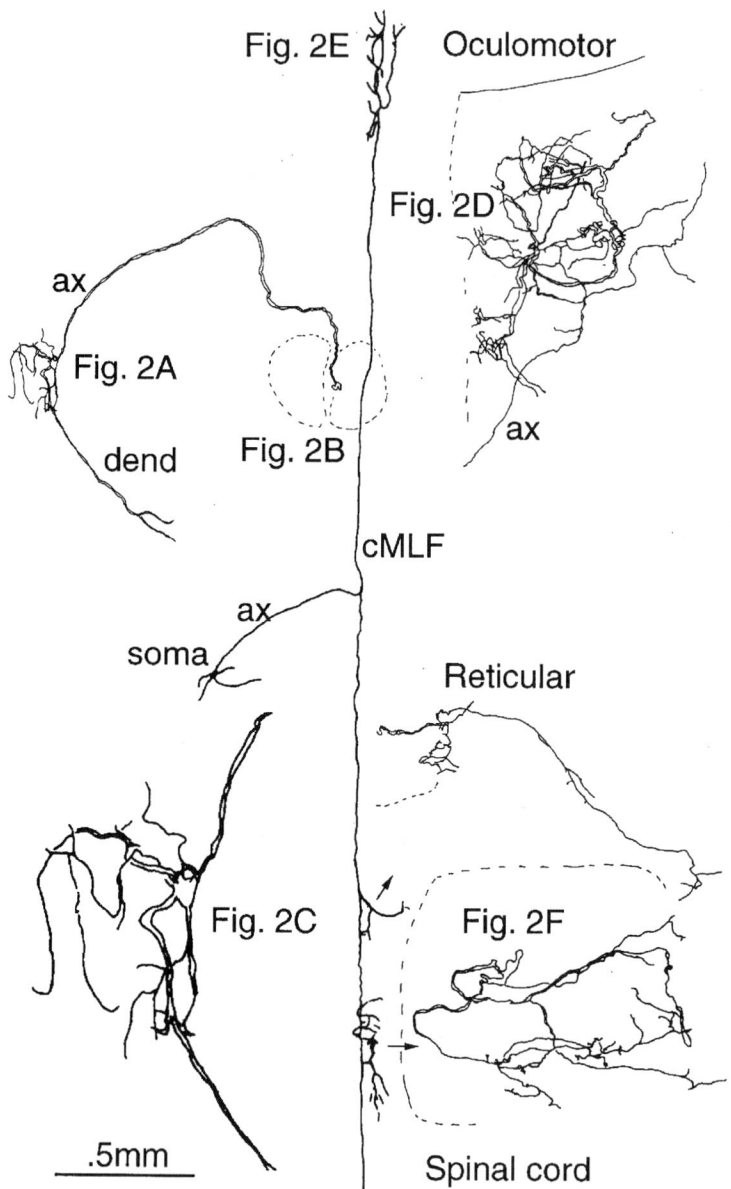

FIGURE 3. Camera lucida drawing of the 45° axis-related tangential neuron illustrated in FIGURE 2. Higher magnification insets correlated with the panels in FIGURE 2 show the termination pattern in the oculomotor nucleus (**2D**), the soma-dendritic arborization in the lateral brainstem (**2A** and **2C**) and two different reticular nuclei (**2F**).

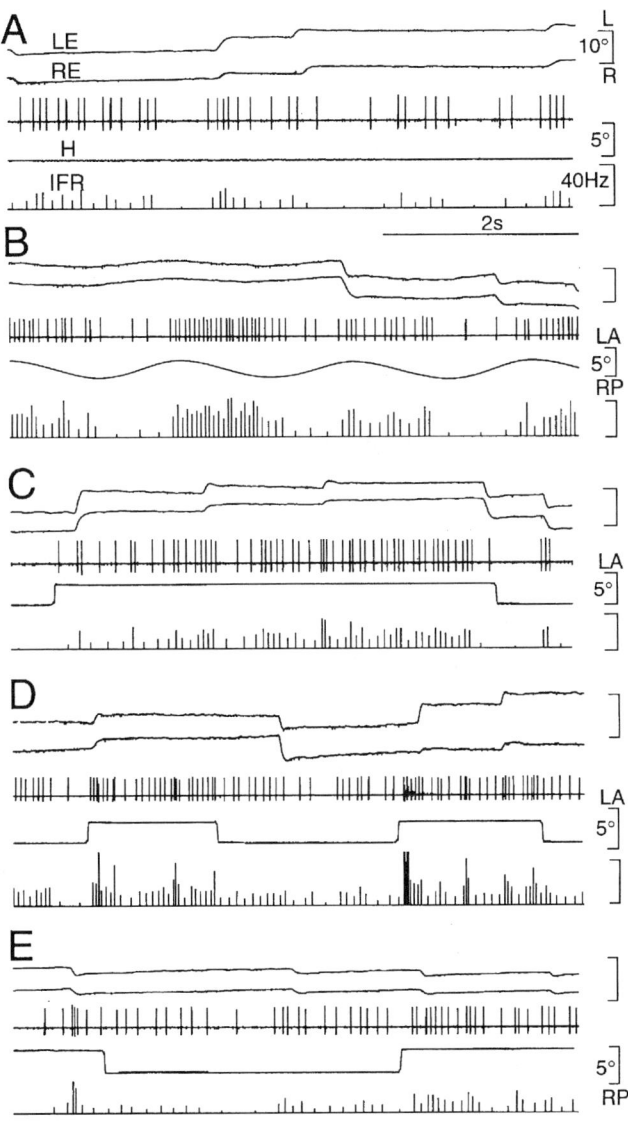

FIGURE 4. Activity of the 45° axis-related tangential neuron show in FIGURES 2 and 3. (**A**) Absence of saccadic and eye position sensitivity during spontaneous eye movements. (**B**) Horizontal/anterior-canal sensitivity with rotation about the 45-deg axis to the ipsilateral (left) side. (**C** and **D**) Static (C) and dynamic (D) sensitivity during combined interaural (left side down) and naso–occipital (nose down) displacements as in B. (**E**) Absence of response to opposite-direction displacements (rightside and tail down). In each set of records, left (LE) and right (RE) eye positions are at the top followed by raw spike discharge, head position (H), and instantaneous firing rate (IFR).

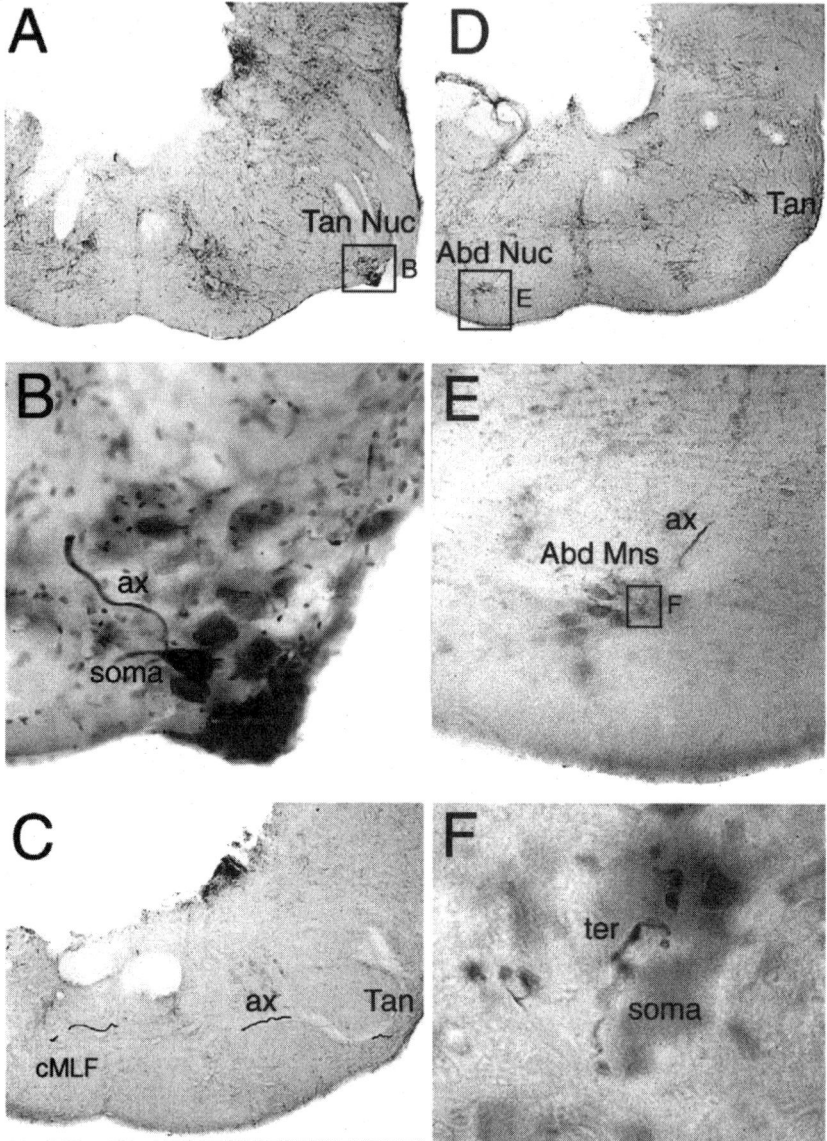

FIGURE 5. Location and projection of a vertical axis-related tangential neuron. (**A** and **B**) Low and high magnification views of the soma in the ventral part of the tangential nucleus. (**C**) The axon projected across the midline (cMLF) to the contralateral abducens nucleus (Abd Nuc in D). (**D–F**) Low and high magnification views of the abducens nucleus (D) and motoneurons (Mns; E), including axonal termination (ter) on an Abd Mn (F). A color version of this figure will be available on the New York Academy of Sciences Web Site (www.nyas.org).

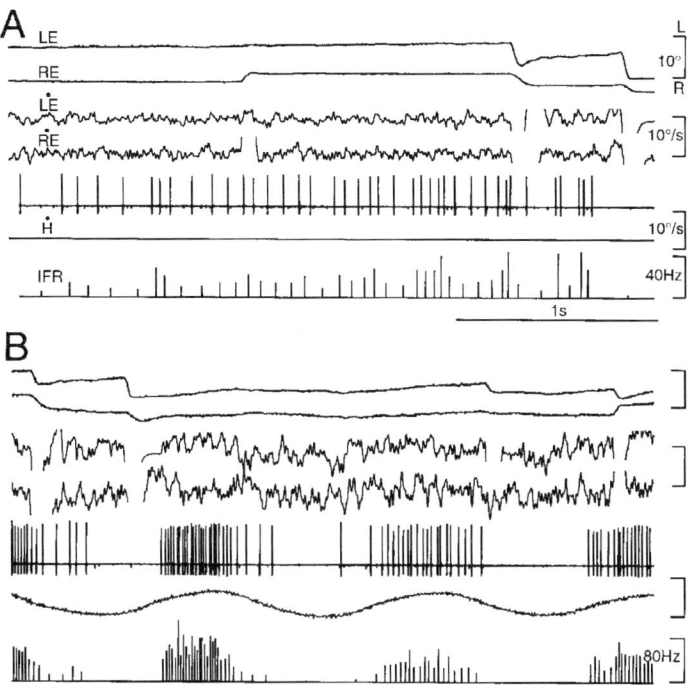

FIGURE 6. Activity of the vertical axis-related tangential neuron shown in FIGURE 5. (**A**) Absence of eye position and velocity sensitivity during spontaneous saccadic and fixation eye movements. (**B**) Low threshold sensitivity to horizontal head rotation. Records are configured as in FIGURE 4, except eye ($\dot{E}$) and head ($\dot{H}$) velocity traces are illustrated below eye position.

Morphology and Physiology of the Tangential Nuclei in Fish

Preliminary structure/function experiments utilized a floating air table with small rotational displacement of ±4° to determine the VA, 45°, or 135° physiological sensitivity of intra-axonal recorded MLF axons before injection of biocytin and subsequent morphological analysis. In the intact goldfish, Tan neurons could be distinguished from those of either the AC (45°) or PC (135°) excitatory and inhibitory vestibular neurons on the basis of signal sensitivity alone. First, neither eye-position sensitivity during static head tilt nor saccadic sensitivity was observed in any of the AC and PC second-order vestibular neurons that subsequently were morphologically identified to be located in the AC or PC subdivisions (FIG. 1A, 1B; FIG. 7). Second, an irregular spontaneous discharge ranging from 10 to 20 sp/s was recorded in all tangential neurons (FIGS. 4A and FIG. 6A). By contrast, AC- and PC-related neurons were only recruited during angular acceleration accompanying either *x* or *y* displacements and never exhibited spontaneous activity; however, some HC-related neurons exhibited a regular background discharge.[19]

Single-cell structure and function analysis of the 18 Tan neurons in the goldfish revealed at least three distinct phenotypes, each exhibiting a similar soma–dendritic architecture (FIGS. 2C, 3, 5B), but with unique axonal properties and physiological signals that appeared to be uniquely correlated to either the VA (FIG. 6), 45° (FIG. 4), or 135° axes (not illustrated). FIGURE 2 shows the morphology of a 45° axis-related tangential neuron whose parent axon projected into the contra-MLF, giving rise to a rostral and caudal axon collateral (inset on right in FIG. 2B). The typical soma, initial segment (ax), and dendritic features are shown in FIGURE 2C, and the low and high magnification documentation of axonal termination in the inferior oblique (IO) and superior rectus (SR) subdivisions of the oculomotor nucleus in FIGURE 2D and 2E. A complete camera lucida drawing of this neuron, along with high-magnification insets correlated with the panels in FIGURE 2 is shown in FIGURE 3. The static and dynamic sensitivity during combined interaural (leftside down) and naso–occitpital (nose down) displacements are shown in FIGURE 4B–4E and the lack of response to opposite-direction displacements (rightside and tail down) in 4F. A similar set of records is shown for a vertical-axis-related (VA) tangential neuron in FIGURES 5 and 6. Low- and high-magnification views of the soma in the tangential nucleus are shown in FIGURE 5A and 5B with parts of the axon projecting across the midline to the contralateral abducens nucleus in FIGURE 5D. Axonal termination onto abducens motoneurons is illustrated in FIGURE 5E and 5F, and the physiological profile is shown in FIGURE 6. Note the absence of eye position and velocity sensitivity during saccadic eye movement and maintained eye position (FIG. 6A) and the low-threshold response to sinusoidal horizontal head rotation (FIG. 6B).

In summary, physiological analysis of VA, 45°, and 135° axis-specific Tan neurons suggest a convergence of first-order canal and otolith-specific afferents onto each subtype (FIGS. 4 and 6). Intracellular biocytin label showed that some neurons in the Tan nucleus have axons that bifurcate to reach oculomotor (FIG. 2D, 2E), hindbrain reticular (FIG. 2F), and spinal targets (FIG. 3); however, in all cases each Tan subtype projected onto the appropriate pair of oculomotor or abducens (internuclear/motor) subnuclei to produce the correct vertical, torsional, or horizontal eye movements (FIG. 7). This analysis of the anatomy and physiology of the Cyprinid tangential and AC/PC canal second-order pathways is summarized in FIGURE 7 to provide a model of reflex neuronal circuitry that can be quantitatively measured and analyzed in respect to the neuronal transformations required for signal processing contributing to vertical and torsional eye movements. Collectively, the observations suggest that the embryonic rhombomeric blueprint for vestibular subnuclei observed in larval and juvenile hindbrains (FIGS. 1 and 7) is retained in the adult (FIGS. 2 and 5), and that each AC, PC, and Tan subgroup is highly segregated in the rhombomeric segments according to direction specificity and oculomotor role.

DISCUSSION

Recent work in Cyprinids, largely goldfish, has delineated most of the anatomy and physiology of brainstem neurons and circuits producing three-dimensional eye motion.[15,20] A major significance of such a blueprint lies in the fact that the genetic and anatomical profiles of each subdivision of vestibular neurons can be correlated with segmented embryonic hindbrain compartments, called rhombomeres, that are homologous in all vertebrates. Preliminary anatomy and physiology in the goldfish (and zebrafish[16]) suggest that the tangential second-order vestibular subnucleus consists of a cluster of neurons largely responsible for vertical and torsional eye motion and position (FIGS. 2–4). Anatomically, the Tan nucleus can be distinguished from adjacent second-order anterior (AC) and posterior (PC) canal subnuclei that provide angular acceleration-specific reflex pathways (FIG. 1 A and 1B; FIG. 7; reference 16). Single-cell morphology of representative neurons from each of the subnuclei in adult fish documents axonal trajectories and termination patterns to the appropriate oculomotor neurons (FIGS. 2, 3, 5). Physiological analysis of individual tangential neurons has established the

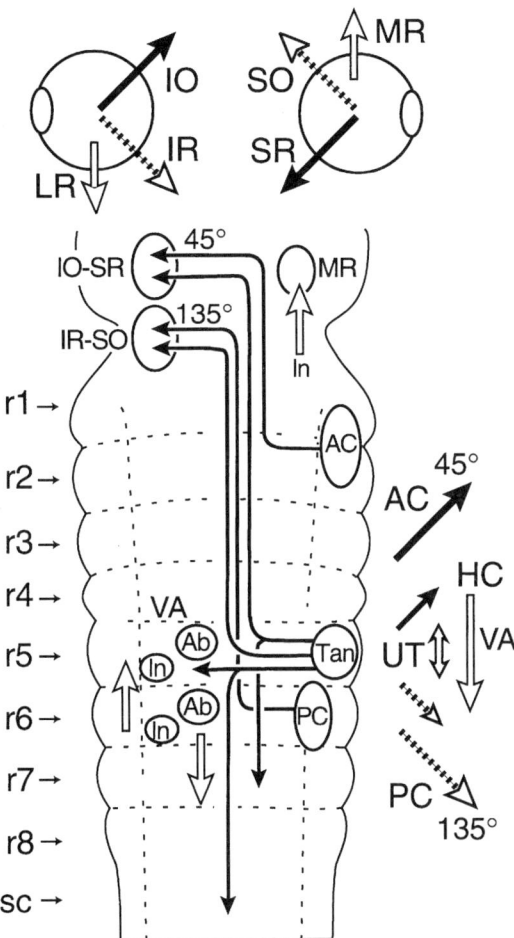

FIGURE 7. Rhombomeric organization of vestibular subnuclei and schematic diagram of canal and otolith-specific vestibulo-oculomotor-related behaviors. The illustration depicts the relationship between semicircular canals, utricle, and extraocular muscles in relationship to the rhombomeric locations of their excitatory vestibulo-ocular subgroups and, in particular, the tangential (Tan) nucleus located in rhombomere segment 5. Appropriately directed, *coded arrows* are used to indicate a notational spatial relationship between the right anterior (AC, 45°, rhs 1–2) and posterior (PC, 135°, rhs 5–6) canal subnuclei with the right superior rectus (SR)/left inferior oblique (IO) and right superior oblique (SO)/left inferior rectus (IR) eye muscles, respectively. In the VA system, abducens (Ab) and internuclear (In) subnuclei are illustrated in rhs 5–6, with suitably directed *arrows* depicting their relationship with the lateral and medial rectus muscles. The proposed rhombomeric origin for vestibulo-ocular subgroups is based on analysis of larval goldfish and zebrafish hindbrains (see FIG. 1).[15–17] Collectively, the observations suggest that the embryonic rhombomeric blueprint for vestibular subnuclei is retained in larval and juvenile hindbrains in respect to individual oculomotor behaviors, and each subgroup is highly segregated in the rhombomeric segments according to their directional specificity.

presence of an appropriately patterned three-dimensional gravitoinertial input from the otolithic end organs with presumed convergence from the semicircular canals (FIGS. 4 and 6). The general argument is that the tangential nucleus represents an anatomically distinct vestibular subgroup responsible for translating patterned gravitoinertial signals from the otolith end organ into ocular-specific reflex behavior (FIG. 7).

Static (instantaneous) gravitoinertial signals are likely directly translated into eye-position displacement by second-order neurons in the tangential nucleus.[21] These neuronal transformations may occur within a reference frame best described by head displacements about a particular set of axes defined by the planes of the horizontal (VA), anterior (45°), and posterior (135°) semicircular canals[18] (FIG. 7). Preliminary evidence supports at least three separate subtypes of axis-specific tangential neurons, each responsive to head displacement about only one axis. Hence, vertebrate VA, 45°, and 135° intrinsic coordinates appear to have evolved as conserved mechanisms by which linear acceleration signals (otoliths) could be combined with angular acceleration (canals) and retinal slip (vision) signals to provide the complete range of response dynamics and performance essential for producing vertical and torsional eye movements.

The vestibular nuclei in all vertebrates, but more noticeably in fish, is discreetly subdivided into coherent groups of cells in which each neuronal cluster is largely subservient to a unique physiological role in oculomotor behavior.[15,22,23] In particular, preliminary evidence provided here suggests that the tangential nucleus acts as a unique gravito-inertial center largely responsible for static vertical and torsional oculomotor reflexes. Since fish are lateral-eyed animals, they may be similar to rabbits and/or rats in which central vestibular neurons may not discriminate head tilt from translation based only on vestibular signals.[2] For example, torsional and vertical eye movements are similar in rabbits in response to dynamic head tilt and/or translation.[24] Moreover, other animals (rodents) generate large vertical and torsional eye movements during translation, including horizontal eye movement.[25] In essence, then, those fish and mammals that largely rely upon nonfoveate vision will likely utilize static otolith-ocular reflexes to exert an orienting function of both eye and posture that keeps the horizon (i.e., visual streak) parallel to earth horizontal. The counterrolling of up to 125° reported in Cyprinids[5] represents the extraordinary range of eye movement produced by this otolith-specific, sub-nucleus in goldfish. Our work suggests that the tangential nucleus might embody such an anatomically distinct vestibular subgroup responsible for translating linear acceleration signals from the otolith end organ into ocular-specific reflex behavior.

In phylogenetic essence, the goldfish tangential nucleus may be best described as a species-specific vestibular subnucleus that stabilizes the static orientation of the globe (each eye) in space. Given the fact that the goldfish has an extensive saccadic system, as well as a highly developed otolith-dependent control of primary eye position, it will be of interest to look at vestibulo-ocular reflexes to see if they distinguish between dynamic head tilt and translation.[26] It has been proposed that otolith signals at low frequencies represent gravitational accelerations, and those at higher frequencies, inertial accelerations (translation), and that in combination with semicircular canal information (angular), the central vestibular complex can distinguish between the two components in respect to gaze control.[2,6] The tangential nucleus in fish might be an excellent experimental site in which to attempt to resolve some of the ambiguities said to exist in "indeterminate" otolith information, since in the aquatic environment with natural three-dimensional buoyancy, both the gravitoinertial and inertial acceleration forces need to be translated into eye-position displacements by presumed second-order vestibular neurons.

In vertebrates, coherent clusters of vestibular neurons occupy unique spatial domains exhibiting specific gene-expression patterns within the rhombomeric scaffold.[15,27,28] As a result, the differentiation of axon trajectory, termination field, and function of each vestibular phenotype is mechanistically linked to the neuromeric blueprint, or alternatively, each vestibular phenotype is shaped by a species-specific evolutionary fate.[15,22] The fish tangential nucleus is of interest in this regard because it appears to originate from a distinct hindbrain morphogenetic segmental unit with a unique developmental and genetic history.[29] As such, the tangential nucleus in the closely related Cyprinid, the zebrafish, offers an excellent site to integrate emerging molecular genetic and developmental tools with physiological correlates that can be quantitatively measured and analyzed in mutational studies. The key issue to be addressed in this future work is whether "genetic networks" and "neuronal networks" can be linked by behavior. If oculomotor function in vertebrates is based on well-conserved anatomy, physiology, and behavioral repertoires, then identifying the "genetics units" responsible for the segmented "rhombomeric units" in the brainstem will allow complementing genetic and molecular approaches to reveal the signaling mechanisms responsible for the development of individual neuronal phenotypes.[22] The first steps toward this goal have been provided for the tangential vestibular subnuclei in the form of a neuronal blueprint that can permit a causal analysis of existing zebrafish mutations.[30]

SUMMARY

Gravitoinertial signals may be directly translated into eye-position displacement by second-order tangential neurons in the goldfish. Preliminary analysis showed that the neuronal transformations are likely to occur within a reference frame best described by head displacements about a particular set of rotational axes defined by the planes of the horizontal, anterior, and posterior semicircular canals. The electrophysiological evidence supports at least three axis-specific subtypes of tangential neurons, each responsive to head displacement about only one axis. Hence, the most parsimonious suggestion is that natural vertical axis, 45°, and 135° intrinsic coordinate reference frames combine linear-acceleration (tilt) signals from the otolith with angular accelerations from the canals to provide a complete range of response dynamics and performance essential for producing three-dimensional reflex eye movements. Because each of the three distinct axis-specific physiological phenotypes target appropriate oculomotor subgroups, we hypothesize that the tangential nucleus is specialized for processing gravito-inertial signals largely responsible for controlling the orientation of the eyes, but also may contribute to pathways maintaining body posture.

REFERENCES

1. HENN, V. 1988. Representation of three-dimensional space in the vestibular, oculomotor, and visual systems. Introduction. Ann. N.Y. Acad. Sci. **545**: 1–9.
2. ANGELAKI, D. E. & B. J. HESS. 1996. Organizational principles of otolith- and semicircular canal-ocular reflexes in rhesus monkeys. Ann. N.Y. Acad. Sci. **781**: 332–347.
3. HENN, V., D. STRAUMANN, B. J. HESS, T. HASLWANTER & N. KAWACHI. 1996. Three-dimensional transformations from vestibular and visual input to oculomotor output. Ann. N.Y. Acad. Sci. **656**: 166–180.
4. PAIGE, G. D. 1991. Linear vestibulo-ocular reflex (LVOR) and modulation by vergence. Acta Otolaryngol. Suppl. Stockh. **481**: 282–286.
5. TRAILL, A. B. & R. F. MARK. 1970. Optic and static contributions to ocular counterrotation in carp. J. Exp. Biol. **52**: 109–124.
6. HESS, B. J. & D. E. ANGELAKI. 1997. Inertial vestibular coding of motion: concepts and evidence. Curr. Opin. Neurobiol. **7**: 860–86.
7. SCHOR, R. H., B. C. STEINBACHER, Jr. & B. J. YATES. 1998. Horizontal linear and angular responses of neurons in the medial vestibular nucleus of the decerebrate cat. J. Vestib. Res. **8**: 107–116.
8. BUSH, G. A., A. A. PERACHIO & D. E. ANGELAKI. 1992. Quantification of different classes of canal–related vestibular nuclei neuron responses to linear acceleration. Ann. N.Y. Acad. Sci. **656**: 917–919.

9. ANGELAKI, D. E., G. A. BUSH & A. A. PERACHIO. 1993. Two-dimensional spatiotemporal coding of linear acceleration in vestibular nuclei neurons. J. Neurosci. **13**: 1403–1417.
10. SUWA, H., E. GILLAND & R. BAKER. 1997. The tangential nucleus of teleost fish is specialized for otolith ocular reflex function. Soc. Neuro. Abst. **23**: 752.
11. GRAF, W. 1988. Motion detection in physical space and its peripheral and central representation. Ann. N.Y. Acad. Sci. **545**: 154–169.
12. PASTOR, A. M., B. TORRES, J. M. DELGADO–GARCIA & R. BAKER. 1991. Discharge characteristics of medial rectus and abducens motoneurons in the goldfish. J. Neurophysiol. **66**: 2125–2140.
13. PASTOR, A. M., R. R. DE LA CRUZ & R. BAKER. 1997. Characterization of Purkinje cells in the goldfish vestibulo–cerebellum during eye movement and adaptive modification of the vestibulo–ocular reflex. *In* The Cerebellum: From Structure to Control, C.I. De Zeeuw, P. Strata, and J. Voogd, Eds.: 359–381. Elsevier. Amsterdam.
14. MARSH, E. & R. BAKER. 1997. Normal and adapted visuooculomotor reflexes in goldfish. J. Neurophysiol. **77**: 1099–1118.
15. BAKER, R. 1998. From genes to behavior in the vestibular system. Otolaryngol.-Head Neck Sur. **119**: 263–275.
16. SUWA, H., E. GILLAND & R. BAKER. 1996. Segmental organization of vestibular and reticular projections to spinal and oculomotor nuclei in the zebrafish and goldfish. Biol. Bull. **191**: 257–259.
17. BAKER, R., H. SUWA, E. MARSH & E. GILLAND. 1996. Segmental arrangement of vestibulo-ocular neurons in the hindbrain of adult zebrafish and closely related cyprinids. Soc. Neurosci. Abst. **22**: 1813.
18. SIMPSON, J. I. & W. GRAF. 1985. The selection of reference frames by nature and its investigators. *In* Adaptive Mechanisms in Gaze Control, A. Berthoz and G. M. Jones, Eds.: 3–20. Elsevier. Amsterdam.
19. GREEN, A., H. SUWA, R. BAKER & H. GALIANA. 1997. Characterization of second order vestibular neurons related to horizontal eye movement in the goldfish. Soc. Neuro. Abst. **23**: 751.
20. BAKER, R., E. GILLAND, A. GREEN, H. STRAKA & H. SUWA. 1998. Rhombomeric organization of horizontal optokinetic and vestibulo–ocular reflexes in goldfish. Soc. Neuro. Abst. **24**: 1411.
21. ANGELAKI, D. E. & B. J. HESS. 1995. Inertial representation of angular motion in the vestibular system of rhesus monkeys. II. Otolith–controlled transformation that depends on an intact cerebellar nodulus. J. Neurophysiol. **73**: 1729–1751.
22. BAKER, R. & E. GILLAND. 1996. The evolution of hindbrain visual and vestibular innovations responsible for oculomotor function. *In* Acquisition of Motor Behavior in Vertebrates, J. R. Bloedel, T. J. Ebner, and S. P. Wise, Eds.: 29–55. MIT Press. Boston.
23. LEE, R. K. K., R. C. EATON & S. J. ZOTTOLI. 1993. Segmental arrangement of reticulospinal neurons in the goldfish hindbrain. J. Comp. Neurol. **329**: 539–556.
24. BAARSMA, E. A. & H. COLLEWIJN. 1975. Eye movements due to linear accelerations in the rabbit. J. Physiol. (Lond.) **245**: 227–249.
25. HESS, B. J. & N. DIERINGER. 1991. Spatial organization of linear vestibuloocular reflexes of the rat: responses during horizontal and vertical linear acceleration. J. Neurophysiol. **66**: 1805–1818.
26. PAIGE, G. D. & D. L. TOMKO. 1991. Eye movement responses to linear head motion in the squirrel monkey. I. Basic characteristics. J. Neurophysiol. **65**: 1170–1182.
27. PRINCE, V. E., A. L. PRICE & R. K. HO. 1998. Hox gene expression reveals regionalization along the anteroposterior axis of the zebrafish notochord. Dev. Genes Evol. **208**: 517–522.
28. GLOVER, J. C. 1996. Development of second-order vestibular projections in the chicken embryo. Ann. N.Y. Acad. Sci. **781**: 13–20.
29. PRINCE, V. E., C. B. MOENS, C. B. KIMMEL & R. K. HO. 1998. Zebrafish hox genes: expression in the hindbrain region of wild–type and mutants of the segmentation gene, valentino. Development. **125**: 393–406.
30. MALICKI, J., A. F. SCHIER, L. SOLNICA–KREZEL, D. L. STEMPLE, S. C. NEUHAUSS, D. Y. STAINIER, S. ABDELILAH, Z. RANGINI, F. ZWARTKRUIS & W. DRIEVER. 1996. Mutations affecting development of the zebrafish ear. Development **123**: 275–283.

Stimulus Processing by Type II Hair Cells in the Mouse Utricle

JEFFREY R. HOLT,[a] MELISSA A. VOLLRATH,[b] AND RUTH ANNE EATOCK[c,d]

[a]*Department of Neurobiology, Harvard Medical School, Massachusetts General Hospital, and Howard Hughes Medical Institute, Boston Massachusetts 02114, USA*

[b]*Division of Neuroscience, Baylor College of Medicine, Houston, Texas 77030, USA*

[c]*The Bobby R. Alford Department of Otorhinolaryngology and Communicative Sciences, Baylor College of Medicine, Houston, Texas 77030, USA*

ABSTRACT: In type II and neonatal hair cells in the mouse utricle, the receptor potentials evoked by low-frequency sinusoidal deflections of the hair bundle are attenuated by adaptation of the mechanoelectrical transduction current and the voltage-dependent activation of a large potassium (K)-selective outwardly rectifying conductance, g_{DR}. These processes may contribute to high-pass filtering of the responses of some utricular afferents to sinusoidal linear accelerations below 2 Hz. Depolarizing receptor potentials are more attenuated by g_{DR} than are hyperpolarizing receptor potentials. It may therefore reduce nonlinear distortion introduced by mechanoelectrical transduction, which generates larger depolarizing currents than hyperpolarizing currents.

The discharge properties of utricular afferents vary according to whether they innervate the striolar or extrastriolar zones of the sensory epithelium. Regional variation in hair-cell properties is likely to contribute. Preliminary results suggest that the outwardly rectifying K conductances of type II cells are slower and larger in the striola than in the extrastriola, consistent with regional variation in the relative numbers of delayed rectifier and A-current K channels.

INTRODUCTION

The mammalian utricle provides information about head position and linear head movements to reflexes that control eye, head, and body position. Robust activity is recorded from utricular afferents in response to linear accelerations in the frequency range from 0 (DC) to 2 Hz.[1] This low-frequency capability distinguishes the mammalian utricle

---

[d]To whom correspondence may be addressed. Phone: 713/798-5145; fax: 713/798-8553; e-mail: eatock@bcm.tmc.edu

from the inner ear organs in which hair-cell properties have been most extensively examined: various auditory organs and the frog saccule, a vibration detector sensitive to frequencies between 10 and 150 Hz.[2] We have been examining whether hair-cell properties in mammalian utricles are specialized in ways that reflect the low-frequency function of these organs, using an *in vitro* preparation of the sensory epithelium of the mouse utricle.

As in the vestibular organs of all amniotes, the hair cells of the mature mouse utricle are classified morphologically as type I or type II. About 60% of the hair cells are type I.[3] Both cell types are present throughout the sensory epithelium. The two cell types can be distinguished during whole-cell recording by the presence or absence of $g_{K,L}$, a delayed rectifier that activates at unusually negative potentials[4] and that is selectively expressed by type I cells.[5] $g_{K,L}$ is acquired by type I hair cells of the mouse utricle during the second half of the first postnatal week.[3] The impact of $g_{K,L}$ on the receptor potential of type I cells has been discussed elsewhere.[6,7] In this report we focus on cells without $g_{K,L}$, that is, neonatal and type II cells.

In hair cells of the frog saccule and turtle, chick, and mouse cochleas, the transduction current through the mechanosensitive conductance, g_{met}, decays during sustained hair-bundle deflections, with time constants ranging from 0.3 to 100 ms.[8-11] Because such adaptation would severely attenuate hair-cell responses to stimuli at low frequencies (<1 Hz), we hypothesized that the process would be slowed, reduced, or eliminated in mammalian vestibular organs. We found, however, that the transduction current in mouse type II hair cells adapts only modestly slower than adaptation in frog saccular hair cells.[12] Here we review the impact of this process on the frequency dependence of the receptor potential in mouse utricular hair cells.

The dominant conductances in most hair cells comprise outwardly rectifying K channels in the basolateral membrane. These may be Ca^{2+}- or voltage-gated, and with the exception of $g_{K,L}$ in type I cells, generally activate positive to resting potential. They provide negative feedback in that they carry outward current in response to the depolarization evoked by inward transduction current, and so act to repolarize the membrane. They therefore represent a second locus of adaptation, one that affects the receptor potential but not the transduction current. Ca^{2+}-gated K channels are also involved in the electrical resonance that tunes the receptor potential in some hair cells, notably those from the turtle cochlea[13] and frog saccule.[14,15] In most vestibular hair cells, however, such electrical resonance is not prominent.[6,7,16] This may be related to the fact that in many vestibular cells voltage-dependent rather than Ca^{2+}-dependent K channels are most numerous.[17] The dominant conductance in many type II cells in the mouse utricle is a voltage-dependent delayed rectifier, g_{DR}, which activates positive to –55 mV.[3] g_{DR} dominates in all mouse utricular hair cells until postnatal day (P) 4, when type I cells begin to acquire $g_{K,L}$.[3] Here we present preliminary data suggesting another possible function for g_{DR}: to compensate at low frequencies for nonlinear properties of the mechanosensitive channels and restore linearity at the level of the receptor potential.

The sensory epithelium of the rodent utricle is regionally organized, both anatomically and in terms of the discharge properties of the primary afferents. Afferent fibers that innervate a cytoarchitectonically distinct strip of epithelium called the striola have different properties from those that innervate the surrounding epithelium, the extrastriola.[18] Striolar afferents have more irregular background discharge and show increasing gain (spikes/s/g of linear acceleration) with frequency between 0 and 2 Hz. Extrastriolar afferents have more regular background discharge and stimulus gains that are independent of stimulus frequency below 2 Hz. Here we compare the frequency dependence of receptor potential data with the afferent data. We also show a regional difference in the kinetics of the outwardly rectifying K conductances of type II hair cells. A possible scheme to explain the observed differences is regional variation in the relative proportions of g_{DR} and a fast inactivating conductance, g_A.

METHODS

Tissue Preparation

Methods were as previously described.[12] Briefly, utricles were excised from young mice (P1–P17; birth = P0; CD-1 outbred strain; timed pregnant females obtained from Charles River, Wilmington, Massachusetts). The utricles were dissected in our standard external solution, containing (in mM) 144 NaCl, 0.7 NaH_2PO_4, 5.8 KCl, 1.3 $CaCl_2$, 0.9 $MgCl_2$, 5.6 D-glucose, 10 HEPES-NaOH, vitamins and amino acids as in Eagle's MEM, at pH 7.4, and 320 mmol kg^{-1}. The epithelium was mounted in an experimental chamber on the stage of a fixed-stage upright microscope (Axioskop FS; Zeiss, Oberkochen, Germany) and viewed with a 40× water-immersion objective with differential interference contrast optics.

Recording

The experimental chamber contained the standard extracellular solution. Recording pipettes contained (in mM): 140 KCl, 0.1 $CaCl_2$, 5 EGTA-KOH, 3.5 $MgCl_2$, 2.5 Na_2ATP, 5 HEPES-KOH, at pH 7.4, and 290 mmol kg^{-1}. Pipette resistances were 3–6 MΩ. Whole-cell recordings were made within the epithelium, as described in reference 12, using an Axopatch 200-A patch-clamp amplifier (Axon Instruments, Foster City, California). Currents were filtered with an 8-pole Bessel filter (Model 902, Frequency Devices, Haverhill, Massachusetts), digitized at ≥ 2 times the corresponding filter frequency using a 12-bit acquisition board (Digidata 1200) and pClamp 6.0 software (Axon Instruments, Foster City, California) and stored on disk. Recordings were obtained at room temperature (22–25°C). Voltages have been corrected for a liquid junction potential of –4 mV. Analysis and fits were done with the program "Origin 5.0" (Microcal Software, Northampton, Massachusetts), which uses a Levenberg-Marquardt least-squares fitting algorithm. Results are presented as means ± standard errors of the mean.

Stimulation

The stimulus was a fluid jet delivered from a glass micropipette and controlled by a fast pressure-clamp system.[12,19] Briefly, the stimulus pipette was pulled to a tip diameter of ~10 µm, filled with the standard extracellular solution, and positioned near the bundle of interest. Stimulus waveforms were steps and sinusoidal bursts controlled by pClamp 6.0 software. The mean rise time for steps was 2.6 ± 0.2 ms (n = 10). The motion was approximately aligned with the hair bundle's orientation axis, defined as the axis of maximum sensitivity.[20] Positive motion indicates deflection toward the tall edge of the bundle.

Hair-bundle deflections were monitored with a Newvicon video camera (Model NC-65, Dage-MTI, Inc., Michigan City, Indiana) and recorded on S-VHS videotape. Deflections were measured off-line from the video image (at 5000×) as the displacement of the bundle at the height of the tallest stereocilia, as viewed from above. A linear relation was obtained between bundle deflection and the output of a pressure transducer that is located at the input to the stimulus pipette. For each cell, the slope of this relation was measured and used to calibrate the output of the pressure monitor for the entire recording. The stimulus traces shown are the output of the pressure monitor calibrated by this method.

RESULTS

Figures 1–4 show results from four hair cells for which we have receptor potential measurements over an extended frequency range. Three of the four cells were from a P4 epithelium, and the fourth was from a P8 epithelium. None of the cells had full calyx endings and all four cells had g_{DR} but not $g_{K,L}$, as shown by whole-cell current responses to voltage steps from a holding potential of –64 mV (FIG. 1A). A cell with $g_{K,L}$ would show appreciable current at potentials negative to –50 mV. As shown by the steady-state current–voltage (I–V) relation in FIGURE 1B, current was very small between –90 and –50 mV. The current at more positive potentials is through g_{DR}, and the small current at potentials negative to –90 mV is through inwardly rectifying channels. At P4, a cell without $g_{K,L}$ and without a calyx may be either a type II cell or an immature type I cell that has not yet acquired $g_{K,L}$ and its calyx.[3] By P8 a cell without $g_{K,L}$ is likely to be a type II cell.[3] In the following paragraphs we relate features of the cells' frequency dependence (FIGS. 2–4) to properties of g_{met} and g_{DR} that are illustrated for a single cell in FIGURE 1.

The cell of FIGURE 1 adapted to step stimuli with a time course and extent close to the mean values for immature and type II cells from this preparation (FIG. 1C). At $P_0 \approx 0.5$,

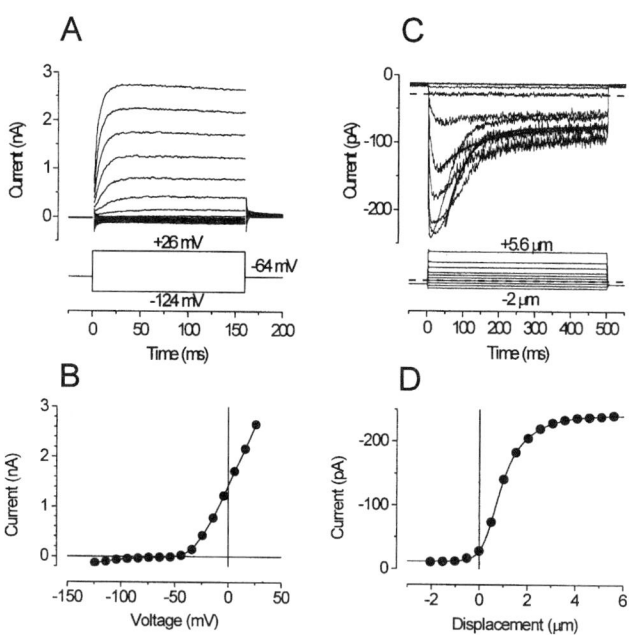

FIGURE 1. Voltage- and mechanically gated currents in a mouse utricular hair cell. The cell was in a P4 epithelium. (**A**) Whole-cell currents evoked by 160-ms voltage steps from the holding potential of –64 mV to potentials between –124 and +26 mV. Capacity transients at the step transitions have been blanked. (**B**) Isochronal I–V relation, taken 150 ms after the start of the iterated voltage steps in part (A). Steps above –50 mV activated an outward current through g_{DR}. A small inwardly rectifying current is evident below –75 mV. (**C**) Transduction currents evoked by a series of step-bundle deflections. *Dashed lines* show resting bundle position and current. (**D**) I(X) function. Peak currents in part (A) were plotted against hair-bundle displacement. *Curve* was obtained by fitting a second-order Boltzmann function to the data.

the major time constant of adaptation (τ_A) was 76 ms (see fit to smaller step in FIG. 1C), and the extent of adaptation was 54% [extent = (peak current − steady-state current)/(peak current − background current)]. Mean values for cells without $g_{K,L}$ (type II and immature cells) were 61 ± 7.6 ms and 62 ± 3.3%.[21] The current-displacement ($I(X)$) function, a plot of the peak transduction current against the displacement of the tip of the tallest stereocilia, is shown in FIGURE 1D. As in other hair cells, this function is well-fit by a second-order Boltzmann (solid curve) and is highly asymmetric. Saturatingly large negative stimuli generated a small fraction of the current evoked by saturatingly large positive stimuli. Moreover, the function's slope changed steeply at the resting position, so that even small positive stimuli generated larger responses than did small negative stimuli.

In FIGURE 2, transduction currents (A) and receptor potentials (B) are shown in response to sinusoidal bursts at various stimulus frequencies, for the same cell as in FIGURE 1. Stimulus amplitude was held constant. In FIGURE 3 the peak–peak (A) and average components (B) of

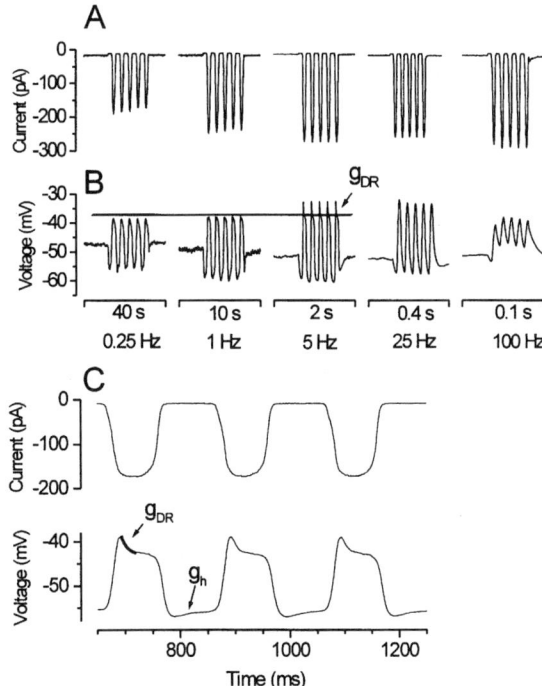

FIGURE 2. (A) Transduction currents and (B) receptor potentials evoked by deflecting the bundle with sinusoidal bursts of constant amplitude and at five frequencies from 0.25 to 100 Hz. Same cell as in FIGURE 1. The straight line in part (B) shows the level to which the peak receptor potential at 5 Hz decayed after activation of g_{DR}. The receptor potential only reached that level at lower frequencies, at which g_{DR} activation was in equilibrium with the slowly changing membrane potential. (C) Transduction current and receptor potential data at 5 Hz on an expanded time scale. From a different cell in the same preparation. The *thick line* is the fit of a monoexponential function to the decay from the depolarizing peak. The time constants from fits of all the peaks at 5 Hz in this cell and in the cell of parts (A) and (B) were between 12 and 20 ms. This time course is consistent with the activation kinetics of g_{DR} in this voltage range.[4] In part (C) the hyperpolarizing phase of the receptor potential also relaxed, reflecting activation of a slow inward rectifier, g_h.

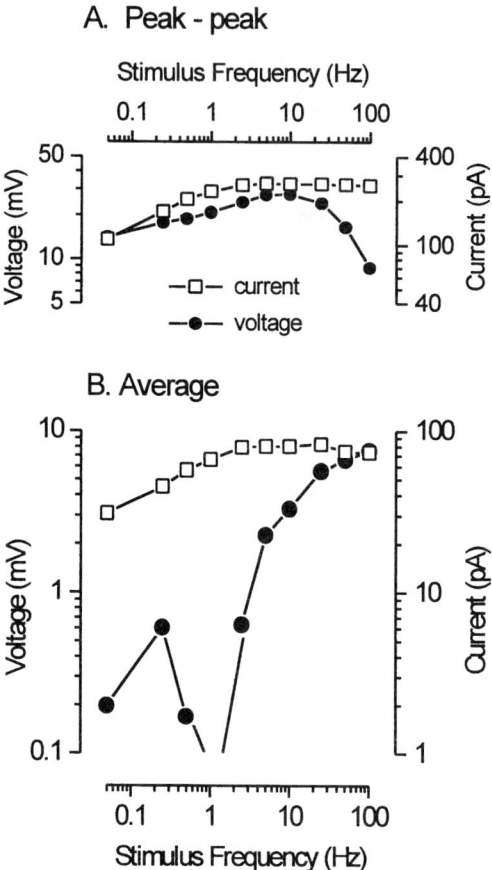

FIGURE 3. The frequency dependence of the transduction current (*open squares*) and receptor potential (*filled circles*) for the cell of FIGURES 1 and 2A,B. (**A**) Peak–peak (AC) responses. The peak–peak response was measured for each of four consecutive stimulus cycles in each burst, and the mean value is plotted. (**B**) Average (DC) responses. At each frequency, the current or voltage during four consecutive stimulus cycles was averaged and the resting level was subtracted.

the transduction currents and receptor potentials are plotted separately as functions of stimulus frequency, again for the cell of FIGURES 1 and 2. The peak–peak component can be compared with the fundamental component of primary afferent responses to linear accelerations (FIG. 4). The average component provides one measure of the nonlinear distortion in the response. The average component in the transduction current, which is always inward, arises because of the highly asymmetric $I(X)$ relation of the hair cell (FIG. 1D). The hair cell generates much larger transduction current for the positive half-cycle of a sinusoidal bundle deflection than it does for the negative half-cycle. As a result, the current averaged over the sinusoidal burst is inward.

Frequency Dependence of the Transduction Current

Both the peak–peak and average components of the transduction current were approximately flat with frequency above 2.5 Hz and fell off at lower frequencies with a similar slope (FIG. 2A and FIG. 3A,B, open squares). This low-frequency roll-off resulted from adaptation of the transduction current. For a positive step of similar amplitude (one-half the peak–peak amplitude of the sine wave, see fit to larger step in FIG. 1C), the response decayed with a major τ_A of 52 ms and an extent of 70%. This time course predicts a high-pass corner frequency, f_c, of 3 Hz, consistent with the sinusoidal data of FIGURES 2 and 3. At the lowest frequency, 0.05 Hz, the peak response was 40% of its value at frequencies above 3 Hz, reasonably close to the 30% of the peak current that remained at the end of a 500-ms adapting step (FIG. 1C).

Frequency Dependence of the Receptor Potential

The frequency dependence of the receptor potential is more complex (FIG. 2B and FIG. 3, filled circles). The roll-off in the peak–peak component above 10 Hz (FIG. 3A) is largely explained by the time required to charge the cell's membrane capacitance, C_m. At high frequencies, there was insufficient time per stimulus cycle to fully charge C_m. Low-pass corner frequencies (f_c) estimated from membrane time constants varied from 30 to 150 Hz for type II and neonatal cells.[21] In the example here f_c was ~30 Hz, consistent with an input resistance of ~1.5 GΩ (calculated as $R_{in} = (2\pi C_m f_c)^{-1}$ and assuming $C_m = 3.5$ pF, the population mean[3]).

The peak–peak component of the receptor potential also fell off at stimulus frequencies below 10 Hz (FIG. 3A). Below 2.5 Hz, the transducer adaptation contributed to this roll-off, but between 2 and 10 Hz the roll-off must reflect voltage-dependent properties of the cell, as it did not occur in the transduction currents. The voltage-dependent properties had a stronger effect on the average component of the response (FIG. 3B). The average component of the receptor potential declined from ~8 mV at 100 Hz to ~0 mV by 1 Hz. The roll-off in the average component means that as frequency decreased, the depolarizing and hyperpolarizing half-cycles became increasingly symmetric. This is evident in the traces in FIGURE 2B.

These differences in the frequency dependence of the transduction current and the receptor potential can be explained by the nonlinear steady-state *I–V* relation of the cell (FIG. 1B) and the activation kinetics of g_{DR}. At low frequencies, depolarization during the positive half-cycle of the stimulus substantially activated g_{DR}. This fed back negatively on the depolarization by producing outward currents that opposed the inward transduction current and by dramatically reducing the cell's input resistance, so that a given inward current evoked a much smaller depolarization. The effect of time-dependent activation of g_{DR} can be seen at the depolarizing peak of the receptor potential at 5 Hz (FIG. 2B,C; data in FIG. 2C are from a different cell). The receptor potential showed a relaxation from the peak that was not evident in the transduction currents (FIG. 2A and top trace, FIG. 2C). Monoexponential fits of the relaxations yielded time constants between 12 and 20 ms, similar to the major time constant of activation of g_{DR} between −30 and −40 mV.[3]

Below 5 Hz, no relaxation occurred at the depolarizing peak of the receptor potential because g_{DR} activation was in equilibrium with the slowly changing membrane potential; note that the receptor potential peaked at about the level to which it relaxed at 5 Hz (line). The strong asymmetry of the transduction current was countered by g_{DR} activation, producing in the example in FIGURE 2B a symmetrical receptor potential with almost zero average component. Relaxations were also sometimes seen at the troughs of the receptor potential (FIG. 1C). These reflected activation of a slow inward rectifier, g_h, by the hyperpolarization.[3,7,22] At 25 and 100 Hz, no relaxation occurred from the depolarizing peak of the receptor potential because g_{DR} activation was too slow to follow the rapid changes in membrane potential.

FIGURE 4 shows the frequency dependence of the peak–peak (A) and average (B) receptor potentials from all four cells for which we have receptor potential data over a broad frequency range. Stimulus amplitude was held constant across frequency for each cell, but varied between cells. Stimuli were chosen to be saturatingly large. The data from the four cells showed similar trends with frequency. The peak–peak component peaked between 2 and 20 Hz and had similar low- and high-frequency slopes, and the average component declined with decreasing frequency to values between 0.2 and 3 mV at 0.05 Hz.

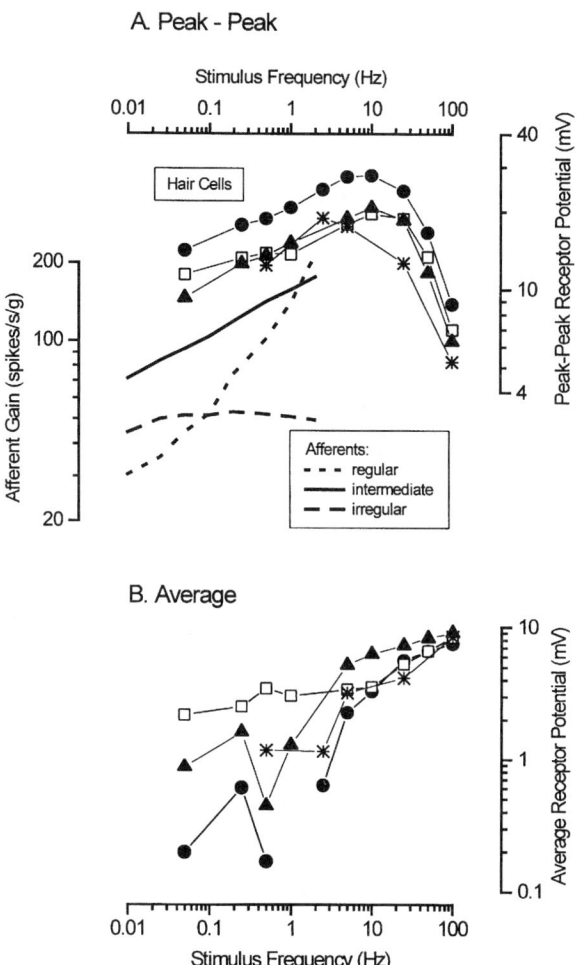

FIGURE 4. The frequency dependence of the peak–peak (**A**) and average (**B**) components of the receptor potential for four hair cells, including the two cells of FIGURES 1–3. Also shown in part (A), on a separate ordinate, are Bode plots of the gain of the fundamental component of primary afferent responses to sinusoidal linear accelerations (replotted from figure 4 in reference 1). The three curves are the mean curves for three groups of afferents, distinguished by the regularity of their background discharge. Linear regression fits (not shown) of these double logarithmic plots of intermediate afferent data (*solid line*) and peak–peak hair-cell data below 3 Hz yielded slopes of ~0.15 log units/log unit.

Comparison with Primary Afferent Data

Also shown in FIGURE 4A are primary afferent data from the utricular nerve of another rodent, the chinchilla, borrowed from reference 1. The fundamental component of the discharge evoked by sinusoidal linear accelerations is plotted as a function of stimulus frequency at and below 2 Hz. The three curves are mean data from three groups of afferents, grouped according to the regularity of their background discharges. The receptor potential data fell off below 3 Hz with a slope similar to that of the data from the intermediate group of utricular afferents.

Regularity of afferent discharge varies with region within the chinchilla utricular macula,[18] increasing with distance from the striola. The intermediate group shown in FIGURE 4 is representative of afferent discharges in between the striola and the periphery of the extrastriola. We did not note the position within the epithelium of the four hair cells shown in FIGURE 4. It therefore remains to be determined whether there is a good match in the frequency dependencies of hair cells and afferent fibers within specific regions of the epithelium. Regional variation in the frequency dependence of receptor potentials might arise from variation in either the mechanosensitive or voltage-dependent conductances. Preliminary evidence for the latter is described next.

Regional Variation in the Kinetics of K Conductances

FIGURE 5 shows whole-cell currents evoked by standard voltage protocols in two sets of neighboring type II cells: (A) striolar and (B) extrastriolar, from two different utricles. The kinetic properties of the large outward K currents evoked by depolarizing voltage steps differ strikingly between regions. On average, the striolar currents activated and inactivated more slowly than the extrastriolar currents, and were larger. Although the two

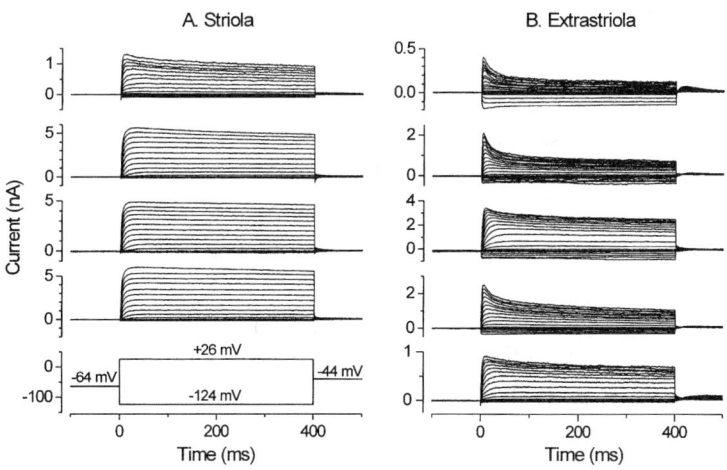

FIGURE 5. Regional variation in voltage-gated currents of neighboring type II cells from (**A**) the striolar region of a P7 utricle (4 cells), and (**B**) the extrastriolar region of a P14 utricle (5 cells). The envelope of the voltage-step series is shown at the bottom of part (A). Cells were immediately adjacent to each other, separated only by supporting cells. Although the current profiles showed some overlap [compare top traces in part (A) and bottom traces in part (B)], the outwardly rectifying currents in the striola were, on average, larger and slower to activate and to inactivate.

utricles shown in FIGURE 5 were of different ages, the difference in currents is related to region and not age. The time to half-peak current was measured for steps to +36 mV from a holding potential of −64 mV in type II cells sampled from various utricles at different ages (P6–P14). It was 1.8 ± 0.2 ms for 10 extrastriolar cells and 2.9 ± 0.3 ms for 6 striolar cells, significantly different at the 0.05 level (Student's t-test). An inactivation index was measured from the same cells by dividing the peak current by the current 390 ms after the onset of the 400-ms step. This index also differed significantly: 1.3 ± 0.1 for striolar cells and 1.9 ± 0.14 for extrastriolar cells (0.05 level, Student's t-test).

In extrastriolar hair cells, the component of the total current that has fast activation and inactivation kinetics fits the description of an A-current. One simple hypothesis to account for the different current profiles in FIGURE 5 is that there is between-cell variation in the absolute numbers of channels belonging to two K conductances, g_A and g_{DR}. The cells in the bottom three panels in A would have mostly g_{DR} channels and the cells in the top two panels in B would have relatively few g_{DR} channels, and intermediate response profiles would reflect more equal mixtures.

DISCUSSION

The receptor potential data presented here, though from a small set of hair cells, can be used as a starting point for a fuller comparison of receptor potential and afferent data. Ultimately such comparisons may reveal how the properties of hair cells in mammalian vestibular organs shape the afferent signals that are fed to postural and gaze reflexes.

Because the receptor potentials were evoked by hair-bundle deflections, their frequency dependence arises within the mechanosensitive and voltage-sensitive conductances of the hair cell, and not at other sites that may affect the afferent data: the mechanical response of the otolith, synaptic mechanisms, or spike-generating mechanisms in the afferent nerve fibers. Here we have noted three sources of frequency dependence affecting the peak–peak receptor potential: high-pass filtering by (1) transducer adaptation and (2) K-channel activation, and low-pass filtering by the membrane capacitance. There is no evidence in mammalian (or avian[16]) vestibular organs for the high-quality electrical resonances exhibited by hair cells of many other organs (see reference 23 for a review).

Over the low-frequency range (≤2 Hz) for which we can compare afferent data, only the high-pass mechanisms have effects. For our set of four hair cells, with similar adaptation and K current profiles, these two mechanisms produced, unsurprisingly, similar attenuation with decreasing frequency below 2 Hz. The slope of the attenuation approximated that of spike discharges from the group of primary afferents with intermediate spontaneous and evoked discharge properties. This preliminary comparison serves only to illustrate that hair-cell properties may contribute substantially to the frequency dependence of afferent discharges. Any similarities may be fortuitous given the large differences in the two preparations. For example: most afferents receive input from both type I and type II hair cells; the *in vitro* preparation lacks several stages that may affect the frequency dependence of the primary afferents; the *in vitro* data were obtained at room temperature rather than body temperature, and with the cells bathed in higher external Ca^{2+} concentration than the hair bundles would face *in vivo*. The lower temperature of the *in vitro* preparation would be expected to slow adaptation and K channel kinetics. The higher external Ca^{2+} concentration would enhance the rate of adaptation. Finally, the hair cells in Figures 1–4 were from utricles during the first postnatal week, an active time for both electrophysiological and morphological development,[3] and the afferent data were from adult chinchillas. The voltage range of activation of g_{DR} is similar in neonatal and older hair cells, but it is possible that features that we did not characterize in these cells, such as long-term inactivation, change with age in ways that influence receptor potentials at low stimulus frequencies.

The variation in frequency dependence within the afferent population, illustrated by the three mean curves in Figure 4A, has a regional basis.[18] We are beginning to uncover regional variation in the voltage-dependent properties of hair cells (FIG. 5). Similar differences have been observed between the central and peripheral zones of the turtle posterior crista.[24] Hair cells of the central zone have larger, slower outward K currents than do hair cells in the periphery. This suggests that regional variation in hair-cell properties is conserved across amniote vestibular organs, as are certain afferent properties. It is not yet clear, however, whether the regional variation in hair-cell properties is relevant to the regional variation in afferent properties. Modeling of data from rat utricular type II cells (M. Saeki and R. A. Eatock, in preparation) suggests that the observed kinetic differences between type II K conductances affect the frequency at which the receptor potential peaks (≥ 10 Hz), but not responses below 2 Hz. As previously noted, we have not determined whether the K conductances experience long-term inactivation, which could affect responses at low frequencies. In the turtle posterior crista, the fast K currents of the peripheral zone show more long-term inactivation in response to small steady depolarizations than do the slow central-zone currents.[25]

The receptor potentials evoked by large bundle deflections were largest around 10 Hz. No afferent data exist above 2 Hz, but recent data on the "translational vestibulo-ocular reflex" suggest that there is appreciable otolith afferent activity at much higher frequencies.[26] Rhesus monkeys with plugged semicircular canals make compensatory eye movements in the dark in response to sinusoidal horizontal body translations at frequencies up to 25 Hz.

The measured responses of otolith afferents to low-frequency head movements have relatively low nonlinear distortion.[1,27] The principal distortion is that arising from the asymmetry of transduction, producing a DC or 0-frequency component. This was 10–20% in squirrel monkey otolith afferents (at 0.1 Hz, 0.4 g), roughly consistent with the average receptor potential (0.1–3 mV) as a percentage of the peak–peak receptor potential (10–20 mV) at 0.1 Hz in FIGURE 4. The data in FIGURES 2 and 3 suggest that voltage-dependent features such as g_{DR} reduce DC distortion relative to that present in the transduction current. Thus, while g_{DR} attenuates the receptor potential at low frequencies, it may at the same time reduce nonlinearity. This effect could be strategic in a system designed to provide input about low-frequency head movements to reflexes controlling gaze and posture.

ACKNOWLEDGMENT

This work was supported by NIDCD Grant DC02290. One of the authors (J. R. H.) is a fellow of the Howard Hughes Medical Institute (HHMI) in the laboratory of D. P. Corey (Investigator, HHMI).

REFERENCES

1. GOLDBERG, J. M., G. DESMADRYL, R. A. BAIRD & C. FERNÁNDEZ. 1990. The vestibular nerve of the chinchilla. IV. Discharge properties of utricular afferents. J. Neurophysiol. **63**: 781–790.
2. LEWIS, E. R., E. L. LEVERENZ & W. S. BIALEK. 1985. The Vertebrate Inner Ear. CRC Press. Boca Raton, Fla.
3. RÜSCH, A., A. LYSAKOWSKI & R. A. EATOCK. 1998. Postnatal development of type I and type II hair cells in the mouse utricle: acquisition of voltage-gated conductances and differentiated morphology. J. Neurosci. **18**: 7487–7501.
4. RÜSCH, A. & R. A. EATOCK. 1996. A delayed rectifier conductance in type I hair cells of the mouse utricle. J. Neurophysiol. **76**: 995–1004.
5. CORREIA, M. J. & D. G. LANG. 1990. An electrophysiological comparison of solitary type I and type II vestibular hair cells. Neurosci. Lett. **116**: 106–111.

6. RENNIE, K. J., A. J. RICCI & M. J. CORREIA. 1996. Electrical filtering in gerbil isolated type I semicircular canal hair cells. J. Neurophysiol. **75**: 2117–2123.
7. RÜSCH, A. & R. A. EATOCK. 1996. Voltage responses of mouse utricular hair cells to injected currents. Ann. N.Y. Acad. Sci. **781**: 71–84.
8. SHEPHERD, G. M. & D. P. COREY. 1994. The extent of adaptation in bullfrog saccular hair cells. J. Neurosci. **14**: 6217–6229.
9. RICCI, A. J. & R. FETTIPLACE. 1997. The effects of calcium buffers on mechanoeletrical transduction in turtle hair cells. Biophys. J. **72**: A266.
10. KIMITSUKI, T. & H. OHMORI. 1992. The effect of caged calcium release on the adaptation of the transduction current in chick hair cells. J. Physiol. **458**: 27–40.
11. KROS, C. J., A. RÜSCH & G. P. RICHARDSON. 1992. Mechano-electrical transducer currents in hair cells of the cultured neonatal mouse cochlea. Proc. R. Soc. Lond. B **249**: 185–193.
12. HOLT, J. R., D. P. COREY & R.A. EATOCK. 1997. Mechanoelectrical transduction and adaptation in hair cells of the mouse utricle, a low-frequency vestibular organ. J. Neurosci. **17**: 8739–8748.
13. ART, J. J. & R. FETTIPLACE. 1987. Variation of membrane properties in hair cells isolated from the turtle cochlea. J. Physiol. **385**: 207–242.
14. HUDSPETH, A. J. & R. S. LEWIS. 1988. Kinetic analysis of voltage- and ion-dependent conductances in saccular hair cells of the bull-frog, *Rana catesbeiana*. J. Physiol. **400**: 237–274.
15. HUDSPETH, A. J. & R. S. LEWIS. 1988. A model for electrical resonance and frequency tuning in saccular hair cells of the bull-frog, *Rana catesbeiana*. J. Physiol. **400**: 275–297.
16. CORREIA, M. J., B. N. CHRISTENSEN, L. E. MOORE & D. G. LANG. 1989. Studies of solitary semicircular canal hair cells in the adult pigeon. I. Frequency- and time-domain analysis of active and passive membrane properties. J. Neurophysiol. **62**: 924–945.
17. ASHMORE, J. F. & D. ATTWELL. 1985. Models for electrical tuning in hair cells. Proc. R. Soc. Lond. B **226**: 325–344.
18. GOLDBERG, J. M., G. DESMADRYL, R. A. BAIRD & C. FERNÁNDEZ. 1990. The vestibular nerve of the chinchilla. V. Relation between afferent discharge properties and peripheral innervation patterns in the utricular macula. J. Neurophysiol. **63**: 791–804.
19. MCBRIDE, D. W. & O. P. HAMILL. 1995. A fast pressure-clamp technique for studying mechanogated channels. *In* Single-Channel Recording. 2nd Ed. B. Sakmann and E. Neher, Eds.: 329–340. Plenum. New York.
20. SHOTWELL, S. L., R. JACOBS & A. J. HUDSPETH. 1981. Directional sensitivity of individual vertebrate hair cells to controlled deflection of their hair bundles. Ann. N.Y. Acad. Sci. **374**: 1–10.
21. HOLT, J. R., A. RÜSCH, M. A. VOLLRATH & R. A. EATOCK. 1998. The frequency dependence of receptor potentials in hair cells of the mouse utricle. Prim. Sensory Neuron **2**: 233–241.
22. HOLT, J. R. & R. A. EATOCK. 1995. The inwardly rectifying currents of saccular hair cells from the leopard frog. J. Neurophysiol. **73**: 1484–1502.
23. EATOCK, R. A., M. SAEKI & M. J. HUTZLER. 1993. Electrical resonance of isolated hair cells does not account for acoustic tuning in the free-standing region of the alligator lizard's cochlea. J. Neurosci. **13**: 1767–1783.
24. BRICHTA, A. M., A. AUBERT, R. A. EATOCK & J. M. GOLDBERG. 1998. Ionic currents of solitary hair cells selectively harvested from different zones of the turtle posterior crista. Assoc. Res. Otolaryng. Abstr. **21**: 24.
25. BRICHTA, A. M. & J. M. GOLDBERG. 1998. Voltage responses of hair cells from the turtle posterior crista to sinusoidal electric currents. Assoc. Res. Otolaryng. Abstr. **21**: 24.
26. ANGELAKI, D. E. 1998. Three-dimensional organization of otolith-ocular reflexes in rhesus monkeys. III. Responses to translation. J. Neurophysiol. **80**: 680–695.
27. FERNÁNDEZ, C. & J. M. GOLDBERG. 1976. Physiology of peripheral neurons innervating otolith organs of the squirrel monkey. III. Response dynamics. J. Neurophysiol. **39**: 996–1008.

The Planes of the Utricular and Saccular Maculae of the Guinea Pig

I. S. CURTHOYS,[a,c] G. A. BETTS,[a] A. M. BURGESS,[a] H. G. MacDOUGALL,[a] A. D. CARTWRIGHT,[a] AND G. M. HALMAGYI[b]

[a]*Vestibular Research Laboratory, Department of Psychology, The University of Sydney, Sydney, New South Wales 2006, Australia*

[b]*Eye and Ear Research Unit, The Department of Neurology, Royal Prince Alfred Hospital, Sydney, New South Wales 2006, Australia*

ABSTRACT: To establish a link between otolith anatomy and function it is necessary to know the regions of the utricular and saccular maculae, which are stimulated by any arbitrary linear acceleration stimulus. That requires accurate information about the location and orientation of the spatially extended maculae in head-fixed coordinates and referred to head-fixed landmarks (such as Reid's line). New data showing the location of the otolithic maculae in the guinea pig with respect to head-fixed stereotaxic coordinates are presented. Guinea pigs were perfused with Karnovsky's fixative and the maculae were exposed while the head was held in a guinea pig stereotaxic device. An electrolytically sharpened fine wire held in a calibrated micromanipulator was touched to points all over the surface of each macula under visual observation with the aid of a high-power operating microscope. The x, y, z coordinates of these points were plotted using a three-dimensional plotting program. Both maculae have pronounced curvature so that dorsoventral shear forces will stimulate regions of both the utricular and saccular maculae.

INTRODUCTION

In the literature there are excellent photographs and illustrations of the gross anatomical characteristics of the utricular and saccular maculae (e.g., reference 1), but there is little information about the relationship of the otolith receptor surface to skull landmarks, which is the information needed for interpreting behavioral studies of otolith function. Recent investigators have approximated the otolithic maculae as being flat surfaces, perpendicular to each other and roughly in the cardinal planes of the head, and this simplification is frequently used to model the response of receptors on these complex structures (e.g., references 2 and 3). However even early anatomical investigations showed that such a simplification was not correct.[4–6] The significant error of the more recent approximations is shown by the fact that the most common schematic representation of the human otolithic maculae[3] depicts the long axis of the human saccular macula to be oriented vertically with respect to head, whereas in fact it is oriented almost horizontally with respect to head (i.e., almost in the human horizontal stereotaxic plane).[4,5]

---

[c]To whom correspondence may be addressed. Phone: 61 2 9351 3570; fax: 61 2 9351 2603; e-mail: ianc@psych.usyd.edu.au

One reason the orientation of these structures with respect to head is not readily available is that in order to carry out the difficult dissection needed to reveal the maculae, the whole temporal bone is usually removed from the skull, thus losing the very skull landmarks that are necessary for orienting the head in experimental studies. The surgical exposure is difficult because the maculae are delicate membranous structures located in the dense petrous temporal bone. In the case of the utricular macula most receptors are located on a thin membrane that is stretched across the fluid-filled vestibule. The saccular macula adheres to the bony inner wall of the saccule of the inner ear.

There have been attempts to obtain the position of these structures by reconstructions, but these attempts have not been successful as shown by the very large discrepancies between different investigators.[5-7] There are very substantial disagreements in the literature about the orientation and configuration of the maculae; for example de Burlet's data[6] is very different from that of Corvera *et al.*[7] Spoendlin has published quasi-quantitative representations of the human utricular and saccular macula topography[8] that appear to be visual estimates of the contours of the structures rather than being obtained by quantitative measurement.

Because of the importance of the anatomical data for understanding otolithic function we sought to obtain quantitative specification of the location of the otolithic maculae in the guinea pig with respect to head landmarks. The guinea pig was studied because the gross and fine anatomy of its vestibular labyrinth is better understood than any other species. Multiple specimens were studied to demonstrate the extent of variation between animals.

METHODS

Guinea pigs were deeply anesthetized with nembutal and perfused with 400 mL of saline containing heparin followed by 500 mL of phosphate-buffered Karnovsky's fixative (4% paraformaldehyde and 5% glutaraldehyde).[9] The head was mounted in a guinea pig stereotaxic apparatus (Kopf) and the inner ear of one labyrinth was dissected to show the maculae. The otoconia and otoconial membrane were very gently removed by a fine jet of saline. Saline was removed from the surface of the utricular macula so that errors due to refraction would not confound the measurements. The macula surface was not allowed to dry—a thin layer of saline always bathed the macula. Methylene blue was used to show the macula surface, staining individual receptor hair cells that could be just discerned at the maximum power of the operating microscope used here. In freshly fixed specimens the hair cells at the striola (the type I hair cells)[1,10,11] stained more darkly than the surrounding hair cells, allowing measurement of the spatial location of the striola.

The x, y, and z coordinates of up to 300 data points per macula were obtained from the surface of the utricular and saccular macula using a Kopf precision micromanipulator that was calibrated in 50-micron steps and was specially modified to be free of backlash. Under very high power, the tip of a fine probe (electrolytically sharpened 40-micron stainless-steel wire) was touched to points on the surface of the macula. In some animals it was possible to obtain data from the utricular macula, the saccular macula, and the horizontal semicircular canal. The latter served as a reference since the orientation of the plane of the horizontal semicircular canal is very well established,[12] and so the plane of the horizontal canal may possibly be used to relate the macula orientation to head position.

The data points were entered into a statistical program, S-Plus,[13] running under Unix, and a program called xgobi[14] allowed the data points to be rotated *en bloc* to show more clearly the projection of the maculae into the major skull planes and to demonstrate the curvature of the macula surfaces.

RESULTS

FIGURES 1 to 4 show views of the guinea pig otolithic maculae in stereotaxic coordinates. The data from a number of animals are included (FIG. 3) to demonstrate the similarity of these data across different animals. FIGURES 1 to 3 show that the utricular macula is a curved structure, with the anterior portion being upturned like the front of a toboggan. To the extent that it is meaningful to assign a "plane" to such a curved structure, then one could describe "the plane" of the utricular macula as being pitched nose down by about 20 deg with respect to the plane of the horizontal semicircular canal. Such a coarse description does not do justice to the curved surface topography of the utricular macula (see FIGS. 1–3).

The pitch angle of the horizontal canal appears to be very large—around 45 deg (FIG. 1), but this is merely a consequence of the rather unusual definition of the stereotaxic horizontal plane in the guinea pig.[12] In fact the natural head position of the guinea pig is pitched down by about 30 deg relative to the stereotaxic horizontal so the horizontal canal is held close to being earth horizontal.[12]

The data from all the specimens are similar (FIG. 3) and that figure shows that there are systematic curves in the planes of both the utricular and saccular maculae. The saccular macula is attached tightly to the curving bony wall of the vestibule and has very close connections to Scarpa's ganglion through the thin cribriform bony plate on which the saccular macula is located. The saccular macula exhibits a complex twisted surface with clear

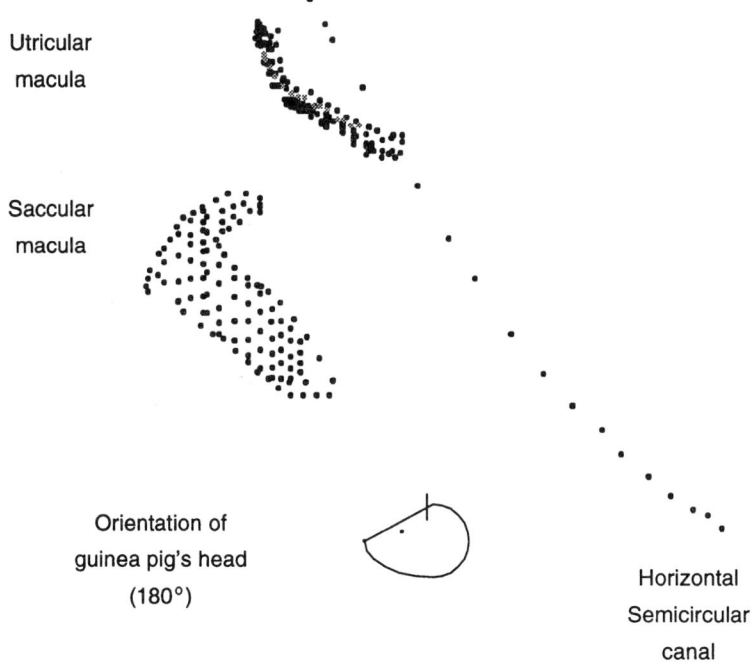

FIGURE 1. View from a directly side-on point of view of points from the surface of the guinea pig utricular macula, saccular macula, and horizontal canal from a left labyrinth in stereotaxic space. The edges of the figure correspond to the horizontal (*X*) and dorsoventral (*Z*) guinea pig stereotaxic axes.

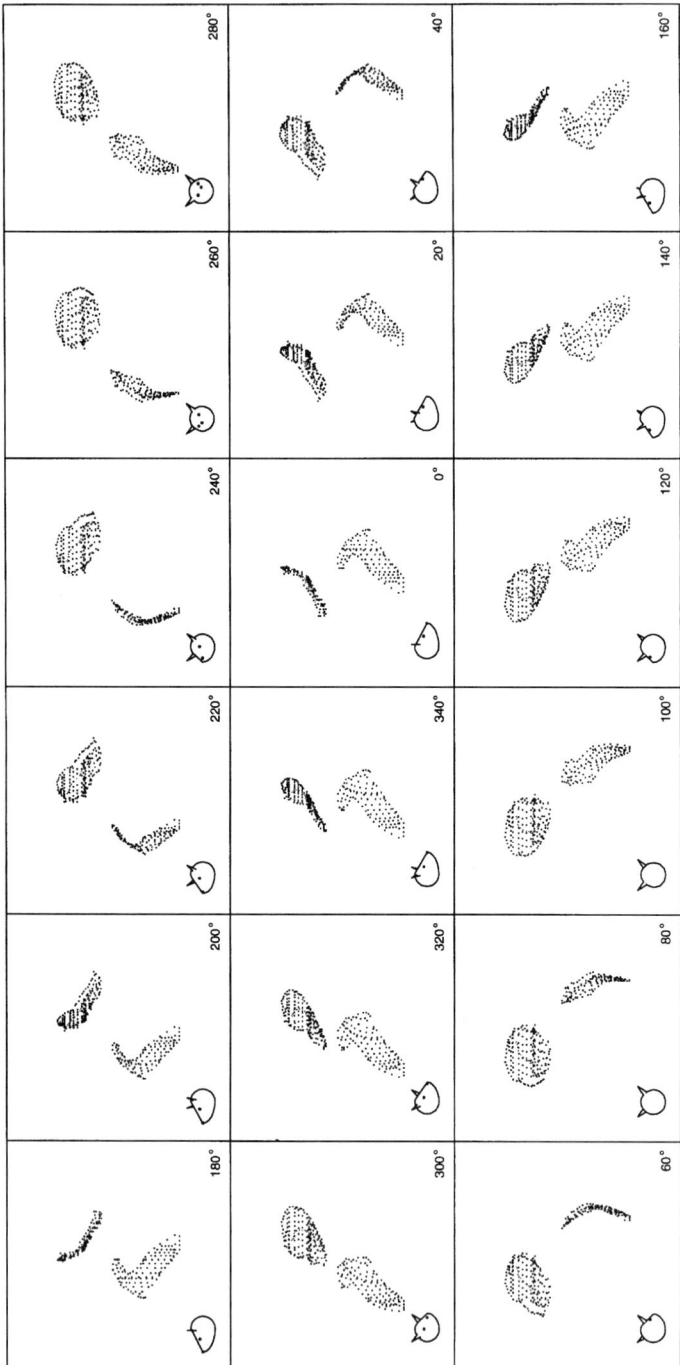

FIGURE 2. Single frames—"still photos"—from the rotation sequence of the guinea pig utricular and saccular macula rotated in 20-degree steps around a Z-axis. The view is from a directly lateral point of view, and the small schematic guinea pig head at the bottom of each inset shows head orientation.

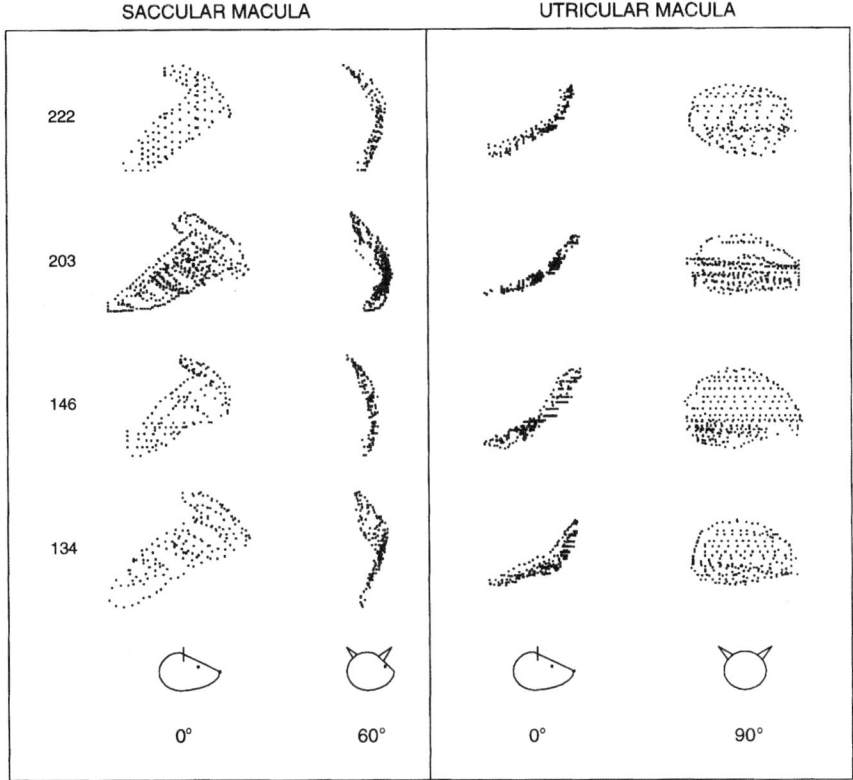

FIGURE 3. Views of the saccular and utricular maculae from 4 animals to show the similarity between animals. The maculae have been rotated to give an "edge-on" view that shows the curvature so clearly. The maculae have been arbitrarily shifted in the dorsoventral dimension to allow comparison between animals.

projections into the frontal plane as well as its predominant sagittal projection. The saccular macula is not simply a sheet of receptors in the sagittal plane but has a considerable projection into the frontal plane (FIG. 4) so that even interaural linear accelerations will stimulate the receptors on the guinea pig saccular maculae. This is clearly seen in direct views looking down on the saccular maculae that show the curved projection of the macula clearly (FIG. 4).

DISCUSSION

Contrary to the simplifications imposed on the maculae by artists and illustrators, it is clear from our data that curvature and asymmetry are more prominent features of the maculae than the flatness and symmetry that have characterized artistic representations. The results confirm other published data concerning the curvature of the utricular macula: Lindeman's photos show this arrangement in the guinea pig,[1] as does Curthoys.[15]

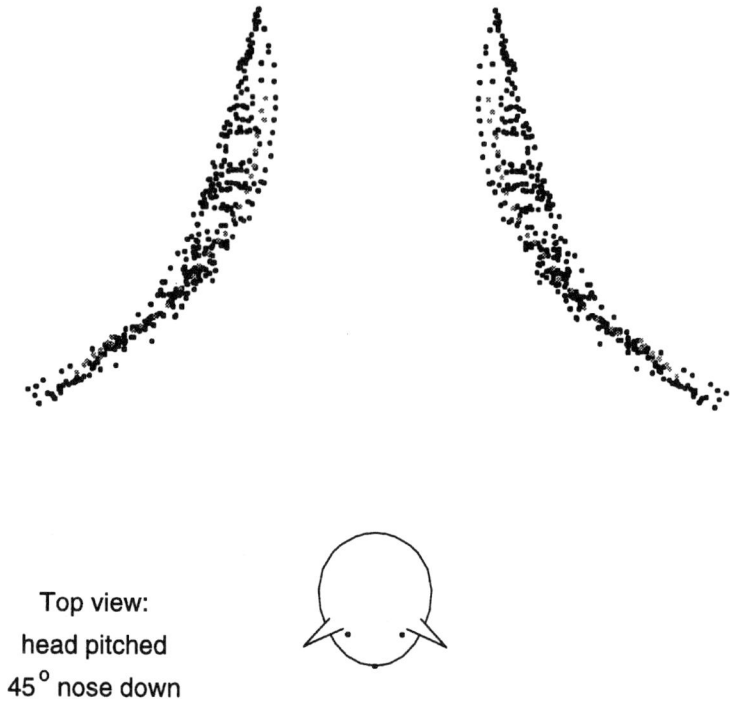

FIGURE 4. To show the curvature of the saccular macula in the guinea pig. This is a directly dorsal view of the data from a single animal with the head pitched nose down by 45 deg (as shown by the inset of the schematic head). The data from a single labyrinth have been reflected around the sagittal plane to represent the maculae in both labyrinths.

Appropriately sectioned temporal bones of the chinchilla also show this curved upturn of the anterior portion of the utricular macula.[10]

The data points were obtained from maculae in animals that had been fixed by perfusion. Could fixation have substantially altered the very characteristics under investigation? Could it have deformed the orientation of the maculae so that the data are invalid? Could the prominent curvature shown in these results be caused by shrinkage or fixation artifact? There are two strong reasons to doubt that fixation affected these results. First, the fixative used (phosphate-buffered Karnovsky's) was specifically chosen because electronmicroscopic evidence has shown that the particular concentration used here caused minimal shrinkage of labyrinth membranes at the electronmicroscopic level.[9] Second, inspection of the maculae in living anesthetized animals confirms the orientation and curvature shown by the quantitative measures. In some anesthetized guinea pigs the inner ear was opened and the utricular and saccular macula were each visualized directly through the endolymph, and they showed similar patterns of curvature to the data presented here.

The following elaborates how this spatial configuration data are important for understanding otolith function. Each macula is covered by receptor hair cells, each with its own preferred direction (polarization). The receptor hair cells are systematically arranged around the surface of the maculae with their polarization directions being exactly opposite on either side of the striola.[1] To relate these structures to function it is necessary to consider the individual hair-cell responses and use that information to determine which hair cells will be stimulated by any linear force stimulus to result in the discharge pattern across fibers in the vestibular nerve.

Physiological studies[15] have shown that for an individual hair cell, excitation is maximal when the direction of linear force corresponds to the polarization direction through the stereocilia toward the kinocilium. As the angle between the direction of this imposed linear force and the direction of the morphological polarization changes, so the magnitude of the excitatory neural response declines. Goldberg *et al.* have shown that in the extrastriola region a number of otolithic receptor cells typically synapse on an individual primary afferent[11] and probably as a result, the breadth of tuning of the preferred direction of primary otolithic afferents may be greater than 180 deg.[16] Shear is the effective stimulus for the otoliths—otolithic receptors are not sensitive to compression forces.[17]

Exact information on the orientation and position of the human macular surfaces is not yet available, so linear force stimuli should be specified by their components in the interaural and dorsoventral axes, with the presumption that the human utricular macula will be mainly stimulated by the interaural shear and the saccular macula predominantly by the dorsoventral shear. On the basis of our anatomical evidence from the guinea pig and the available anatomical data in the human,[5–7] this is only a very approximate assignment, and there are regions of the utricular macula that will be stimulated by dorsoventral shear forces.

ACKNOWLEDGEMENTS

We gratefully acknowledge the support of NHMRC of Australia and the partial support from Research Grant 5 R01 DC 02372-02 from the National Institute on Deafness and Other Communication Disorders, National Institutes of Health. The assistance of Warren Davies is gratefully acknowledged.

REFERENCES

1. LINDEMAN, H. H. 1969. Studies on the morphology of the sensory regions of the vestibular apparatus. Adv. Anat. Embryol. Cell Biol. **42**: 1–113.
2. MALCOLM, R. & G. MELVILL JONES. 1974. Erroneous perception of vertical motion by humans seated in the upright position. Acta Otolaryngol. **77**: 274–283.
3. HOWARD, I. P. 1982. Human Visual Orientation. Wiley. New York.
4. ROSENHALL, U. 1972. Vestibular macular mapping in man. Ann. Otol. Rhinol. Laryngol. **81**: 339–351.
5. QUIX, F. H. 1925. The function of the vestibular organ and the clinical examination of the otolithic apparatus. J. Laryngol. Otol. **50**: 425–443.
6. DE BURLET, H. M. 1930. Die Stellung der Maculae acusticae im Schädel des Menschen und einiger Säugetiere. Morphol. Jahrb. **64**: 377–393.
7. CORVERA, J., C. S. HALLPIKE & E. H. J. SCHUSTER. 1958. A new method for the anatomical reconstruction of the human macular planes. Acta Otolaryngol. **49**: 4–16.
8. SPOENDLIN, H. 1965. Strukturelle Eigenschaften der vestibularen Rezeptoren. Schweiz. Arch. Neurol. Neurochir. Psychiatry **96**: 219–230.
9. ANNIKO, M. & P.-G. LUNDQUIST. 1977. The influence of different fixatives and osmolarity on the ultrastructure of the cochlear neuroepithelium. Arch. Otolaryngol. **218**: 67–73.
10. FERNANDEZ, C., R. A. BAIRD & J. M. GOLDBERG. 1988. The vestibular nerve of the Chinchilla. I. Peripheral innervation patterns in the horizontal and superior semicircular canals. J. Neurophysiol. **60**: 167–203.
11. GOLDBERG, J. M., G. DESMADRYL, R. A. BAIRD & C. FERNANDEZ. 1990. The vestibular nerve of the chinchilla. V. Relation between afferent discharge properties and peripheral innervation patterns in the utricular macula. J. Neurophysiol. **63**: 791–804.

12. CURTHOYS, I. S., E. J. CURTHOYS, R. H. I. BLANKS & C. H. MARKHAM. 1975. The orientation of the semicircular canals in the guinea pig. Acta Otolaryngol. **80**: 197–205.
13. BECKER, R. A., J. M. CHAMBERS & A. R. WILKES. 1988. The new S language: a programming environment for data analysis and graphics. Brooks/Cole. Pacific Grove, Calif.
14. SWAYNE, D. F., D. COOK & A. BUJA. 1998. Xgobi: interactive dynamic data visualization in the X-window system. J. Comput. Graph. Stat. **7**: 113–130.
15. CURTHOYS, I. S. 1987. Eye movements produced by utricular and saccular stimulation. Aviat. Space Environ. Med. (Suppl.). **58**: A192–A197.
16. FERNANDEZ, C. & J. M. GOLDBERG. 1976. Physiology of peripheral neurons innervating otolith organs of the squirrel monkey. I. Response to static tilts and to long-duration centrifugal force. J. Neurophysiol. **39**: 970–984.
17. SHOTWELL, S. L., R. JACOBS & A. J. HUDSPETH. 1981. Directional sensitivity of individual vertebrate hair cells to controlled deflection of their hair bundles. Ann. N.Y. Acad. Sci. **374**: 1–10.

Miniature EPSPs and Sensory Encoding in the Primary Afferents of the Vestibular Lagena of the Toadfish, *Opsanus tau*

RACHEL LOCKE,[a] JEAN VAUTRIN,[b] AND STEPHEN HIGHSTEIN[a,c]

[a]*Washington University School of Medicine, Department of Otolaryngology, Box 8115, 4566 Scott Avenue, St. Louis, Missouri 63110, USA*

[b]*INSERM 432, Université Montpellier II, Visiting Laboratory of Neurophysiology, NINDS, NIH, Building 36, Room 2C02, 9000 Rockville Pike, Bethesda, Maryland 20892, USA*

ABSTRACT: The synaptic activity transmitted from vestibular hair cells of the lagena to primary afferent neurons was recorded *in vitro* using sharp, intracellular microelectrodes. At rest, the activity was composed of miniature excitatory postsynaptic potentials (mEPSPs) at frequencies from 5 to 20/s and action potentials (APs) at frequencies between 0 and 10/s. mEPSPs recorded from a single fiber displayed a large variability. For mEPSPs not triggering APs, amplitudes exhibited an average coefficient of variance (CV) of 0.323 and rise times an average CV of 0.516. APs were only triggered by mEPSPs with larger amplitudes (estimated 4–6 mV) and/or steeper maximum rate of rise (10.9 mV/ms, ± 3.7 SD, $n = 4$ experiments) compared to (3.50 mV/ms, ±0.07 SD, $n = 6$ experiments) for nontriggering mEPSPs. The smallest mEPSPs showed a fast rise time (0.99 ms between 10% and 90% of peak amplitude) and limited variability across fibers (CV: 0.18) confirming that they were not attenuated signals, but rather represented single-transmitter discharges (TDs). The mEPSP amplitude and rise-time relationship suggests that many mEPSPs represented several, rather than a single pulse of secretion or TDs. According to the estimated overall TD frequency, the coincidence of TDs contributing to the same mEPSP were not statistically independent, indicating a positive interaction between TDs that is reminiscent of the way subminiature signals group to form minature signals at the neuromuscular junction. Depending on the duration and intensity of efferent stimulation, a complete block of AP initiation occurred either immediately or after a delay of a few seconds. Efferent stimulation did not significantly change AP threshold level, but abruptly decreased mEPSP frequency to a near-complete block that followed the block of APs. Maximum mEPSP rate of rise decreased during, and recovered progressively after, efferent stimulation. After termination of efferent stimulation, mEPSP amplitude did not recover instantly and for a few seconds the amplitude distribution of synaptic events showed fewer large-amplitude events than during the control period. This confirms that mEPSP amplitude and rate of rise properties, which are critical for triggering afferent APs, are modified by efferent activity. The depression of afferent AP firing during efferent stimulation corresponded to a decrease in mEPSP frequency and, to a lesser extent, a decrease in mEPSP amplitude and rate of rise, suggesting, a decrease in the level of interaction among TDs contributing to a mEPSP.

[c]To whom correspondence may be addressed.

INTRODUCTION

The labyrinth of the vestibular system transduces head position and movement and transfers this information to the brain via a frequency code of complex dynamic signals carried on the VIIIth cranial nerve. Boyle *et al.*[1] identified five potential sites in the transduction cascade from head movement to nerve signal that might contribute to creating and shaping the dynamics of the afferent message. One of these sites is the synapse between the hair cell and its innervated primary vestibular afferent fiber.

Of the six labyrinthine endorgans, the lagena of the toadfish, *Opsanus tau*, was most favorable for the study of miniature synaptic potentials because they could be routinely recorded with high fidelity.[2] The lagena senses linear acceleration and is extremely sensitive,[3] perhaps actually functioning as a stationarity detector. In the absence of movement, spontaneous activity can be recorded in lagenar primary afferents as miniature excitatory postsynaptic potentials (mEPSPs) and action potentials (APs), indicating that there is ongoing transmitter release (most likely glutamate[4]) from hair cells. Movement of hair-cell stereocilia generates receptor potentials that modify this ongoing transmitter release. APs appear to be initiated by summation of mEPSPs in the afferent fiber. Thus, features of acceleration of the animal are encoded by the firing patterns of lagenar afferents.

The lagena is also endowed with a powerful central nervous system efferent innervation (FIG. 1) that appears to modify signal transmission to the brain.[5] In toadfish semicircular canal afferents, efferent activity has been documented to modify the message (AP frequency code) carried by canal afferents.[5-8] Although mEPSPs could not be recorded in semicircular canal afferents in this species, mEPSPs can routinely be recorded in lagenar afferents. The efferent synapse is commonly believed to be cholinergic (reviewed by Warr[9]), and inhibitory postsynaptic potentials have been recorded in vestibular hair cells following efferent stimulation (Boyle *et al.*[1]).

The present experiments were performed to study the effects of efferent stimulation upon AP initiation by mEPSPs in lagenar afferents. Results presented indicate that inhibition of the afferent message following efferent stimulation is due to a decrease in mEPSP frequency together with a decrease in mEPSP amplitude and rise time.

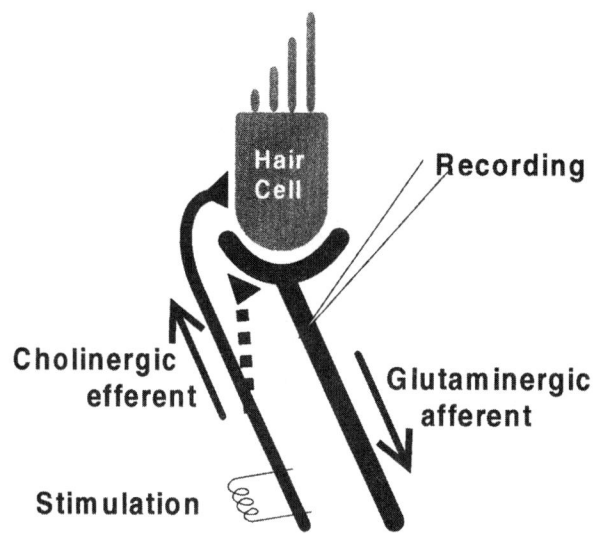

FIGURE 1. Diagrammatic representation of lagena epithelium innervation.

Early studies on neuromuscular transmission have suggested that miniature postsynaptic signals reflect the presynaptic release of invariable quanta because when neuromuscular junctions (NMJs) are mature and in resting condition, miniature signals exhibit a median modal amplitude (large-mode) and a low variability in amplitude and rise time.[10] However, subminiature signals and small-mode, skew-distributed miniature signals were also reported.[11,12] When recordings with improved resolving power are performed, large-mode and small-mode miniatures both appear to be composed of subminiature discharges of transmitter (TDs).[13–15] (In the absence of evidence indicating that subminiatures correspond to either partial or complete release of vesicle contents or another form of transient release, we conservatively use the term transmitter discharge or TD, originally introduced by Fatt and Katz,[10] to refer to the presynaptic events producing brief transient postsynaptic signals.)

The amplitude distribution of miniature synaptic potentials recorded at central synapses is usually skewed and it has frequently been proposed that mEPSPs that are several times the modal amplitude correspond to the simultaneous release of several TDs.[16,17] Subminiature TDs within a miniature cannot always be resolved, however, the only interpretation known for the stereotypical increase in rise time and the rise-time variability with the amplitude of small-mode miniatures at central and peripheral synapses is an extension of the duration of transmitter secretion. Thus it was suggested that small-mode miniatures are due to a short-term interaction between TDs; in other words, TDs might occur in bursts.[18,19] That the form of the amplitude distribution at a given NMJ changes dynamically with the experimental conditions indicated that the variability in miniature signal amplitude reflects variations in the number of TDs and their synchronization.[19–21]

The present study takes into account that lagenar mEPSP amplitude and rise-time exhibit the same type of relationship, fluctuations, and rapid changes as the amplitude and rising phase of small-mode miniature signals recorded at central synapses and challenged NMJs. Results indicate that inhibition of the afferent message following efferent stimulation is due a decrease in mEPSP frequency together with an alteration of mEPSP amplitude and rise time. We interpreted the rapid decrease and recovery of mEPSP peak amplitude and rise time following efferent stimulation as a possible rearrangement of the number and synchronization of TDs.

Preliminary reports have been published.[2]

METHODS

Preparation

Adult toadfish *Opsanus tau* of either sex (ca. 500 g) were supplied by the Marine Biological Laboratory, Woods Hole, Massachusetts, where they were maintained at the Marine Resources Center. Fish were lightly anesthetized by immersion in MS222 (3-aminobenzoic acid ethyl ester; Sigma, St. Louis, Missouri) and partially immobilized by an intramuscular injection of Pavulon (pancuronium bromide, 0.05 mg/kg; Sigma). Fish were secured in a plastic tank filled with sea water. A dorsal craniotomy exposed the vestibular labyrinth on one side leaving its innervation intact. The lagena and its afferent and efferent connections together with the medulla containing the efferent vestibular nuclei were removed and placed in a temperature-controlled, Sylguard-coated experimental chamber. The medulla, containing the efferent vestibular nuclei was inverted and carefully placed so that two minutin stainless-steel pins (Carolina Biological) sticking vertically out of the Sylguard and spaced 1 mm apart pierced the efferent nuclei. The chamber contained toadfish teleost Ringer, concentrations in mM Na^+: 165; K^+: 5; Ca^{2+}: 1.5; Mg^{2+}: 0.5.

Stimulation and Recording

All experiments were performed at 15°C. Microelectrodes were fabricated from standard wall borosilicate glass tubing and had resistances of 50–80 MΩ when filled with 3 M KCl. They were inserted into fibers at the base of the lagena where the nerve enters the epithelium, which was probably no more than 100 µm from the sensory epithelium. The recording amplifier was standard and had a bandwidth of DC-10 kHz. Electric pulse train stimulus parameters were, pulse width, 100 µs, frequency, 100 Hz, and amplitude adjusted for maximum effectiveness with a limit of several hundred µA. APs and mEPSPs were stored on magnetic videotape using a PCM-VCR system (VR100 from Instrutech, 37-kHz bandwidth) for future analysis.

Analysis

The electrophysiological signals were played back from magnetic tape and digitized at 50 kHz by a 1401 plus (CED) and a personal computer using Spike 2 software and/or SCAN and PAT software (Courtesy of John Dempster, Strathclyde University). The threshold level of the SCAN program detecting postsynaptic mEPSPs was set 0.5–1 mV above the resting potential level. Series of 40 to 700 (typically several hundreds) mEPSPs were automatically acquired. Each event captured was manually selected before being entered into the database individually. Comparing chart recording and events entered in the database we found that typically a few percent (rarely up to 10%) of possible events are missing from the statistics because they occurred in the dead time of the sampling program and were not captured. MEPSPs often overlapped and events occurring at an interval shorter than an individual rise time appeared integrated into a single-peak potential transient in which the individual time course could not always be differentiated. The absence of any transient negative slope during the rising phase was the objective criterion established to qualify an event as one mEPSP. This did not guarantee the discrimination between one and two presynaptic events when these occurred at a short interval; however, it preserved the consistency in analysis criteria. Positions of AP threshold were identified as the inflection point between the EPSP and AP rising phase using a high magnification time and voltage display and by referring to and tracing the differentiated signal (see FIG. 5). The maximum rising slope of AP-triggering mEPSPs was measured from the onset to the AP-threshold initiation point. The maximum rising slope of nontriggering mEPSPs was measured between the onset and peak amplitude. Peak amplitude and rise-time variability were expressed as coefficients of variance: $CV = \sigma/\mu$ (where σ is the standard deviation and μ the mean). All numbers are given as mean ± standard deviation.

RESULTS

Miniature Synaptic Potentials Recorded in Lagenar Primary Afferents

Glass microelectrodes were visually guided into the nerve root as close to its exit from the lagenar *macula* as possible. Entry into an afferent was signaled by an abrupt decrease in the baseline potential (afferent resting potential, mean = 70.5 ± 7.6 mV, $n=10$) and, in the unstimulated preparation the appearance of spontaneous depolarizations superimposed upon the hyperpolarized baseline. These spontaneous depolarizations were interpreted to be mEPSPs and occurred at frequencies from 5 to 20/s. Spontaneous APs were also regularly observed at frequencies up to 10/s. A break on the rising phase of most APs suggested that they were triggered by mEPSPs. These data taken together suggest that up to 50% of mEPSPs triggered APs. Ten afferents with stable resting potentials and stable levels of spontaneous activity were chosen for analysis. FIGURE 2A illustrates typical records of 14 consecutive signals from one fiber. Two AP-triggering and 12 nontriggering mEPSPs were superimposed by aligning them on their onset. Visual inspection indicated that steeply rising

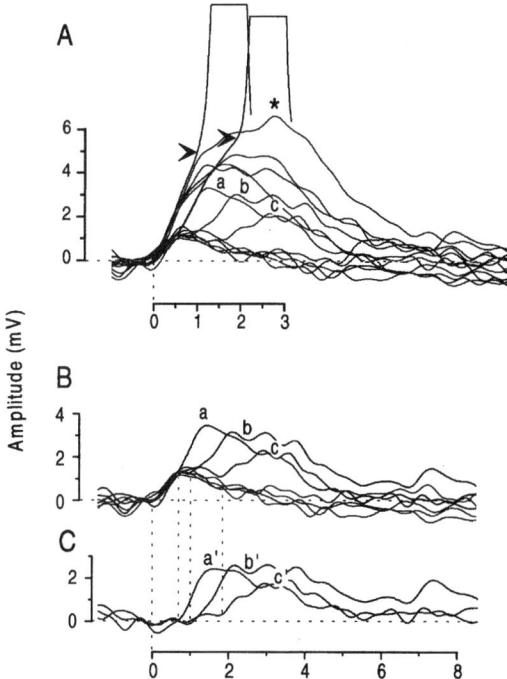

FIGURE 2. Representative series of consecutive mEPSPs recorded in terminals of primary neurons *in vitro*. (**A**) mEPSPs are superimposed, their onsets aligned. Two APs (*thicker lines*) were triggered during that series and are truncated on the figure. *Arrowheads* indicate the abrupt increase in slope corresponding to AP onset. The * points to an mEPSP that peaks at a potential above the threshold potential reached by other mEPSPs. The slope of this mEPSP was insufficient to trigger an AP. Before the onset of an AP, AP triggering mEPSPs are congruent with the time course of nontriggering mEPSPs. Some large mEPSPs are congruent with some smaller mEPSPs. Slow rising mEPSPs are congruent with the smallest mEPSPs. (**B**) Seven smallest mEPSPs from the series shown in part A. Signal *c* shows two peaks, and its rising phase exhibits temporarily a negative slope strongly suggesting that *c* corresponds to two presynaptic events of secretion. Such signals were considered as two mEPSPs. Signal *b* shows a constantly rising phase to a unique peak. The break on its rising phase and the congruency of its time course before the break with smaller mEPSP's time course strongly suggest that *b* also corresponds to two presynaptic events. However, due to the difficulty in appreciating less obvious breaks, such signals were classified as one mEPSP and incorporated in mEPSP statistics (see Methods). Signal *a* has no break on the rising phase; however, it could also result from two presynaptic events occurring at an interval < 1 ms as shown in part C. (**C**) Signals *a'*, *b'*, and *c'* were obtained from signals *a*, *b*, and *c* in part B after subtracting the average of the 4 smallest mEPSPs of the series shown in part B. Differentiated signals *a'*, *b'*, and *c'* exhibit time courses of putative mEPSPs with onset occurring at various time delays after the onset of the averaged mEPSPs. The time interval between signals *a* and *a'* is about 0.75 ms, between *b* and *b'* about 1 ms, and between *c* and *c'* about 1.8 ms.

mEPSPs were more likely to trigger APs than were their more slowly rising counterparts. Additionally, the voltage threshold at which an individual AP was initiated (horizontal arrows) also appeared to be dependent upon the slope of the rising phase of the triggering mEPSP; AP thresholds occurred at a more negative potential when the mEPSP rising phase was steeper (lower of two horizontal arrows). Note that nontriggering mEPSPs with slow rising phases (* on FIG. 2A) could be larger than mEPSPs that triggered an AP. It was therefore concluded that APs were triggered by mEPSPs that were relatively large and had relatively steep rising phases.

Characteristics of mEPSP

To facilitate the analysis of mEPSP characteristics, seven mEPSPs from FIGURE 2A are displayed again in FIGURE 2B. The three largest are labeled a, b, c. The average of the four smallest unlabeled mEPSPs was subtracted from a, b, and c to give the three voltages labeled a', b', c' in FIGURE 2C. The rising phase of signal c (FIG. 2B) shows two obvious peaks and a distinct region of negative slope, suggesting that c might correspond to two (or more) presynaptic events. In the analysis conducted below, such signals were considered to be two mEPSPs. In contrast to c, signal b shows a constantly rising phase to a unique peak. The break on the rising phase, and the congruency of the time course before the break with the smaller mEPSPs, suggests that b might also correspond to two presynaptic events. However, due to the subtlety of this less obvious break compared to the one in c, and in order to preserve consistency, such events were considered as one mEPSP. The signal a has no apparent break on its rising phase, and although it could also have arisen from two presynaptic events occurring simultaneously, signals such as a were also analyzed as unitary mEPSPs.

The peak amplitude and 10% to 90% rise time of several hundred non-AP triggering mEPSPs were measured from six fibers. These statistics exclude mEPSPs that triggered APs because peak amplitude and rise time could not be measured in those cases. Thus, true mEPSP amplitude and variability are likely to be larger than determined from the analysis below. Mean peak amplitude was found to be $3.03, \pm 0.57$ mV; mean CV = 0.324 ± 0.053 A typical frequency distribution of peak amplitudes is shown in FIGURE 3A. The distribution is skewed to the left, with small amplitudes being predominant. There are no events in the distribution with amplitudes less than 1 mV because such events could not be routinely discriminated from the baseline. Even if these events could have been measured, the distribution would obviously have retained its asymmetry. Individual mEPSP rise times ranged from 0.25 to 4 ms; mean = 1.44 ms, ± 0.25, with an average CV of 0.516 ± 0.085 mEPSP rise times were also asymmetrically distributed with a predominance of short rise times (FIG. 3B).

A typical relationship between the amplitude and rise time from a single fiber of mEPSPs (FIG. 3C) revealed that the smallest mEPSPs had the shortest rise times. For the six fibers analyzed, rising phases were restricted between two slopes; the mean minimum slope varied between 0.25 and 0.50 mV/ms and the mean maximum slope varied between 3 and 8 mV/ms. Further, because peak amplitude and rise time were positively, and not negatively, correlated, it is suggested that mEPSPs were not significantly attenuated electrotonically. Because the slope of the mEPSP rising phase may play a role in triggering APs, the maximum rising slope of the 411 mEPSPs analyzed in FIGURE 3A and 3B was evaluated (FIG. 3D). Maximum slope did not depend significantly on the rise time, but was better correlated with the peak amplitude (FIG. 3E and 3F, respectively).

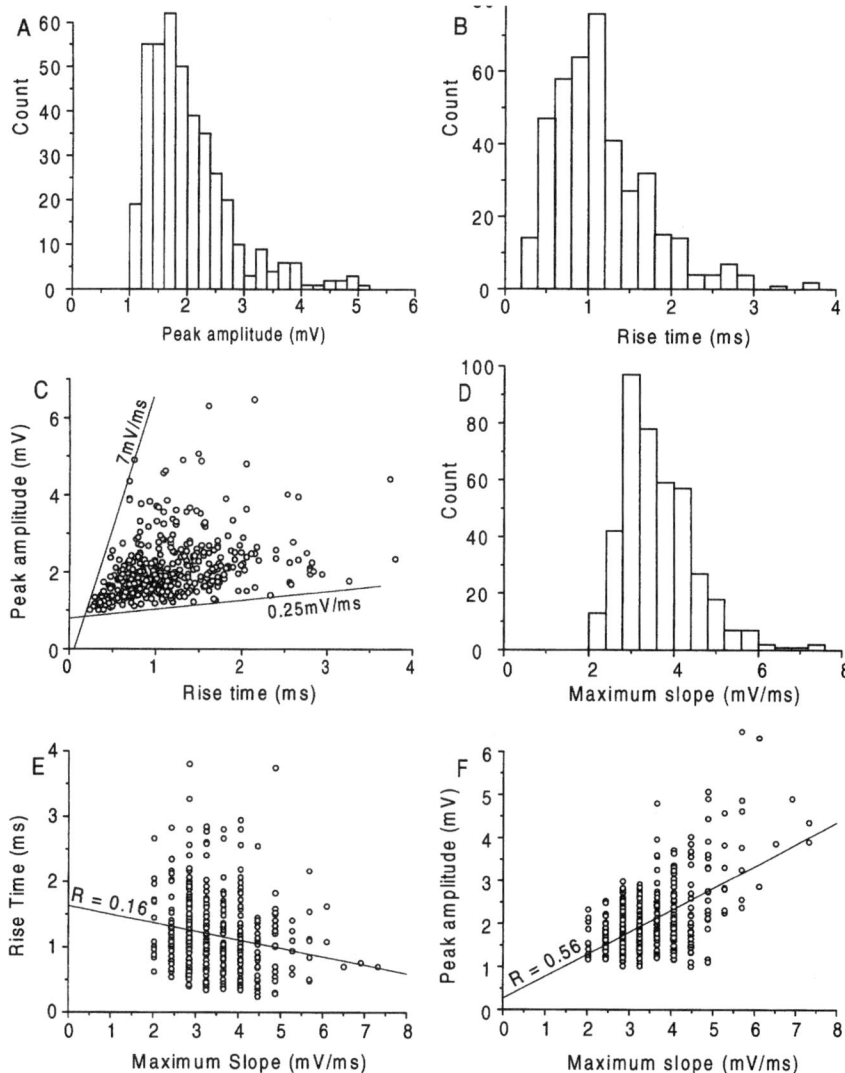

FIGURE 3. Typical amplitude and rise-time properties of 411 mEPSPs that did not trigger an AP. In this experiment 1.7% of mEPSPs triggered an AP. (**A**) Frequency distribution of peak amplitude. The mean (2.05 mV) is larger than the mode (~1.7 mV). Amplitude variability (CV=39%) is height. (**B**) Frequency distribution of rise time. The mean is 1.18 ms and standard deviation 0.72 mV. (**C**) Peak amplitude vs. rise time plot. Points are confined between two slopes. The 0.25-mV/ms overall slope corresponds to the minimal rate (about 1 kHz) of the smaller mEPSPs for which there is still a single peak. The maximum overall slope varied from one experiment to another and is here 7 ms/mV. (**D**) Frequency distribution of the maximal slope measured on each mEPSP. The mean (3.48 mV/ms) exceeds the mode (~3 mV/ms). (**E**) Rise time vs. maximum slope plot. There is no correlation, as indicated by the value of the linear correlation coefficient R. (**F**) Peak amplitude vs. maximum slope plot. R value suggests some correlation.

Characterization of the Smallest mEPSPs Detectable

To analyze the smallest mEPSPs the detection threshold was set as low as possible (0.5–1 mV above the mean level of the baseline noise) so that non-mEPSPs were only occasionally captured (estimated at approximately 2–10% of the sample). After manual elimination of these false triggers, the 20 smallest signals were averaged ($n = 3$ fibers; FIG. 4 and TABLE 1). The time course of the averaged signals had a rapid rising, and an exponential falling phase, suggesting that most of the smallest events captured were indeed mEPSPs. The falling phase was well described by a single exponential function having a time constant of about 2 ms. For the three fibers on which this analysis was completed, the average signal exhibited a peak amplitude of 1.42 ± 0.49 mV with a rise time of 0.99 ± 0.18 ms. A foot consistently appeared along the rising phase of the averaged smallest mEPSP approximately 0.25 mV above the baseline, suggesting that these mEPSPs might have been preceded by an even smaller event. Another interpretation is that this foot reflects some oscillating component in the baseline noise and that only mEPSPs coinciding with the positive phases of these baseline oscillations of the postsynaptic potential hit the capture trigger.

MEPSP Slope and Action-Potential Triggering

The data presented in FIGURE 2 suggested that mEPSPs with steeper rising phases were more likely to trigger APs than were mEPSPs with slower rising slopes. To further

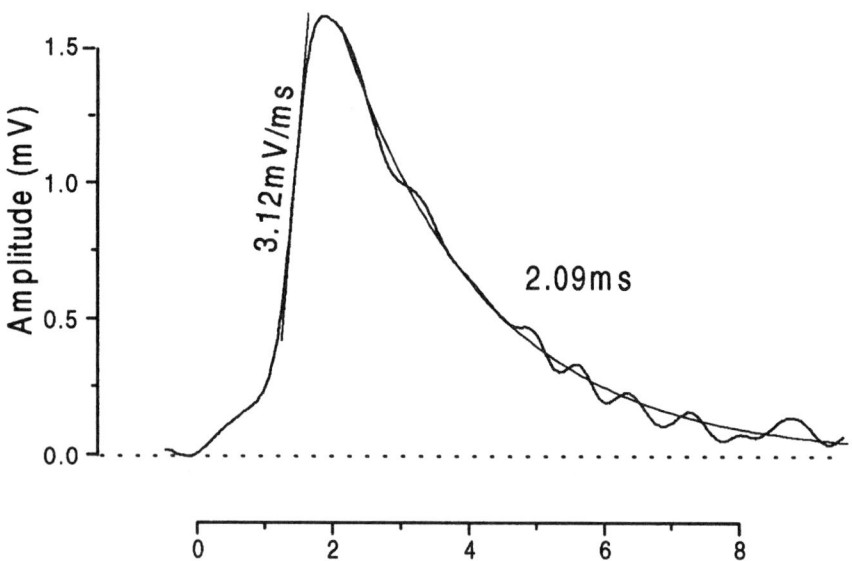

FIGURE 4. Averaging of 20 smallest mEPSPs of an experimental series. Peak amplitude is 1.61 mV, rise time (10–90% of peak) is 0.92 ms, maximum slope of rising phase is 2.24 mV/ms, and falling phase was accurately described by a 2.09-ms time-constant-decaying single exponential. The early rising phase shows a slower time course generating a break at ~0.25 mV above the baseline. See TABLE 1 for statistics.

TABLE 1. Maximum Slope of Nontriggering and AP-Triggering mEPSP Rising Phases

	Nontriggering Slopes			AP-Triggering Slopes			% Triggered
	Mean	SD	n	Mean	SD	n	
Lagena 22	3.48 mV/ms	± 0.88	411	7.79 mV/ms	± 3.25	7	1.7
Lagena 35	4.71 mV/ms	± 1.65	545	15.09 mV/ms	± 8.07	255	46.8

Note: SD = Standard deviation; n = sample size.

investigate this, a quantitative evaluation of these data is now presented. Because the peak amplitude and rise time of AP-triggering mEPSPs could not be determined, the maximum slope of AP-triggering and non-triggering mEPSPs was instead measured. The maximum slope of nontriggering mEPSPs was measured between 10 and 90% peak amplitude, and the maximum slope of AP-triggering mEPSPs was measured before the sharp increase in the slope corresponding to the AP initiation (cf. FIG. 5A). Data for two fibers with different proportions of AP-triggering and nontriggering mEPSPs are reported in TABLE 1. In these two fibers, AP-triggering and nontriggering mEPSPs exhibited significantly different maximum slopes, confirming the visual impression suggested by FIGURE 2. FIGURE 5 shows the distribution of these maximum slopes of triggering (FIG. 4B) and nontriggering (FIG. 4C) mEPSPs in the same fiber (lagena 35, TABLE 1).

Efferent Activity Inhibits Action-Potential Firing and Alters mEPSPs Properties

The effects of electric pulse activation of the efferent vestibular system upon mEPSPs and AP initiation in lagenar afferents were studied. It was previously shown that individual efferent neurons were self-activated by alert fish at a maximum frequency of 100 Hz.[1] In the experiments described here, 100-Hz electric pulse train stimuli (pulse width 100 μs) at varying intensities below 500 μA were employed. The evolution of an average experiment is shown in FIGURE 6. As stimulation intensity (pulse amplitude) increased (FIG. 6A), both AP firing rate and mEPSP (6B, C) frequency rapidly decreased. Following the cessation of efferent stimulation, AP frequency usually rapidly recovered. In the experiment illustrated, however, AP frequency remained below that of the prestimulus period (FIG. 6C). Note that the decrease in mEPSPs (FIG. 6B) was delayed with respect to the decrease in AP frequency. This suggests that during efferent stimulation the largest mEPSPs and/or those with the steepest rising phases were eliminated first. This suggestion is borne out by the data in FIGURE 6D and 6F. Interestingly, for a few seconds following the cessation of stimulation mEPSP amplitude remained suppressed (FIG. 6D). MEPSP amplitude subsequently increased beyond the prestimulus level (FIG. 6D). However, mEPSP rise time was not affected by efferent stimulation (FIG. 6F). The depression of the slope of all mEPPs (AP-triggering and nontriggering) by the efferent stimulation and its recovery at the termination of the stimulation were both delayed by a few seconds. Similar behavior was observed in all other experiments. AP threshold (FIG. 6E) and time to reach threshold (FIG. 6G)

TABLE 2. Changes in Peak Amplitude and Variability After Efferent Stimulation[a]

	Mean	CV	n	SD
Control	1.000	0.5670	327	0.567
From 0s to 5s	0.971	0.329	123	0.320
From 5s to 10s	1.040	0.519	132	0.540
From 10s to 15s	1.073	1.062	108	1.140

[a]The results of 6 efferent stimulation experiments from 3 lagenas are pooled and expressed relative to control amplitude before stimulations set arbitrarily at 1.000. CV = Coefficient of variance.

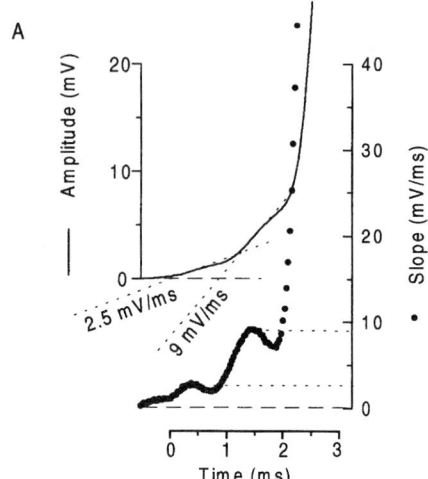

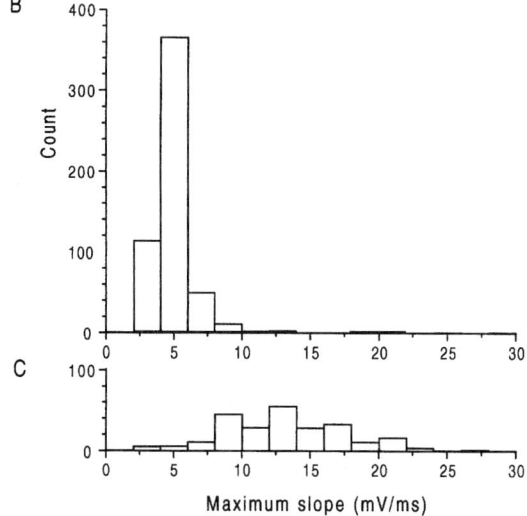

FIGURE 5. mEPSP maximum slopes distribute in two classes. (**A**) The line (left abscissa) represents the membrane potential during the rising phases of a mEPSP and of the AP (truncated) that the mEPSP triggered. *Dotted lines* are tangents to the potential amplitude when the slope reaches a local maximum. *Filled circles* (right abscissa) represents the slope of the mEPSP and the AP. The slope was calculated differentiating the successive digital samples describing the postsynaptic signal. As often (see FIG. 2A), the mEPSP rising phase shows a break separating two distinct time courses. The slope exhibits two maxima during the mEPSP rising phase. (**B**) Frequency distribution of maximum slope of nontriggering mEPSPs. (**C**) Frequency distribution of maximum slope of AP-triggering mEPSPs and AP-triggering (B) in the same afferent fiber. Means are, respectively, 4.74 mV/ms ± 2.24 S.D. and 12.81 mV/ms ± 8.97 S.D.

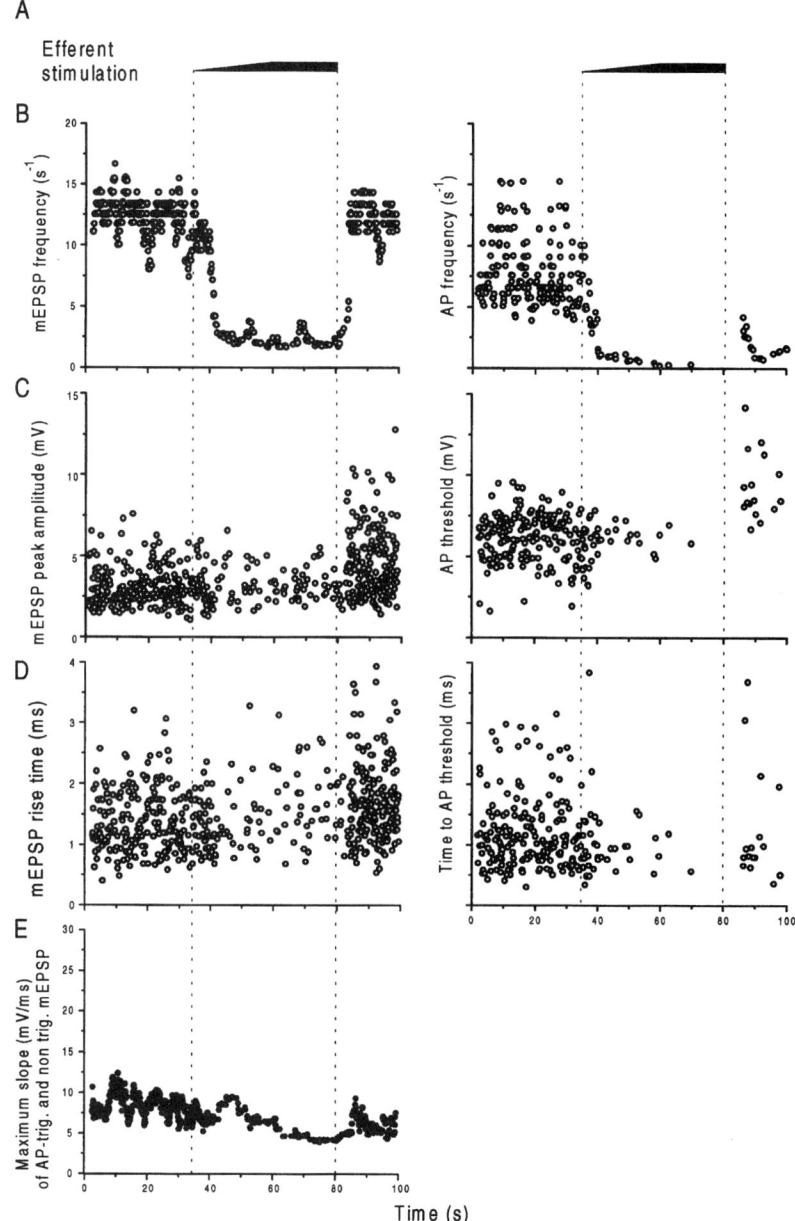

FIGURE 6. Afferent AP firing and mEPSP properties before, during, and after efferent stimulation *in vivo*. The effects were particularly progressive in this experiment. (**A**) Efferent stimulation. (**B**) Evolution of mEPSP frequency. (**C**) AP frequency. (**D**) Peak amplitudes of mEPSPs. (**E**) Threshold of APs. (**F**) mEPSP rise time measured between 10 and 90% of peak amplitude. (**G**) Time to AP trigger instant measured between 10 and 90% of threshold level. (**H**) Maximum slope of nontriggering and AP-triggering mEPSPs.

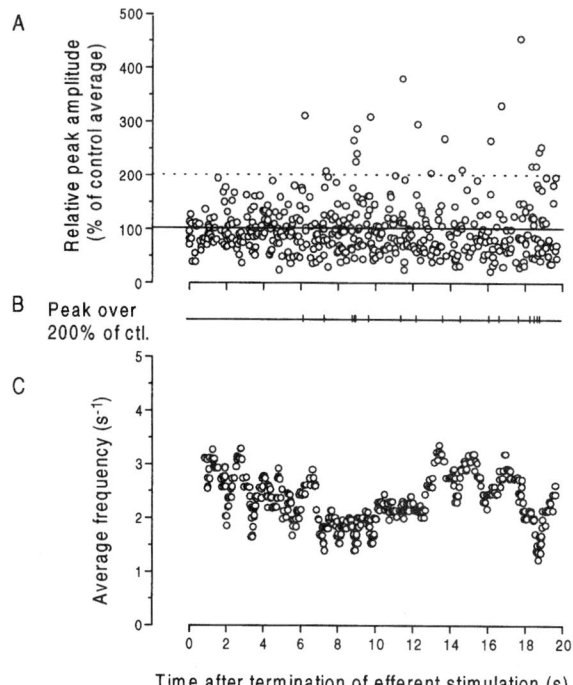

FIGURE 7. Evolution of mEPSP properties after termination of efferent stimulation. The results of 6 stimulation experiments were pooled after standardization. Stimulation lasted 30 to 45 s and were performed on 3 lagena preparations. (**A**) mEPSP peak amplitudes. Amplitudes are expressed relative to the average control amplitude before the efferent stimulation. Only two APs occurred during the period reported on the plots for all 6 experiments. (**B**) *Ticks* mark the position in time of the mEPSPs that were twice or more the control amplitude. (**C**) mEPSP frequency during the same period of time.

remained steady during efferent stimulation, but AP threshold and, more transiently, time to reach threshold increased beyond their prestimulus levels following stimulation. Taken together these data suggest that the decrease in AP frequency observed during efferent stimulation is due to changes in mEPSP frequency, amplitude, and slope.

FIGURE 7 illustrates the evolution of mEPSP amplitude and frequency after the the termination of efferent stimulation. In six fibers, efferent stimulation lasted between 30 and 45 s. mEPSP amplitudes were normalized to their average control amplitude in each cell and are plotted as percent of control amplitude. Note that mEPSPs twice or more of the control amplitude (tick marks, FIG. 7B) were absent during the first 6 s following the cessation of efferent stimulation. The frequency of these large mEPSPs gradually increased thereafter. These results are quantified in TABLE 2.

DISCUSSION

The amplitude, slope, and rise time of mEPSPs measured in spontaneously firing lagenar afferents were found to vary. Analysis of mEPSP properties demonstrated a vague positive correlation between mEPSP amplitude and rise time and between amplitude and maximum slope. AP triggering depended upon mEPSP rising slope as well as mEPSP amplitude. Further, efferent stimulation decreased and ultimately eliminated the largest mEPSPs, thereby decreasing AP frequency.

Electrotonic Attenuation

Electrotonic attenuation is known to reduce the amplitude and increase the time course of transient potentials occurring in neurites at sites remote from the recording electrode. Thus, it is often proposed that variability in mEPSP amplitude and rise time reflects variable attenuation of signals occurring at remote sites. As generally found here and elsewhere, however, the smallest mEPSPs had the shortest rise times, and amplitude and rise time were not negatively but weakly positively correlated, suggesting that mEPSP variability does not reflect variable electrotonic attenuation. Furthermore APs were only triggered by mEPSPs with a large amplitude and/or short rise time, indicating that mEPSPs with apparent lower amplitude and longer rise times were indeed either too small and/or too slow to trigger APs at the site where they were generated and that their amplitude and rise-time properties did not follow principally from differential electrotonic attenuation.

Quantal Nature of Synaptic Transmission

Since the pioneering work on the neuromuscular junction,[10] the recording of spontaneous, miniature, asynchronous postsynaptic signals (in the absence of APs) is generally considered evidence of a quantal form of synaptic transmission. Thus the recording of more or less random small EPSPs in primary afferent terminals in the goldfish sacculus[22] and frog vestibular semicircular canals[23,24] is generally considered sufficient evidence of a quantal form of transmission. However, the quantal nature of neuromuscular transmission was defined on the basis that all miniature endplate potentials; either asynchronous and miniature or unitary evoked events show the same amplitude, time course, and pharmacological sensitivities. Indeed, the small variability in large-mode miniature endplate potential amplitude (CV ~20%) and rise time has been classically assigned to unidentified causes or to measurement errors. In fact mEPSPs recorded in lagenar afferents exhibit a variability in amplitude ($\geq$ 30%) and rise time that is much larger and more comparable to the variability in the characteristics of small-mode miniature signals recorded at developing or challenged neuromuscular junctions and central nervous system synapses.[25] (Note that the quantal composition of evoked synaptic signals cannot be characterized if individual events exhibit a CV $\geq$ 0.50 because of blurring of the histogram peaks; cf.[26] For this reason nontriggering lagenar mEPSPs should be comparable to small-mode miniature synaptic signals.)

Variability in mEPSPs

There are several presynaptic sites on each hair cell and each afferent fiber makes synaptic contact with several hair cells. Therefore, variability in mEPSP properties might reflect variability in receptor density or quantal size across these synaptic contacts. At defined central synapses, however, the variability in miniature signals (small-mode type) within a single synapse has been shown to be as large as that across synapses,[25] suggesting that so-called quantal size might not actually be fixed. Indeed miniature signals simultaneously recorded at double-sided synapses (postsynaptic receptors on both sides of the cleft[27]) show correlated variability,[17] indicating that the miniature signal is not produced by a fixed transmitter quantum. In fact amplitude variability is always accompanied by rise-time variability. Distance between release site and postsynaptic receptors was proposed to explain longer rise times. However, transmitter diffusion from a remote release site might increase the rise time but would also dilute the transmitter, causing a negative, rather than positive correlation, between amplitude and rise time. Transmitter diffusion across the synaptic cleft, binding to postsynaptic receptors and channel opening are rapid events,[28] such that only extended release might explain rise times lasting several milliseconds.[29] If the fusion pore of a vesicle remains open longer the duration of the secretion and would explain the correlation between amplitude and rise time. Alternatively, brief TDs occurring at intervals shorter than individual responses would also extend the rise time, of the signal. The predominance of small amplitude and short rise time then, suggests that single discharges predominate.

Subminiature Transmitter Discharges

At the neuromuscular junction it has been demonstrated that several subminiature TDs contribute to a single spontaneous postsynaptic signal.[13–15,30] Similar findings were reported at central synapses.[16,31,32] At the neuromuscular junction, TDs were found to be generally better synchronized and in larger numbers than at central synapses. It was shown that during development, hyperosmotic shock, or poisoning of the presynaptic terminal, miniature endplate potentials can show the same variability as miniature signals at central synapses, with a predominance of small amplitudes and short rise times.[18,20] Normal and small-amplitude miniature endplate currents were shown to originate from the same release sites[33] and the release process to reversibly switch dynamically between the two types of mEPSPs.[21,20] One explanation for this phenomenon could be a change in the number and level of synchronization between TDs.[19] If this were the case, it would be appropriate to investigate the possibility that mEPSPs in lagenar afferents exhibit the same variable TD composition as do miniature signals at other synapses.

Coincidence Between Transmitter Discharges

Long and complex rising phases together with the positive correlation between amplitude and rise time suggests that lagenar mEPSPs might follow from the coincidence of several TDs, and that only the smallest mEPSPs may be the actual elementary events. It therefore seems appropriate to characterize the probability of coincident TDs. This probability may be estimated from TD frequency, which in turn may be estimated from the mean (3 mV) and the mode (1.5 mV) of the mEPSP amplitude distribution, assuming that the mode reflects the first Gaussian mode and the mean amplitude of mEPSPs produced by a single TD. This leads to an estimated average of two TDs per mEPSPs in the present experiment. Since mEPSP frequency ranged between 10 s^{-1} and 20 s^{-1} (mean = 14.1 s^{-1}; ± 3.5 SD; $n = 11$), the TD frequency should be between 20 s^{-1} and 40 s^{-1}. Therefore, the probability for a TD to occur by chance during the rising phase of the response to another TD is 0.02 to 0.04. MEPSPs produced by three TDs coinciding is extremely unlikely (< 1/1000). The number of mEPSPs two or three times the mode far exceeds the estimated probability of two or three random TD coincidences. With a 4% incidence of two-TD mEPSPs, the mean EPSP amplitude would be just above 104% of the single-TD-mEPSP amplitude, whereas the mean EPSP amplitude is actually much larger (~3 mV).

Rossi and colleagues[24] concluded that frog labyrinthine mEPSPs do not exhibit interaction and that coincidences between presynaptic events are at a chance level. However, their time interval distribution was based on 2.5-ms-wide bins and thus would have overlooked 1-ms or shorter time interactions. Furthermore, Rossi *et al.*[24] described the amplitude distribution of their most elementary postsynaptic event as a log-normal distribution, without providing any rational for using such a distribution. A log-normal distribution is skewed with the predominance of small amplitudes and is similar to the amplitude distribution of the small-mode miniature endplate potentials described at the neuromuscular junction and other synapses. We verified that Rossi's published distributions of mEPSP amplitude can be well described by two Gaussian functions. The best fit was with the mode of the second Gaussian at 1.79 times and 1.90 times the mode of the first Gaussian (for Rossi et al.,[24] figs. 2D and 6A, respectively). This is consistent with larger mEPSPs resulting from two TDs occurring at a short interval. Thus, in the frog labyrinth, as in the toadfish lagena, many mEPSPs originally identified as single events may actually represent several TDs; the coincidence of these TDs was not likely to be random. It is not currently clear what underlies the coincidence between TDs.

MEPSP Slope and AP Triggering

APs were preferentially triggered by mEPSPs having a large amplitude and a steep rising slope, probably because with faster rise time there is less inactivation of Na$^+$ channels and activation of K$^+$ channels. mEPSP slopes distributed into two classes, representing triggering and nontriggering mEPSPs. From the perspective of the synchronization of TD hypothesis, this could be explained by differences in the number and synchronization of the TDs composing triggering and nontriggering mEPSPs. Although we have no evidence, it is possible that triggering mEPSPs are composed of more and better synchronized TD-like large-mode mEPPs compared to small-mode mEPPs. Regarding the interaction between TDs, either there are several classes of mEPSPs with different levels of TD interaction, or the level of interaction increases abruptly when a certain threshold is reached, as was shown to be the case for mEPSPs.[18,19]

Efferent Action upon mEPSPs and AP Generation

During the first 10 s after the termination of efferent stimulation, the recovery of large-amplitude mEPSPs and AP frequency was progressive, suggesting an incremental increase in the number of coincident TDs. Therefore, TD interactions, as well as mEPSP frequency, participate in the encoding of the lagenar sensory message. Identifying the nature of TD interactions could contribute to the understanding of the encoding of the lagenar sensory message.

Thus we have shown that the rise time and amplitude of the mEPSPs transmitted from lagenar hair cells to primary afferents determine the spike-train parameters that encode the sensory message originating in this endorgan. The efferent vestibular system modulates these PSP properties, thereby increasing or decreasing the strength and content of this sensory message from endorgan to brain.

REFERENCES

1. BOYLE, R., J. P. CAREY & S. M. HIGHSTEIN. 1991. Morphological correlates of response dynamics and efferent stimulation in horizontal semicircular canal afferents of the toadfish, *Opsanus tau*. J. Neurophysiol. **66**: 1504–1521.
2. LOCKE, R. E. & S. M. HIGHSTEIN. 1990. Efferent modulation of synaptic noise in lagenar afferents of the toadfish. Soc. Neurosci. Abstr. **16**: 735.
3. LEWIS, E. R., R. A. BAIRD, E. L. LEVERENZ & H. KOYAMA 1982. Inner ear: dye injection reveals peripheral origins of specific sensitivities. Science **215**: 1641–1643.
4. KATAOKA, Y. & H. OHMORI. 1996. Of known neurotransmitters, glutamate is the most likely to be released from chick cochlear hair cells. J. Neurophysiol. **76**: 1870–1879.
5. HIGHSTEIN, S. M. & R. BAKER. 1986. Organization of the efferent vestibular nuclei and nerves of the toadfish, *Opsanus tau*. J. Comp. Neurol. **243**: 309–325.
6. HIGHSTEIN, S. M. & R. BAKER. 1985. Action of the efferent vestibular system on primary afferents of the toadfish, *Opsanus tau*. J. Neurophysiol. **54**: 370–384.

7. BOYLE, R. & S. M. HIGHSTEIN. 1990. Resting discharge and response dynamics of horizontal semicircular canal afferents of the toadfish, *Opsanus tau*. J. Neurosci. **10**: 1557–1569.
8. BOYLE, R. & S. M. HIGHSTEIN. 1990. Efferent vestibular system in the toadfish: action upon horizontal semicircular canal afferents. J. Neurosci. **10**: 1570–1582.
9. WARR, W. B. 1992. Organization of olivocochlear systems in mammals. *In* The Mammalian Auditory Pathway: Neuroanatomy, D. B. Webster, A. N. Popper, and R. R. Fay, Eds.: 440–448. Springer-Verlag. New York.
10. FATT, B. & B. KATZ. 1952. Spontaneous subthreshold activity at motor nerve endings. J. Physiol. **117**: 109–128.
11. BEVAN, S. 1976. Sub-miniature end-plate potentials at untreated frog neuromuscular junctions. J. Physiol. (Lond.) **258**: 145–155.
12. KRIEBEL, M. E. 1988. The neuromuscular junction. *In* Handbook of Experimental Pharmacology, V. P. Whittaker, Ed.: 537–566. Spinger-Verlag. Berlin.
13. LILEY, A. W. 1957. Spontaneous transmitter release of transmitter substance in multiquantal units. J. Physiol. **136**: 595–605.
14. WERNIG, A. & H. STIRNER. 1977. Quantum amplitude distributions points to functional unity of the synaptic "active zone." Nature **269**: 820–822.
15. ERXLEBEN, C. & M. E. KRIEBEL. 1988. Subunit composition of the spontaneous miniature end-plate current at the mouse neuromuscular junction. J. Physiol. **400**: 659–676.
16. KORN, H., C. SUR, S. CHARPIER, P. LEGENDRE & D. S. FABER. 1994. The one-vesicle hypothesis and multivesicular release. Adv. Second Messenger Phosphoprotein Res. **29**: 301–322.
17. FRERKING, M., S. BORGES & M. WILSON. 1995. Variation in GABA mini amplitude is the consequence of variation in transmitter concentration. Neuron **15**: 885–895.
18. VAUTRIN, J. & M. E. KRIEBEL. 1991. Characteristics of slow-miniature end-plate currents show a subunit composition. Neuroscience **41**: 71–88.
19. VAUTRIN, J. & J. L. BARKER. 1995. How can exocytosis account for the actual properties of miniature synaptic signals? Synapse **19**: 144–149.
20. VAUTRIN, J. 1992. Miniature endplate potentials induced by ammonium chloride, hypertonic shock and botulinum toxin. J. Neurosci. Res. **31**: 318–326.
21. KRIEBEL, M. E., F. LLADOS & J. VAUTRIN. 1996. Hypertonic treatment reversibly increases the ratio of giant skew-miniature endplate potentials to bell-miniature endplate potentials. Neuroscience **71**: 101–117.
22. ISHII, Y., S. MATSUURA & T. FURUKAWA. 1971. Quantal nature of transmission at the synapse between hair cells and eighth nerve fibers. Jpn. J. Physiol. **21**: 79–89.
23. SCHESSEL, D.A. & S.M. HIGHSTEIN. 1981. Is transmission between the vestibular type I hair cell and its primary afferent chemical? Ann. N.Y. Acad. Sci. **374**: 210–214.
24. ROSSI, M. L., M. MARTINI, B. PELUCCHI & R. FESCE. 1994. Quantal nature of synaptic transmission at the cytoneural junction in the frog labyrinth. J. Physiol. **478** (1): 17–35.
25. STEVENS, C. 1993. Quantal release of neurotransmitter and long-term potentiation. Cell 70/Neuron **10**: 55–63.
26. VAUTRIN, J., A. SCHAFFNER, B. FONTAS & J. L. BARKER. 1993. Frequency modulation of transmitter release. J. Physiol. (Paris) **87**: 51–73.
27. VAUTRIN, J., A. SCHAFFNER & J. L. BARKER 1994. Fast presynaptic $GABA_A$ receptor-mediated Cl-conductance in cultured rat hippocampal neurones. J. Physiol. **479** (1): 53–63.
28. CLEMENTS, J. D. 1996. Transmitter time course in the synaptic cleft; its role in central synaptic function. Trends Neurosci. **19**: 163–171.
29. VAN DER KLOOT, W. 1995. The rise times of miniature endplate currents suggest that acetylcholine may be release over a period of time. Biophys. J. **69**: 148–154.
30. VAUTRIN, J. 1986. Subunits in quantal transmission at the mouse neuromuscular junction: test of peak intervals in amplitude distributions. J. Theor. Biol. **120**: 363–370.
31. EDWARDS, F. A., A. KONNERTH & B. SAKMANN. 1990. Quantal analysis of inhibitory synaptic transmission in the dentate gyrus of rat hippocampal slices: a patch-clamp study. J. Physiol. **430**: 213–249.
32. MAPLE, B. R., F. S. WERBLIN & S. M. WU. 1994. Miniature excitatory postsynaptic currents in bipolar cells of the tiger salamander retina. Vision Res. **34**: 2357–2362.
33. VAUTRIN, J. & M. E. KRIEBEL. 1992. Focal, extracellular recording of slow miniature junctional potentials at the mouse neuromuscular junction. J. Neurosci. Res. **31**: 502–506.

A Review of Otolith Pathways to Brainstem and Cerebellum

J. A. BÜTTNER-ENNEVER[a]

Institute of Anatomy, Ludwig-Maximilian-University of Munich, Pettenkoferstrasse 11, D-80336 Munich, Germany

ABSTRACT: Our knowledge of otolith pathways is developing rapidly, but is still far from complete. Primary afferents from the sacculus and utricle terminate mainly in the lateral, inferior and caudal superior vestibular nuclei, and the ventral cerebellum, in particular the nodulus. Otolith signals descend via reticulo- and vestibulospinal pathways in the spinal cord to influence neck motoneurons and ascending proprioceptive afferents. Utricular information can reach the extraocular eye muscles via mono-, di-, and multisynaptic pathways, but saccular afferents probably only by multisynaptic pathways.

The otolith signals are relayed from the vestibular nuclei, medullary reticular formation, inferior olive, and lateral reticular nucleus to sagittal zones in the caudal cerebellar vermis (nodulus and uvula), and influence the deep cerebellar nuclei. The graviceptive information could be channeled by the cerebellar efferents back to the vestibular and inferior olive complex, or fed into ascending pathways that would innervate the mescencephalon, the thalamus, and cerebral cortex.

INTRODUCTION

The vestibular system senses the movement and position of the head in space, and uses the information to stabilize vision, control posture, and register the orientation of the body.[1-3] The vestibular signals are generated in the labyrinth, which contains two different types of sensory detectors: the semicircular canals and the otoliths. The three semicircular canals register *angular* acceleration of the head in space: the two otolith organs, the utricle and the saccule, detect *linear* accelerations of the head. The sensory epithelium of the utricle, the *macula utriculi*, lies in the horizontal plane. It detects earth-horizontal linear acceleration in any direction, and registers changes in head position with respect to gravity, such as a static tilt of the head. The saccule sensory epithelium is oriented vertically (see Curthoys *et al.*, this volume). It detects acceleration in the vertical plane, and its signals may play an important role in stabilizing vision by compensating for the vertical head movements that occur during locomotion. In this article we review the current knowledge on the brainstem pathways carrying otolith signals.

Neurons that encode otolith signals usually have a sensitivity to angular acceleration as well as tilt, implying the frequent convergence of otolith and canal signals.[4,5] It

---

[a]Address for telecommunication: Phone: 0049 89 5160 4880; fax: 0049 89 5160 4857; e-mail: buettner@anat.med.uni-muenchen.de

has been suggested that if otolith stimulation alone is insufficient to induce compensatory eye movements, it must be combined with another sensory cue, for example, from the semicircular canals, or an acoustic or optokinetic signal.[6] The regions in which otolith signals have been recorded are the vestibular nuclei, the inferior olive and the lateral reticular nucleus in the medulla oblongata, the nodulus and ventral uvula of the cerebellum, the cerebellar nuclei, and the motoneurons for oculomotor and neck movements. There is also some evidence that otolith activity may reach the thalamus and the cortex. Finally, one should keep in mind that the otoliths are not the sole source of graviceptive information to the body, as evidenced by residual faculties of bilateral labyrinthectomized patients. Mittelstaedt[7] describes an extravestibular graviceptor system of vascular origin, and a second system possibly associated with the kidneys.

PRIMARY AFFERENTS FROM THE OTOLITH ORGANS

In man there are about 6000 utricular and 4000 saccular axons from the otolith maculae that combine with the canal afferents to form the vestibular nerves of about 20,000 fibers.[8,9] The cells are bipolar neurons with their cell bodies in the Scarpa's ganglion. The utricular primary sensory afferents travel to the brainstem via the superior division of the vestibular (VIIIth) cranial nerve, those of the sacculus in the inferior division along with the posterior canal afferents. In several species, it has been found that the otolith and canal primary afferents enter the medulla at the level of the lateral vestibular nucleus (LVN), where they divide into an ascending and descending branch. The descending branch of all primary vestibular afferents innervates a central region of the vestibular complex.[10,11] The ascending axon collaterals terminate in SVN and project further to the anterior and posterior vermis of the cerebellum, mainly in the nodulus (lobule X) and the ventral uvula (lobule IXd). Barmack *et al.*[12] estimated that in rabbit 70% of primary afferents project directly to the cerebellum, almost exclusively to the nodulus and uvula, but with sparse terminations in the anterior lobe (FIGS.1 and 2).

Earlier studies using degeneration, or bulk-tracers, have reported some differences between the terminals of canals and otoliths in the vestibular nuclei, in spite of major inconsistencies in the results (for review see references 10 and 11). For example, only canal afferents were found in the superior vestibular nucleus (SVN) and the interstitial nucleus of the vestibular nerve: a saccule input but no canal afferents terminated in the LVN and the y-group.[10,11,13–15] Recent single-cell reconstructions of identified otolith[16,17] and canal primary afferents[18,19] terminating in the vestibular nuclei have revealed important detailed differences that were not clear from the previous studies. The utricular afferents terminate mainly in the inferior (or descending) vestibular nucleus (IVN), innervating the lateral, medial (MVN), and superior vestibular nuclei, and even the abducens nucleus, sparsely.[17,20] Additional electrophysiological evidence was put forward for a direct projection to the whole of the abducens nucleus from the utricle,[21–23] see below. In comparison, saccule fibers terminate mainly in LVN and IVN, sometimes in SVN. A few collaterals enter the adjacent reticular formation and the spinal trigeminal nucleus.[16] Although no otolith terminals were found in the y-group or the cerebellum in these studies, this could be a consequence of low sample numbers. In the pigeon[24] and gerbil[15] otolith, afferents were reported in the posterior cerebellar cortex and cerebellar nuclei, although canal afferents to these regions were more numerous.

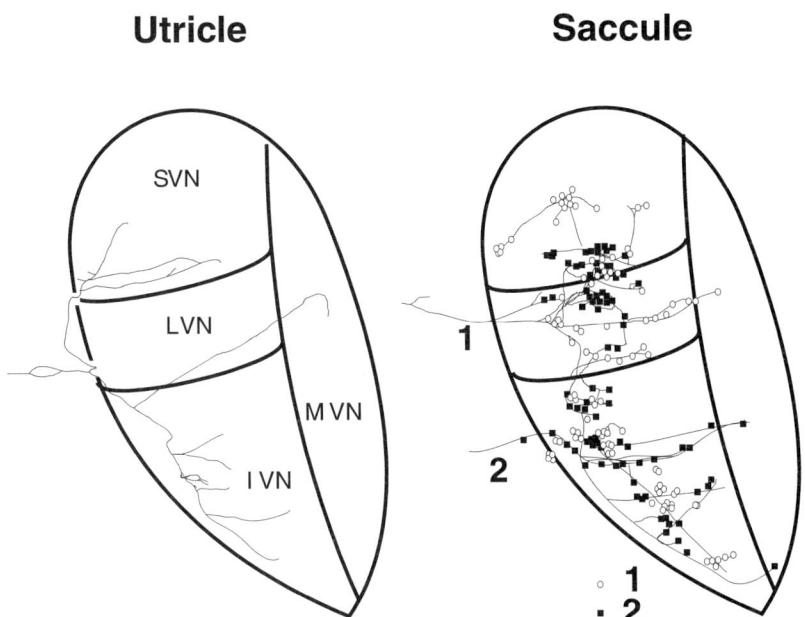

FIGURE 1. Reconstruction of the axonal arborization and terminals of identified otolith primary afferents, plotted on a schematic drawing of the vestibulonuclear complex. **Left**: One utricular afferent; **right**, two saccule primary afferents. (Redrawn from Imagawa et al.,[16,17] with permission.)

INTRANUCLEAR CONNECTIONS OF THE VESTIBULAR NUCLEAR COMPLEX

Within the vestibular complex there are many pathways that interconnect one region with another[25] (see reviews in references 10 and 11). The pattern is complex, but has been usefully interpreted by Epema et al.[25] They show that the central region of the vestibular complex, the magnocellular region, comprises the main output route of the vestibular complex. The surrounding parts of the vestibular complex consist mostly of smaller cells, which are intimately interconnected with other vestibular nuclei, often with many commissural or cerebellar connections, but they project into the magnocellular output area, and through its efferent pathways—for example, the medial longitudinal fasciculus (MLF), the ascending Deiters' tract (ATD), or the brachium conjunctivum (BC)—may reach extraocular motoneurons.[10,11]

Evidence for specific otolith local circuits within the vestibular nuclei has recently been reported by Uchino et al.[26] Saccular afferents from one population of hair cells were shown to activate vestibular nuclei (VN) neurons monosynaptically, while afferents from another population of hair cells *located on the opposite side of the saccule striola*, project to the same VN neurons disynaptically via an inhibitory neuron. This was called cross-striola inhibition.

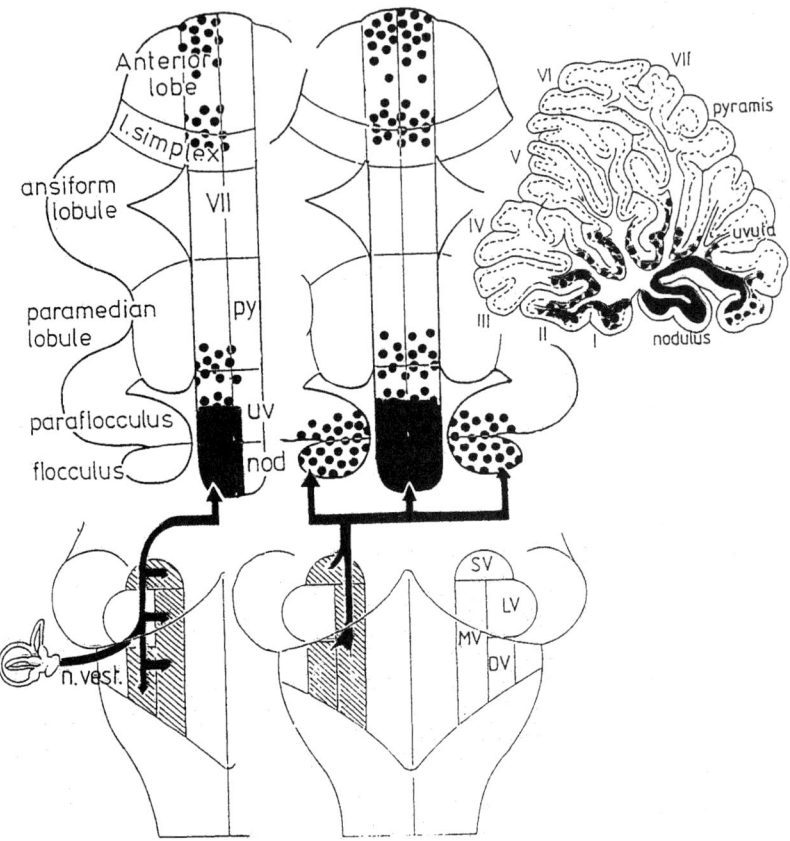

FIGURE 2. Pattern of input to the cerebellum from primary vestibular nerve fibers (**left**), and the secondary vestibular nuclei efferents (**right**). Notice the similarity in location of primary and secondary vestibular projections mainly in the nodulus, shown on a dorsal and sagittal view of the cerebellum. (Taken from Voogd et al.[62] Reproduced with permission.)

VESTIBULAR-OCULAR PATHWAYS

When the head is tilted to the side there is a compensatory counterrolling of the eyes, which varies in size according to species. Counterrolling can reach about 15 deg in man and up to 60 deg in rabbit. This phenomenon is a clear indication that static signals from the otoliths reach extraocular motoneurons, in particular those of the superior and inferior oblique.[5] The pathways have usually been assumed to be disynaptic through the vestibular nuclei, which are analogous to the canal VOR pathways, although there is no reason why this should necessarily be the case.[27] A large amount of convergence of otolith and canal signals has been recorded at the level of the vestibular nuclei,[21–23,28] although this may often be concealed by anesthesia.[6] The amount of precise anatomical evidence for the pathways by which the signals reach the motoneurons, especially with respect to the saccule, is small.

A *disynaptic* input from the utricle to ipsilateral *abducens motoneurons* has been known for some time.[29–31] The results are supported by those of Uchino et al.,[21] who stimulated the utricular nerve and localized secondary ipsilateral abducens-projecting cells to LVN and rostral IVN. There is mounting evidence for an additional *monosynaptic* excitatory utriculo-abducens input[17,21–23,32] accompanied by a long latency hyperpolarizing input to contra- and ipsilateral medial rectus motoneurons.[22] The pathway carrying the disynaptic utricular signal to the medial rectus motoneurons in the oculomotor nucleus was suggested to be the ATD.[33] Chen-Huang and McCrea[34] later confirmed that the ATD carried utricular inputs combined with a canal signal to ipsilateral medial rectus motoneurons, and generated vergence of the eyes during linear acceleration. The size of the utricular signal depended on the fixation distance, implying an important neural multiplier system in the vestibular nuclei, and not a simple disynaptic utricle–oculomotor neuron pathway.

In a re-examination of utriculo-ocular pathways Sasaki et al.[30] found no evidence for a direct monosynaptic input to trochlear motoneurons to account for the phenomenon of counterrolling of the eyes during static tilt; implying that it is multisynaptic. The experiments were designed to specifically limit spread of the stimulating current to canals, which could be the explanation for the disynaptic response found in previous studies.

These are the only studies that have been found that support a mono- or disynaptic otolith–extraocular motoneuron pathway, and the possibility remains that multisynaptic pathways may play an important role in otolith–extraocular motoneuron responses, such as counterrolling. At present there is much less evidence for the involvement of the saccule, as opposed to the utricle, in evoking reflex eye movements. A new and clinically useful test of otolith function, an intense click to one ear, is known to synchronously activate saccule afferents and produce neck myogenic potentials.[35–37] Interestingly the application of the loud click does not generate detectable reflex eye movements (Curthoys, personal communication). In contrast, there is overwhelming experimental evidence that both the saccule and utricle contribute directly and powerfully to neck movements, as described below.

DESCENDING OTOLITH PATHWAYS

Stimulation of the saccule and utricle lead to different patterns of activation of neck muscles, and could serve to compensate for the head perturbations during locomotion.[2] *Utricular* stimulation activates monosynaptically vestibulospinal neurons mainly in ventrocaudal LVN and rostromedial IVN, whose axons descend predominantly (73%) in the ipsilateral lateral vestibulospinal tract (i-LVST) to the cervicothoracic level.[20,21,30] The inputs were effective in tilting the head to the *ipsilateral* side by activating ipsilateral neck extensors and flexors, and inhibiting (trisynaptically) the neck muscles on the contralateral side.[28,38]

The saccule activated a different group of vestibulospinal cells.[39,40] They lay mostly in ventrocaudal LVN and rostral IVN, and 63% possessed axons that descend in the medial vestibulospinal tract (MVST) to the upper cervical spinal segments. Subsequent studies demonstrated a *bilaterally* organized excitatory saccule input to neck extensors in i-LVST and c-MVST, and an inhibitory effect on neck flexors in MVST or possibly reticulospinal pathways.[40] Thus the functional result of saccule activation would be head movements in the pitch plane, forward and backward.[28] These authors concluded that the saccule plays an important role in maintaining the relative position of head and body against vertical linear acceleration. The bilaterality of axons terminals from vestibulospinal fibers carrying saccular information has not yet been confirmed in other studies.

The nucleus reticularis gigantocellularis lies beneath the vestibular nuclei in the medulla, and is known to participate in the generation of head movements.[41] Vestibular-responsive neurons were found in the caudal parts of this region. These cells were modulated mainly by otolithic stimulation, and in addition possibly by vertical semicircular canal stimulation, in about half of this population.[42] The cells lie medially, in a region known to project strongly to the spinal cord, the nodulus and lobus simplex of the cerebellum. The nucleus reticularis gigantocellularis may be important for the modulation of neck- and postural-muscle activity, during the interaction of head movements with gravity.[42]

Another pathway to the spinal cord that carries otolith signals, presumably serotinergic, originates in the medial medullary reticular formation,[43] perhaps from nucleus raphe pallidus.[44] The descending fibers from nucleus raphe pallidus terminate strongly throughout the whole length of the spinal cord in the ventral horn.[45] Otolith inputs to such a pathway must have a widespread tonic influence on somatomotor output. Ascending spinal pathways to the vestibular nuclei from neck muscle afferents, also carry otolith information that feeds back into the vestibular complex.[46]

THE LATERAL RETICULAR NUCLEUS

Neurons of the lateral reticular nucleus (LRN) lie in the lateral medulla oblongata, and can be divided into a magnocellular and parvicellular division. Units responsive to tilt lie mostly in the parvicellular LRN.[47] In part, the otolith response may come from direct vestibulo-LRN projections originating in LVN and SVN,[48,49] but also from vestibulospinal excitation of spinoreticular neurons terminating in LRN.[50] The LRN receives spinal afferents from ascending axon collaterals of propriospinal neurons of C3 and C4, which also activate forelimb motoneurons monosynaptically.[51] Information from the sensorimotor cortex, the tectum, and red nucleus also converge on the LRN as do the propriospinal neurons. These signals, combined with otolith information in LRN, are carried by mossy fibers to the cerebellar cortex, and terminate in a zonal fashion, similar to the sagittal organization of inferior olive inputs to the cerebellum.[52,53] The regions of the vermis that receive projections from the LRN include the pyramis, uvula, nodulus, and floccular region.[54] The more lateral parts of the paramedian (VIII) lobe also receive afferents, but the anterior lobe receives the most dense projections. The LRN afferents also contact the cerebellar nuclei, as do all cerebellar cortical afferents.[55] The LRN pathways must serve to coordinate postural limb, neck, and otolith information via cerebellar pathways.

INFERIOR OLIVE

Otolith sensitivity can be recorded from neurons of the inferior olive (IO) in the β-group (β-nucleus) and the dorsal medial cell column of the medial accessory olive (DMCC) in the rabbit and rat (FIG. 3). These two subnuclei are particularly rich in GABAergic terminals, and, according to Barmack,[13] are not depleted of GABA by lesions of the cerebellar nuclei (the main source of GABA afferents to the inferior olive). The cell group considered to be the origin of otolith afferents to the inferior olive is outlined in Paxinos and Baxter's atlas of the rat brain[56] as the parasolitary nucleus (PSol). It is part of the vestibular complex, which lies at the caudal confluence of MVN and IVN, adjacent to the solitary tract and extending dorsally to the floor of the forth ventricle.[57] The PSol receives a vestibular primary afferent projection from the ipsilateral ganglion of Scarpa, and is composed exclusively of small GABAergic neurons whose axons project to the ipsilateral DMCC and β-nucleus of the inferior olive.[57] Neurons in the PSol have been retrogradely labeled by injections of WGA-HRP into either the ipsilateral β-nucleus or DMCC. Lesions of the PSol deplete the ipsilateral β-nucleus and the DMCC of GABAergic boutons.[57]

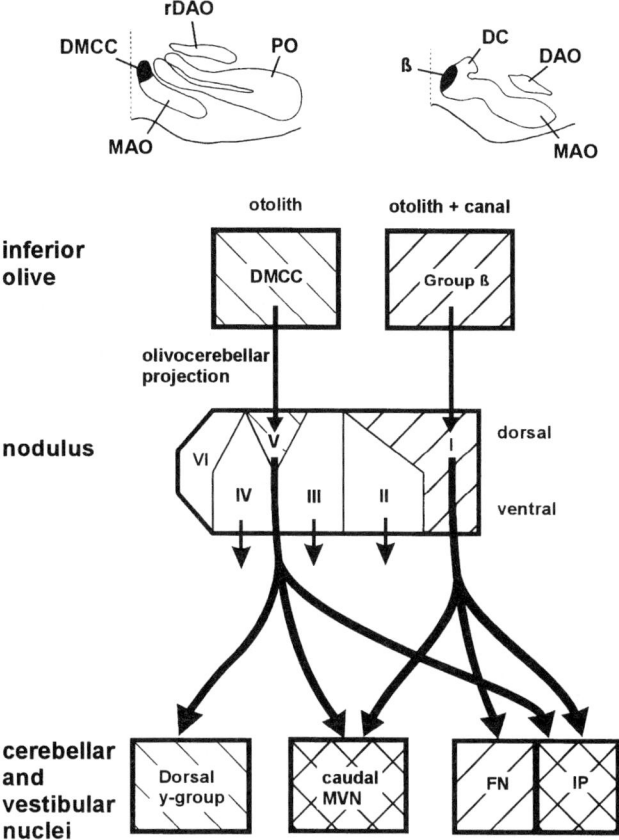

FIGURE 3. Summary diagram of projections from cell groups in the inferior carrying otolith signals (DMCC and Group β) to specific sagittal zones (I medial, and dorsal V lateral) of the nodulus (and uvula; see text). These otolith receiving zones of the vermis project onto the cerebellar and vestibular nuclei. (Adapted from Voogd et al.,[62] with permission.)

Electrophysiological recording from single cells in the DMCC demonstrate that these neurons encode signals attributed to stimulation of the ipsilateral utricle, evoked by natural static roll-tilt about the longitudinal axis.[13] Single-cell recordings from the β-nucleus showed that most of these neurons had optimal planes of sensitivity that aligned with either the ipsilateral anterior or posterior semicircular canal.[13,58] Single-cell recordings from the cells of the PSol show that these small GABAergic cells contain both otolithic and semicircular canal information (Barmack, personal communication). Interestingly, vestibular stimulation in the plane of the horizontal semicircular canal fails to modulate the activity of cells in PSol, the β-nucleus, and DMCC. Furthermore, horizontal vestibular stimulation fails to evoke any climbing fiber activity in the uvula and nodulus.[59]

Olivary efferents exit from the hilum of the olive, cross the midline, and enter the cerebellum via the inferior cerebellar peduncle. Efferents from the β-group and DMCC terminate as climbing fibers within specific sagittal zones of the nodulus (FIG. 3), extending into both the dorsal and ventral uvula, and possibly the pyramis of lobule VIII.[60] Some regions of the inferior olive are also connected reciprocally to the cerebellar nuclei (see below).[61,62] However, this may not extend to either the β-nucleus or DMCC.

AFFERENTS AND EFFERENTS OF THE NODULUS AND UVULA

In the nodulus and uvula, both vestibularly modulated climbing fiber afferents from group β and the DMCC, and optokinetically modulated climbing fiber afferents from the dorsal cap (DC) and the ventrolateral outgrowth of the principal olive, terminate on the cerebellar Purkinje cells within sagittal modules or zones (review reference 62; monkey;[63] rabbit,[64] rat[65]). Throughout the pathway that originates from the PSol and terminates in the uvula–nodulus as climbing fibers, there is a strict topography, with sagittal strips of climbing fibers representing functional zones on the surface of the nodulus (FIG. 3). These sagittal strips encode in mediolateral distance from the midline a full 180 degrees of posssible vestibular stimulation. Activation of the ipsilateral posterior semicircular canal evokes climbing fiber responses in a medial strip. Activation of the anterior semicircular canal evokes climbing fiber responses in a more laterally located strip. Inserted between these two sagittal vestibular climbing fiber strips is an "optokinetic climbing fibre strip" that is activated by posterior–anterior stimulation of the ipsilateral eye.[62,63,66]

While the termination of vestibular primary afferents is largely confined to the nodulus and ventral uvula, the termination of vestibularly modulated climbing fibers extends into the dorsal uvula where they interdigitate with climbing fibers that encode somatosensory afferents.[60] The pattern of termination of vestibular climbing fibers in sagittal zones is reminiscent of the zonal pattern generated by immunohistochemical staining for Zebrin (aldolase c).[62] This similarity does not extend to function, since the zebrin strips are not coextensive with the vestibular climbing fiber zones.

The nodulus and ventral uvula receive both primary and secondary vestibulocerebellar mossy fibers, which terminate diffusely, that is, not in sagittal zones, throughout the caudal vermis. The secondary vestibular mossy fiber afferents carry otolith signals as well as the climbing fibers.[59] The afferents arise bilaterally from the vestibular complex, mainly from the caudal MVN and from IVN.[54,60,67–70]

The efferents of the nodulus and ventral uvula project in a topographic fashion back onto the vestibular nuclei, such that the lateral nodulus projects to the caudal MVN, and the medial nodulus to the middle part of MVN.[60,62,71] Tracing experiments using biocytin in the rabbit followed ventral nodulus Purkinje cell efferents of the most medial zone (responsive to otolith inputs) to the fastigial nucleus and the white matter around both fastigial and the interposed nuclei: the axons continued into the brainstem to innervate the caudal and parvocellular parts of MVN region, and IVN.[72] Other zones did not innervate the fastigial nucleus. Less selective studies followed nodulus efferents to the SVN, but Wylie and colleagues found no projection from the otolith-responsive zone, only the horizontal optokinetic-sensitive zone projected to SVN.

AFFERENTS AND EFFERENTS OF THE CEREBELLAR NUCLEI

Otolith activity in the fastigial nucleus is not well known, but has been recently recorded in behaving monkeys by Büttner and colleagues, mainly in rostral fastigial nucleus[73] (see this volume). The activity may come in part from nodulus Purkinje cell efferents of the median sagittal zone carrying otolith and canal signals. Single-cell reconstruction of the efferents have demonstrated terminals in the fastigial nucleus and the periinterpositus white matter.[62,72] FIGURE 3 is taken from Voogd[62] and summarizes the results of several studies on the rabbit: in other species there is an analogous organization of the inputs and outputs in the different sagittal zones of the nodulus. The medial zone (with a

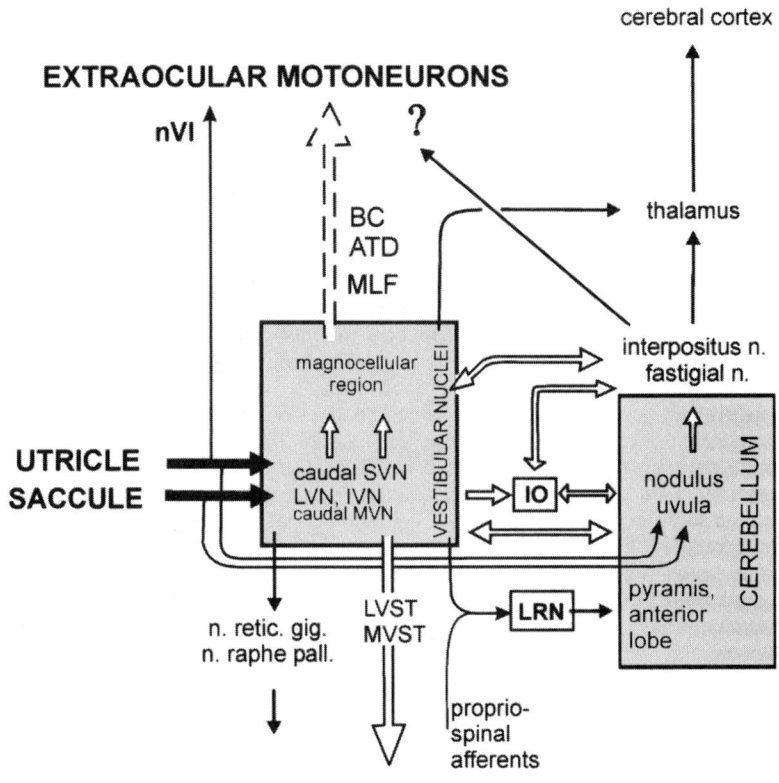

FIGURE 4. Summary diagram of structures and pathways carrying otolith signals. *Shaded areas* represent the vestibular nuclear complex and the cerebellum.

β-group input) supplies terminals to the fastigial and interpositus nuclei and to the parvicellular caudal MVN, from where the signals could reach the output zones of the vestibular nucleus in the MVN/SVN magnocellular region. The DMCC olivary afferents project to a dorsal nodulus zone (V) and the uvula, which in turn project to the nucleus interpositus, the dorsal y-group, and the caudal MVN, according to Bernard.[65] With these connections in mind, it is likely that otolith signals could be carried to other parts of the brain by cerebellar nuclear efferents or from the output pathways of the vestibular nuclei.

The efferent pathways from the fastigial nucleus,[74] include: first, the uncinate fasciculus, which crosses through the contralateral fastigial nucleus (FN), to terminate in the LVN and IVN ventrally. An ascending branch travels rostrally beneath the MLF, through the fields of Forel to the contralateral thalamus, to terminate in the VPLo area. A second fastigiobulbar pathway projects ipsilaterally to LVN and IVN, and a small group of fibers cluster at the medial border of the brachium conjunctivum and ascend to thalamic levels, crossing in the midpons with the main brachium decussation. Afferents to the pretectal areas, midline thalamus, and VPLo regions from the fastigial nuclei have been documented.[74–77]

OTOLITH SIGNALS IN THE THALAMUS AND CEREBRAL CORTEX

It is likely that the internal representation of "verticality," also called the subjective visual vertical (SVV), depends at least in part on otolith sensory afferents to thalamic and cortical structures. According to the studies quoted earlier and summarized in FIGURE 4, otolith information could reach the thalamus through two main afferent routes: from the cerebellar nuclei and the vestibular nuclei.[74–78] In an analysis of over 100 patients Dieterich and Brandt[79–80] plotted the location of acute lesions throughout the length of the brainstem; these lesions led to a permanent tilt of the SVV, and cyclorotation of the eyes. Dieterich and Brandt demonstrated a decussation at the pontine level, the anatomical basis of which could arise from pathways carrying otolith information crossing the midline; but which pathways is not clear. Lesions of the posterior thalamus were also associated with a permanent tilt of the SVV,[81] and the infarcts in these patients were thought to lie in the regions roughly correlating with the "vestibular-receiving" thalamus in monkeys.[78] How the region relates to the cerebellar thalamic territory is not clear.[74–77]

Thalamic information is relayed to the cerebral cortex,[82] but there are very few reports of otolith responses in the cerebral cortex (see Dieterich *et al.*, and Hoffmann *et al.*, in this volume). The vestibular cortical circuits described by Guldin and colleagues[83–85] are based, at present, on *canal information*, and not otolith signals. In one study, electrical stimulation of both saccule or utricle gave rise to evoked potentials in the same areas of the cat cerebral cortex (3a, 2v, and SII) as canal stimulation.[86]

Sophisticated multiaxis turntables and sledges in several laboratories around the world, including the one designed by Volker Henn, who we are at present celebrating with this conference in Zurich, have recently enabled scientists to make great progress in understanding "otolith-signal processing" in the central nervous system. We are only just beginning to be able to outline the neuroanatomical pathways carrying the otolith information.

ACKNOWLEDGMENTS

The author is indebted to Dr. Neal Barmack for providing valuable comments. The excellent technical assistance of Ursula Schneider and Ahmed Messoudi is gratefully acknowledged. This work was supported by a grant from the German Research Council SFB 462/B3.

REFERENCES

1. COHEN, B. 1974. The vestibulo-ocular reflex arc. *In* Handbook of Sensory Physiology, Vol. 6, H. H. Kornhuber, Ed.: 477–540. Springer-Velag. New York.
2. GRAF, W. M., E. A. KESHNER, F. J. R. RICHMOND, Y. SHINODA, K. STATLER & Y. UCHINO. 1997. How to construct and move a cat's neck. J. Vestib. Res. **7**: 219–237.
3. PRECHT, W. 1979. Vestibular mechanisms. Annu. Rev. Neurosci. **2**: 265–289.
4. FUKUSHIMA, K., S. I. PERLMUTTER, J. F. BAKER & B. W. PETERSON. 1990. Spatial properties of second-order vestibulo-ocular relay neurons in the alert cat. Exp. Brain Res. **81**: 462–478.
5. PRECHT, W., J. H. ANDERSON & R. H. I. BLANKS. 1979. Canal-otolith convergence on cat ocular motoneurons. Prog. Brain Res. **50**: 459–468.
6. FUKUSHIMA, K. & J. FUKUSHIMA. 1991. Otolith-visual interaction in the control of eye movement produced by sinusoidal vertical linear acceleration in alert cats. Exp. Brain Res. **85**: 36–44.
7. MITTELSTAEDT, H. 1992. Somatic versus vestibular gravity reception in man. Ann. N.Y. Acad. Sci. **656**: 124–139.
8. BERGSTROM, B. 1973. Morphology of the vestibular nerve. II The number of myelinated vestibular nerve fibers in man at various ages. Acta Otolaryngol. (Stockh) **76**: 173–179.

9. MARKHAM, C. H. 1989. Anatomy and physiology of otolith-controlled ocular counterrolling. Acta Oto-Laryngol. **468**: 263–266.
10. BÜTTNER-ENNEVER, J. A. 1992. Patterns of connectivity in the vestibular nuclei. Ann. N. Y. Acad. Sci. **656**: 363–378.
11. GERRITS, N. M. 1990. Vestibular nuclear complex. *In* The Human Nervous System, G. Paxinos, Ed.: 863–88. Academic Press. San Diego.
12. BARMACK, N. H., R. W. BAUGHMAN, P. ERRICO & H. SHOJAKU. 1993. Vestibular primary afferent projection to the cerebellum of the rabbit. J. Comp. Neurol. **327**: 521–534.
13. BARMACK, N. H. 1996. GABAergic pathways convey vestibular information to the beta nucleus and dorsomedial cell column of the inferior olive. Ann. N.Y. Acad. Sci. **781**: 541–552.
14. GACEK, R. R. 1969. The course and central termination of first order neurons supplying vestibular end organs in the cat. Acta Oto-Laryngol. **254**: 1–66.
15. KEVETTER, G. A. & A. A. PERACHIO. 1986. Distribution of vestibular afferents that innervate the sacculus and posterior canal in gerbil. J. Comp. Neurol. **254**: 410–424.
16. IMAGAWA, M., W. M. GRAF, H. SATO, H. SUWA, N. ISU, R. IZUMI & Y. UCHINO. 1998. Morphology of single afferents of the saccular macula in cats. Neurosci. Lett. **240**: 127–130.
17. IMAGAWA, M., N. ISU, M. SASAKI, K. ENDO, H. IKEGAMI & Y. UCHINO. 1995. Axonal projections of utricular afferents to the vestibular nuclei and the abducens nucleus in cats. Neurosci. Lett. **186**: 87–90.
18. ISHIZUKA, N., S. SASAKI & H. MANNEN. 1982. Central course and terminal arborizations of single primary vestibular afferent fibers from the horizontal canal in the cat. Neurosci. Lett. **33**: 135–139.
19. SATO, F., H. SASAKI, N. ISHIZUKA, S. SASAKI & H. MANNEN, 1989. Morphology of single primary vestibular afferents originating from the horizontal semicircular canal in the cat. J. Comp. Neurol. **290**: 423–439.
20. SATO, H., K. ENDO, H. IKEGAMI, M. IMAGAWA, M. SASAKI & Y. UCHINO, 1996. Properties of utricular nerve-activated vestibulospinal neurons in cats. Exp. Brain Res. **112**: 197–202.
21. UCHINO, Y., H. IKEGAMI, M. SASAKI, K. ENDO, M. IMAGAWA & N. ISU. 1994. Monosynaptic and disynaptic connections in the utriculo-ocular reflex arc of the cat. J. Neurophysiol. **71**: 950–958.
22. UCHINO, Y., M. SASAKI, H. SATO, M. IMAGAWA, H. SUWA & N. ISU. 1996. Utriculoocular reflex arc of the cat. J. Neurophysiol. **76**: 1896–1903.
23. UCHINO, Y., M. SASAKI, H. SATO, M. IMAGAWA, H. SUWA & N. ISU. 1997. Utricular input to cat extraocular motoneurons. Acta Oto-Laryngol. **528**: 44–48.
24. DICKMAN, J. D. & Q. FANG. 1996. Differential central projections of vestibular afferents in pigeons. J. Comp. Neurol. **367**: 110–131.
25. EPEMA, A. H., N. M. GERRITS & J. VOOGD. 1988. Commissural and intrinsic connections of the vestibular nuclei in the rabbit: a retrograde labeling study. Exp. Brain Res. **71**: 129–146.
26. UCHINO, Y., H. SATO & H. SUWA. 1997. Excitatory and inhibitory inputs from saccular afferents to single vestibular neurons in the cat. J. Neurophysiol. **78**: 2186–2192.
27. HWANG, J. C. & W. F. POON. 1975. An electrophysiological study of the sacculo-ocular pathways in cats. Jpn. J. Physiol. **25**: 241–251.
28. UCHINO, Y. 1997. Connections between otolith receptors and neck motoneurons. Acta Oto-Laryngol. **528**: 49–51.
29. BAKER, R., W. PRECHT & A. BERTHOZ. 1973. Synaptic connections to trochlear motoneurons determined by individual vestibular nerve branch stimulation. Brain Res. **64**: 402–406.
30. SASAKI, M., K. HIRANUMA, N. ISU & Y. UCHINO. 1991. Is there a three neuron arc in the cat utriculo-trochlear pathway? Exp. Brain Res. **86**: 421–425.
31. SCHWINDT, P. C., A. RICHTER & W. PRECHT. 1973. Short latency utricular and canal input to ipsilateral abducens motoneurons. Brain Res. **60**: 259–262.
32. LANG, W. & S. KUBIK. 1979. Primary vestibular afferent projections to the ipsilateral abducens nucleus in the cat. Exp. Brain Res. **37**: 177–181.
33. HESS, B. J. M. & N. DIERINGER. 1991. Spatial organization of linear vestibuloocular reflexes of the rat-responses during horizontal and vertical linear acceleration. J. Neurophysiol. **66**: 1805–1818.
34. CHEN-HUANG, C. & R. A. MCCREA. 1998. Viewing distance related sensory processing in the ascending tract of deiters vestibulo-ocular reflex pathway. J. Vestib. Res. **8**: 175–184.

35. MUROFUSHI, T. & I. S. CURTHOYS. 1997. Physiological and anatomical study of click-sensitive primary vestibular afferents in the guinea pig. Acta Otolaryngol. (Stockh.) **117**: 66–72.
36. MUROFUSHI, T., I. S. CURTHOYS & D. P. GILCHRIST. 1996. Response of guinea pig vestibular nucleus neurons to clicks. Exp. Brain Res. **111**: 149–152.
37. MUROFUSHI, T., I. S. CURTHOYS, A. N. TOPPLE, J. G. COLEBATCH & G. M. HALMAGYI. 1995. Responses of guinea pig primary vestibular neurons to clicks. Exp. Brain Res. **103**: 174–178.
38. IKEGAMI, H., M. SASAKI & Y. UCHINO. 1994. Connections between utricular nerve and neck flexor motoneurons of decerebrate cats. Exp. Brain Res. **98**: 373–378.
39. SATO, H., M. IMAGAWA, N. ISU & Y. UCHINO. 1997. Properties of saccular nerve-activated vestibulospinal neurons in cats. Exp. Brain Res. **116**: 381–388.
40. UCHINO, Y., H. SATO, M. SASAKI, M. IMAGAWA, H. IKEGAMI, N. ISU & W. M. GRAF. 1997. Sacculocollic reflex arcs in cats. J. Neurophysiol. **77**: 3003–3012.
41. COWIE, R. J., M. K. SMITH & D. L. ROBINSON. 1994. Subcortical contributions to head movements in macaques. 2. Connections of a medial pontomedullary head-movement region. J. Neurophysiol. **72**: 2665–2682.
42. FAGERSON, M. H. & N. H. BARMACK. 1995. Responses to vertical vestibular stimulation of neurons in the nucleus reticularis gigantocellularis in rabbits. J. Neurophysiol. **73**: 2378–2391.
43. CHAN, Y. S., C. W. CHEN & C. H. LAI. 1996. Response of medullary reteicular neurons to otolith stimulation during bidirectional off-vertical axis rotation of the cat. Brain Res. **732**: 159–168.
44. YATES, B. J., T. GOTO, I. KERMAN & P. S. BOLTON. 1993. Responses of caudal medullary raphe neurons to natural vestibular stimulation. J. Neurophysiol. **70**: 938–946.
45. HOLSTEGE, G. 1991. Descending motor pathways and the spinal motor system: limbic aand nonlimbic components. Prog. Brain Res. **87**: 307–421.
46. SATO, H., T. OHKAWA, Y. UCHINO & V. J. WILSON. 1997. Excitatory connections between neurons of the cervical nucleus and vestibular neurons in the cat. Exp. Brain Res. **115**: 381–386.
47. POMPEIANO, O. & K. HOSHINO. 1977. Responses to static tilts of lateral reticular neurons mediated by contralateral labyrinthine receptors. Arch. Ital. Biol. **115**: 211–236.
48. CARLETON, S. C. & M. B. CARPENTER. 1983. Afferent and efferent connections of the medial, inferior and lateral vestibular nuclei in the cat and monkey. Brain Res. **278**: 29–51.
49. LADPLI, R. & A. BRODAL. 1968. Experimental studies of commissural and reticular formation projections from the vestibular nuclei in the cat. Brain. Res. **8**: 65–96.
50. POMPEIANO, O. 1979. Neck and macula labyrinthine influences on the cervical spinoreticulocerebellar pathway. Prog. Brain Res. **50**: 501–514.
51. EKEROT, C.-F., B. LARSON & O. OSCARSSON. 1979. Information carried by the spinocerebellar paths. Prog. Brain Res. **50**: 79–90.
52. APPS, R. & J. R. TROTT. 1997. Topographical organization within the lateral reticular nucleus mossy fibre projection to the c1 and c2 zones in the rostral paramedian lobule of the cat cerebellum. J. Comp. Neurol. **381**: 175–187.
53. KÜNZLE, H. 1975. Autoradiographic tracing of the cerebellar projections from the lateral reticular nucleus in the cat. Exp. Brain Res. **22**: 255–266.
54. BARMACK, N. H., R. W. BAUGHMAN & F. P. ECKENSTEIN. 1992. Cholinergic innervation of the cerebellum of the rat by secondary vestibular afferents. Ann. N.Y. Acad. Sci. **656**: 566-579.
55. PARENTI, R., F. CIRCIRATA, M. R. PANTO & M. F. SERAPIDE. 1996. The projections of the lateral reticular nucleus to the deep caerebellar nuclei. An experimental analysis in the rat. Eur. J. Neurosci. **8**: 2157–2167.
56. PAXINOS, G. & C. WATSON. 1986. The Rat Brain in Stereotaxic Coordinates. Academic Press. Orlando.
57. BARMACK, N. H., B. J. FREDETTE & E. MUGNAINI. 1998. Parasolitary nucleus: a source of GABAergic vestibular information to the inferior olive of rat and rabbit. J. Comp. Neurol. **392**: 352–372.
58. BARMACK, N. H., M. H. FAGERSON, B. J. FREDETTE, E. MUGNAINI & H. SHOJAKU. 1993. Activity of neurons in the beta nucleus of the inferior olive of the rabbit evoked by natural vestibular stimulation. Exp. Brain Res. **94**: 203–215.
59. BARMACK, N. H. & H. SHOJAKU. 1995. Vestibular and visual climbing fiber signals evoked in the uvula-nodulus of rabbit cerebellum by natural stimulation. J. Neurophysiol. **74**: 2573–2589.

60. FUSHIKI, H. & N. H. BARMACK. 1997. Topography and reciprocal activity of cerebellar purkinje cells in the uvula-nodulus modulated by vestibular stimulation. J. Neurophysiol. **78**: 3083–3094.
61. IKEDA, Y., H. NODA & S. SUGITA. 1989. Olivocerebellar and cerebelloolivary connections of the oculomotor region of the fastigial nucleus in the macaque monkey. J. Comp. Neurol. **284**: 463–488.
62. VOOGD, J., N. M. GERRITS & T. J. H. RUIGROK. 1996. Organization of the vestibulocerebellum. Ann. N.Y. Acad. Sci. **781**: 553–579.
63. WHITWORTH, R. H., D. E. HAINES & G. W. PATRICK. 1983. The inferior olive of a prosimian primate, galago senegalensis. II. olivocerebellar projections to the vestibulocerebellum. J. Comp. Neurol. **219**: 228–240.
64. SATO, Y. & N. H. BARMACK. 1985. Zonal organization of the olivocerebellar projections to the uvula in rabbits. Brain Res. **359**: 281–292.
65. BERNARD, J. F. 1987. Topographical organization of olivocerebellar and corticonuclear connections in the rat. An WGA-HRP study. I. Lobules IX, X, and the flocculus. J. Comp. Neurol. **263**: 241–258.
66. BRODAL, P. & A. BRODAL. 1981. The olivocerebellar projection in the monkey. Experimental studies with the method of retrograde tracing of horseradish peroxidase. J. Comp. Neurol. **201**: 375–393.
67. BARMACK, N. H., R. W. BAUGHMAN, F. P. ECKENSTEIN & H. SHOJAKU. 1992. Secondary vestibular cholinergic projection to the cerebellum of rabbit and rat as revealed by choline acetyltransferase immunohistochemistry, retrograde and orthograde tracers. J. Comp. Neurol. **317**: 250–270.
68. EPEMA, A. H., N. M. GERRITS & J. VOOGD. 1990. Secondary vestibulocerebellar projections to the flocculus and uvulonodular lobule of the rabbit: a study using HRP and double fluroescent tracer techniques. Exp. Brain Res. **80**: 72–82.
69. SATO, Y., K. KANDA, K. IKARASHI & T. KAWASAKI. 1989. Differential mossy fiber projections to the dorsal and ventral uvula in the cat. J. Comp. Neurol. **279**: 149–164.
70. THUNNISSEN, I. E., A. H. EPEMA & N. M. GERRITS. 1989. Secondary vestibulocerebellar mossy fiber projection to the caudal vermis in the rabbit. J. Comp. Neurol. **290**: 262–277.
71. SHOJAKU, H., K. SATO, K. IKARASHI & K. KAWASAKI. 1987. Topographical distribution of Purkinje cells in the uvula and the nodulus projecting to the vestibular nuclei in cats. Brain Res. **416**: 100–112.
72. WYLIE, D. R. W., C. I. DE ZEEUW, P. L. DIGIORGI & J. I. SIMPSON. 1994. Projections of individual Purkinje cells of identified zones in the ventral nodulus to the vestibular and cerebellar nuclei in the rabbit. J. Comp. Neurol. **349**: 448–463.
73. SIEBOLD, C., L. GLONTI, S. GLASAUER & U. BÜTTNER. Rostral fastigial nucleus activity in the alert monkey during three-dimensional passive head movements. J. Neurophysiol. **77**: 1432–1446.
74. ASANUMA, C., W. T. THACH & E. G. JONES. 1983. Brainstem and spinal projections of the deep cerebellar nuclei in the monkey with observations on the brainstem projections of the dorsal column nuclei. Brain Res. Rev. **5**: 299–322.
75. STERIADE, M. 1995. Two channels in the cerebellothalamocortical system. J. Comp. Neurol. **354**: 57–70.
76. MIDDLETON, F. A. & P. L. STRICK. 1997. Cerebellar output channels. Int. Rev. Neurobiol. **41**: 61–81.
77. PERCHERON, G., C. FRANCOIS, B. TALBI, J. YELNIK & G. FÈNELON. 1996. The primate motor thalamus. Brain Res. Rev. **22**: 93–181.
78. LANG, W., J. A. BÜTTNER-ENNEVER & U. BÜTTNER. 1979. Vestibular projections to the monkey thalamus: an autoradiographic study. Brain Res. **177**: 3–17.
79. BRANDT, T. & M. DIETERICH. 1995. Central vestibular syndromes in roll, pitch, and yaw planes—Topographic diagnosis of brainstem disorders. Neuro-Ophthalmology **15**: 291–303.
80. DIETERICH, M. & T. BRANDT. 1993. Ocular torsion and tilt of subjective visual vertical are sensitive brainstem signs. Ann. Neurol. **33**: 292–299.
81. DIETERICH, M. & T. BRANDT. 1993. Thalamic infarctions: Differential effects on vestibular function in the roll plane (35 patients). Neurology **43**: 1732–1740.

82. SCHMAHMANN, J. D. & D. N. PANDYA. 1990. Anatomical investigation of projections from thalamus to posterior parietal cortex in the rhesus monkey: a WGA-HRP and fluorescent tracer study. J. Comp. Neurol. **295**: 299–326.
83. AKBARIAN, S., O. J. GRÜSSER & W. O. GULDIN. 1992. Thalamic connections of the vestibular cortical fields in the squirrel monkey (saimiri sciureus). J. Comp. Neurol. **326**: 423–441.
84. GULDIN, W. & O. J. GRÜSSER. 1998. Is there a vestibular cortex? TINS **21**: 254–259.
85. GULDIN, W. O., S. AKBARIAN & O. J. GRÜSSER. 1992. Cortico-cortical connections and cytoarchitectonics of the primate vestibular cortex: a study in squirrel monkeys (Saimiri sciureus). J. Comp. Neurol. **326**: 375–401.
86. JIJIWA, H., T. KAWAGUCHI, S. WATANABE & H. MIYATA. 1991. Cortical projections of otilith organs in the cat. Acta. Otolaryngol. (Stockh.) **481**: 69–72.

Signal Processing Related to the Vestibulo-Ocular Reflex during Combined Angular Rotation and Linear Translation of the Head

ROBERT A. MCCREA[a] AND CHIJU CHEN-HUANG

Department of Neurobiology, Pharmacology, and Physiology, University of Chicago, 5807 South Ellis Avenue, Chicago, Illinois 60637, USA

ABSTRACT: The contributions of vestibular nerve afferents and central vestibular pathways to the angular (AVOR) and linear (LVOR) vestibulo-ocular reflex were studied in squirrel monkeys during fixation of near and far targets. Irregular vestibular afferents did not appear to be necessary for the LVOR, since when they were selectively silenced with galvanic currents the LVOR was essentially unaffected during both far- and near-target viewing. The linear translation signals generated by secondary AVOR neurons in the vestibular nuclei were, on average, in phase with head velocity, inversely related to viewing distance, and were nearly as strong as AVOR-related signals. We suggest that spatial–temporal transformation of linear head translation signals to angular eye velocity commands is accomplished primarily by the addition of viewing distance multiplied, centrally integrated, otolith regular afferent signals to angular VOR pathways.

INTRODUCTION

Movements of the head in space are often a complex combination of angular rotation and linear translation. The vestibulo-ocular reflex (VOR) functions to stabilize images on the retina during both types of head movement. Angular head acceleration stimulates the labyrinthine semicircular canals and evokes an angular vestibulo-ocular reflex (AVOR). Linear, or translational head acceleration stimulates vestibular otolith receptors and evokes the linear vestibulo-ocular reflex (LVOR). One complicating factor is that the rotational eye velocity required to stabilize a visual target on the retina when the head and eyes are translated is inversely related to the distance of the target from the eyes.[1] When the distance of a visual target from the eyes is large, the eye movement required to stabilize the image is negligible. But as the distance of the image from eyes decreases, the eye rotational velocity required to maintain image stability for even small head translations can be considerable. Thus in order to produce a compensatory eye movement when the head is simultaneously rotated and translated the central nervous system has to temporally transform sensory afferent signals from different labyrinthine endorgans into eye movement commands that have appropriate spatial and dynamic characteristics at different viewing

[a]To whom correspondence may be addressed. Phone: 773/702-6374; fax: 773/702-6374; e-mail: ramccrea@midway.uchicago.edu

distances. This spatial and temporal computation must be carried out rapidly and accurately in order for clear vision to be maintained.

Head acceleration is the necessary stimulus for producing sensory signals in vestibular afferents that innervate both the hair cells in the semicircular canal crista and the otolith macula, but there are important differences in the signals carried by the vestibular primary afferent fibers that innervate each endorgan. In squirrel monkeys, the afferents that innervate the semicircular canals generate signals that are roughly in phase with, or slightly lead, angular head velocity during mid- to high-frequency head movements.[2] On the other hand, the head translation signals generated by otolith afferents are roughly in phase with, or slightly lead, linear head acceleration.[3] In each endorgan, the afferents that phase lead velocity or acceleration tend to be the irregularly discharging afferents that innervate hair cells in the most sensitive regions of the sensory epithelium.[4]

The premotor and motor signals that are required to produce the VOR are generally thought to be a combination of angular position and velocity commands.[5] Consequently, transformation of sensory vestibular signals into appropriate oculomotor commands involves partial integration of the angular head velocity signals generated by semicircular canal afferents and nearly double integration of the linear head acceleration signals carried by otolith afferents. Otolith signals related to linear translation also have to be converted into appropriate angular velocity-position oculomotor commands whose amplitude varies as a function of viewing distance (or vergence angle) in order for the LVOR to work properly. Since the AVOR and the LVOR are evoked by stimulation of different vestibular receptors and clearly require different processing in the CNS, the question arises whether these two reflexes share common premotor pathways.

There is evidence for the existence of separate otolith-ocular pathways in mammals[6] (see also Uchino *et al.*, this volume). However, the results of recent physiological studies[7–9] suggest that the central processing of signals related to the AVOR and combined angular and linear VOR (CVOR) are not entirely separate. One question is whether, and to what extent, central VOR pathways in the squirrel monkey related to the horizontal semicircular canal AVOR also carry signals related to the translational LVOR.

A second question is whether different classes of vestibular afferents are specifically involved in generating the translational LVOR. In the squirrel monkey, the signals generated by regularly discharging semicircular canal afferents are sufficient to produce the AVOR evoked in the dark or during far-target viewing, but irregular canal afferents appear to be necessary for producing viewing distance-related changes in the AVOR.[10] It has been suggested that the phase lead in the linear responses of irregular otolith afferents may provide an important part of the physiological substrate for producing spatial and temporal transformation of otolith signals related to the LVOR[11] (see Angelaki, this volume).

In this paper we will review the results of single unit and behavioral studies in the squirrel monkey that were designed to determine the contribution of irregular afferents and central AVOR pathways to the LVOR during near- and far-target viewing. The results of these studies suggest that at midband frequencies of linear stimulation the central AVOR pathways contribute significantly to the translational LVOR, and that irregular afferents are not necessary for producing the LVOR.

METHODS

Squirrel monkeys were prepared for bilateral eye movement and single-unit recordings. Eye movements were measured using the magnetic search-coil technique. Single-unit recordings were obtained with tungsten microelectrodes introduced into the vestibular nuclei. Labyrinthine stimulating electrodes were chronically implanted into the perilymphatic space bilaterally so that secondary vestibular units could be orthodromically identified and so that vestibular irregular afferents could be silenced by bilateral application of anodal currents. Short cathodal shocks (100 μs, 50–300 μA) were used for identifying secondary vestibular neurons in the vestibular nuclei. DC anodal galvanic currents 50–100 μA were applied bilaterally in a calibrated manner so that irregular vestibular afferents could be selectively silenced.[10,12] A few (6) units were antidromically activated following electrical stimulation of the ipsilateral ascending tract of Deiters.[9] Records of single unit firing rate and eye movements were selected on the basis of behavioral criteria (accurate fixation of targets), desaccaded and averaged. The gain and phase of sinusoidal responses were determined by fitting sinusoidal functions to an average of concatenated, desaccaded, averaged records.

FIGURE 1A is an illustration of the experimental setup used. During experiments squirrel monkeys were seated on a position servocontrolled vestibular turntable. Recordings were made with the monkey's head held in the plane of the horizontal semicircular canal at three different positions with respect to the axis of rotation. In the head-center (HC) position, the monkeys interaural plane was centered on the axis of turntable rotation, and turntable rotation stimulated primarily the horizontal semicircular canals, evoking the AVOR (FIG. 1B, center traces). The monkeyís head was then positioned 20 cm in front of the axis of turntable rotation with the nose out (NO) such that turntable rotation produced a combination of angular and linear head movements that had a synergistic effect on the CVOR (FIG. 1B, top traces). The monkey's head was then positioned 20 cm off the axis of rotation with its nose in (NI) such that turntable rotation produced a combination of angular and linear head movements that had a synergistic effect on the CVOR (FIG. 1B, bottom traces). The contribution of the translational LVOR to the CVOR was assessed by subtracting responses evoked in the head-center position from responses evoked in the off-axis position (FIG. 1B, right column).

Monkeys were trained to fixate and pursue small laser spots projected onto a cylindrical tangent screen and to fixate small LED targets that were positioned 10-cm (near-target) or 1.3–1.7-cm (far-target) distance from the eyes. In most experiments single-unit and eye-movement responses were recorded at three frequencies of turntable rotation (0.7, 1.9, and 4.0 Hz).

Comments on The Use of Galvanic Currents for Silencing Irregular Afferents

Irregular afferents are significantly more sensitive to galvanic currents than regularly discharging afferents.[13] However, the differential sensitivity of irregular afferents to galvanic currents is relative, and anodal galvanic currents reduce the firing rate of both regular and irregular afferents. In squirrel monkeys small anodal currents (50–100 μA) can be used to completely silence most irregular afferents without silencing most regular afferents. Thus it is possible to study the effects of selective silencing of irregular afferents on the head-movement responses of central vestibular neurons and reflexes, provided care is taken to use small-amplitude rotational or translational stimuli that do not recruit silenced irregular afferents or silence regular afferents whose firing rate is already much lower than normal.

Bilaterally applied anodal galvanic currents dramatically decrease the firing rate of most regular and irregular afferents, so the tonic vestibular inputs to the brain provided both classes of afferent are dramatically affected. Central neurons that do not receive symmetric excitatory and inhibitory labyrinthine inputs derived from similar afferent inputs would be expected to exhibit a tonic change in firing rate during bilateral application of galvanic currents; regardless of the ultimate source of their vestibular inputs (see FIGS. 5 and 6 below). The tonic change in firing rate could be interpreted as head tilt or constant velocity head rotation by other circuits in the brain. Consequently, the technique may be inappropriate for assessing the contribution of irregular afferents to steady-state nystagmus, posture, or VOR

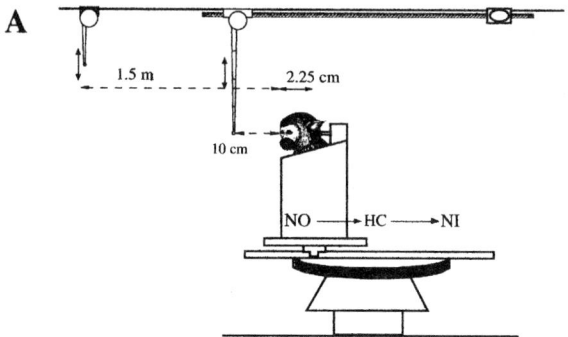

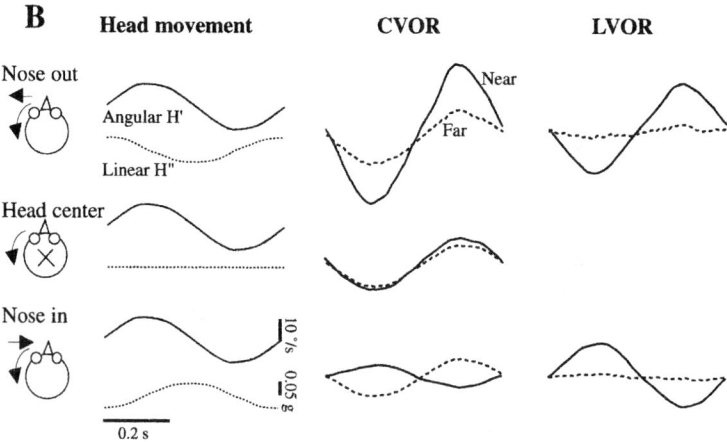

FIGURE 1. Methods used for studying vestibular signal processing during combined linear and angular head movement. (**A**) Experimental setup. The monkey was seated on a linear track mounted on the top of a position servocontrolled vestibular turntable. Animals were trained to fixate small, retractable targets 1.5 m and 0.1 m from the eyes. The near target was attached to a motorized track on the ceiling that allowed its position to be adjusted as the monkey was moved to different off-axis positions on the turntable. (**B**) Head and eye velocity evoked during table rotation with the monkey placed 20 cm in front of the axis of rotation (nose out, *top panel*), on the axis of rotation (head center, *center panel*), and 20 cm behind the axis of rotation (nose in, *bottom panel*). The CVOR eye movements evoked when the monkey was fixating a far target (*dashed traces*) and near target (*solid traces*) are superimposed in the *center column*. The *right-hand column* illustrates the calculated LVOR response. The illustrated records were selected when the vergence angle (*dashed lines*; negative values are convergence) and eye position were within 2° of the target, and are averages of 30–60 cycles at 1.9 Hz.

velocity storage. In any case, it is prudent to examine the responses of peripheral and central vestibular pathways related to the VOR or other vestibular reflexes during application of galvanic currents before assuming that the currents produced a behavioral change that was attributable to irregular afferents.

Bilaterally applied cathodal currents increase the firing rate of both regular and irregular vestibular afferents, and increase the tonic firing rate of most central vestibular neurons; therefore, DC cathodal currents are probably not useful for differentiating the contributions of regular and irregular afferents to central physiology or behavior. The most easily interpreted effect of galvanic currents is the lack of an effect. If bilaterally applied anodal currents have little effect on a behavior, it is reasonable to conclude that irregular afferents are not necessary to produce it.

In sum, the effects of galvanic currents on vestibular reflexes need to be interpreted with caution, after examining the effects of the procedure on the physiology of relevant central pathways and considering the consequences of a dramatic reduction in the spontaneous firing rate of central vestibular pathways.

RESULTS

The gain of the AVOR and CVOR evoked by turntable rotation during near-target viewing was significantly larger than during far-target viewing (FIG. 1B). In the head-center position, the gain of the AVOR increased 23% during near-target viewing. During off-axis rotation, the CVOR gain increased nearly threefold during near-target viewing in the NO position. In the NI off-axis position, linear and angular head velocity were oppositely directed, and the CVOR response reversed in direction during near-target viewing. The gain of the LVOR was typically small during far-target viewing, but increased dramatically during near-target viewing.

Most horizontal canal-related secondary vestibular neurons were sensitive to the LVOR. FIGURE 2 illustrates the averaged responses of two secondary units that were likely to have

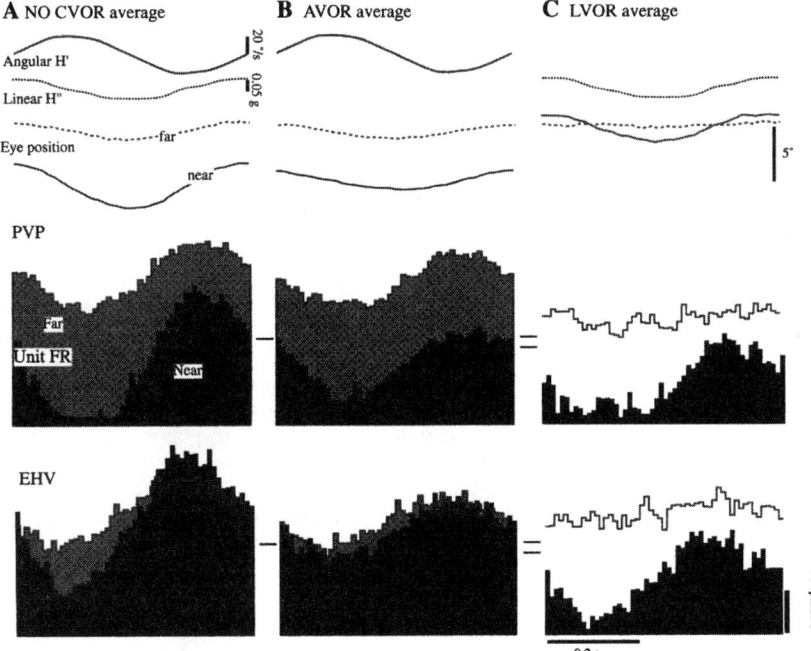

FIGURE 2. PVP and EHV unit CVOR, AVOR, and LVOR responses during near and far target viewing. (**A, B**) : Averaged head, eye, and unit responses during 1.9-Hz rotation in the NO off-axis position (A) and when the head was centered on the axis of rotation (B). The calculated LVOR responses are illustrated in (**C**). Examples of the responses of a PVP unit and an EHV unit are illustrated.

been involved in producing the angular and linear vestibulo-ocular reflex. The top traces illustrate records of angular head velocity, linear head acceleration, and horizontal eye movements recorded in the NO off-axis position and HC position during rotation at 1.9 Hz while the monkey fixated targets that were far and near to the eyes. Each unit represents an example of two important classes of secondary vestibular units that are involved in generating the AVOR. Each unit was activated at monosynaptic latency following short shock cathodal stimulation of the vestibular nerve. In the squirrel monkey, position-vestibular-pause (PVP) units project both to medial and lateral rectus motoneurons.[9,14] These units are sensitive to ipsilateral angular head velocity contralateral eye movements, pause during saccades, and exhibit a reduced sensitivity to head movements during VOR cancellation.

A second class of units that participate in the direct AVOR pathways are eye-head-vestibular (EHV) units. Most of these units are sensitive to ipsilateral angular head velocity during the VOR, but to contralateral head velocity during VOR cancellation (EHV units) and to contralateral eye velocity during ocular pursuit. The EHV unit illustrated in FIGURE 2 was activated antidromically following electrical stimulation of the ipsilateral ascending tract of Deiters, which suggests that it is part of the direct "three-neuron" vestibulo-ocular pathway from the horizontal semicircular canal to the medial rectus muscle of the ipsilateral eye. Both PVP and EHV units were sensitive to ipsilateral head velocity during the AVOR (FIG. 2B). The responses increased in amplitude during rotation in the NO off-axis position, particularly during near-target viewing (FIG. 2A, filled histograms). The difference in response amplitude is related to the LVOR (FIG. 2C). The modulation in both units and in eye movement attributable to the LVOR were small during far-target viewing, but large during near-target viewing. In both units, the LVOR signals had nearly the same phase relationship to linear head velocity as they had to angular head velocity.

The LVOR-related responses of all of the PVP and EHV units that were sensitive to ipsilateral head velocity during the AVOR are illustrated in FIGURE 3. FIGURE 3A is a polar plot of the far- (solid small circles) and near- (open large circles) target LVOR responses of PVP units. The right-hand dashed line represents signals that were in phase with ipsilateral linear head velocity. Clockwise (downward) points phase lag ipsilateral head velocity. Some PVP units were not sensitive to the LVOR, but the responses of most units were modulated roughly in phase with ipsilateral linear head velocity. The mean PVP unit LVOR response in the NO position at 1.9 Hz was 5.1 sp/s/cm/s during near-target viewing and phase-lagged ipsilateral head velocity by 24 deg (thick line with open circle in FIG. 3A). EHV units tended to be more sensitive to the LVOR than PVP units. Their mean LVOR response in the NO position at 1.9 Hz was 11.6 sp/s/cm/s during near-target viewing, and phase-led ipsilateral head velocity by 25 deg (thick line with open square in FIG. 3B). The vector sum of the population responses of PVP and EHV units is plotted in FIGURE 3C. The phase of the LVOR eye-movement response and regular vestibular afferents are also illustrated in the polar plot. The phase of the response of these two major classes of secondary VOR neurons is roughly the same as the response phase of the eye movements evoked by the LVOR, and lags the signals generated by regular vestibular otolith afferents by 90 degrees. This 90° phase lag re otolith regular afferents was remarkably constant during 0.7-, 1.9-, and 4.0-Hz offaxis rotations, which suggests that the otolith signals carried by central AVOR pathways had been temporally integrated.

Contribution of Regular and Irregular Afferents to the LVOR

The LVOR responses of secondary VOR neurons lagged regularly discharging otolith afferents by 90°, and irregularly discharging afferents by an even larger amount. If otolith irregular afferents contribute significantly to the LVOR signals carried by secondary VOR

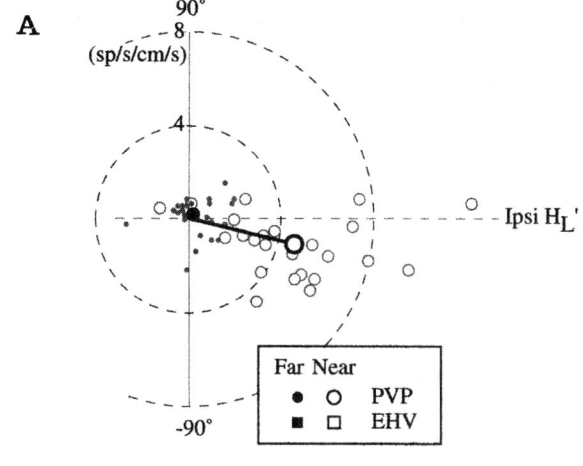
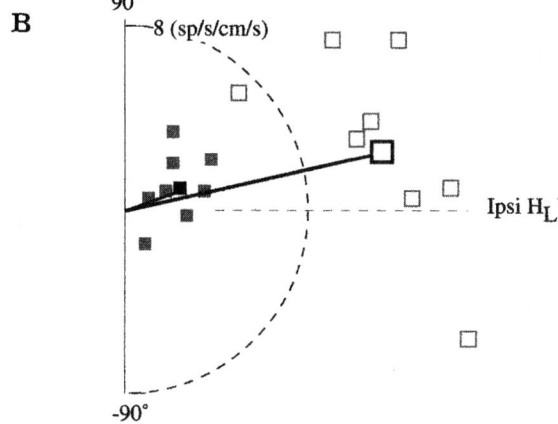
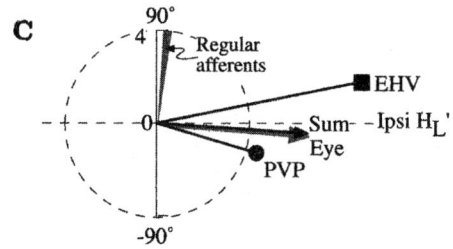

FIGURE 3. Polar plots of the LVOR responses of PVP (**A**) and EHV (**B**) units recorded in the 20-cm NO off-axis position during 1.9-Hz rotation. Only units sensitive to ipsilateral head velocity during the AVOR are illustrated. Phase lead ipsilateral linear head velocity is counterclockwise. *Small filled symbols* are responses recorded during far-target viewing. *Large open symbols* are responses during near-target viewing. The *thick lines* indicate vector average responses for each population of cells. These averages are compared to the phase of the LVOR evoked by the same stimulus and the response phase of regular afferents in (**C**).

pathways, their signals would have had to have been at least partially integrated a second time. On the other hand, if regularly discharging vestibular afferents contribute the bulk of the otolith inputs to LVOR pathways, their signals need to have been integrated only once. Although it would seem to be simpler for the CNS to use the signals of regular afferents to generate the LVOR signals carried by central VOR pathways, the brain may not use Occam's razor as a rule for constructing the central connectivity of the LVOR network during development. If the translational LVOR evoked during off-axis rotation relied primarily on regular afferents, then functional ablation of irregular afferent signals should have no effect on the LVOR or on the signals carried by secondary VOR pathways.

Galvanic currents affect the firing behavior of both regular and irregular afferents. Bilateral application of anodal currents inhibits the firing rate of both regular and irregular afferents. In squirrel monkeys, the effects of anodal currents on vestibular nerve afferent discharge, secondary VOR pathways, and on eye movements have been carefully studied (see the Methods section). The results of these studies suggest that most squirrel-monkey irregular afferents can be selectively silenced and prevented from firing during head rotation provided that small-amplitude stimuli are used (FIG. 4A). The typical effect of 50-µA anodal currents bilaterally applied for 5s on the spontaneous firing rate and rotational response of a secondary PVP unit is illustrated in FIGURE 4B. The amount of current was approximately four times the amount required to produce a threshold response in the incoming fiber volley field potential in the vestibular nuclei, and its application reduced the amplitude of the irregular afferent contribution to this potential by more than 90%.[12] During the onset and offset of the anodal currents, PVP units generated a transient response; consequently, only responses recorded 0.5 s after the onset of currents were analyzed.

During galvanic ablation of irregular afferents (GA) the spontaneous firing rate of PVP units was significantly reduced (FIG. 4B), although most units were not silenced and continued to be modulated by head rotation. The reduction in spontaneous firing rate was due to the inhibitory effects of currents on both regular and irregular afferents. Superimposed averaged responses of two PVP units in the presence (shaded histograms) and absence (dark histograms control traces) of bilaterally applied 50-µA anodal currents are illustrated in FIGURE 5. In both units, GA reduced the spontaneous discharge rate, and the rotational responses in the off direction of the unit illustrated in Figure 5B were clipped during GA when the unit's firing rate was driven into inhibitory saturation. Fortunately, most secondary VOR units in the squirrel monkey were not silenced by GA with the stimuli used in these studies.

The head-movement responses of PVP neurons were affected by GA in several ways. The rotational responses of some units were decreased in gain (e.g., FIG. 5A) while the responses of others increased in gain (FIG. 5B). On average, the angular rotational responses of eye-movement units and the AVOR in the dark were unaffected by GA.[10,15] The effects of GA on the CVOR responses of PVP units were also variable. The CVOR responses of the secondary PVP unit illustrated in FIGURE 6 were slightly enhanced during GA, and GA had little effect on response phase. In other units, GA produced a small decrease in response gain during the CVOR.

Galvanic ablation of irregular afferents had little effect on the LVOR component of the CVOR. The effects of GA on the CVOR and LVOR evoked during NO off-axis sinusoidal rotation is illustrated in FIGURE 7. FIGURE 8 illustrates the effects of GA on the CVOR and LVOR produced by steps in head acceleration in the dark. Galvanic ablation of irregular afferents reduced the gain of the CVOR by a small amount during near-target viewing (FIGS. 7A and 8A), but this reduction was largely attributable to the effects of GA on the viewing-distance-enhanced component of the AVOR. When AVOR responses were subtracted out from CVOR responses, the remaining eye movement was remarkably unaffected by GA. In FIGURE 7B LVOR responses recorded in the presence and absence of GA were superimposed during 1.9-Hz NO rotation. The average responses recorded at three frequencies of rotation are plotted in FIGURE 7C.

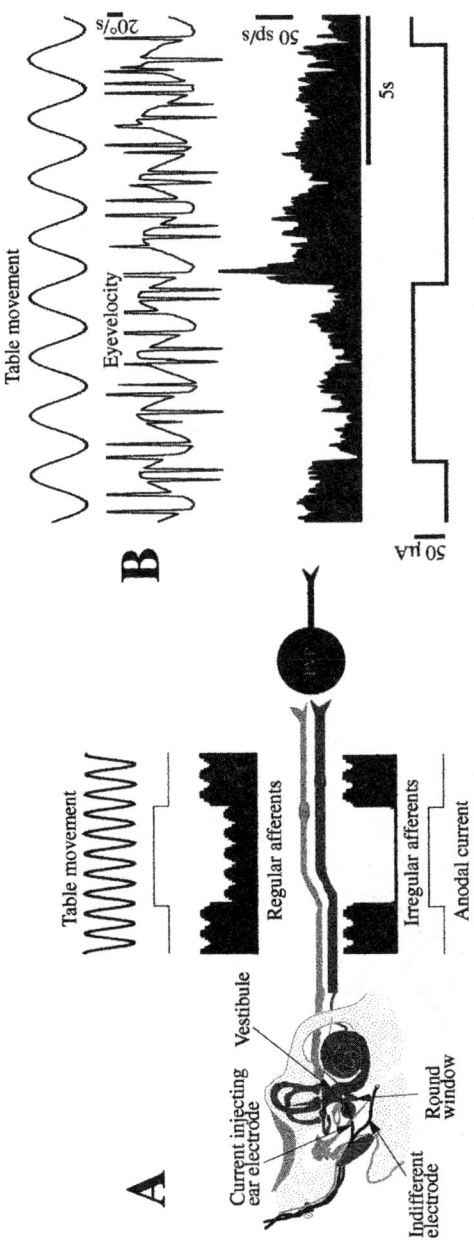

FIGURE 4. Effects of anodal currents on the firing behavior of primary and secondary vestibular neurons. (**A**) Schematic diagram of the experimental galvanic ablation (GA) of irregular afferent discharge modulation during head movements. Perilymphatic anodal currents inhibit both regular and irregular primary afferent fibers, but appropriately calibrated currents can selectively silence irregular afferents and remove their discharge modulation during head movements. (**B**) Effect of GA on the firing behavior of a typical secondary PVP unit. Records of turntable velocity and eye velocity are shown above the record of unit firing rate. The bottom trace indicates the onset and offset of anodal current. Inhibition of regular and irregular afferent inputs produces a decrease in spontaneous firing rate. The effects of irregular afferents on the modulation of PVP unit firing rate can then be assessed. Note the transient responses during the onset and offset of currents, and the lack of inhibitory saturation during rotation.

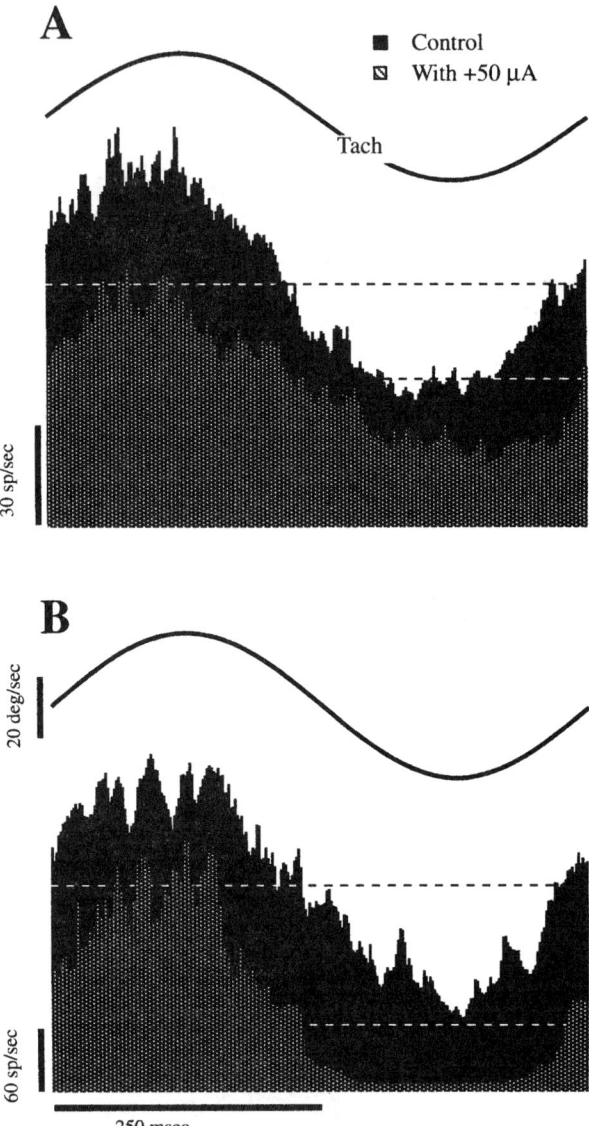

FIGURE 5. Averaged responses of two PVP units during irregular afferent ablation. (**A**) Example of a PVP unit whose rotational gain decreased during GA. (**B**) Example of a PVP unit whose rotational gain increased during GA. Responses recorded during GA (*hatched histograms*) are superimposed on control responses recorded during interleaved periods when GA was not applied (*filled histograms*). Note that the spontaneous firing rate of each unit (*dashed lines*) decreased during GA, and that the response of the unit illustrated in part (B) was clipped by this reduction. Such clipping occurred relatively infrequently with the stimuli used in this study, and only marginally affected the population response of PVP units.

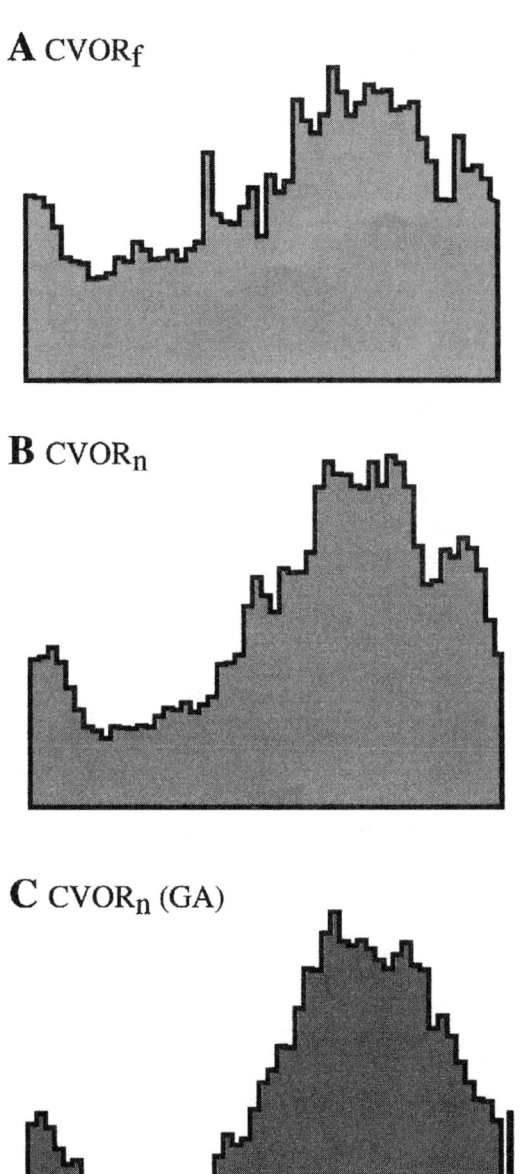

FIGURE 6. CVOR responses of a PVP unit during galvanic ablation of irregular afferents. (**A**) Unit response to 1.9-Hz rotation recorded in the NO position during far-target viewing. (**B**) Unit responses recorded during near-target viewing. (**C**) The response recorded during galvanic ablation of irregular afferents. Note that the response in part (C) is larger than the response in either part (A) or (B).

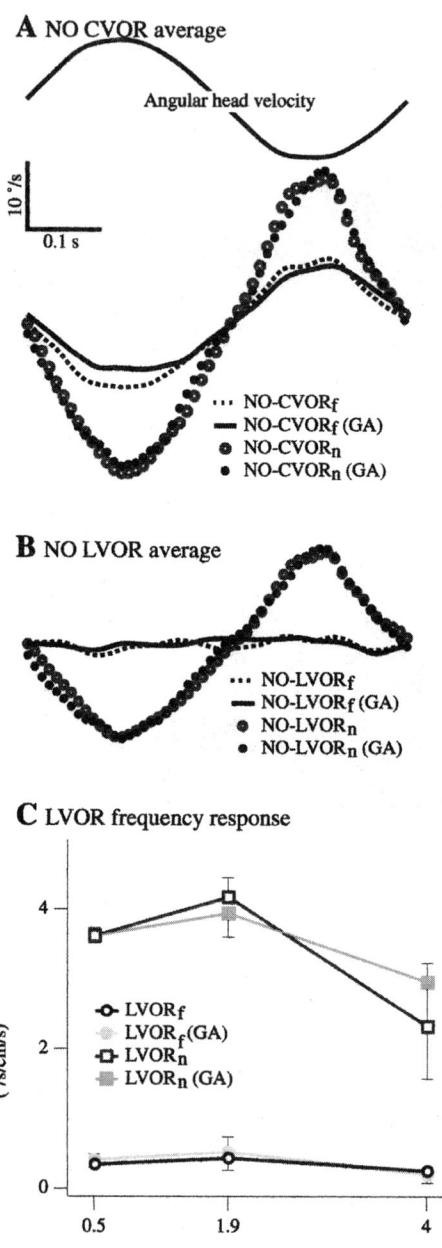

FIGURE 7. Lack of effect of galvanic ablation of irregular afferents on the sinusoidal CVOR and LVOR. (**A**) Effects of GA on the CVOR during far (CVORf) and near (CVORn) target viewing. (**B**) Calculated effect of GA on the LVOR during far (LVORf) and near (LVORn) target viewing. (**C**). LVOR gain plotted as a function of frequency. Filled symbols indicate records obtained during GA; open symbols indicate responses obtained during interleaved epochs in which current was not applied.

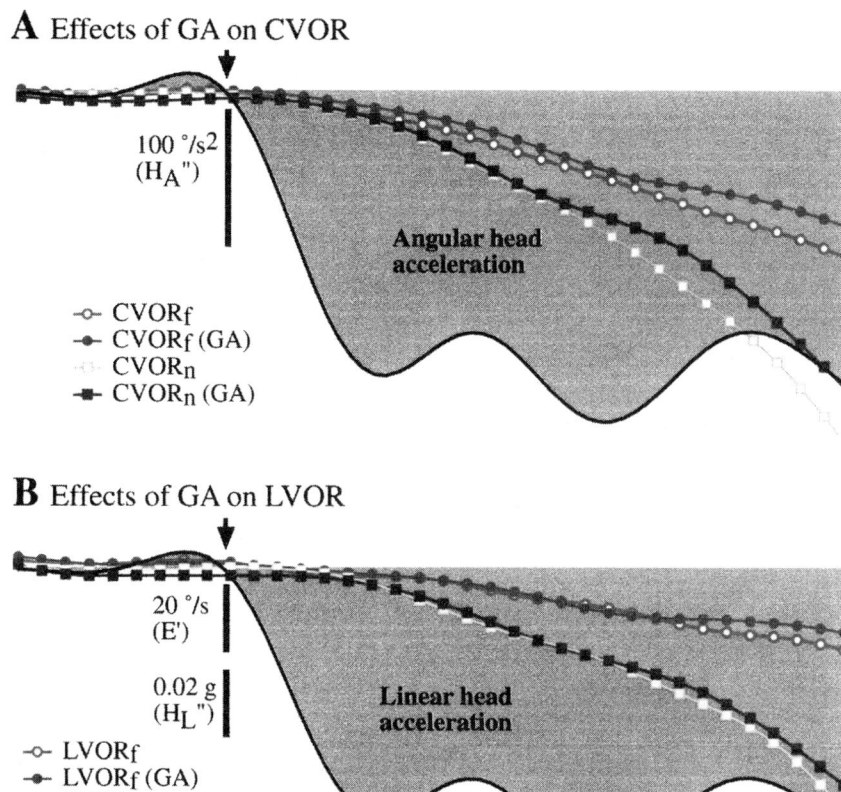

FIGURE 8. Effect of galvanic ablation of irregular afferents on the CVOR and LVOR evoked in the dark by a step in turntable acceleration with the monkey in the NO position. The monkey was required to fixate an earth stationary target that was near or far from its head. Steps in head acceleration were triggered in the dark, shortly after extinguishing the target. (**A**) Effects of GA on CVORf and CVORn responses. The small reduction in CVOR gain was almost entirely attributable to the effects of GA on the AVOR (not shown). (**B**) Estimated effect of GA on the LVOR evoked by steps in linear head acceleration in the NO off-axis position. Response were calculated by subtracting AVOR responses recorded in the head center position from the CVOR records shown in part (A). Note that GA had little effect on the response to this high-frequency stimulus.

GA also had no significant effect on the LVOR evoked by steps in head acceleration. The LVOR eye-movement responses illustrated in FIGURE 8B are averaged records calculated by subtracting AVOR responses from the CVOR responses illustrated in FIGURE 8A. The step in linear head acceleration was generated in the dark briefly after the target LED was extinguished. Although the currents used in the illustrated experiment were strong enough to produce a small decrease in the gain of the AVOR, they had no effect on the LVOR, either when the eyes had been fixated on a far (LVORf) or near (LVORn) target. Thus, the signals generated by regularly discharging afferents are sufficient to produce the LVOR during mid-to high-frequency (0.5–6-Hz) off-axis rotation.

DISCUSSION

During walking, running, or leaping from one tree branch to another, complex, changing, combinations of linear head translation and angular head rotation stimulate the otolith organs and semicircular canals. These vestibular sensory signals must be converted into accurate oculomotor commands that keep images of interest in a stable position on both retinas if visual acuity is to be maintained. Otolith signals are transformed into appropriate angular oculomotor commands and then combined with semicircular canal signals on secondary VOR pathways to produce the translational LVOR. Specifically, the linear acceleration signals carried by otolith regular afferents are dynamically transformed by a central integrator and multiplied by a factor inversely related to viewing distance before being sent to AVOR pathways, which then anatomically perform the spatial transformation of viewing-distance-adjusted linear velocity signals into angular oculomotor commands.

The speed with which the VOR neural network performs the computation previously described above is impressive. The latency of the LVOR illustrated in FIGURE 8B is 20–30 ms, which is only slightly longer than the 10-ms latency of the AVOR. This rapid transformation of otolith signals into oculomotor commands is accomplished in part by keeping the matrix algebra associated with this transformation simple and linear. It is possible that integration and viewing-distance multiplication are carried out in a single step simply by setting the gain of a central neural LVOR integrator with a signal proportional to viewing distance, such as vergence angle. The spatial transformation of linear velocity signals to angular oculomotor commands would be done effortlessly by using the same anatomical pathways that accurately produce the AVOR to produce the translational LVOR.

FIGURE 9 summarizes how we think signals from horizontal canal afferents and utricular afferents are combined during off-axis rotation. Both the AVOR and the LVOR components of the CVOR are produced primarily by inputs from regularly discharging vestibular afferents. The utricular signals are multiplied by a factor inversely related to viewing distance and integrated before being sent to secondary VOR pathways. Viewing-distance adjustments in the AVOR are mediated primarily by irregular afferent signals that have been multiplied by a factor inversely related to viewing distance and integrated. The viewing-distance adjusted output of the AVOR and LVOR integrators is sent to extraocular motoneurons via direct pathways and via inputs to secondary VOR pathways. In this scheme the temporal transformation and parametric gain adjustment of otolith signals is carried out in region(s) of the brain related to the VOR neural integrator, such as the prepositus nucleus, while the spatial transformation of the these signals is accomplished by anatomically distributing these transformed signals to premotor AVOR pathways, including secondary canal ocular pathways.

It is unlikely that the preceding simple description of the central organization of otolith-ocular pathways is sufficient. First of all, static head tilt produces torsional counterrolling eye movements, while the interaural translation produces horizontal eye movements; yet the utricular signals produced by the two stimuli are comparable. It seems likely that the pathways that mediate torsional counterrolling are separate from those that produce oculomotor responses to linear translation. Current behavioral evidence suggests that the static tilt pathways carry otolith signals that are low-pass filtered rather than high-pass filtered (see Paige, this volume). A second complication is that it seems likely that separate integrators are required for the AVOR and translational LVOR, since the LVOR often requires disconjugate eye movements, particularly during naso–occipital translation or when eccentric near targets are being viewed, while the AVOR produces primarily conjugate eye movements. Thus the LVOR requires separate premotor pathways to the motoneurons that innervate each muscle on each eye, and central integrators related to each muscle.[16]

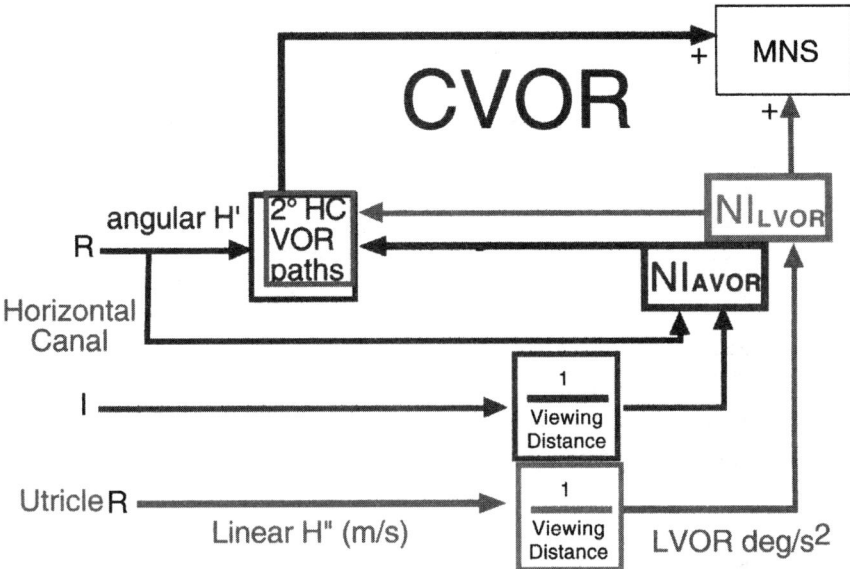

FIGURE 9. Schematic diagram summarizing how horizontal semicircular canal and utricular afferent signals are combined to produce the CVOR during off-axis rotation. The anatomy of secondary horizontal canal pathways is used to transform linear head-velocity signals into angular eye velocity motor commands. Viewing distance adjustments in the CVOR are accomplished by integration of horizontal canal irregular-afferent (I) signals and utricular regular-afferent (R) signals that have been multiplied by a factor inversely related to viewing distance. Thus the signals carried by central VOR pathways code the angular eye velocity required to produce stable vision during different combinations of linear and angular head movement.

A third problem is that the addition of integrated irregular horizontal canal afferent signals to AVOR is not sufficient to produce accurate compensatory eye movements. The use of integrated regular otolith afferent acceleration signals is precisely what is needed to produce a compensatory LVOR eye movement, once the signals have been multiplied by a central estimate of viewing distance, but a similar algorithm fails to work well for the AVOR, particularly at high frequencies of head rotation. In fact, the gain of squirrel monkey AVOR eye movements is typically too low to maintain image stability during near-target viewing, particularly at high frequencies of head rotation.[10]

A heuristic explanation for the better dynamic response of viewing distance adjustments in the LVOR is that the amplitude of translational head movements is potentially much larger in natural circumstances than the amplitude of high-frequency rotational head movements. A mechanism for rapid adjustment of the gain of the LVOR as a function of viewing distance was probably necessary early in vertebrate evolution and is clearly an essential part of the neural substrate that functions to stabilize images on the retina. There may have been less selective pressure to develop a specialized central mechanism to compensate for the small errors in the AVOR produced by changes in viewing distance. The approximate change in the AVOR produced by the selective addition of viewing distance multiplied, and integrated signals carried by the canal afferents most closely related to head acceleration may have proved to be sufficient.

CONCLUSION

The problem of producing compensatory angular eye movements that maintain stable vision of objects in three-dimensional extrapersonal space during head translation was probably encountered early in vertebrate evolution. The simplest solution to the problem appears to be to multiply an integrated vestibular estimate of head acceleration by a factor inversely related to viewing distance and to use existing semicircular canal vestibulo-ocular reflex pathways to produce compensatory angular eye velocity. It is likely that this simple spatial–temporal transformation is a fundamental feature of the architecture of the brainstem neural network that produces the vestibulo-ocular reflex.

REFERENCES

1. VIIRRE, E., D. TWEED, K. MILNER & T. VILIS. 1986. A reexamination of the gain of the vestibuloocular reflex. J. Neurophysiol. **56**: 439–450.
2. GOLDBERG, J. M. & C. FERNANDEZ. 1971. Physiology of peripheral neurons innervating semicircular canals of the squirrel monkey. II. Response to sinusoidal stimulation and dynamics of peripheral vestibular system. J. Neurophysiol. **34**: 661–675.
3. FERNANDEZ, C. & J. M. GOLDBERG. 1976. Physiology of peripheral neurons innervating otolith organs of the squirrel monkey. III. Response dynamics. J. Neurophysiol. **39**: 996–1008.
4. LYSAKOWSKI, A., L. B. MINOR, C. FERNANDEZ & J. M. GOLDBERG. 1995. Physiological identification of morphologically distinct afferent classes innervating the cristae ampullares of the squirrel monkey. J. Neurophysiol. **73**: 1270–1281.
5. ROBINSON, D. A. 1981. The use of control systems analysis in the neurophysiology of eye movements. Ann. Rev. Neurosci. **4**: 463–503.
6. UCHINO, Y., H. IKEGAMI, M. SASAKI, K. ENDO, M. IMAGAWA & N. ISU. 1994. Monosynaptic and disynaptic connections in the utriculo-ocular reflex arc of the cat. J. Neurophysiol. **71**: 950–958.
7. MCCONVILLE K. M. V., R. D TOMLINSON & E.-Q. NA. 1996. Behavior of eye-movement-related cells in the vestibular nuclei during combined rotational and translational stimuli. J Neurophysiol. **76**: 3136–3148.
8. TOMLINSON, R. D., K. M. V. MCCONVILLE & E.-Q. NA. 1996. Behavior of cells without eye movement sensitivity in the vestibular nuclei during combined rotational and translational stimuli. J. Vest. Res. **6**: 145–158.
9. CHEN-HUANG, C. & R. A. MCCREA. 1998 Viewing distance related sensory processing in the ascending tract of Deiters vestibulo-ocular reflex pathway. J. Vestibular. Res. **8**: 175–184.
10. CHEN-HUANG, C. & R. A. MCCREA. 1998. Contribution of vestibular nerve irregular afferents to viewing distance-related changes in the vestibulo-ocular reflex. Exp. Brain Res. **119**: 116–130.
11. BUSH, G. A., A. A. PERACHIO & D. E. ANGELAKI. 1993. Encoding of head acceleration in vestibular neurons. I. Spatiotemporal response properties to linear acceleration. J. Neurophysiol. **69**: 2039–2055.
12. CHEN-HUANG, C., R. A. MCCREA & J. M. GOLDBERG. 1997. Contributions of regularly and irregularly discharging vestibular-nerve inputs to the discharge of central vestibular neurons in the alert squirrel monkey. Exp. Brain Res. **114**: 405–422.
13. SMITH C.E. & J. M. GOLDBERG. A stochastic afterhyperpolarization model of repetitive activity in vestibular afferents. Biol. Cybern. **54**: 41–51.
14. MCCREA, R. A., A. STRASSMAN, E. MAY & S. M. HIGHSTEIN. 1987. Anatomical and physiological characteristics of vestibular neurons mediating the horizontal vestibulo-ocular reflex of the squirrel monkey. J. Comp. Neurol. **264**: 547–570.
15. MINOR L.B. & J. M GOLDBERG. 1991. Vestibular-nerve inputs to the vestibulo-ocular reflex: a functional-ablation study in the squirrel monkey. J. Neurosci. **11**: 1636–1648.
16. MCCONVILLE, K., R. D. TOMLINSON, W. M. KING, G. PAIGE & E.-Q. NA. 1994. Eye position signals in the vestibular nuclei: consequences for models of integrator function. J. Vestibular Res. **4**: 391–400.

Otolith Processing in the Deep Cerebellar Nuclei

U. BÜTTNER,[a] S. GLASAUER, L. GLONTI, J. F. KLEINE, AND C. SIEBOLD

Department of Neurology, Ludwig Maximilians University, D 81377 Munich, Germany

ABSTRACT: To investigate the otolith contribution to the responses of "vestibular only" neurons in the rostral fastigial nucleus (FN), single-unit activity was recorded in the alert monkey with the head fixed during static and dynamic stimulation (± 15 deg, 0.06–1.4 Hz) around an earth-fixed horizontal axis. Head orientation could be altered allowing for roll, pitch, and intermediate planes of orientation. For the vast majority of neurons a response vector orientation (RVO) with an optimal response and a null-response at a head orientation 90 deg apart could be determined. Presumably more than 30% of the vestibular only neurons had an otolith input, as indicated by responses to static tilt, head-position-related activity, large phase changes (> 100 deg) of neuronal activity between 0.06 and 1.4 Hz, changes of the RVO at different frequencies and complex responses (spatio–temporal convergence). Thus, neurons in FN reflecting an otolith or a combined canal–otolith input are much more common than up to now thought. Vestibular-only neurons are most likely involved in vestibulospinal mechanisms. Their precise functional role has yet to be determined.

INTRODUCTION

The cerebellum is intimately related to the vestibular system. Primary vestibular afferents project directly to the cerebellum, mainly to the nodulus/uvula[3] and to a lesser extent to the anterior vermis.[40] Occasional collaterals to other structures, including the fastigial nucleus[31] and the flocculus,[3] have been reported. The vestibulocerebellum (flocculus, nodulus, uvula) has intense, reciprocal connections with the vestibular nuclei.[40] But cerebellar cortical structures not belonging to the vestibulocerebellum such as the entire vermis, also receive a vestibular nuclei input.[25] This also applies to the deep cerebellar nuclei, most prominently to the fastigial nucleus (FN). Mossy fiber (MF) projections derive from all vestibular nuclei (except the lateral vestibular nuclei) and are bilateral.[28] Most efferent connections from FN cross within the cerebellum, traverse within the contralateral FN, and leave the cerebellum via the uncinate fasciculus. The projections from the caudal FN are bilateral and mainly to the lateral and inferior vestibular nucleus. More rostral areas of FN project to the contralateral superior and medial vestibular nucleus.[28]

Projections from the interpositus nuclear complex (NI) to the vestibular nuclei have been reported in the cat,[23] monkey,[18] and rat.[9] They reach each of the main vestibular nuclei on the ipsilateral side with the most prominent projection to the lateral vestibular nucleus.

---

[a]Address for communication: Professor Dr. U. Büttner, Neuologische Klinik, Klinikum Grosshadern, Marchioninistrasse 15, D 81377 München. Phone: 89-7095-2560; fax: 89-7095-8883; e-mail: ubuettner@brain.nefo.med.uni-muenchen.de

A few fibers also reach the contralateral vestibular nuclei. There is also a significant projection from the vestibular nuclei to NI.[9]

This anatomical evidence of strong vestibular connections of the cerebellum, however, does not allow for any functional considerations, particularly with regard to the question whether canal or otolith-related information is carried in these pathways. Staining of the individual canal,[31] utricular,[21] and saccular[22] afferents could give answers to these questions. So far only canal afferents have been shown to have a direct input to the cerebellum. However, the number of stained utricular and saccular neurons reported so far is small. Considering that about 70% of vestibular afferents project to nodulus/uvula[3] and approximately 50% of the vestibular responsive mossy fibers (MF) in the nodulus/uvula have otolith-related static sensitivity,[4] as well as the fact that about 50% of the primary vestibular afferents derive from the otolith organs,[41] it would be surprising if there is no direct otolith–afferent input, at least to the nodulus/uvula.

Recently, individual branches of the vestibular nerve that innervate semicircular canal and otolith organs were injected with tracer substances and the terminal distribution was investigated in the brainstem and the cerebellum of the pigeon.[10] With this method utricular and saccular afferents were traced to the deep cerebellar nuclei. Using the same method in the squirrel monkey, Naito et al.[27] reported saccular and utricular projections to the cerebellum without further specification. With electrophysiological methods one can stimulate individual semicircular canal ampullae and record in cerebellar structures.[42] This approach has also been used to stimulate the utricular[32] and saccular[33] nerve, and to record from vestibulospinal neurons. However, it appears not to have been used so far to determine an otolith input to cerebellar structures.

In the cerebellar cortex otolith-related natural stimuli have been applied mainly to neurons in the nodulus/uvula.[4] Approximately 50% of the vestibularly driven neurons (Purkinje cells, mossy fibers) respond to static tilt. The remaining neurons were modulated by dynamic vertical vestibular stimulation. In the vermis (lobulus V–VII) a few neurons have been reported to respond to static tilt.[30] In a more recent study a high percentage of neurons in the anterior vermis was shown to display spatio–temporal convergence, indicating at least a partial otolith input.[29]

There appears to be no report of neurons in the flocculus responding to static tilt. However, gaze-velocity Purkinje cells in the monkey are modulated differently during the rotational VOR and during eccentric rotation, which induces eye translation and otolith stimulation.[36] In the medial deep cerebellar nucleus, the FN, tilt-responsive neurons have been encountered in the cat.[15,16] The majority of neurons was located in the rostral part of FN and showed an increase of neuronal activity during ipsilateral tilt and a decrease during contralateral tilt. It is not known, whether vestibular neurons in the interposed nucleus carry an otolith-related signal.

In the monkey, FN can be divided into a rostral and a caudal part. The caudal part receives an input from lobulus VI and VII of the dorsal vermis.[28] This part of the vermis is considered as pontocerebellum, since it receives strong projections from the pons.[43] Neurons are modulated with saccades[19] or smooth-pursuit eye movements.[37] Consequently this area has been labeled oculomotor vermis.[43] Neurons in the caudal FN are also modulated with saccades[12,20] or smooth-pursuit eye movements,[8,13] and have therefore been called the fastigial oculomotor region (FOR).[28] In contrast, neurons in rostral FN are not modulated with individual eye movements.[8] Neurons here show a robust modulation with natural vestibular stimulation in the horizontal plane[8] and in vertical planes.[7,35] These "vestibular only" neurons are unlikely to be involved in eye-movement control, but rather in vestibulo-spinal mechanisms (neck, gait, posture). Vestibular-only neurons respond to constant-velocity optokinetic stimulation in a direction-specific manner.[8] The slow gradual activity changes are related to the "velocity storage" mechanism, which is known to also influence vestibulospinal pathways.[39]

Previous work in the vestibular nerve and vestibular nuclei shows that a response to static tilt is certainly related to an otolith input and does not reflect a vertical canal input. With dynamic stimulation, a number of criteria have to be considered. At a given stimulus frequency the response modulation varies with head orientation. A vertical-canal-related neuron has its optimal response when the canal is in the plane of rotation, that is, in the right anterior–left posterior (RALP) or in the left anterior–right posterior (LARP) canal plane (FIG. 1). The orientation for the optimal response is also called the response vector orientation (RVO). Deviations from RVO lead to a continuous decrease of modulation and a null-response (no modulation) at an orientation of 90 deg apart. This modulation can be fitted by a cosine function.[1] For neurons receiving inputs from different canals ("canal convergence") RVOs can range from roll to pitch stimulation. The phase for these neurons receiving an input from one or several vertical canals is close to head velocity, except a phase shift of 180 deg around the null-response. Also for canal-related neurons the phase at a given orientation remains close to head velocity over the frequency range of 0.01–2 Hz.[1,24] And finally the RVO does not alter at different stimulus frequencies.[24] For regular and irregular otolith afferents responses are similar in many aspects. They also have an RVO and a null-response at 90 deg apart. The RVO does not alter with stimulus frequency, but in contrast to canal-related neurons responses are close to head position.[11] This has also been found for a number of neurons in the vestibular nuclei. Other neurons in the vestibular nuclei show a continuously increasing phase lag up to 180 deg over the frequency range of 0.01–1.0 Hz.[34] This implies that an otolith-related neuron can also have a head-velocity-related response at a given frequency.

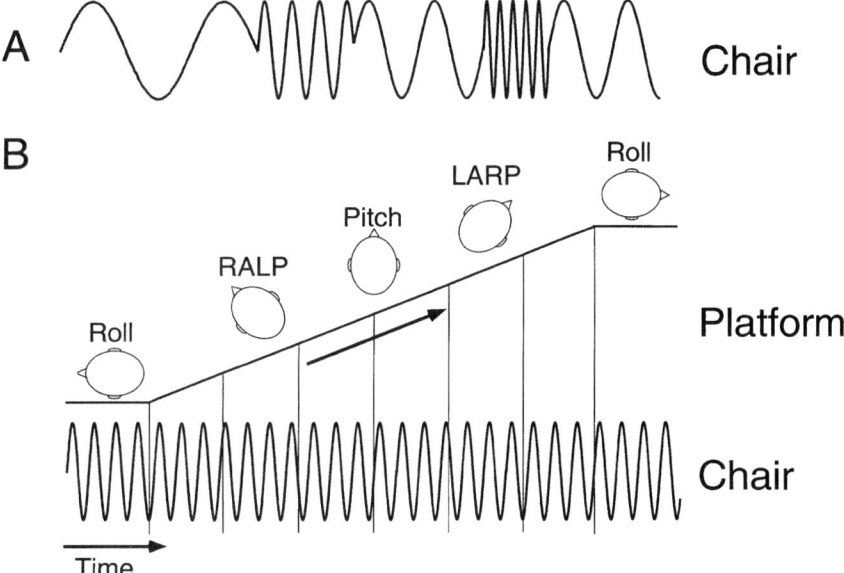

FIGURE 1. Schematic representation of the applied stimuli. Sinusoidal stimuli (**A**),(**B**) consisted of vertical vestibular stimulation around an earth-fixed horizontal axis. In (A) different frequencies (0.06–1.4 Hz) were applied at the same head orientation, which was the response vector orientation (RVO) obtained at 0.6 Hz. In (B) the sinusoidal stimulation was combined with a slow platform rotation continuously changing the orientation of the monkey's head.

All these characteristics refer to neurons that receive afferent inputs only from canals or otoliths. Deviations from these rules indicate canal–otolith convergence. For such neurons it is assumed that the canal afferents and the otolith afferents have different spatial orientations (RVO) and different temporal response characteristics. This spatio–temporal convergence leads to complex responses.[2] With a complex response the modulation during different head orientations at a given stimulus frequency can often not be fitted by a cosine function. The RVO can also alter with different stimulus frequencies. For neurons in the vestibular nuclei canal–otolith convergence is also assumed, when the gain (relative position) at frequencies of <0.1 Hz is flat and phase close to head position, and at higher frequencies (0.1–1.0 Hz) gain and phase lead increased.[24]

The purpose of the present study was to investigate whether otolith-related signals are carried by neurons in the FN of the alert monkey during natural static and dynamic stimulation around an earth fixed horizontal axis, and if so, what kind.

METHODS

Monkeys (*Macaca mulatta*) were chronically prepared for single-unit recordings (for details see Boyle *et al.*[6]). During the experiment the head was immobilized by a head holder and the monkey sat with the head erect (stereotaxic horizontal) in a primate chair. Three-dimensional eye position was recorded by a dual search-coil system,[5] and neuronal activity was recorded with varnished tungsten microelectrodes. The FN was approached perpendicularly.

Vestibular stimulation consisted of sinusoidal movements (0.06–1.4 Hz; ±7.5–±15 deg) around an earth-fixed horizontal axis. An independent motor allowed additional slow rotations around the monkey's Z-axis. This leads to different head orientations allowing for roll ($\beta = \pm 90$ deg), pitch ($\beta = 0$ deg), RALP ($\beta = 45$ deg), LARP ($\beta = -45$ deg), and intermediate planes of rotation (FIG. 1). Neuronal activity was also recorded during static tilt (up to ± 20 deg) at different head orientations. A third independent motor allowed rotations around an earth vertical axis, which was used for horizontal canal stimulation.

Data Analysis

All data (neuronal activity, 3-dimensional head and eye position) were monitored and stored for further off-line computer analysis. Signals were digitized with real-time occurrence of neuronal activity (spikes) and a sampling rate of 200 Hz for the remaining channels. For a given stimulus condition neuronal activity was averaged for 3–20 cycles, depending on the stimulus frequency. When the head orientation altered continuously (FIG. 1B), responses at neighboring orientations were included for averaging. Averaged neuronal responses were fitted by least-square best sine wave function, thereby determining sensitivity (imp·s^{-1}/deg) and phase in relation to head position. Positive values of phase indicate that neuronal activity leads head position. A least-square best sine wave function was fitted for the amplitude of the neuronal responses at different head orientations, which allowed us to calculate the optimal response (RVO).

RESULTS

General Characteristics

Neurons were recorded in the fastigial nucleus (FN) on both sides of the cerebellum. About 15–20% of the neurons in FN can be driven by dynamic vertical vestibular stimulation.[35] The vast majority of vestibular-only neurons was encountered in the rostral two-thirds of FN. There was no specific distribution in this area of vertical canal- and otolith-related vestibular-only neurons. Vestibular-only neurons were spontaneously active (range 20–100 imp/s; average 58 imp/s). The activity was irregular with an average coefficient of variation of the interspike interval of 0.73 and values up to 1.4.[35] Vestibular responses were the same in the light or in darkness. Responses were also not affected by the level of alertness as judged by the eye-movement recordings. During sinusoidal vestibular stimulation, about 50% of the neurons were silenced during part of the cycle ("cut off"), which is known to be a common phenomenon for vestibular neurons in FN.[8,14]

STIMULATION AT 0.6 HZ

The vast majority of vestibular-only neurons in FN responded in a uniform manner during sinusoidal vertical vestibular stimulation (0.6 Hz, ±15 deg) around an earth-fixed horizontal axis. The depth of modulation changed systematically with head orientation (FIGS. 2 and 3). At a given orientation neurons had an optimal response (RVO), and at an orientation 90 deg apart a null-response with no modulation (FIGS. 2 and 3). The neuronal modulation at different orientations could be fitted by a cosine function. Phases were constant except for an 180 deg phase shift around the null-response (FIG. 3). For neurons with a phase close to head velocity a canal-related response was assumed, just as an otolith-related response was assumed for neurons with a phase ±45 deg relative to head position. Based on these criteria, 20–25% of the vestibular-only neurons had an otolith-related

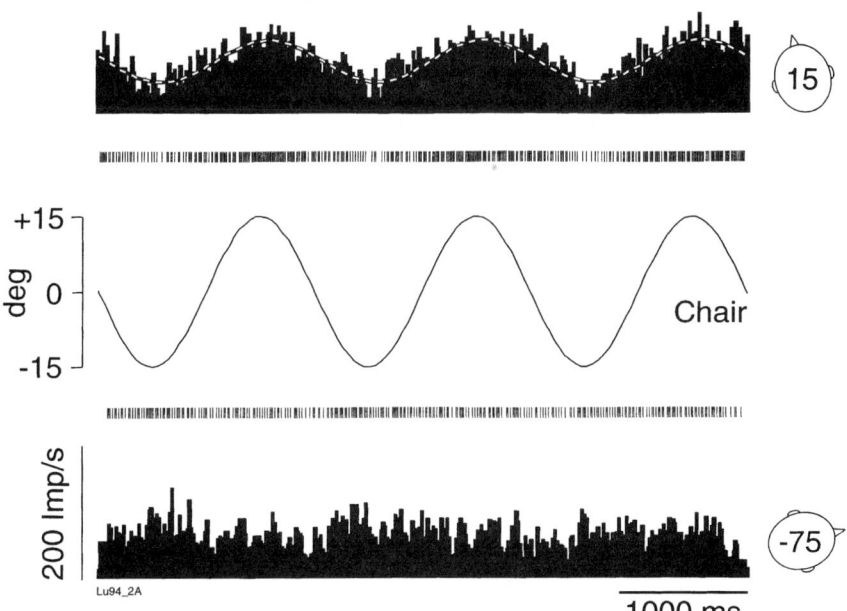

FIGURE 2. Otolith-related "vestibular only" neuron in the left FN during vertical stimulation (0.6 Hz ± 15 deg) at different head orientations (top: β = + 15 deg; bottom: β = –75 deg). *Chair* indicates head position. At the top head orientation is close to the RVO and neuronal activity is in phase with head position. Activity increases with nose-up (NV). *Stippled line* (**top**) indicates best sine wave fit of neuronal activity. At an orientation 90 deg apart (**bottom**) there is virtually no modulation.

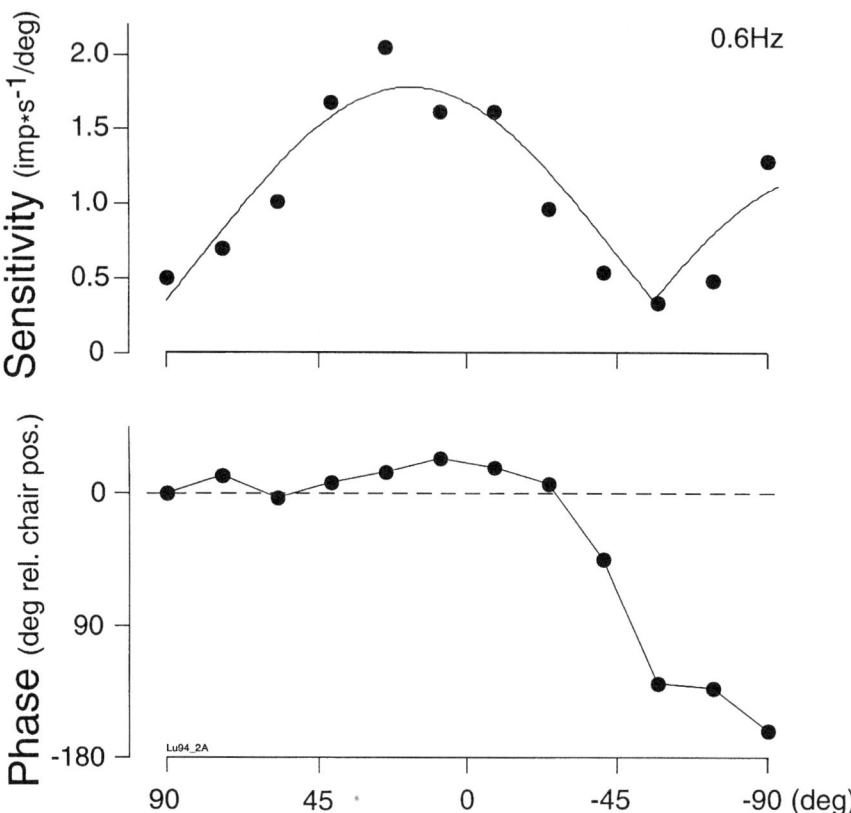

FIGURE 3. Sensitivity and phase of an otolith-related neuron (same as in FIG. 2) during different head orientations. Sensitivity decreases monotonically from the head orientation with the optimal response (RVO) to a head orientation 90 deg apart with no neuronal modulation (null-response). Phase is close to head position at all orientations except a phase shift of 180 deg around the null-response.

response, while the majority of neurons reflected a canal response.[35] The sensitivity at the RVO for the otolith-related neurons ranged from 1.02 to 3.66 imp·s^{-1}/deg, with an average of 2.34 imp·s^{-1}/deg. There was no preferred orientation for the RVOs. They were directed to the ipsilateral as well as to the contralateral side. A few neurons also had their RVO in the pitch plane, during nose-up or nose-down orientation. There was also no specific distribution of the RVOs within the FN.

In addition to the otolith-related response, many of the neurons showed a modulation during stimulation of the horizontal semicircular canal (yaw stimulation).[35] Thus convergence of otolith- and horizontal canal-related responses were quite common. However, complex responses,[2] which reflect convergence of otolith and (vertical) canal-related afferents, were less often encountered. The example in FIGURE 4 shows that the sensitivity was constant at all orientations except for a steep decrease around the phase shift of 180 deg. Such a response pattern does not reflect a single or combined vertical canal or a single otolith input. It could result from some central nonlinear processing or convergence of canal and otolith inputs (spatio–temporal convergence).

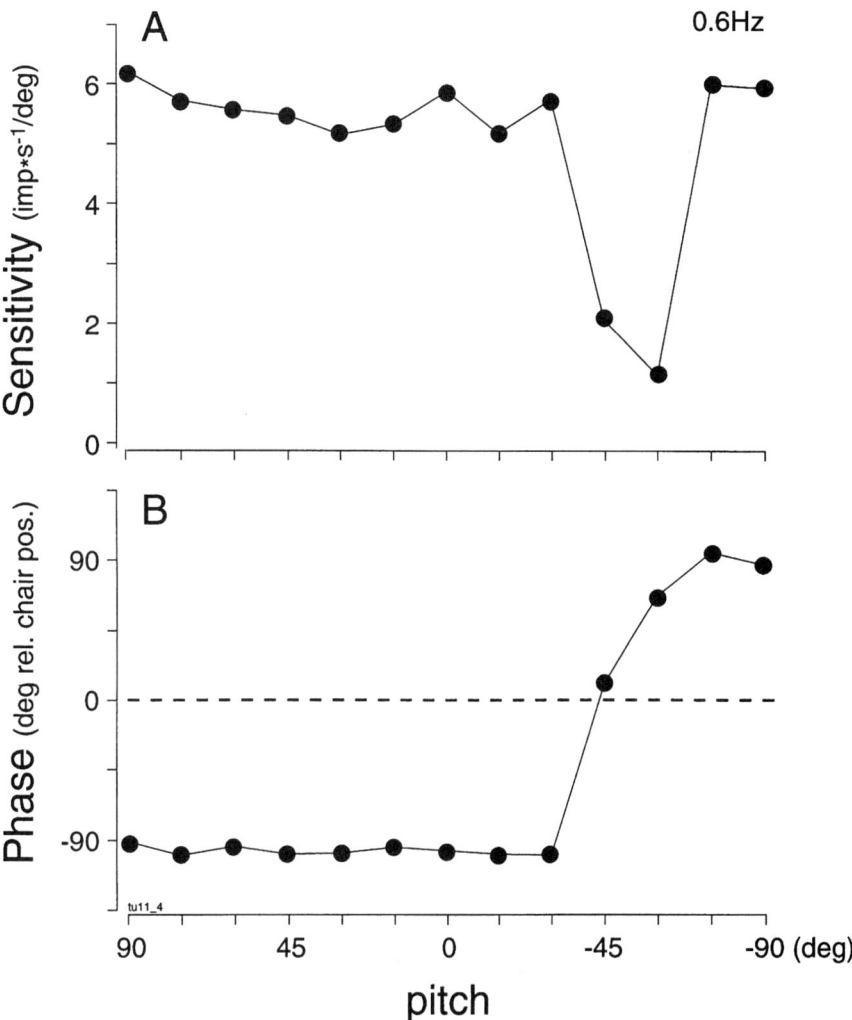

FIGURE 4. Complex "vestibular only" neuron recorded in the right FN during sinusoidal vertical vestibular stimulation (0.6 Hz; ±15 deg). The sensitivity (**A**) remains constant at all head orientations (abscissa), except a sharp drop in sensitivity around the head orientation, at which phase (**B**) reverses by 180 deg.

STATIC RESPONSES

Whereas neurons with a canal-related response were not modulated during static tilt, most neurons with an otolith-related response during dynamic stimulation were.[35] The RVOs obtained from static tilt in different orientations and from dynamic stimulation at 0.6 Hz were similar for the same neurons with differences of the RVO not exceeding 15 deg. The dynamic sensitivity (expressed in imp·s^{-1}/deg) was always higher than the position-related static sensitivity. This is also known to be the case for many otolith afferents.[11] A few neurons responding during dynamic stimulation did not respond during static tilt. This might reflect the higher sensitivity during dynamic stimulation. It probably also relates to the rather irregular activity of FN neurons and the fact that only tilts up to ±20 deg were applied, which makes it difficult to establish smaller activity changes.

FREQUENCY RESPONSES AT 0.06–1.4 HZ

Within the frequency range of 0.06–1.4 Hz, the phase of vertical canal-related neurons stays close to head velocity for vestibular afferents[17] and vestibular nuclei neurons.[24] For otolith-related neurons the phase is close to head position. This applies for otolith afferents[11] and many otolith-related vestibular nuclei neurons.[24,34] However, Schor et al.[34] could demonstrate in the cat with all semicircular canals plugged that a high percentage of vestibular nuclei neurons show an increasing phase lag of up to 180 deg over the frequency range of 0.01–1 Hz. Thus at a given frequency a head-velocity-related response can also reflect an otolith input.

To test this possibility, vestibular-only neurons in FN were investigated over a wide frequency range (0.06–1.4 Hz). Initially the RVO was determined at 0.6 Hz. All other frequencies were investigated at this orientation (RVO). More than 80% of the neurons responded at all frequencies. A few head-velocity- (canal-) related neurons did not respond clearly at the lowest frequency (0.06 Hz), which was probably due to the low stimulus velocity (5.7 deg/s) at this frequency. Based on the phase and gain analysis, three different response patterns can be distinguished: (1) neurons with head velocity-related phases; (2) neurons with head-position-related phases; and (3) neurons with large phase changes at different frequencies. Neurons with head-velocity-related phases probably reflect a canal-related input. For these neurons the phase was close to head velocity over the whole frequency range. Furthermore, the sensitivity (imp·s^{-1}/deg) showed a steep increase.[24] An otolith-related response can be assumed for those neurons, for which the phase is close to head position over the whole frequency range (FIG. 5C, 5D). In agreement with this, sensitivity (imp·s^{-1}/deg) remained constant or showed small increases with higher frequencies. The largest group of neurons had phase changes exceeding more than 50 deg over the frequency range investigated. For some neurons these changes could be as high as 180 deg (FIG. 5A, 5B). For most neurons the phase advanced continuously with higher frequencies. Other neurons showed an increasing phase lag only with frequencies above 0.5 Hz. A continuously increasing phase lag, as seen with vestibular nuclei neurons,[34] was generally not observed.

The neuron in FIGURE 5B had a head-velocity-related response around 0.2 Hz. However, this neuron most likely had an (at least additional) otolith-related input. These findings clearly show that a head velocity related response at a given stimulus frequency does not rule out an otolith input.

RESPONSE VECTOR ORIENTATION AT DIFFERENT FREQUENCIES

For canal[17] and otolith[11] afferents the RVO remains the same at different stimulus frequencies. This is also the case for most vestibular nuclei neurons receiving a canal-[24] or otolith-related[24,34] input. Only central neurons with complex responses (spatio–temporal convergence) due to canal–otolith convergence exhibit a shift of the RVO and the null-response at different frequencies. Often a null-response cannot be determined.[2] Since this is another paradigm to reveal otolith-related responses in the FN, for some vestibular-only neurons the head orientation was systematically altered at different frequencies. This rather lengthy stimulus protocol could not be completed for all neurons.

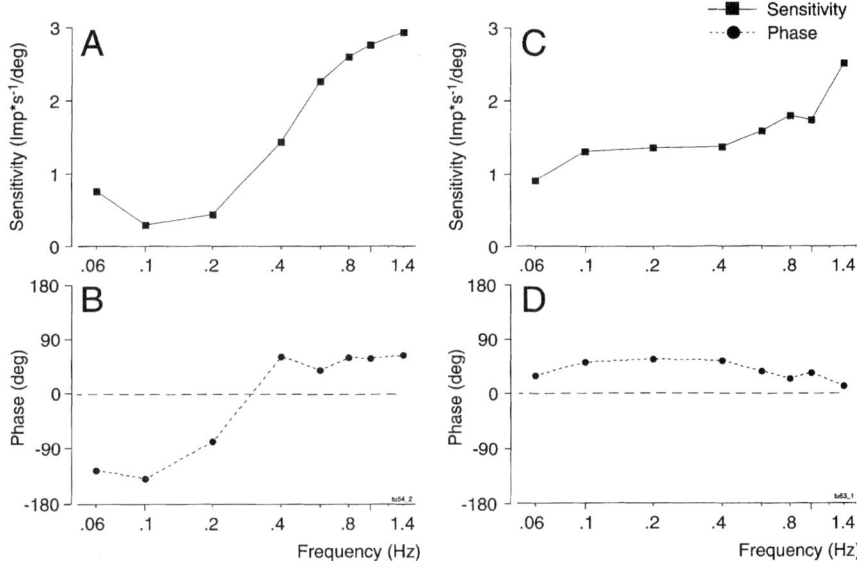

FIGURE 5. Neuronal responses of two FN (**A, B**) and (**C, D**) neurons at different stimulus frequencies (abscissa). The head orientation was kept constant and corresponded to the RVO obtained at 0.6 Hz. For the neuron on the left (A, B), phase changed by more than 130 deg over the frequency range tested. In contrast, for the neuron on the right (C, D), phase remained constant, slightly leading head position. For both neurons sensitivity increased with frequency more pronounced for the neuron on the left (A).

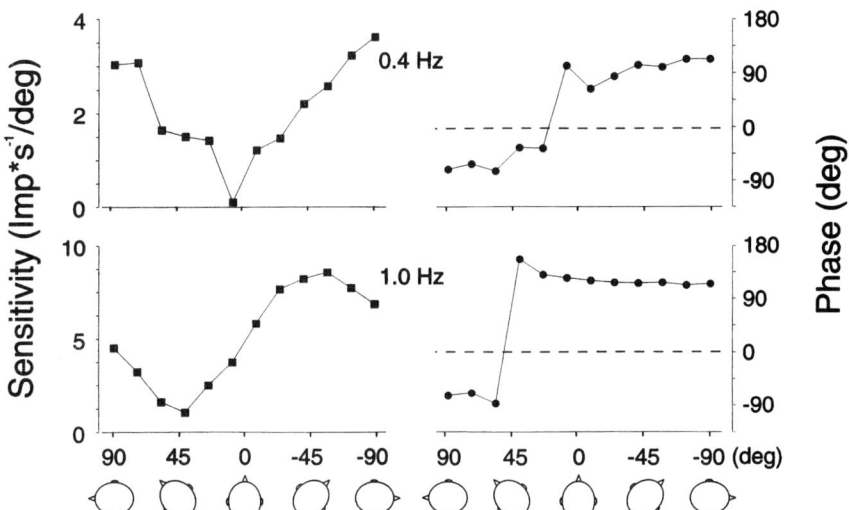

FIGURE 6. "Vestibular only" neuron in FN during different stimulus frequencies (0.4 and 1.0 Hz) and head orientations. The RVO and the null-response occur at different head orientations for different frequencies. At 0.4 Hz the null-response is close to $\beta = 0$ (pitch), whereas at 1.0 Hz it is shifted to $\beta = 45$ deg. Phase is close to head velocity, except a reversal around the head orientation with the null-response.

Most neurons had a constant phase (except a 180-deg phase shift around the null-response) at all head orientations for a given frequency. Accordingly a null-response and an optimal response (RVO) could be determined and sensitivity values at different head orientations could be fitted by a cosine function (FIG. 6). For a few neurons, particularly those with low sensitivities, more gradual phase changes around the null-response were observed at frequencies below 0.4 Hz. This was not the rule, however, and was difficult to interpret in view of the irregular activity and the low sensitivity of these neurons. Many neurons had a consistent RVO (and null-response) at different frequencies with differences not exceeding 20 deg. For most of these neurons the phase of the response was also related to head velocity at all stimulus frequencies. This suggests that these neurons reflect a canal-related input either from individual canals or canal convergence.

Other neurons clearly exhibited a shift of the RVO (and the null-response) at different frequencies (FIG. 6) occasionally exceeding 100 deg. Independent of the phase behavior at different stimulus frequencies (see earlier) these frequency-dependent shifts of the RVO reflect some spatio–temporal convergence, which requires an otolith input.

COMMENT

Many cerebellar structures contain otolith-related information. In the cerebellar cortex the most prominent and obvious representation is in the nodulus/uvula region,[4] but there is also ample evidence for otolith-related activity in the anterior vermis.[29] With appropriate stimulation an effect of otolith activation can also be demonstrated in the floccular region.[36] From the deep cerebellar nuclei, the FN has the strongest connections with the vestibular nuclei. With natural vestibular stimuli it can be shown that neurons in the FN carry a precise and robust vestibular signal in spite of their irregular spontaneous activity. With stimulation around an earth-fixed horizontal axis, changes of head orientation systematically alter the neuronal modulation, allowing for the determination of an optimal response (RVO) and a null-response.[35]

As described in detail earlier, canal-related neurons in the nerve and vestibular nuclei respond in a rather uniform manner. Specifically, phase is close to head velocity over the frequency range of 0.05–1.0 Hz, at a given stimulus frequency the neuronal modulation at different head orientations can be fitted by a cosine function, and an RVO can be determined. Finally, the RVO is independent of stimulus frequency. Many vestibular-only neurons in FN show deviations from these rules, indicating an otolith-related input. A considerable percentage of these neurons had a clear RVO and phase related to head position at all frequencies (FIG. 5C, 5D). They probably reflect those, that only receive an otolith input. Others reflect canal–otolith convergence. These are neurons with a complex response pattern at a given stimulus frequency (FIG. 4), and a shift of the RVO at different frequencies (FIG. 6). It should be emphasized that these signs of spatio–temporal convergence do not necessarily indicate canal–otolith interaction, but could also derive from otoliths with different spatial orientation and temporal response characteristics.

Quite a number of vestibular-only neurons had an increasing phase *lead,* often exceeding 130 deg over the frequency range of 0.06–1.4 Hz (FIG. 5A, 5B). Such responses have not been described for otolith-related vestibular nuclei neurons. Interestingly otolith-related neurons in the vestibular nuclei can show an increasing phase *lag* up to 180 deg, a pattern so far not observed in the FN. Since the otolith nerve afferents do not show such large phase leads or lags, it will be of interest to see whether the FN might play a role in the generation of the central phase changes seen in the vestibular nuclei.

Based on the just-mentioned criteria, it can be estimated that more than 30% of vestibular-only neurons carry an otolith-related signal. These neurons are distributed mainly in the rostral FN without specific clusters. Their RVOs include all orientations, which has also been noted for Purkinje cells (PCs) in the anterior vermis.[29]

The activity of these neurons in the FN is mainly determined by their vestibular nuclei input. They are modulated by the inhibitory PC input, which, for the vestibular-only neurons, derives from the anterior vermis but also from the nodulus/uvula. So far nothing is known about functional differences of neurons receiving an anterior vermis or a nodulus/uvula input.

It was mentioned earlier that vestibular-only neurons in the FN are probably related to vestibulospinal mechanisms (neck, gait, postures). Lesions in the FN lead to a falling tendency to the ipsilateral side.[26,38] Single-unit recordings during natural vestibular stimulation described here reveal that neurons with a large variety of RVOs and frequency-dependent phase and gain characteristics are present. It will be a challenge to determine how these signals precisely interact with the cerebellar cortex and the vestibular nuclei to optimize the vestibulospinal motor output.

ACKNOWLEDGMENTS

This work was supported by the Deutsche Forschungsgemeinschaft (DFG). We thank B. Pfreundner and I. Wendl for preparing the manuscript, and M. Seiche for editing it.

REFERENCES

1. BAKER, J., J. GOLDBERG, G. HERMANN & B. PETERSON. 1984. Optimal response planes and canal convergence in secondary neurons in vestibular nuclei of alert cats. Brain Res. **294**: 133–137.
2. BAKER, J., J. GOLDBERG, G. HERMANN & B. PETERSON. 1984. Spatial and temporal response properties of secondary neurons that receive convergent input in vestibular nuclei of alert cats. Brain Res. **294**: 138–143.
3. BARMACK, N. H., R. W. BAUGHMAN, P. ERRICO & H. SHOJAKU. 1993. Vestibular primary afferent projection to the cerebellum of the rabbit. J. Comp. Neurol. **327**: 521–534.
4. BARMACK, N. H & H. SHOJAKU. 1995. Vestibular and visual climbing fiber signals evoked in the uvula-nodulus of the rabbit cerebellum by natural stimulation. J. Neurophysiol. **74**: 2573–2589.
5. BARTL, K., C. SIEBOLD, S. GLASAUER, C. HELMCHEN & U. BÜTTNER. 1996. A simplified calibration method for three-dimensional eye movement recordings using search-coils. Vision Res. **36**: 997–1006.
6. BOYLE, R., U. BÜTTNER & G. MARKERT. 1985. Vestibular nuclei activity and eye movements in the alert monkey during sinusoidal optokinetic stimulation. Exp. Brain Res. **57**: 362–369.
7. BÜTTNER, U., C. SIEBOLD & L. GLONTI. 1996. Vestibular signals in the fastigial nucleus of the alert monkey. Ann. N.Y. Acad. Sci. **781**: 304–13.
8. BÜTTNER, U., A. F. FUCHS, G. MARKERT-SCHWAB & P. BUCKMASTER. 1991. Fastigial nucleus activity in the alert monkey during slow eye and head movements. J. Neurophysiol. **65**: 1360–1371.
9. COMPOINT, C., C. BUISSERET-DELMAS, M. DIAGNE, P. BUISSERET & P. ANGAUT. 1997. Connections between the cerebellar nucleus interpositus and the vestibular nuclei: an anatomical study in the rat. Neurosci. Lett. **238**: 91–94.
10. DICKMAN, J. D. & Q. FANG. 1996. Differential central projections of vestibular afferents in pigeons. J. Comp. Neurol. **367**: 110–131.
11. FERNANDEZ, C. & J. M. GOLDBERG. 1976. Physiology of peripheral neurons innervating otolith organs of the squirrel monkey. III. Response dynamics. J. Neurophysiol. **39**: 996-1008.
12. FUCHS, A. F., F. R. ROBINSON & A. STRAUBE. 1993. Role of the caudal fastigial nucleus in saccade generation.1. Neuronal discharge patterns. J. Neurophysiol. **70**: 1723–1740.
13. FUCHS, A. F., F. R. ROBINSON & A. STRAUBE. 1994. Participation of the caudal fastigial nucleus in smooth-pursuit eye movements. I. Neuronal activity. J. Neurophysiol. **72**: 2714–2728.

14. GARDNER, E. P. & A. F. FUCHS. 1975. Single-unit responses to natural vestibular stimuli and eye movements in deep cerebellar nuclei of the alert rhesus monkey. J. Neurophysiol. **38**: 627–649.
15. GHELARDUCCI, B. 1973. Responses of the cerebellar fastigial neurones to tilt. Pflügers Arch. **344**: 195–206.
16. GHELARDUCCI, B., O. POMPEIANO & K. M. SPYER. 1974. Distribution of the neuronal responses to static tilts within the cerebellar fastigial nucleus. Arch. Ital. Biol. **112**: 126–141.
17. GOLDBERG, J. M. & C. FERNANDEZ. 1984. The vestibular system. *In* Handbook of Physiology. The Nervous System. Sensory Processes, Vol. III, Sec. 1: 977–1022. American Physiological Society. Bethesda, Md.
18. GONZALO-RUIZ, A. & C. LEICHNETZ. 1990. Connections of the caudal cerebellar interpositus complex in a new world monkey (Cebus apella). Brain Res. **25**: 919–927.
19. HELMCHEN, C. & U. BÜTTNER. 1995. Saccade-related Purkinje cell activity in the oculomotor vermis during spontaneous eye movements in light and darkness. Exp. Brain Res. **103**: 198–208
20. HELMCHEN, C., A. STRAUBE & U. BÜTTNER. 1994. Saccade-related activity in the fastigial oculomotor region of the macaque monkey during spontaneous eye movements in light and darkness. Exp. Brain Res. **98**: 474–482.
21. IMAGAWA, M., N. ISU, M. SASAKI, K. ENDO, H. IKEGAMI & Y. UCHINO. 1995. Axonal projections of utricular afferents to the vestibular nuclei and the abducens nucleus in cats. Neurosci. Lett. **186**: 87–90.
22. IMAGAWA, M., W. GRAF, H. SATO, N. SUWA, N. ISU, R. IZUMI & Y. UCHINO. 1998. Morphology of single afferents of the saccular macula in cats. Neurosci. Lett. **240**: 127–130.
23. ITO, J., I. MATSUOKA, M. SASA, S. TAKAORI & M. MORIMOTO. 1983. Input to lateral vestibular nucleus as revealed by retrograde horseradish peroxidase technique. Oto-Rhino-Laryngol. **30**: 64–70.
24. KASPER, J., R. H. SCHOR & V. J. WILSON. 1988. Response of vestibular neurons to head rotations in vertical planes. I. response to vestibular stimulation. J. Neurophysiol. **60**: 1753–1764.
25. KOTCHABHAKDI, N. & F. WALBERG. 1978. Cerebellar afferent projections from the vestibular nuclei in the cat: an experimental study with the method of retrograde axonal transport of horseradish peroxidase. Exp. Brain Res. **31**: 591–604.
26. KURZAN, R., A. STRAUBE & U. BÜTTNER. 1993. The effect of muscimol micro-injections into the fastigial nucleus on the optokinetic response and the vestibulo-ocular reflex in the alert monkey. Exp. Brain Res. **94**: 252–260.
27. NAITO, Y., A. NEWMAN, W. S. LEE, K. BEYKIRCH & V. HONRUBIA. 1998. Projections of the individual vestibular end-organs in the brain stem of the squirrel monkey. Hear. Res. **87**: 141–156.
28. NODA, H., S. SUGITA & Y. IKEDA. 1990. Afferent and efferent connections of the oculomotor region of the fastigial nucleus in the macaque monkey. J. Comp. Neurol. **302**: 330–348.
29. POMPEIANO, O., P. ANDRE & D. MANZONI. 1997. Spatiotemporal response properties of cerebellar Purkinje cells to animal displacement: a population analysis. Neuroscience **81**: 609–626.
30. PRECHT, W., R. VOLKIND & R. H. BLANKS. 1977. Functional organization of the vestibular input to the anterior and posterior cerebellar vermis of cat. Exp. Brain Res. **27**: 143–160.
31. SATO, F., H. SASAKI, N. Ishizuka, S. SASAKI & H. MANNEN. 1989. Morphology of single primary vestibular afferents originating from the horizontal semicircular canal in the cat. J. Comp. Neurol. **290**: 423–439.
32. SATO, H., K. ENDO, H. IKEGAMI, M. IMAGAWA, M. SASAKI & Y. UCHINO. 1996. Properties of utricular nerve-activated vestibulospinal neurons in cats. Exp. Brain Res. **112**: 197–202.
33. SATO, H., M. IMAGAWA & N. ISU. 1997. Properties of saccular nerve-activated vestibulospinal neurons in cats. Exp. Brain Res. **116**: 381–388.
34. SCHOR, R. H., A. D. MILLER, S. J. TIMERICK & D. L. TOMKO. 1985. Responses to head tilt in cat central vestibular neurons. II. Frequency dependence of neural response vectors. J. Neurophysiol. **53**: 1444–1452.
35. SIEBOLD, C., L. GLONTI, S. GLASAUER & U. BÜTTNER. 1997. Rostral fastigial nucleus activity in the alert monkey during three dimensional passive head movements. J. Neurophysiol. **77**: 1432–1446.

36. SNYDER, L. H. & W. M. KING. 1996. Behavior and physiology of the macaque vestibulo ocular reflex response to sudden off axis rotation: computing eye translation. Brain Res. Bull. **40**: 293–301.
37. SUZUKI, D. A. & E. L. KELLER. 1988. The role of the posterior vermis of monkey cerebellum in smooth-pursuit eye movement control. II. Target velocity-related Purkinje cell activity. J. Neurophysiol. **59**: 19–40.
38. THACH, W. T., H. P. GOODKIN & J. G. KEATING. 1992. The cerebellum and the adaptive coordination of movement. Annu. Rev. Neurosci. **15**: 403–442.
39. THODEN, U., J. DICHGANS & T. SAVIDIS. 1977. Direction-specific optokinetic modulation of monosynaptic hind limb reflexes in cats. Exp. Brain Res. **30**: 155–160.
40. VOOGD, J., N. M. GERRITS & T. J. RUIGROK. 1996. Organization of the vestibulocerebellum. Ann. N.Y. Acad. Sci. **781**: 553–579.
41. WILLIAMS, P. L. 1995. Gray's Anatomy, 38th ed. Churchill Livingstone. New York.
42. WILSON, V. J., J. A. ANDERSON & D. FELIX. 1974. Unit and field potential activity evoked in the pigeon vestibulocerebellum by stimulation of individual semicircular canals. Exp. Brain Res. **19**: 142–157.
43. YAMADA, J. & H. NODA. 1987. Afferent and efferent connections of the oculomotor cerebellar vermis in the macaque monkey. J. Comp. Neurol. **265**: 224–241.

Control of Spatial Orientation of the Angular Vestibulo-Ocular Reflex by the Nodulus and Uvula of the Vestibulocerebellum

BORIS M. SHELIGA,[a] SERGEI B. YAKUSHIN,[a] ADAM SILVERS,[b] THEODORE RAPHAN,[c] AND BERNARD COHEN[a,d]

[a]*Department of Neurology, Mount Sinai School of Medicine, 1 Gustave L. Levy Place, New York, New York 10029, USA*

[b]*Department of Radiology, Mount Sinai School of Medicine, 1 Gutave L. Levy Place, New York, New York 10029, USA*

[c]*Department of Computer Information Sciences, Brooklyn College of the City University of New York, Brooklyn, New York, 11210 USA*

ABSTRACT: Eye velocity produced by the angular vestibulo-ocular reflex (aVOR) tends to align with the summed vector of gravity and other linear accelerations [gravito-inertial acceleration (GIA)]. Defined as "spatial orientation of the aVOR," we propose that it is controlled by the nodulus and uvula of the vestibulocerebellum. Here, electrical stimulation, injections of the $GABA_A$ agonist, muscimol, and single-cell recordings were utilized to investigate this spatial orientation. Stimulation, injection, and recording sites in the nodulus were determined *in vivo* by MRI and verified in histological sections. MRI proved to be a sensitive, reliable way to localize electrode placements. Electrical stimulation at sites in the nodulus and sublobule d of the uvula produced nystagmus whose slow-phase eye-velocity vectors were either head centric or spatially invariant. When head centric, the eye velocity vector remained within ± 45° of the vector obtained with the animal upright, regardless of head position with respect to gravity. When spatially oriented, the vector remained relatively constant in space in one on-side position, with respect to the vector determined with the animal upright. A majority of induced movements from the nodulus were spatially oriented. Spatially oriented movements were generally followed by after-nystagmus, which had the characteristics of optokinetic after-nystagmus (OKAN), including orientation to the GIA. After muscimol injections, horizontal-to-vertical cross-coupling was lost or reduced during OKAN in tilted positions. This supports the hypothesis that the nodulus mediates yaw-to-vertical or roll cross-coupling. The injections also shortened the yaw-axis time constant and produced contralateral horizontal spontaneous nystagmus, whose velocity varied as a function of head position with regard to gravity. Nodulus units were tested with static head tilt, sinusoidal oscillation around a spatial horizontal axis with the head in different orientations relative to the pitching plane, and off-vertical axis rotation (OVAR). The direction of the response vectors of the otolith-recipient units in the nodulus, determined from static and/or dynamic head tilts, were confirmed by OVAR. These

[d]To whom correspondence may be addressed. Phone: 212/241-7068; fax: 212/831-1610; e-mail: bcohen@smtplink.mssm.edu

vector directions lay close to the planes of the vertical canals in 7/10 units; many units also had convergent input from the vertical canals. It is postulated that the orientation properties of the aVOR result from a transfer of otolith input regarding head tilt along canal planes to canal-related zones of the nodulus. In turn, Purkinje cells in these zones project to vestibular nuclei neurons to control eye velocity around axes normal to these same canal planes.

INTRODUCTION

The angular VOR (aVOR) exhibits the property of spatial orientation. During nystagmus induced by rotation of the subject or surround, the yaw-axis component of slow phase velocity tends to align with gravity or with tilts of the gravito–inertial acceleration vector (GIA) with regard to the head.[e,1-12] We have shown that the central velocity storage mechanism is responsible for spatial orientation of the aVOR, and that the direct visual and vestibular pathways do not orient eye velocity to the GIA.[12] Thus, the study of the spatial orientation of the aVOR is essentially a study of the three-dimensional characteristics of velocity storage. Alignment of the eye-velocity vector to the GIA has been modeled by three processes: a reduction in the time constant of the dominant horizontal component; an increase in the torsional and/or vertical time constants; and the appearance of orthogonal vertical or torsional "cross-coupled" eye velocity.[1,2,6-8,11,12] From the frequency characteristics of velocity storage, and from the finding that velocity storage is activated by visual, vestibular, and somatosensory input in both monkeys[13] and humans,[14] it has been inferred that this type of spatial orientation is likely to be important for postural and gaze stabilization during the centripetal accelerations experienced in circular locomotion[15] and when passively transported around curves on bicycles, motorcycles, cars, and so on.

Behavioral, stimulation and lesion studies have shown that the nodulus and uvula control the horizontal time constant of velocity storage, one element in determining its spatial orientation. The horizontal aVOR time constant is reduced when the head is reoriented with respect to gravity from the upright position during postrotatory nystagmus (tilt dumping), both in humans[16,17] and in monkeys.[9,18-23] The horizontal time constant is also reduced when subjects view a relative stationary visual surround during vestibular nystagmus or OKAN (light dumping).[21,24-26] Electrical stimulation of the nodulus and lobule 9d of the ventral uvula reproduces this reduction.[23] Both tilt dumping and light dumping are lost after ablation of the nodulus and uvula.[21] Electrical stimulation of the nodulus also produces after-nystagmus that has the characteristics of optokinetic after-nystagmus (OKAN).[27] This suggests that it is not only possible to discharge velocity storage, but also to induce it by nodulus stimulation.

The nodulus and ventral uvula are also involved in producing the changes in vertical and/or torsional aVOR components that tilt the eye-velocity vector during spatial orientation. After complete nodulo-uvulectomy, there is a loss of spatial orientation of the aVOR in the monkey.[1,2,9,10] Removal of only medial portions of the nodulus and uvula causes a loss of yaw-to-vertical and yaw-to-roll cross-coupling, and control of the horizontal aVOR time constant is maintained.[2] From this, it is likely that different portions of the vermal cortex of the nodulus and uvula participate in different parts of the process of spatial orientation.[2]

Much of the information about the representation of spatial orientation in the nodulus and uvula has come from studies utilizing visual motion around different axes. A striking feature of the cortex of the nodulus and uvula is its division into parasagittal zones that are

---

[e]We define spatial orientation of the a VOR as alignment of the eye-velocity vector with gravito–inertial acceleration. Gravito–inertial acceleration (GIA) is the vector sum of gravitational and inertial accelerations. Cross-coupling is the appearance of vertical and/or torsional eye velocities in response to tilts of the GIA during yaw-axis stimulation.

innervated by different subnuclei of the inferior olive.[28-31] In the rabbit, activity induced by optokinetic stimulation about a vertical axis during yaw or horizontal optokinetic nystagmus (OKN) is processed in the caudal dorsal cap (cdc), while activity related to optokinetic stimulation about a horizontal axis parallel to the axis of the ipsilateral anterior canal (135°) reaches the cerebellar cortex through the rostral dorsal cap (rdc) and ventrolateral outgrowth (vlo).[32-39] Vestibular signals aligned approximately with the anterior- and posterior-canal axes project from the ß nucleus of the inferior olive.[32,33,36,40-43] Utricular signals project from the dorsomedial cell column (DMCC)[44] and the β nucleus.[41,45] As yet, no signals related to lateral canal activation have been found in the inferior olive or in climbing fiber responses.[33,42] From these data, it has been proposed that the three reciprocal planes of the semicircular canals are represented across the vermis of the nodulus in zones defined by inferior olive input.[33,46] From midline to the left, the right-anterior-left-posterior-canal (RALP) plane is represented in the medial 1.5 mm, yaw-axis visual movement (horizontal OKN) in the next 0.5–1.0 mm, and the left-anterior-right-posterior (LARP) plane 1.0–2.0 mm more laterally.[42,47] These zones project to appropriate regions of the superior and medial vestibular nuclei where appropriate canal plane responses could be generated.[46] This could provide a cellular mechanism for realizing spatial orientation of the aVOR.[2] Since the major transmitter of Purkinje cells is GABA,[48] presumably the eye velocity changes produced by the direct projections of the nodular cortex to the vestibular nuclei occur through inhibition and disinhibition.

Four sagittal zones have been distinguished in the monkey, based on four white-matter compartments innervated by discrete subnuclei of the contralateral inferior olive.[31] Olivocerebellar projections to the different zones in monkey have been inferred by analogy with the organization in rabbit. The most medial zone (Zone 1) extends from the midline to 0.8-mm lateral in the nodulus. It receives climbing fiber input from the ß nucleus medially and the cdc laterally. The second zone (Zone 2) extends from 0.8 to 2.4 mm in the nodulus and uvula, and forms the lateral limits of the vermis over the posterior surface of the cerebellum. Zone 2 was labeled from injections including the vlo, rostral ß, and rostral medial accessory olive (rMAO). Similar to the rabbit, the vlo projection is probably restricted to Zone 2 in the nodulus, while axons from the rostral β nucleus may extend into the uvula. Zone 3, innervated by the cdc, extends from 2.4 to 3.2 mm and is restricted to the nodulus. An auxiliary projection to this zone from the ß nucleus cannot be excluded, but is unlikely in the nodulus. Zone 4 extends from 3.2 to 4mm lateral. Its olivocerebellar afferents are currently unidentified in the rhesus monkey, but by analogy with the lateral zone in rabbit, probably include the DMCC and rMAO.[32,36,49] If it is assumed that the same canal-, otolith- and OKN-related information is present in the inferior olive of the monkey as in the rabbit, Zone 1 has horizontal axis anterior canal (135°) VOR and otolith (ß), as well as yaw-axis representation from the cdc (yaw OKN). Input to Zone 2 would be related to horizontal axis (135°) OKN and VOR and the utricle. Zone 3, extending laterally beyond the vermis, would receive information about yaw-axis OKN. This canal-related organization has been postulated as the basis for spatial orientation in the rabbit,[33] and we have proposed that it is also the basis for the spatial orientation of the aVOR in the monkey.[1,2] Recently, a canal-based orientation system was also found in the pigeon nodulus in response to visual flow along canal planes.[82]

As noted in the preceding paragraphs, electrical stimulation of the nodulus and rostroventral uvula (lobule d), caused a reduction in the horizontal aVOR time constant in the monkey,[23] one component of spatial orientation of the aVOR.[1,2,6,7] When prolonged, such stimulation also produced after-nystagmus, which had the characteristics of OKAN.[27] Only horizontal eye movements were recorded, however, either for short[23] or longer stimuli,[27] and it is not known whether vertical and torsional components would also be produced by electrical stimulation of the nodulus or uvula or whether the OKAN-like after-nystagmus also has the property of alignment to the spatial vertical as does OKAN.[7] There is also a paucity of information about whether the nodulus and uvula receive otolith input in the monkey. Otolith information, which codes tilts of the GIA with respect to the head, and which would be critical in orienting the aVOR, has been found in the cat.[50] Such cells have not been studied in the monkey, and the directions of their response vectors have not been determined. The purpose of this study was to determine how the nodulus organizes spatial orientation of the aVOR in the monkey. We wished to establish whether trajectories of eye velocity in response to electrical stimulation of the nodulus had spatial components, and how spatial orientation of the aVOR was altered after inactivation of the nodulus with the $GABA_A$ agonist, muscimol. We also wished to determine the spatial organization of otolith-related units in the nodulus.

METHODS

Two rhesus and one cynomolgus monkey were used in this study. The experiments conformed to the Guide for the Care and Use of Laboratory Animals (National Research Council, 1996), and were approved by the Institutional Animal Care and Use Committee. Under sterile surgical conditions and anesthesia, an acrylic ring was implanted that held the animals' heads painlessly during experiments.[51] A Delrin bar, attached to this ring, permitted introduction of microelectrodes into deep brain structures. Two three-turn coils were implanted on one eye to measure horizontal, vertical, and torsional eye position.[22,52] To calibrate eye movements, the animals were rotated in light at 30°/s about the pitch, roll, and yaw axes. It was assumed that horizontal and vertical gains were close to unity in this condition. Roll gains were assumed to be 0.6 when rotation was around a naso–occipital axis aligned with the spatial vertical.[53] Eye velocities to the left, down, and counterclockwise from the animal's point of view, are represented by downward deflections in the velocity traces in the figures.

During testing animals sat in a primate chair in a three-axis vestibular stimulator surrounded by an optokinetic drum. Each axis went through the center of rotation of the head. The stimulator has been described in detail in previous publications.[20,54,55] With the monkey upright, the yaw axis was aligned with gravity, and the horizontal stereotaxic plane was aligned with the spatial horizontal. Thus, the lateral semicircular canals were tilted up approximately 30° from the earth horizontal plane during the experiments.[53,56] Amphetamine sulfate (0.3 mg/kg) was given 30 min before testing to maintain alertness.

Electrical stimuli were 0.5-ms constant current, cathodal pulses of 40–60 μA given at a frequency of 200 Hz for 10–30 s. The pulses were monopolar, with the electrode tip referenced to the bolts implanted on the skull. For injection, a stainless-steel guide tube was inserted into the brain, aimed at the region of interest. Injections were made using a Hamilton pressure syringe inserted through the guide tube toward the designated injection site. Muscimol (1.2–1.4 μL of a 1-μg/μL solution) was injected over a period of 2 min. The syringe was left in place for an additional 3 min. Testing began 10 min after the start of injection and lasted for 2–3 h. In a control experiment, conducted on a separate day, 1.2–1.4 μL of 0.9% NaCl was injected into one of the effective locations.

Units were recorded extracellularly using 80-μ tungsten electrodes, varnished to the tip. Electrodes had a resistance of ≈ 1 mΩ tested at 1 kHz. Unit amplifiers had a bandwidth from 200 Hz–10 kHz. The location of the recording electrodes in the nodulus was based on stereotaxic coordinates and the location of characteristic activity related to eye movements in the abducens and rostral fastigial nuclei. The accuracy of electrode placement was confirmed in the MRI. Because the bodies of the animals were not restrained during recording, sinusoidal oscillation in unit activity during yaw oscillation of the body could reflect neck proprioceptive as well as semicircular-canal input. Vertical-canal input, on the other hand, could be separated from neck input on the basis of temporal phase of response. Therefore, testing during unit recording was limited to determining otolith sensitivity and vertical canal input, and we did not study potential lateral canal input.

Unit sensitivity to tilts of the GIA was determined in three conditions. The animal was statically tilted at 30° and/or 60° for 30 s with the head in different orientations to the plane of tilt. After each tilt, the animal was returned to the upright position for about 20 s. The animal was also sinusoidally oscillated around a spatial horizontal axis at ± 18° and with the head in different orientations to the plane of oscillation. Third, some of the units were tested during off-vertical axis rotation (OVAR). In our coordinate system, 0° corresponded to nose-down, 180° to nose-up, 90° to right side down (RSD), and 270° to left side down (LSD) tilts. Unit sensitivity is presented as $imp \cdot s^{-1} \cdot g^{-1}$. This was calculated by assuming that the resting discharge was the mean value of all temporal gains ($imp \cdot s^{-1}$) obtained from a variety of head orientations to the plane of tilt or the plane of oscillation around a spatial horizontal axis. Unit sensitivity during static or dynamic tilts was plotted as a function of head orientation relative to the tilt plane. Sensitivity was positive when it was within ± 90° relative to head position, and as negative in other cases.

With the exception of optokinetic nystagmus (OKN), all testing was done in darkness. Per- and postrotatory nystagmus were induced by rotation in yaw around a vertical axis at a constant velocity at speeds of 30°/s and 60°/s ($270°/s^2$ acceleration and deceleration). OKN was induced by rotating the visual surround about the animal's yaw axis at 60°/s for 30 s with the animal upright, left side down (LSD), or right side down (RSD). Optokinetic after-nystagmus (OKAN) was then recorded in darkness. The animals were also tilted statically toward the LSD or RSD positions at angles of 0° to 90° in 15° steps. Each angle of tilt was maintained for 60 s. Finally, animals were rotated in yaw at velocities of 30–120°/s about axes tilted from the vertical (off-vertical axis rotation, OVAR) at angles of 0° to 90° in 15° steps. Each axis of tilt was maintained for at least 40 s.

Eye-position voltages and voltages related to the velocity or position of the axes of rotation were recorded by amplifiers with a bandpass of DC to 40 Hz. Voltages were digitized at 600 Hz/channel with 12-bit resolution, and stored on disk. Eye-position voltages were digitally differentiated and saccades were removed. Slow phase velocities were analyzed from the onset of OKAN during optokinetic stimulation or from the initial jump in velocity during per-rotatory and postrotatory nystagmus to the point where the yaw-axis eye velocity decayed to zero. In all cases, the direction of nystagmus is indicated by the direction of the slow phase eye velocity. Time constants were determined by dividing the area under the slow phase velocity envelope by the initial jump in velocity.[57] Vectors of slow phase velocity of nystagmus induced by stimulation were determined in two dimensions, yaw and pitch, from responses that were greater than 3°/s in each of the three tested positions (upright, left, and right side down; FIGS. 3 and 4). A mean vector was calculated for the average horizontal and vertical components of eye velocity during stimulation after subtracting any spontaneous velocities in the period just before stimulation.

Electrode locations were identified *in vivo* with MRI. Under ketamine/xylazine anesthesia, sharpened 120-μ tungsten wires were introduced as markers through the Delrin plate, and moved to positions 3–5 mm above the positions where the stimulating/recording electrodes or pipettes were located during experiments. Scans were performed on a whole-body 1.5-T scanner (GE Medical Systems, Milwaukee, WI). The sedated animal with the electrode array in place was put in the scanner in a supine position. A quadrature knee coil was placed around the animal's head. The knee coil was used to maximize signal to noise. A T1-weighted, 3-D SPGR volume-acquisition scan was performed (TR = 24, TE = 5) using a 256 × 256 matrix and 12-cm × 12-cm field of view. The acquired images were 1.2 mm thick, with 60 images in total. Scan time for this sequence was approximately 6 min. The acquired images were transferred to an independent workstation (GE Medical Systems). Using the interactive image-analysis program, the acquired axial images were reformatted into sagittal and coronal images, and interactively adjusted to be tangential to the plane of the electrodes (FIG. 1).

Upon completion of the experiments, seven electrolytic lesions (cathodal DC, 60 μA for 30 s) were placed at the same locations where electrical stimulation and muscimol injections were delivered. Several days later, the animals were deeply anesthetized and perfused intracardially with saline and 10% formalin. The brains were removed, embedded in gelatin, serially sectioned in sagittal stereotaxic planes, stained with cresyl violet, and the location of the recording and stimulation sites was determined.

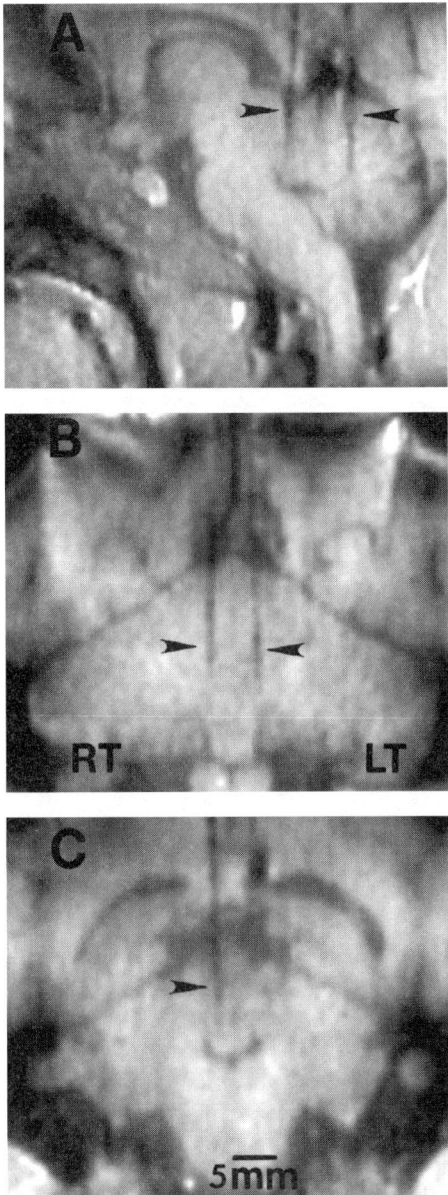

FIGURE 1. MRI of brainstem and cerebellum of monkey used for identification of electrode tracks. (**A**) Sagittal section 2.5 mm from midline on right showing two electrodes (*arrows*) flanking region of interest in cerebellum. The electrode tracks were separated by 8 mm. (**B**) Coronal section through posterior electrode pair. The electrode tracks (*arrows*), which were 5 mm apart, lay on either side of the vermis. (**C**) Coronal section about 2-mm caudal to abducens nucleus. *Arrow* points to electrode track 1.5-mm lateral on the right, at a location similar to the tracks shown in FIGURE 2. See text for details.

RESULTS

Location of Sites of Stimulus and Injection

The MRI, taken to identify the location of the microelectrodes *in vivo*, was of good quality (FIG. 1A–1C), except in the immediate vicinity of the stainless-steel bolts implanted in the skull. The implanted tungsten electrodes, used as markers, did not cause significant distortion of the surrounding brain tissue. From the implanted electrodes, we were able to determine the rostral/caudal and medial/lateral limits of the regions that were explored, as well as the depth of the relevant structures. The most anterior electrodes were just rostral to lobules 1–3 of the anterior cerebellum, and the most posterior electrodes were close to the interface between sublobules c and d of the uvula (arrows, FIG. 1A). At the level of the cerebellum, they were separated by 8 mm. Electrodes straddling the area of interest in the posterior vermis were 5 mm apart (arrows, FIG. 1B). An electrode that penetrated the nodulus 1.8 mm to the right of the midline is shown in FIGURE 1C.

Microelectrode tracks similar to the track shown in FIGURE 1C were identified in sagittal histological sections about 1.5 mm to the right of the midline (FIG. 2A). This section also included the abducens nucleus (not shown) and the medial vestibular nucleus (MVN; FIG 2B). Dorsally, the electrode tracks traversed the fastigial nucleus (FN), and terminated at the sites marked by the lesions in the first and second folia of the posterior bank of the nodulus (FIG. 2A) or at a, d, and f in the diagram of FIGURE 2B. Additional sites, which were electrically stimulated and where injections of muscimol were placed, were located more laterally in the nodulus (c, e, and g) and more medially in the anterior bank of sublobule d of the uvula (b). These sites are shown projected onto the diagram of FIGURE 2B, and their laterality is noted in TABLE 1. Four of the seven identified electrode locations were in folia 1 and 6 of the nodulus, two were in folia 2 and 5 of the nodulus, and 1 was in folia 1 of sublobule d of the uvula.

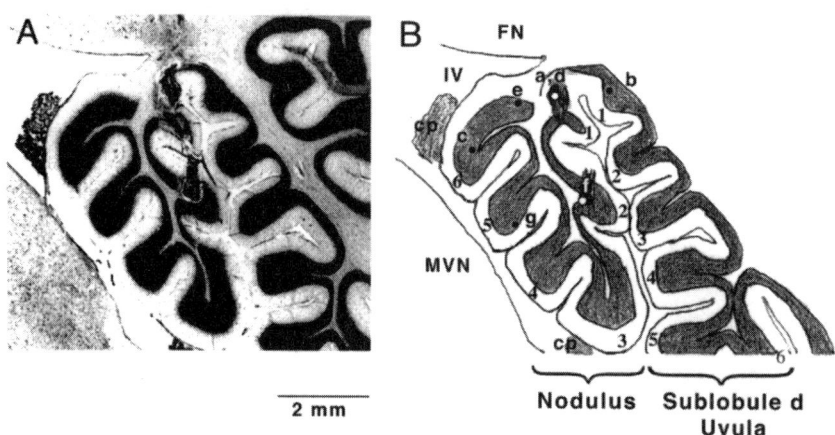

FIGURE 2. (**A**) Sagittal histological section, stained with cresyl violet, of nodulus and sublobule d of the uvula. Two lesions were made at sites of injection, one in folium 1 and the second in folium 2 of the posterior bank of the nodulus. (**B**) Diagram of section in part (A). Both the folia of the nodulus and of sublobule d of the uvula are numbered from 1 to 6. Sites of injection are labeled with letters and are marked by the open and *closed circles*. a, d, and f lay in this section, b was located more medially, and c, e, and g were located more laterally. All sites are projected onto this diagram (see TABLE 1 for laterality). *Abbreviations*: MVN, medial vestibular nucleus; IV, fourth ventricle; FN, fastigial nucleus; cp, choroid plexus.

TABLE 1. Effects of Stimulation and Muscimol Injections in Nodulus and Uvula

Cerebellar Structure	Location of Injection		Orientation of Evoked Response	Change in Vertical Component of OKAN After Muscimol Injection (%)						Change in Horizontal aVOR Tc (%)	
	Site	Distance from the Midline (mm)		Tilt Ipsi SD		Tilt Contra SD				Ipsi aVOR	Contra aVOR
				Ipsi OKAN	Contra OKAN	Ipsi OKAN	Contra OKAN				
				Right Side							
Post. Nod. Folium 1	a	2.1	HC	−58	−85	−93	−96			−71	−44
Post. Nod. Folium 1	d	2.2	SO-lSD	−52	−59	−10	−73			−67	−54
Post. Nod. Folium 2	f	2.2	SO-lSD	−47	−74	−20	−61			−43	+22
				Left Side							
Ant. Uvula Folium 1	b	0.75	SO-rSD	−59	−18	−54	−61			−21	−38
Ant. Nod. Folium 6	c	2.2	SO-rSD	−33	+23	−8	−64			−2	−39
Ant. Nod. Folium 6	e	2.2	SO-rSD	−12	−18	+50	−41			−8	−29
Ant. Nod. Folium 5	g	3.15	Not done	−66	−17	−45	−54			−8	−19
			Means ± (SD)	−47 ± 19	−35 ± 38	−26 ± 45	−64 ± 17			−31 ± 29	−29 ± 25

Note: Columns 1–3: Sites of stimulation and injection with laterality; column 4: Orientation of evoked nystagmus, head-centered (HC) or spatially oriented (SO), right or left side down (rSD or lSD); columns 5–8: Percent increase (+) or decrease (−) in vertical cross-coupled components of OKAN after muscimol injection; columns 9–10: Percent increase (+) or decrease (−) in horizontal aVOR time constant after muscimol. Means and ± 1 standard deviation below.

Stimulation

Nystagmus with horizontal, vertical, and torsional components was commonly induced by electrical stimulation in and around the nodulus. From some sites, the vectors of the slow phase eye velocity of the induced nystagmus were relatively fixed in head coordinates regardless of head position with respect to gravity. We termed these responses as "head centric." Generally, there was no after-nystagmus associated with head-centric responses, and eye velocity promptly fell toward zero after stimulation. Similar responses have previously been induced from the uvula.[23,27] A sample response, induced from site a in the first folium of the nodulus, 2.1 mm to the right of the midline (FIG. 2B), is shown in FIGURE 3. With the animal upright (FIG. 3A), stimulation caused nystagmus whose slow phase eye velocity was predominantly along the yaw axis to the right, first with an upward and then a downward component. There was also a small counterclockwise roll component. The direction of the mean slow phase velocity vector relative to the head is shown by the arrow under the head of the upright monkey in the inset and by the arrow drawn in the two-dimensional (yaw and pitch) phase plane plot on the right. The vector was in the downward direction, being predominantly in yaw to the right. When the animal was on its left (FIG. 3B) or right side (FIG. 3C), the characteristics of the trajectory with respect to the head were similar to those of the upright trajectory. That is, the eye velocity vector was initially down with respect to the head along the yaw right and pitch up directions, and then curved back toward the yaw axis. Despite the downward eye velocity induced when the animal was left side down (FIG. 3B) and an upward velocity induced when it was right side down (FIG. 3C), the mean vectors for all three conditions were maintained within ±45° of the vector obtained with the animal upright. Thus, in each case the dominant vector of the induced nystagmus moved with the head, regardless of the animal's position with respect to gravity.

In contrast, at other sites in the nodulus, spatially dependent responses were evoked. These had a slow rise in slow phase velocity, followed by a substantial period of after-nystagmus at the end of stimulation. The per-stimulus response induced from site c in folium 1, 3.3 mm to the left of the midline, will be considered first (FIG. 4). With the animal upright (FIG. 4A), the slow phase velocity was predominantly to the left during stimulation, but a downward velocity was also induced that declined rapidly toward the end of stimulation. As a result, there was an upward velocity vector that was close to the yaw axis (inset FIG. 4A, and phase plane plot). When the animal was tilted onto its left side (FIG. 4B), eye velocity was also predominantly to the left along the yaw axis during stimulation. Thus, the per-stimulus velocity vector was again upward with respect to the head and it shifted with respect to gravity. In contrast, when the animal was stimulated

FIGURE 3. Stimulation at site a in nodulus, 2.1 mm to the right of the midline (see FIG. 2) with the animal (**A**) upright (**B**) left side down (LSD), or (**C**) right side down (RSD) that produced a head-centric response. There was a rapid rise in eye velocity at the onset of stimulation. The induced movement was predominantly horizontal to the right, but there were also vertical and roll components. Eye velocity fell rapidly at the end of stimulation and there was no after-nystagmus. (HEV, horizontal eye velocity; VEV, vertical eye velocity, REV, roll eye velocity.) Eye position traces are not shown. Eye velocities to the right, up and clockwise from the animal's point of view produce upward deflections in the velocity traces. On the right are horizontal/vertical phase-plane plots. The plots start from positions close to the origin at the onset of stimulation and proceed eccentrically. They terminate at the end of stimulation. The *arrows* give the mean direction and amplitude of the per-stimulus eye-velocity vector. Note that the direction of the eye-velocity vector in upright and tilted positions remained relatively fixed with regard to the head, despite the change in head position with regard to the GIA.

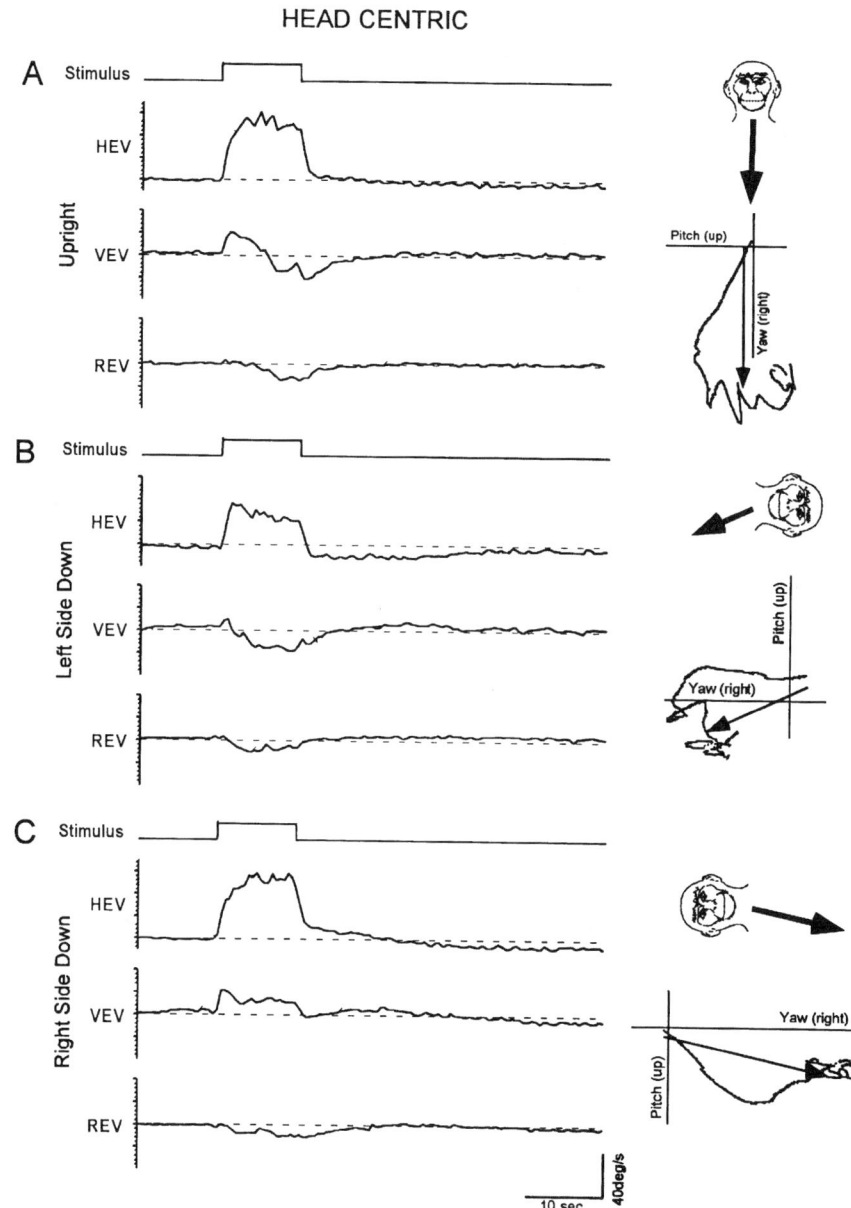

in the right-side-down-position (FIG. 4C), the yaw-axis component disappeared, and the induced nystagmus was predominantly in the pitch and roll directions. As a result, the mean-eye velocity vector in the pitch/yaw plane was spatially upward. Thus, it was maintained relatively constant in space for the upright and right-side-down positions (compare insets, FIG. 4A, and 4C). We termed per-stimulus responses that tended to maintain their orientation in space within ±45° of the vector obtained with the animal upright, irrespective of the animal's head position with regard to gravity, as "spatially oriented." In all cases, per-stimulus, spatially oriented vectors were obtained only when animals were on one side, and in no case was spatial orientation maintained for both the right- and left-side-down positions. In some instances, there was no spatial constancy, but rather the eye-velocity vector changed as a function of head orientation, although it was not maintained along a constant direction. We called these responses "spatially dependent."

At 11 sites in the nodulus, responses were elicited with sufficient magnitude (> 3°/s) for analysis in at least two positions. For stimulation at five sites on the left side of the nodulus and uvula, the four responses from the nodulus were spatially oriented during right-side-down tilts and the response from the uvula was spatially oriented for left-side-down tilt. For the responses to stimulation at six sites on the right side, one was spatially oriented and two had spatially dependent responses when tilted left side down. Therefore, spatial constancy was mainly elicited when the animal was on the side contralateral to stimulation. Head-centric responses to either side were elicited from the remaining three sites on the right side. Spatially oriented and spatially dependent per-stimulus nystagmus accounted for 72% of the responses induced from the region of the nodulus and 39% of the responses induced from the region of the fastigial nucleus. In contrast, only 28% of responses induced from the region of the nodulus were classified as head centric, whereas 61% of the responses from the region of the fastigial nucleus had such characteristics. Thus, electrical stimulation of the nodulus most commonly induced per-stimulus responses that were either spatially oriented or had spatial dependence.

As in a previous study,[27] after-nystagmus was elicited by nodulus stimulation with the characteristics of OKAN. This included its spatial orientation properties, as well.[6,7] With the animal upright (FIG. 4A), the slow phase velocity of the after-nystagmus was predominantly to the left, producing a velocity vector that was upward along the yaw and gravitational axes, which in this case were coincident. In the LSD position (FIG. 4B), yaw velocity to the left was induced during stimulation, and a cross-coupled upward vertical velocity appeared during the after-nystagmus. Consequently, the velocity vector of the

---

FIGURE 4. Simulation at site c in nodulus, 2.2 mm to the left of the midline in nodulus produced a spatially oriented response with the animal RSD (C). (**A**) Stimulation with the animal upright produced a slow rise in eye velocity to the left and down, followed by after-nystagmus that declined along the time constant of OKAN. The vector of the induced per-stimulus response was up and to the right both spatially and with respect to the head (phase-plane graph on right). The relationship of the induced vector to the monkey's head is shown in the inset above. (**B**) With the animal LSD, the per-stimulus nystagmus was to the left, but the vertical component disappeared. The direction of the per-stimulus vector was upward with regard to the head and tilted with regard to gravity (phase-plane plot and inset on right). There was a cross-coupled, upward after-response that tended to swing the vector of the induced eye velocity toward the GIA. (**C**) When RSD, the horizontal component disappeared and down/roll eye velocities appeared. This produced a per-stimulus vector that was oriented spatially in the upward direction, similar to that induced with the animal upright (A). No after-nystagmus was induced in the RSD position.

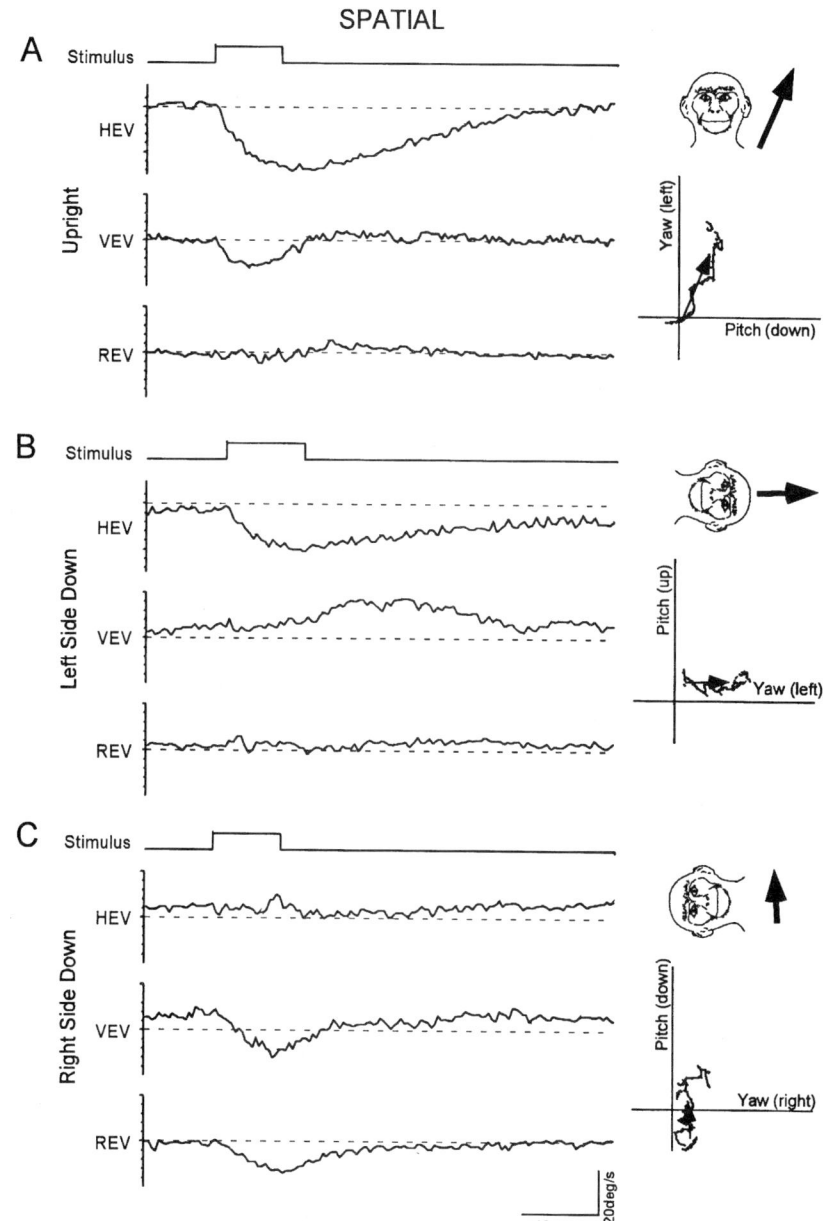

after-nystagmus swung toward alignment with the gravity vector. An orientation vector of the after-nystagmus was calculated from the declining yaw and pitch velocities ($y = -3 - 1.2 \cdot x$; $r = 0.71$, $n = 85$).[6,7] The resultant angle of the vector ($-51°$) was close to the angle of the orientation vector computed for this animal's left OKAN in the LSD position ($-49°$; FIG. 5C, I). In the RSD position (FIG. 4C), no yaw velocity was induced, and there was no cross-coupling from vertical and roll to yaw. The yaw-axis time constant in the upright position was 23 s, and it fell to 14 s in the LSD position. The orientation of the eye-velocity vector to the GIA in tilted positions, as in FIG. 4B, the absence of cross-coupling from vertical or roll eye velocities to yaw, as in FIG. 4C, and the reduction of the yaw-axis time constant in side-down positions, as in FIG. 4A and, 4B, are essential features of the orientation properties of velocity storage.[6,7] This strengthens the conclusion that the after-nystagmus induced by stimulation of the nodulus was similar to OKAN,[27] and that it was due to activation of velocity storage.

Muscimol Injections

Optokinetic Stimulation and Cross-Coupling

Seven muscimol injections and a control injection of saline were made into the region of the nodulus. Typical OKN and OKAN in upright and tilted positions before muscimol are shown in FIGS. 5A–5C and 6A–6C. Only horizontal and vertical slow-phase velocities (HEV and VEV) are displayed. With the animal upright (FIGS. 5A, and 6A), the OKN and OKAN were predominantly horizontal. A small upward slow phase velocity developed at the onset of OKAN when the animal went into darkness, reflecting the weak upward spontaneous nystagmus present in darkness in most monkeys.[7] Similar to previous results,[1,2,7,12] horizontal OKAN time constants were shorter in tilted than upright positions, and a prominent cross-coupled vertical component appeared (VEV, FIG. 5B, 5C; FIG. 6B, 6C). It was upward during OKAN to the left when the animal was in the LSD position (FIG. 5C), and during OKAN to the right when the animal was RSD (FIG. 6B). A smaller downward component appeared during leftward OKAN in the RSD position (FIG. 5B) and during rightward OKAN in the LSD position (FIG. 6C).

With the animal in the upright position, muscimol inactivation of the nodulus did not affect yaw-axis OKN (FIGS. 5D, 6D). The nystagmus remained purely horizontal during OKN, and the orientation vector derived from the subsequent OKAN remained vertical, although the time constant was generally reduced (see below). With the animal in a tilted position, however, cross-coupled OKAN velocities were substantially reduced or abolished. The largest reduction occurred after injection at site a, 2.1 mm to the right of the midline (FIG. 2B). Cross-coupling was essentially abolished for left OKAN, with the animal either left (FIG. 5F) or right side down (FIG. 5E). In this instance, it was also lost for right OKAN with the animal LSD (FIG. 6F), and reduced for right OKAN with the animal RSD (FIG. 6E).

Changes in cross-coupled eye velocities for this site before and after muscimol are summarized in the pitch/yaw phase plane plots of OKAN (FIGS. 5G–5J; 6G–6J). In the control condition for OKN and OKAN with upward cross-coupled eye velocity before muscimol (FIG. 5I, 6G), the eye-velocity vector was tilted $-49°$ and $-42°$. For OKN and OKAN that induced downward cross-coupled velocities (FIGS. 5G, 6I), the orientation vectors were tilted 50° and 36°. After muscimol injection into the right nodulus, there was no significant slope for the orientation vector when the animal was left side down for either left (FIG. 5J) or right (FIG. 6J) OKAN velocities. The RSD vector was diminished from 50° to 13° when the slow phase velocity was to the left (FIG. 5H), but the OKAN continued to orient to gravity for velocities to the right (FIG. 6H). Thus, after injecting muscimol at this site in the nodulus on the right, the vector stayed with the body during LSD tilts, regardless of the direction of the nystagmus velocity. This also occurred for the RSD position when the eyes moved to the right.

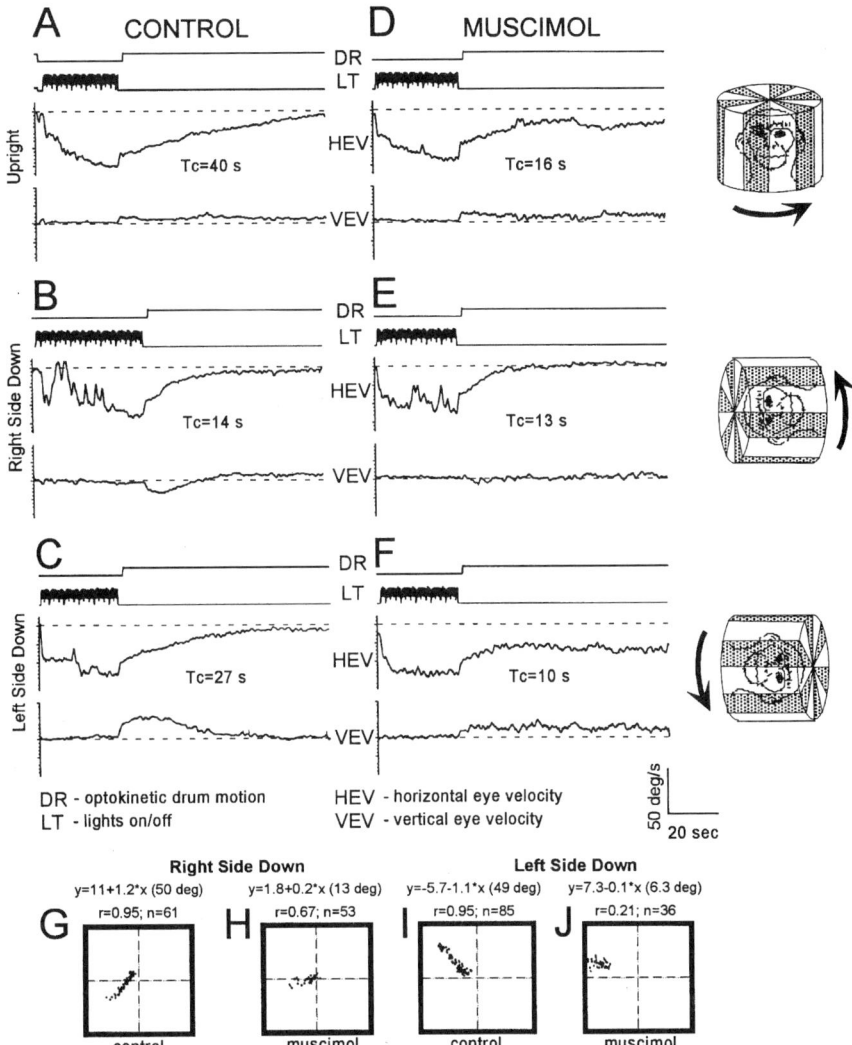

FIGURE 5. Effects of muscimol injection at site a in nodulus, 2.1 mm to right of midline on OKN and OKAN with left slow phase velocities. (**A**)–(**C**) Horizontal and vertical OKN and OKAN velocities before injection with the animal (A) upright (B) RSD, and (C) LSD. Note the reduction in the horizontal time constant with the animal in side-down positions, and the appearance of vertical cross-coupled velocities during OKAN. (**D**)–(**F**) After muscimol, the horizontal time constant was reduced in the upright position and the cross-coupled OKAN velocities disappeared. (**G**)–(**J**) Phase-plane plots of the declines in yaw and pitch eye velocity from the onset to the end of OKAN, before (G, I) and after (H, J) muscimol. Each plot starts farthest from the origin at the onset of OKAN and approaches the origin along the orientation vector of the system.[1,2,6,7] Horizontal eye velocity is on the abscissa, and vertical eye velocity on the ordinate of each graph. The equations, which describe the least-square linear approximation of the data points, are given above each graph together with the correlation coefficient and number of data points. The slope of the linear approximation is in parentheses. Before injection, the velocity vector of the OKAN was tilted 50° (G) and –49° (I) in side-down positions. After injection the slope was reduced (H) or abolished (J).

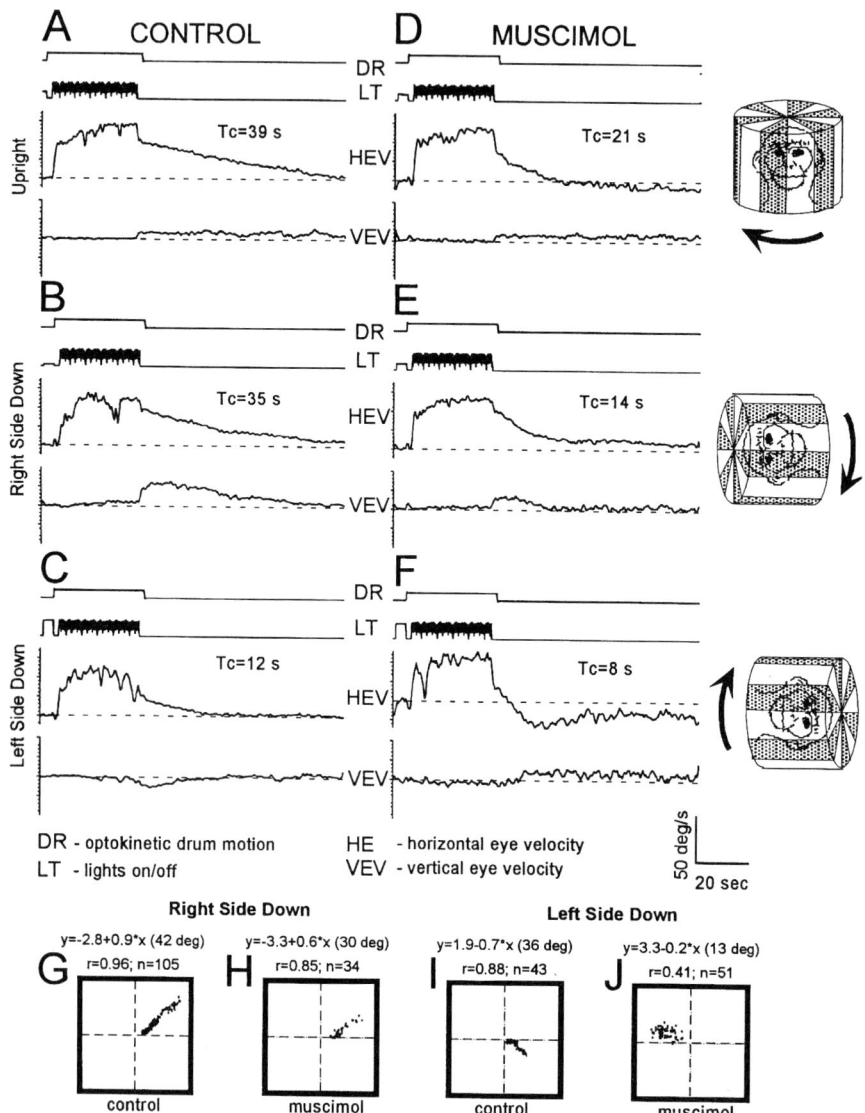

FIGURE 6. Scheme as in FIGURE 5 except that OKN and OKAN had right slow phase velocities. Muscimol caused a reduction in the horizontal time constant (D), a reduction in upward cross-coupling (E), and a loss of down cross-coupling (F). Note the maintained slope in (H) and the loss of slope in (J) for the orientation vectors.

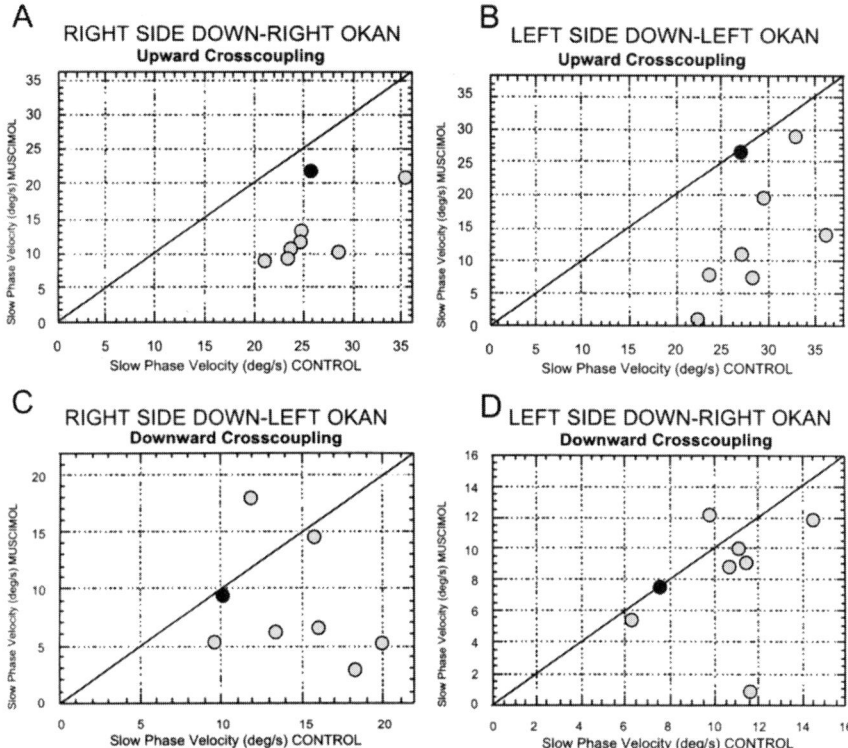

FIGURE 7. Graphs of cross-coupled vertical velocity before (*abscissa*) and after (*ordinate*) muscimol injections at the 7 sites shown in FIGURE 2B. Values from each experiment are shown by a *single dot* for (**A**) right OKAN, RSD; (**B**) left OKAN, LSD; (**C**) left OKAN, RSD, and (**D**) right OKAN, LSD. The *heavy diagonal line* shows where the values would fall if there were no effect of the injections. Values obtained after a saline injection at one of the positive sites (*filled circles*) lay close to this line. Reduction in cross-coupling was greatest in (A)–(C).

Maximal upward and downward cross-coupled slow phase velocities before (abscissa) and after (ordinate) muscimol injections were calculated for all sites (FIG. 7) as well as the percent of the cross-coupling after muscimol relative to the original cross-coupling (TABLE 1). The heavy lines in FIGURE 7 show where points would have fallen if there had been no effect of injection. Following saline injections, the values lay on or close to this line (filled circles). In contrast, cross-coupled velocities were significantly reduced after muscimol injections. Upward cross-coupling was reduced for both right- (A) and left-side-down (B) positions, whereas downward coupling was diminished more for right- (C) than left-side-down (D) positions. In comparisons made of cross-coupled velocities before and after injection, the largest reductions (≈ 65%) occurred when the side contralateral to injection was down and the OKAN slow eye velocity was to the contralateral side, as in FIGURE 5F (TABLE 1; contra OKAN, contra SD). Reductions in other sites were less, ranging between 40 and 47%. Despite the variability, which is probably related, at least in part, to the different sites of injection, there was some reduction in cross-coupled slow phase velocity at most sites, including one in folium 1 of sublobule d of the uvula.

OKAN and Vestibular Horizontal Time Constants

Muscimol injection also affected the horizontal OKAN time constant, which was reduced at most of the injection sites. Injection at site a, in folium 1, 2.1 mm to the right, for example, caused the horizontal OKAN time constants to become shorter on both sides with the animal upright (FIGS. 5A, 5D and 6A, 6D). The horizontal OKAN time constants were also shorter in the side-down positions after muscimol at this site. At other sites, there was less reduction or no effect. Changes in time constant of vestibular nystagmus, induced with velocity steps of 60°/s with the animal upright in darkness, had similar characteristics to those described for OKAN in that the time constants generally became shorter after muscimol injection (FIG. 8A, 8B; TABLE 1). Before injection, mean vestibular time constants from all sites were 31 ± 6 s for left slow phase velocity and 39 ± 8 s for right velocities. After injection, time constants fell to 23 ± 8 s and 23 ± 5 s for left and right slow phase velocities, respectively.

Spontaneous Nystagmus

Spontaneous contralateral horizontal nystagmus was induced after each of the muscimol injections (FIG. 8B). Slow phase velocities ranged from 4°/s to 45°/s (mean 21 ± 15°/s). Rising time constants of the horizontal nystagmus, tested after the animal was put into darkness, ranged from 10 s to 33 s. An example of a 14-s rising time constant to a steady-state spontaneous nystagmus level of 20°/s to the left is shown in FIG. 8B. At all six sites in the nodulus, the slow phase velocity was contralateral to the side of injection. It was ipsilateral from the site b in lobule d of the uvula. At all sites, the nystagmus was present only in darkness, and was suppressed in light. Upward vertical components were less than 10°/s, and were most likely a reflection of the animal's spontaneous upward nystagmus in darkness.

There was a marked effect of head position re gravity on the horizontal slow phase velocity of the spontaneous nystagmus (FIG. 9). After muscimol, head positions to the left caused an increase in slow phase velocity to the left and a decrease in slow phase velocity to the right, and vice versa. This could cause a reversal of the direction of spontaneous nystagmus recorded in the upright position, so that the animal had an apogeotropic, direction-changing positional nystagmus, if considered with regard to the quick phase direction. The nystagmus was most symmetrical at site g, but there was reversal of slow phase velocity at other sites as well (a, b, c, and e). The velocities induced by static tilt at each 15° for 6/7 sites after muscimol injection could be approximated by straight lines with high correlation coefficients ($r \geq 0.95$). Interestingly, the slopes from the 6/7 sites were approximately parallel, with a mean change in slow phase velocity of $\approx 18°*s*g^{-1} \pm 2°/s$ (SD). There was no increase in vertical or torsional eye velocity during the horizontal nystagmus in side-down positions. Therefore, the vector or the positional nystagmus was not spatially oriented. No positional nystagmus was induced in control tilts (FIG. 9, dotted line) or after saline injection (dashed line).

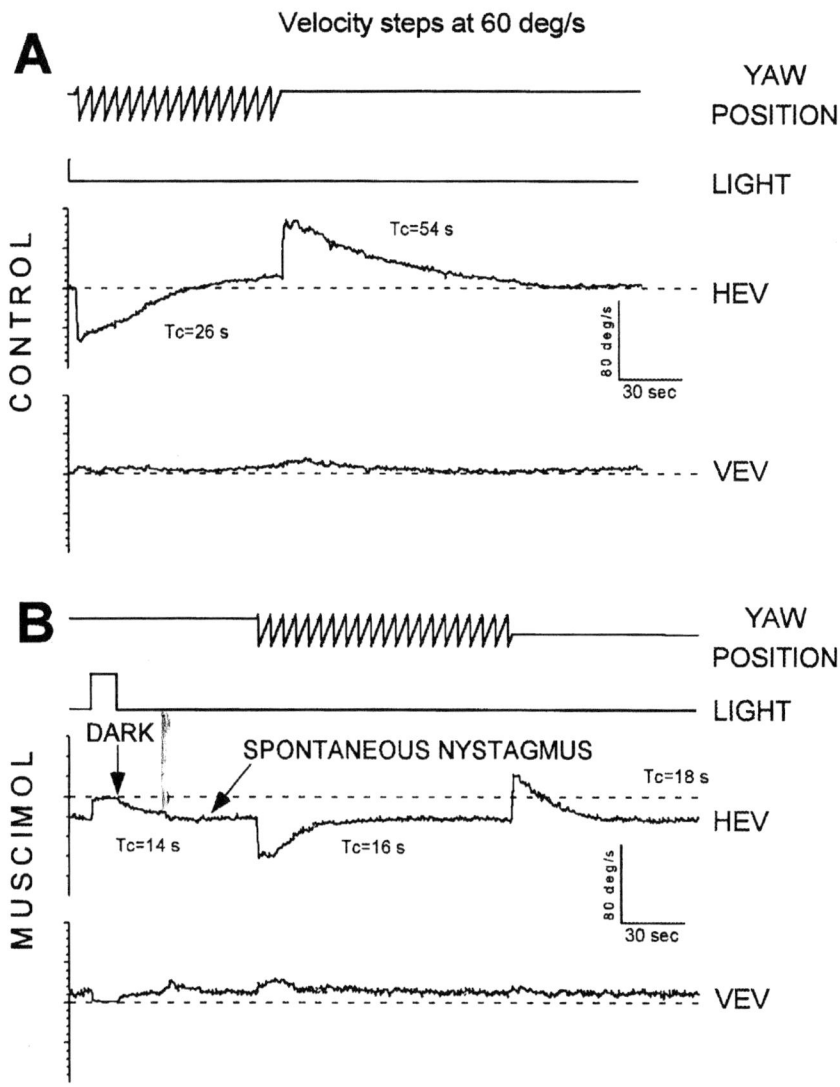

FIGURE 8. Horizontal (HEV) and vertical (VEV) slow phase velocity during per- and post-rotatory nystagmus in response to angular rotation at 60°/s, recorded in darkness. (**A**) Before injection at site a, 2.1 mm to the right of the midline, the time constants of per- and postrotatory nystagmus were 26 s (left) and 54 s (right) for the per- and post-rotatory nystagmus, respectively. (**B**) After injection, time constants of the per- and postrotatory nystagmus were reduced over those obtained before injection. Spontaneous contralateral (*left*) horizontal nystagmus appeared, rising to a steady state after the animal had been in light with a time constant of 14 s. There was no change in the animal's spontaneous vertical nystagmus before and after injection.

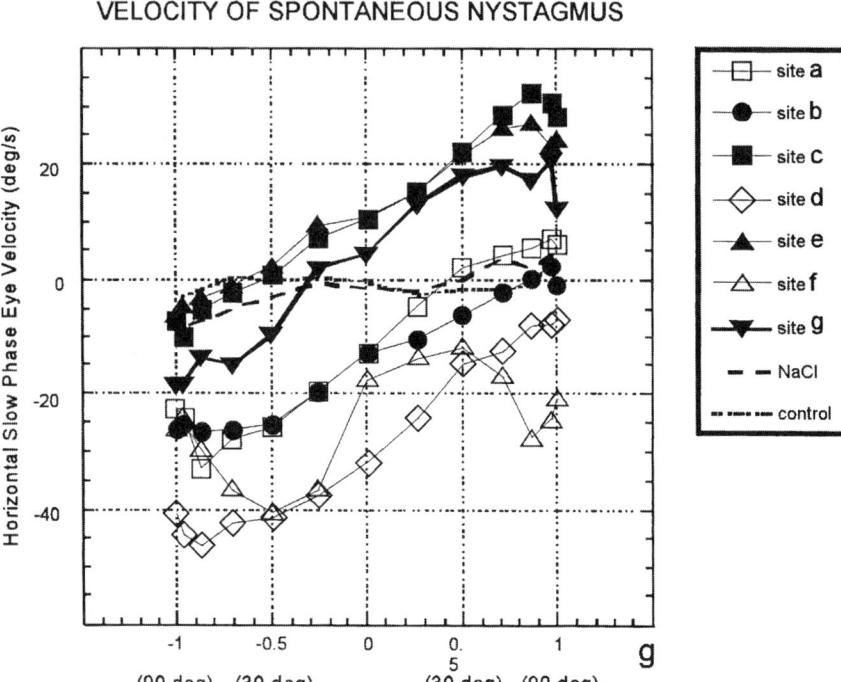

FIGURE 9. Alterations in the velocity of the horizontal spontaneous nystagmus produced by injection of muscimol at various sites in the nodulus and uvula as a function of head position with regard to gravity. *Filled symbols* are for left-sided and *open symbols* for right-sided injections. The sites are identified in the box on the right. With the exception of the results at site f, a *straight line* approximated the data in each set well. The average slopes for these six sites was $18.3°/g \pm 2.2$ ($n = 6$) with a mean correlation coefficient of 0.98 (0.01. No positional nystagmus was induced in tilted positions in controls or after NaCl injection.

Off-Vertical Axis Rotation

The animals were rotated in yaw at 60°/s about axes tilted at angles from 0° to 90° in 15° steps. The steady-state (bias) velocities, the depth of modulation as a function of head position with regard to gravity, and the phase of these modulations were analyzed for both the horizontal and vertical components. Bias velocities were induced both before and after muscimol injection. When the slow phases of the nystagmus induced by OVAR and the spontaneous nystagmus were oppositely directed, the bias velocity was reduced by the amount of the spontaneous nystagmus. Before injection, horizontal bias velocities decreased slightly as the angle of tilt was increased. After injection, this tendency was retained for some sites and lost at others. The amplitude of the modulations in horizontal slow phase velocity during each cycle of OVAR increased after injection. This increase in modulation could be attributed to the reduction in the horizontal time constant associated with muscimol injection (FIG. 8, TABLE 1), similar to the effects of habituation.[52] The phase of modulation of horizontal slow phase velocity was unaffected by muscimol injections.

TABLE 2. Directions of Response Vectors for Ten Otolith-Related Units Obtained with Static Tilts, Dynamic Tilts, and with OVAR[a]

Unit No.	CV	Spontan. (imp*s^{-1})	Static Tilt Gain (imp*s^{-1}*g)	Static Tilt Phase (deg)	Dynamic Tilt Gain (imp*s^{-1}*g)	Dynamic Tilt Phase (deg)	OVAR Phase CW (deg)	OVAR Phase CCW (deg)	OVAR Phase Aver. (deg)	Response Plane
					Otolith Only—Related					
4	0.44	56.8 ± 6.6	11.4	93	17.5	69	147 ± 45	8 ± 42	78	Contra-SD
					Otolith & Vertical Canal Related					
3	0.33	59.5 ± 3.3	6.8	124	42.0	38	—	—	—	Contra-AC
6	0.41	32.4 ± 3.1	—	—	54.2	145	—	—	—	Contra-PC
12	0.42	42.5 ± 7.5	—	206	12.4	139	—	—	—	Ipsi-PC
13	0.32	62.7 ± 2.4	19.0	222	20.1	226	—	—	—	Contra-PC
14	0.41	54.9 ± 3.8	—	—	91.2	68	—	—	—	Ipsi-SD
15	0.31	73.1 ± 3.2	—	—	11.3	131	290 ± 32	226 ±41	258	Contra-SD
					Spatial-Temporal Convergence					
2	0.44	57.8 ± 6.5	1.0	335	7.7	311	228	45	316	Ipsi-AC
16	0.39	51.2 ± 8.7	13.8	244	7.6	277	305 ± 27	210 ± 31	258	Contra-AC
18	0.38	62.6 ± 10.3	33.2	306	25.0	341	205 ± 29	23 ± 5	294	Contra-AC

[a]The directions of the response vectors (phases) obtained with the different techniques were generally comparable.

Abbreviations: Ipsi, Ipsilateral; contra, contralatral; SD, side down; AC, anterior canal; PC, posterior canal; imp., impulse; deg., degree.

Single-Unit Recording

Ten units, which were sensitive to head orientation with regard to gravity, were recorded in the rostral 1/3 of the nodulus. The mean coefficient of variation of the units was 0.33 ± 0.05, identifying them as "irregular" units[58] (TABLE 2). The mean spontaneous activity of these cells was 55.4 ± 11.3 imp·s^{-1}. All of these units received convergent input from neck-muscle proprioceptors, determined by causing oscillations in unit activity while pressing on the back of the neck. Unit sensitivity to tilts of the GIA, determined during static tilt, dynamic tilt, and OVAR are listed in TABLE 2.

Recordings from a typical cell during static tilt are shown in FIGURE 10A. The unit was recorded while the head was tilted 60° about a horizontal axis in various orientations relative to the plane of tilt. Each time the animal was tilted, there was a phasic increase in activity (downward arrows). We attribute this to neck-muscle activation due to the body-position change. The mean firing rate in tilt was determined for each head orientation in the period shown by the heavy horizontal lines (FIG. 10A). The sensitivity of the unit (imp·s^{-1}·g^{-1}) varied as a function of the orientation of the head to the plane of tilt (FIG. 10B). The data were well fit by a cosine function ($r=0.760$) that had a peak sensitivity of about 14 imp·s^{-1}·g^{-1} with a head orientation of 244° and a zero crossing at 154°. Thus, this unit was polarized close to the plane of the left posterior canal (225°).

All units were tested with sinusoidal oscillation around a spatial horizontal axis with the head in similar orientation relative to the plane of tilt as for static tilt. The orientation of the sensitivity vectors for these 10 units was similar for static tilt as for dynamic oscillation about a spatial horizontal axis. One of the 10 units had activity modulated in phase with head position, indicating no convergent input from the vertical canals (TABLE 2). Six of the tested units had modulation in phase with head velocity. The spatial response of one of them is shown in FIGURE 10D. This unit had a spatial phase of 145°, suggesting convergent input from the right posterior canal. Its peak sensitivity was about 54 imp·s^{-1}·g^{-1}. Three of the 10 units had temporal phases that changed as a function of orientation of the head to the plane of tilt (FIG. 10E).

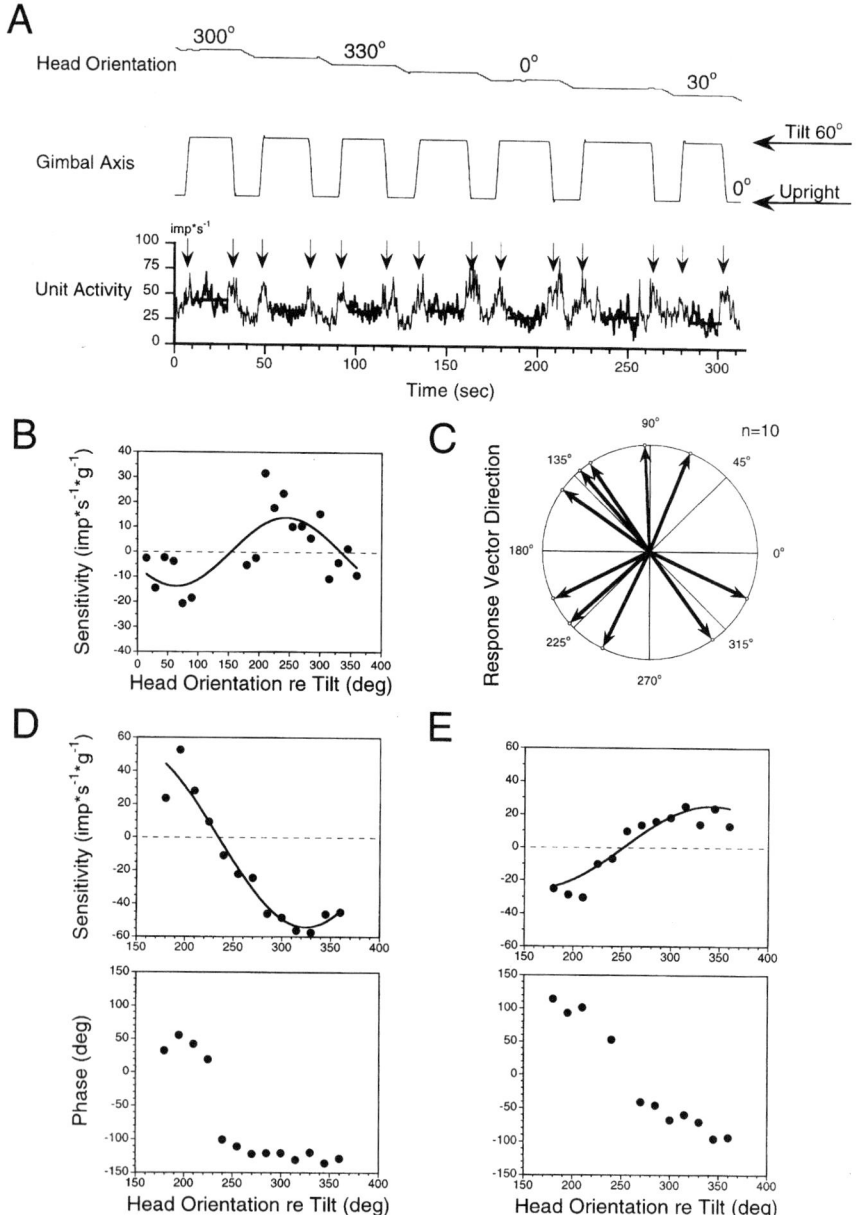

These units have been described in the vestibular nuclei as "spatial-temporal convergence units."[59,60] This could also indicate possible convergent input from the vertical canals.[61] Five of the 10 recorded units were also tested during OVAR. Response vectors led head orientation in all cases. If the average response vector was defined based on the sum of the responses during clockwise and counterclockwise rotation, then the orientation of the response vectors for each unit was close to that based on static or dynamic tilt data.

The average direction of the response vectors for the 10 units are shown in FIGURE 10C. For seven units, the response vectors were determined from static tilts. For three units, not tested with static tilt, response vectors were based on dynamic data. We determined that units were related to particular canal planes if the direction of the response vector was within ± 22.5° from the canal plane. Maximal activation of 7 of the 10 units occurred for tilts that were close to the planes of the right (135°) and left (225°) posterior canals and the left (315°) and right anterior (45°) canal, respectively. Three units were ≈ 90° from the midline. Thus, a majority of nodulus units that were recorded tended to have their maximal activation vectors aligned with vertical-canal planes, and some of them also received vertical-canal input.

DISCUSSION

This study supports the postulate that the nodulus is a critical structure for determining alignment of the eye-velocity vector with the GIA, which we define as spatial orientation of the aVOR. When the head is tilted relative to the GIA, this spatial orientation is achieved by an alteration of the horizontal, vertical, and roll time constants of velocity storage and by the generation of cross-coupled vertical and/or roll velocities.[1,2,6–8,12] While the nodulus has been shown to play a critical role in controlling the horizontal time constant of the aVOR from stimulation[23,27] and lesion studies,[1,2,21,62] the present study demonstrates that the nodulus can also produce spatially oriented cross-coupled responses when electrically activated. The spatially oriented eye-velocity responses induced from the nodulus behaved as if the animal were tilted, and the nodulus was generating a response to an altered orientation in space. When the head was actually tilted, generally to the contralateral side, the response was held constant in space, despite the change in the head position with respect to gravity. Consistent with previous data,[27] poststimulus eye velocities that had the appropriate temporal and spatial characteristics of OKAN and of velocity storage, including orientation to the GIA, were also elicited from these sites.

Stimulation sites that caused cross-coupled components of eye velocity produced a concomitant reduction of yaw-to-vertical cross-coupling of horizontal OKAN when inactivated with muscimol. Although the effect was variable, inactivation occurred at all seven

◀

FIGURE 10. Effect of static tilts (A)–(C) and dynamic tilts (D), (E) on single units recorded in the nodulus. Coordinate system: tilt forward (0°/360°), LSD (90°), tilt back (180°), RSD (270°). (**A**) Variations in firing frequency produced by tilting the animal 60° with the head oriented differently to the plane of tilt (*top trace*). The *arrows* point to the phasic increases in firing, presumably caused by activation of neck proprioceptors. *Heavy horizontal lines* indicate the mean activity during the periods used for analysis. (**B**) Sensitivity of this unit plotted as a function of head orientation with regard to the plane of tilt. The data were fitted with a sinusoid whose maximum represents the direction of the response vector, which in this instance, was 244°. (**C**) Polar plot representing direction of response vectors for all recorded units. One unit was oriented along a 90° axis, and two other units fell within ± 22.5° of 90°/270° plane. The other units fell close to the LARP (135°, 315°) or RALP (45°, 225°) planes. (**D**) Spatial responses of otolith-related nodulus units tested with dynamic tilt. (D) Non-STC unit; (E) STC unit. Gains (*top graph*) and phases (*bottom graph*) are plotted as a function of head orientation with regard to the plane of tilt.

sites. This is strong support for the postulate that cross-coupling is controlled by the nodulus, and is consistent with effects of lesions that caused a loss of spatial orientation of the aVOR for eye velocity around all axes.[1,2,63,64] Since there is direct input from the labyrinth to the nodulus, the effects of stimulation could conceivably have been mediated by back activation of collateral branches through the vestibular nuclei. The injection data rule out this possibility, because it was local inactivity of cerebellar circuits that caused the reduction in cross-coupling.

The per-stimulus eye velocities that were oriented in space did so with the head in only one lateral direction, and moved with the head when the animal was tilted in the opposite direction. In general, it appeared that the direction of spatial constancy was for contralateral eye velocities. This is probably a reflection of the fact that the predominant pathways from the nodulus to the vestibular nuclei are ipsilateral,[46] to the side that produces contralateral slow phase velocities when excited. Taken together, with the results of stimulation, these findings extend the conclusions from bilateral lesion experiments,[2] in that they suggest that sites in the nodulus control spatial orientation in a specific direction. This is compatible with findings that there was only unilateral, never bilateral, dumping after nodulus stimulation.[23] The muscimol injections that abolished cross-coupling with or without changes in horizontal time constant are consistent with the idea that the processes that control cross-coupling and the aVOR time constant are separately controlled by different portions of the cerebellar cortex.[2]

The idea that spatial orientation should be represented in the cerebellum is not new. It was implicit in the findings that the cerebellum is more highly developed in Cetacea than in terrestrial animals.[65] Since whales have little or no fine limb coordination, it was surmised that their well-developed cerebella could be used for spatial orientation and navigation. The cerebellum of the weakly electric fish, *Eigenmanni*, is also involved in location in space.[66] A similar conclusion was reached in the rat.[67] More recently, Pompeiano and coworkers have come to a similar conclusion.[68] Investigating spatiotemporal-response properties of Purkinje cells in the anterior vermis (lobules 1–3), they found a population of Purkinje cells that coded the direction of head tilt in space. "For each selected time in the tilt cycle, the direction of the population vector closely corresponded to that of the head tilt, while its amplitude was related to that of the stimulus." They conclude that this could provide a substrate for the "spatial organization of vestibulospinal reflexes induced by otolith receptors," and that "the Purkinje cells of the cerebellar cortex are expected to show prominent responses to head rotation, which could affect the spatially organized postural responses by utilizing vestibular and reticular targets."[67–72]

The regions that were explored in this study lay within a relatively restricted range, and the MRI proved to be of value in pinpointing the sites of recording in the cerebellum. Particularly when recording from regions that do not have characteristic activity, as from the nodulus, it may be difficult to determine depth and laterality during electrode penetrations in the live animal. It was of interest that the anatomic structure defined by the MRI was consistent with the postmortem histology: the mean separation of the electrode tracks on the right and left side of the nodulus were 5.4 mm (TABLE 1), which is close to the 5-mm separation of the electrodes shown in FIGURE 1B. Thus, MRI provided an efficient technique for establishing electrode placement. Initially, there was concern that the stainless-steel bolts would become overheated when subjected to the magnetic fields, and that the artifacts from the bolts would obscure the MRI. Neither of these concerns proved relevant, nor was it necessary to have special equipment to hold the head in a stereotaxic plane when the MRI was taken, since the plane of section was readily adjusted in the computer reconstruction of the images.

The regions that were injected in the nodulus extended from 2.1 mm to 3.3 mm from the midline (mean 2.7 ± 0.5 mm, $n = 6$). This encompasses the lateral border of Zone 2 at 2.4 mm, which contains input from the vlo nucleus, the rostral medial accessory olive (rMAO), and the rostral ß nucleus. In the rabbit these regions carry climbing fiber activity from the ipsilateral anterior canal, the contralateral posterior canal, and the utricle. Zone 3 receives climbing fiber input from cdc of Kooy, which in turn gets information from the nucleus of the optic tract (NOT) about ipsilateral horizontal retinal slip for movements about a vertical axis. This is a response to movement in the same plane as the lateral canals.[73] Assuming that the spread of the electrical current for a 40-µA current[23] and for a muscimol injection of about 1 µL was approximately 1.0 mm,[74,75] then a 1.2–1.4-µL injection would extend somewhat farther. Thus, we presume that the vertical rotatory responses that were elicited by stimulation and the cross-coupling that was abolished by injection were probably largely due to activation or inactivation of Zone 2, and alterations in horizontal time constant were probably due to activation or inactivation of Zone 3. The nodulus is a large structure, approximately 7–8 mm wide, 2 mm thick, and 5 mm from dorsal to ventral surface, and only limited regions were explored in our study. As a result, conclusions about nodulus function must necessarily be tentative.

An interesting aspect of muscimol inactivation of the nodulus was the occurrence of contralateral spontaneous nystagmus from all sites of injection. The origin of this nystagmus was likely due to a reduction in GABAergic inhibition of neurons in the ipsilateral vestibular nuclei. The fact that the nystagmus had contralateral slow phase velocities is probably explained by the fact that the output connections of the nodulus are predominantly to the ipsilateral vestibular nuclei. Muscimol would reduce inhibition in this vestibular nucleus, thereby producing contralateral velocities. Two findings suggest that the relation between head position and velocity of spontaneous nystagmus is a general property of the vestibular system. The alteration in eye velocity is similar to the effect of head position on spontaneous nystagmus that was observed after muscimol injections in the vestibular nuclei (Yokota and Cohen, unpublished observations). Second, the slope of the response was similar after all injections, and was independent of the site of injection.

The positional nystagmus after muscimol was apogeotropic. That is, the slow phases were toward gravity and the quick phases were away from gravity, and this reversed when the monkey's head position was reversed. Similar nystagmus is often observed after cerebellar and/or brainstem lesions (unpublished observations). Since no vertical or roll spontaneous nystagmus was generated as a function of head orientation, the vector of spontaneous eye velocity was directed along the head yaw and remained invariant. Therefore, this aspect of the response was not related to spatial orientation, apparently precluding involvement by velocity storage. On the other hand, the rising time constants of the slow phase velocity were in the range of 10 to 33 s, which was in the range of the aVOR-dominant time constant, and the peak velocities ranged up to 40°/s, which is close to the saturation velocity of velocity storage,[57,76] as well as to the saturation level of lateral canal-related, horizontal vestibular-only (VO) neurons to optokinetic stimulation.[77,78] Thus, the nystagmus could have been produced by reduction of Purkinje cell inhibition of VO neurons in MVN, which are believed responsible for production of velocity storage.[54,79,80] The source of the nystagmus as well as the process that rather precisely sets the level of spontaneous nystagmus, as a function of head position with regard to gravity, are subjects for further study.

There were only relatively minor effects of nodulus inactivation on the nystagmus associated with OVAR, namely an increase in the modulation in horizontal eye velocity associated with head position with regard to gravity, and a slight decrease in the peak steady-state yaw velocity. This is consistent with the previous lesion studies,[2] in which OVAR nystagmus was produced with relatively normal characteristics after the nodulus and uvula were removed. The implication of this is that processing of otolith information to produce slow phase velocity during OVAR is probably not done in these regions of the vestibulocerebellum.

Single-unit recording also supported the idea that the nodulus subserves a spatial orientation function by the broad presence of otolith-related activity. Such activity would be a prerequisite to determining the tilt of the head with regard to the GIA. Interestingly, the sensitivity vectors of the nodulus units were polarized close to the same anterior- and posterior-canal planes that are represented in Zone 2 of the vermis of the nodulus,[31] where injections of muscimol caused a loss of cross-coupling. Recently, we demonstrated that otolith-related units in the vestibular nuclei had their maximal sensitivity close to the planes of the vertical canals from which they had convergent input.[61] This otolith–canal convergence is likely to have specific functional significance: it provides a mechanism for sensing head orientation along particular planes of movement.

From the clustering of the unit sensitivities in or close to canal planes in the nodulus, we utilize a previous model[1,2] to propose how system time constants that account for spatial orientation of the aVOR might be dynamically altered. If the incoming otolith activity to individual zones were organized along canal planes, as is suggested in FIGURE 10C and TABLE 2, and if the canal-related Purkinje cells located in these canal-related zones were activated by otolith neurons whose sensitivity vectors lay close to these same canal planes, then the Purkinje cells could provide a feedback mechanism to accomplish the changes in time constants and the production of cross-coupling that are the basis for spatial orientation. This is shown schematically in FIGURE 11. A broad distribution of spatial vectors in the vestibular periphery, that is, in the utricle and saccule, is mapped onto the nodulus in three functional canal planes, which then influence the eye movements produced around the axis of that canal plane. Purkinje cells in the nodulus canal-related zones project to regions of MVN and SVN where eye movements in these canal planes are produced.[46] The result of otolith activation due to a tilt of the head with regard to gravity would first be to activate the otolith neurons that sensed the tilt. These in turn would project to the canal-related zones in the nodulus and possibly sublobule d of the uvula via mossy fiber input, in a canal-related organization. In turn, Purkinje cells in these zones would project to specific canal-recipient areas of the vestibular nuclei to produce eye velocity along the axes of each of the reciprocal canal pairs that were associated with that tilt. Presumably, this would be done by inhibition of lateral canal-related VO neurons in MVN and by disinhibition of vertical canal-related VO neurons in SVN.[1,2] If the subject was in angular motion, this input could then alter the axis of eye rotation to align with the tilt of the GIA with regard to the head.

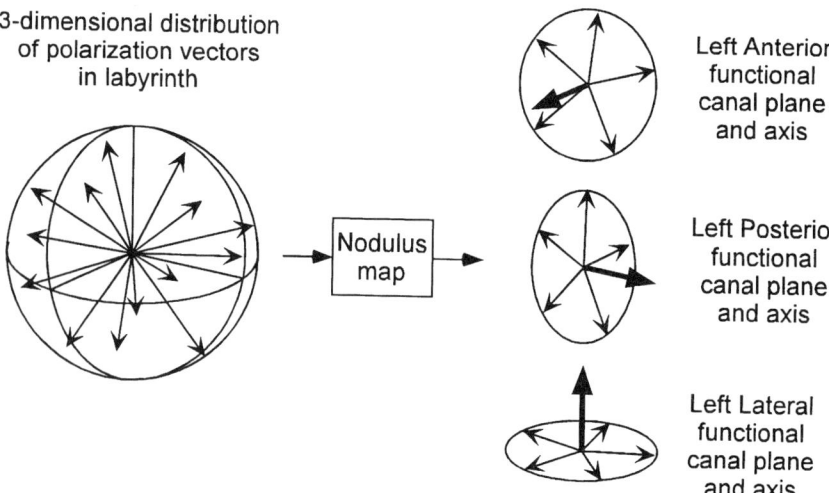

FIGURE 11. Schematic representation of scheme whereby units with polarization vectors in all directions in the labyrinth are mapped into the semicircular-canal coordinate scheme in the nodulus. See text for details.

Thus, the neural machinery to implement spatial orientation functions during a wide range of motions could simply be a honing of otolith-related polarization vectors that control spatial orientation, which are widely distributed in the utricular and sacular maculae, to more confined planes related to the semicircular canal organization in the nodulus. Such a mechanism has previously been suggested as a means for estimating three-dimensional head velocity during rotation about axes that are not aligned with the GIA.[81] This mechanism could then be utilized for controlling spatial orientation of semicircular canal-based systems utilized for estimating spatial orientation during circular locomotion,[15] as well as during passive circular motion.

ACKNOWLEDGMENTS

This work was supported by the following grants from the NIH: DC03284, EY11812, NS00294, DC03787, EY04148, and EY01867. We thank Victor Rodriguez and Jun Maruta for technical assistance and John Abela of Radiology for performing the MRI's.

REFERENCES

1. WEARNE, S., T. RAPHAN & B. COHEN. 1996. Nodulo-uvular control of central vestibular dynamics determines spatial orientation of the angular vestibulo-ocular reflex. Ann. N.Y. Acad. Sci. **781**: 364–384.
2. WEARNE, S., T. RAPHAN & B. COHEN. 1998. Control of spatial orientation of the angular vestibulo-ocular reflex by the nodulus and uvula. J. Neurophysiol. **79**: 2690–2715.
3. RAPHAN, T. & B. COHEN. 1986. Multidimensional organization of the vestibulo-ocular reflex (VOR). In Adaptive Processes in Visual and Oculomotor System, E. L. Keller and D. S. Zee, Eds.: 285–292. Pergamon Press. New York.
4. HARRIS, L. R. & G. R. BARNES. 1987. Orientation of vestibular nystagmus is modified by head tilt. In The Vestibular System: Neurophysiologic and Clinical Research, M. D. Graham and J. L. Kemink, Eds.: 539–548. Raven Press. New York.
5. HARRIS, L. R. 1987. Vestibular and optokinetic eye movements evoked in the cat by rotation about a tilted axis. Exp. Brain Res. **66**: 522–532.
6. RAPHAN, T. & D. STURM. 1991. Modelling the spatiotemporal organization of velocity storage in the vestibuloocular reflex by optokinetic studies. J. Neurophysiol. **66**(4): 1410–1420.
7. DAI, M., T. RAPHAN & B. COHEN. 1991. Spatial orientation of the vestibular system: dependence of optokinetic after nystagmus on gravity. J. Neurophysiol. **66**(4): 1422–1438.
8. RAPHAN, T., M. DAI & B. COHEN. 1992. Spatial orientation of the vestibular system. Ann. N.Y. Acad. Sci. **656**: 140–157.
9. ANGELAKI, D. E. & B. J. M. HESS. 1994. Inertial representation of angular motion in the vestibular system of rhesus monkeys. I. Vestibuloocular reflex. J. Neurophysiol. **71**: 1222–1249.
10. ANGELAKI, D. E. & B. J. M. HESS. 1995. Inertial representation of angular motion in the vestibular system of rhesus monkeys. II. Otolith-controlled transformation that depends on an intact cerebellar nodulus. J. Neurophysiol. **73**: 1729–1751.
11. RAPHAN, T. & B. COHEN. 1996. How does the vestibulo-ocular reflex work? In Disorders of the Vestibular System, R. W. Baloh & G. M. Halmagyi, Eds: 20–47. Oxford University Press. New York.
12. WEARNE, S., T. RAPHAN & B. COHEN. 1997. Contribution of vestibular commissural pathways to velocity storage and spatial orientation of the angular vestibulo-ocular reflex. J. Neurophysiol.
13. SOLOMON, D. & B. COHEN. 1992. Stabilization of gaze during circular locomotion in darkness; II. Contribution of velocity storage to compensatory eye and head nystagmus in the running monkey. J. Neurophysiol. **67**(5): 1158–1170.
14. BLES, W. & S. KOTAKA. 1986. Stepping around: nystagmus, self-motion perception and Coriolis effects. In Adaptive Processes in Visual and Oculomotor Systems, E. L. Keller and D. S. Zee, Eds.: 465–471. Pergamon. Oxford.
15. COHEN, B., S. WEARNE, M. DAI & T. RAPHAN. 1999. Spatial orientation of the angular vestibulo-ocular reflex. J. Vest. Res. In press.

16. BENSON, A. J. 1974. Modification of the response to angular accelerations by linear accelerations. In Handbook of Sensory Physiology, Vol. VI, H. Kornhuber, Ed.: 281–320. Springer. Berlin.
17. FETTER, M., D. TWEED, B. HERMANN, B. WOHLAND-BRAUN & E. KOENIG. 1992. The influence of head position and head reorientation on the axis of eye rotation and the vestibular time constant during postrotatory nystagmus. Exp. Brain Res. **91**: 121–128.
18. MERFELD, D. M., L. R. YOUNG, G. D. PAIGE & D. L. TOMKO. 1993. Three dimensional eye movements of squirrel monkeys following postrotatory tilt. J. Vest. Res. **3**(2): 123–141.
19. MERFELD, D. M. & L. R. YOUNG. 1995. The vestibulo-ocular reflex of the squirrel monkey during eccentric rotation and roll tilt. Exp. Brain Res. **106**(1): 111–122.
20. RAPHAN, T., B. COHEN & V. HENN. 1981. Effects of gravity on rotatory nystagmus in monkeys. Ann. N.Y. Acad. Sci. **374**: 44–55.
21. WAESPE, W., B. COHEN & T. RAPHAN. 1985. Dynamic modification of the vestibulo-ocular reflex by the nodulus and uvula. Science **228**: 199–201.
22. DAI, M., L. MCGARVIE, I. B. KOZLOVSKAYA, T. RAPHAN & B. COHEN. 1994. Effects of spaceflight on ocular counterrolling and spatial orientaion of the vestibular system. Exp. Brain Res. **102**: 45–56.
23. SOLOMON, D. & B. COHEN. 1994. Stimulation of the nodulus and uvula discharges velocity storage in the vestibulo-ocular reflex. Exp. Brain Res. **102**: 57–68.
24. COHEN, B., V. MATSUO & T. RAPHAN. 1977. Quantitative analysis of the velocity characteristics of optokinetic nystagmus and optokinetic after-nystagmus. J. Physiol. (Lond.). **270**: 321–344.
25. COHEN, B., V. HENN, T. RAPHAN & D. DENNETT. 1981. Velocity storage, nystagmus, and visual vestibular interactions in humans. Ann. N.Y. Acad. Sci. **374**: 421–433.
26. WAESPE, W. & B. COHEN. 1983. Flocculectomy and unit activity in the vestibular interaction. Exp. Brain Research. **51**: 23–35.
27. HEINEN, S. J., D. K. OH & E. L. KELLER. 1992. Characteristics of nystagmus evoked by electrical stimulation of the uvular/nodular lobules of the cerebellum in monkey. J. Vest. Res. **2**: 233–245.
28. VOOGD, J., 1964. The Cerebellum of the Cat. Structure and Fiber Connections. F. A. Davis. Philadelphia.
29. VOOGD, J. 1969. The importance of fiber connections in the comparative anatomy of the mammalian cerebellum. In Neurobiology of Cerebellar Evolution and Development, R. Llinas, Ed.: 493–514. A.M.A. Chicago.
30. VOOGD, J. & F. BIGARÉ. 1980. Topographical distribution of olivary and corticonuclear fibers in the cerebellum. A review. In The Inferior Olivary Nucleus: Anatomy and Physiology, J. Courville, C. C. d. Montigny, and Y. Lamarre, Eds.: 207–234. Raven Press. New York.
31. VOOGD, J., N. M. GERRITS & T. J. H. RUIGROK. 1996. The organization of the vestibulocerebellum. Ann. N.Y. Acad. Sci. **781**: 553–579.
32. BALABAN, C. D. & R. T. HENRY. 1988. Zonal organization of olivo-nodulus projections in albino rabbits. Neurosci. Res. **5**(5): 409–423.
33. BARMACK, N. H. & H. SHOJAKU. 1992. Representation of a postural coordinate system in the nodulus of the rabbit cerebellum by vestibular climbing fiber signals. In Vestibular and Brain Stem Control of Eye, Head, and Body Movement, H. Shimazu and Y. Shinoda, Eds.: 331–338. Japan Scientific Societies Press. Tokyo.
34. GRAF, W. 1988. Motion detection in physical space and its peripheral and central representation. Ann. N.Y. Acad. Sci. 545: 154–169.
35. KANO, M. S., M. KANO & K. MAEKAWA. 1990. Receptive field organization of climbing fiber afferents responding to optokinetic stimulation in the cerebellar nodulus and flocculus of the pigmented rabbit. Exp. Brain Res. **82**(3): 499–512.
36. KATAYAMA, S. & N. NISIMARU. 1988. Parasagittal zonal pattern of olivo-nodular projection in rabbit cerebellum. Neurosci. Res. **5**(5): 424–438.
37. LEONARD, C. S., J. I. SIMPSON & W. GRAF. 1988. The spatial organization of visual messages in the flocculus of the rabbit's cerebellum. I. Typology of inferior olive neurons of the dorsal cap of Kooy. J. Neurophysiol. **60**: 2073–2090.
38. SIMPSON, J. I., W. GRAF & C. LEONARD. 1981. The coordinate system of visual climbing fibers to the flocculus. In Progress in Oculomotor Research, A. F. Fuchs and W. Becker, Eds.: 475–484. Elsevier. New York.

39. TAKEDA, T. & K. MAEKAWA. 1984. Collateralized projection of visual climbing fibers to the flocculus and nodulus of the rabbit. Neurosci. Res. **2**: 125–132.
40. ALLEY, K., R. BAKER & J. I. SIMPSON. 1975. Afferents to the vestibulo-cerebellum and the origin of the visual climbing fibers in the rabbit. Brain Res. **98**: 582–589.
41. BARMACK, N. H., M. FAGERSON, B. J. FREDETTE, E. MUGNAINI & H. SHOJAKU. 1993. Activity of neurons in the beta nucleus of the inferior olive of the rabbit evoked by natural vestibular stimulation. Exp. Brain Res. **94**: 203–215.
42. BARMACK, N. H. & H. SHOJAKU. 1995. Vestibular and visual signals evoked in the uvula-nodulus of the rabbit cerebellum by natural stimulation. J. Neurophysiol. **74**: 2573–2589.
43. SHOJAKU, H., Y. SATO, K. IKARASHI & T. KAWASAKI. 1987. Topographical distribution of Purkinje cells in the uvula and the nodulus projecting to the vestibular nuclei in cats. Brain Res. **416**: 100–112.
44. BARMACK, N. H. & M. H. FAGERSON. 1994. Vestibularly-evoked activity of single neurons in the dorsomedial cell column of the inferior olive in rabbit. Soc. Neurosci. Abstr. **20**: 1190.
45. BARMACK, N. H. & H. SHOJAKU. 1989. Topography and analysis of vestibular-visual climbing fibre signals in the rabbit cerebellar nodulus. Soc. Neurosci. Abstr. **15**: 180.
46. WYLIE, D. R., C. I. DE ZEEUW, P. L. DIGIORGI & J. I. SIMPSON. 1994. Projections of individual Purkinje cells of identified zones in the ventral nodulus to the vestibular and cerebellar nuclei in the rabbit. J. Comp. Neurol. **349**: 448–463.
47. FUSHIKI, H. & N. H. BARMACK. 1997. Topography and reciprocal activity of cerebellar Purkinje cells in the uvula-nodulus modulated by vestibular stimulation. J. Neurophysiol. **78**: 3083–3094.
48. ITO, M., S. M. HIGHSTEIN & J. FUKUDA. 1970. Cerebellar inhibition of the vestibulo-ocular reflex in rabbit and cat and its blockade by picrotoxin. Brain Res. **17**: 524–526.
49. TAN, J., N. M. GERRITS, R. NANHOE, J. I. SIMPSON & J. VOOGD. 1995. Zonal organization of the climbing fiber projection to the flocculus and nodulus of the rabbit. A combined axonal tracing and acetylcholinesterase histochemical study. J. Comp. Neurol. **356**: 23–50.
50. MARINI, G., L. PROVINI & A. ROSINA. 1975. Macular input to the cerebellar nodulus. Brain Res. **99**: 367–371.
51. SIROTA, M. G., B. M. BABAEV, I. N. BELOOZEROVA, A. N. NYROVA, S. B. YAKUSHIN & I. B. KOZLOVSKAYA. 1988. Neuronal activity of nucleus vestibularis during coordinated movement of eyes and head in microgravitation. The Physiologist. **31**(Suppl.1): 8–9.
52. COHEN, B., *et al.* 1992. Vestibuloocular reflex of rhesus monkeys after spaceflight. J. Appl. Physiol. **73**(2): 121S–131S.
53. YAKUSHIN, S. B., M. J. DAI, J.-I. SUZUKI, T. RAPHAN & B. COHEN. 1995. Semicircular canal contribution to the three-dimensional vestibulo-ocular reflex: a model-based approach. J. Neurophysiol. **74**: 2722–2738.
54. DAI, M. J., T. RAPHAN, B. COHEN & C. SCHNABOLK. 1991. Spatial orientation of velocity storage during post-rotatory nystagmus. Soc. Neurosci. Abstr. **17**: 314.
55. REISINE, H. & T. RAPHAN. 1992. Neural basis for eye velocity generation in the vestibular nuclei during off-vertical axis rotation. Exp. Brain Res. **92**: 209–226.
56. BLANKS, R. H. I., I. S. CURTHOYS, M. BENNET & C. H. MARKHAM. 1985. Planar relationships of the semicircular canals in rhesus and squirrel monkeys. Brain Res. **340**: 315–324.
57. RAPHAN, T., V. MATSUO & B. COHEN. 1979. Velocity storage in the vestibulo-ocular reflex arc (VOR). Exp. Brain Res. **35**: 229–248.
58. GOLDBERG, J. M. & C. FERNANDEZ. 1971. Physiology of peripheral neurons innervating semicircular canals of the squirrel monkey. I. Resting discharge and response to angular accelerations. J. Neurophysiol. **34**: 635–660.
59. SCHOR, R. H., A. D. MILLER, & D. L. TOMKO. 1984. Responses to head tilt in cat central vestibular neurons. I. Direction of maximum sensitivity. J. Neurophysiol. **51**: 136–146.
60. BAKER, J., J. GOLDGERG, G. HERMANN & B. PATERSON. 1984. Spatial and temporal response properties of secondary neurons that receive convergent input in vestibular nuclei of alert cats. Brain Res. **294**:138–143.
61. YAKUSHIN, S. B., T. RAPHAN & B. COHEN. 1999. Spatial properties of otolith units recorded in the vestibular nuclei. This issue.
62. SINGLETON, G. T. 1967. Relationships of the cerebellar nodulus to vestibular function: A study of the effects of nodulectomy on habituation. Laryngoscope **77**: 1579–1620.

63. ANGELAKI, D. E. & B. J. M. HESS. 1994. The cerebellar nodulus and ventral uvula control the torsional vestibulo-ocular reflex. J. Neurophysiol. **72**: 1443–1447.
64. ANGELAKI, D. E. &. B. J. M. HESS. 1996. Organizational principles of otolith and semicircular canal-ocular reflexes in rhesus monkeys. Ann. N.Y. Acad. Sci. **781**: 332–347.
65. JANSEN, J. & A. BRODAL, 1954. Aspects of Cerebellar Anatomy: 423. Johan Grundt Tanum Forlag. Oslo, Norway.
66. FENG, A. S. & T. H. J. BULLOCK. 1977. Neuronal mechanisms for object discrimination in the weakly electric fish Eigenmannia virescens. J. Exp. Biol. **66**: 141–158.
67. DAHHAOUI, M., J. LANNOU, T. STELZ, J. CASTON & J. M. GUASTAVINO. 1992. Role of the cerebellum in spatial orientation in the rat. Behav. Neural. Biol. **58**: 180–189.
68. POMPEIANO, O., P. ANDRE & D. MANZONI. 1997. Spatiotemporal response properties of cerebellar Purkinje cells to animal displacement: a population analysis. Neuroscience **81**: 609–626.
69. MANZONI, D., P. ANDRE & O. POMPEIANO. 1995. Responses of Purkinje cells in the cerebellar anterior vermis to off-vertical axis rotation. Pflug. Arch. **431**: 141–154.
70. MANZONI, D., O. POMPEIANO & P. ANDRE. 1998. Neck influences on the spatial properties of vestibulospinal reflexes in decerebrate cats: role of the cerebellar anterior vermis. J. Vestib. Res. **8**: 283–297.
71. ANDRE, P., D. MANZONI & O. POMPEIANO. 1998. Spatiotemporal response properties of cerebellar Purkinje cells to neck displacement. Neuroscience **84**: 1041–1058.
72. MANZONI, D., O. POMPEIANO & P. ANDRE. 1998. Convergence of directional vestibular and neck signals on cerebellar Purkinje cells. Pflueg. Arch. **435**: 617–630.
73. SIMPSON, J. I. 1984. The accessory optic system. Annu. Rev. Neurosci. **7**: 13–41.
74. SANDKUEHLER, J., B. MAISCH & M. ZIMMERMANN. 1987. The use of local anaesthetic microinjections to identify central pathways: a quantitative evaluation of the time course and extent of the neuronal block. Brain Res. **68**: 168–178.
75. STRAUBE, A., R. KURZAN & U. BUETTNER. 1991. Differential effects of bicuculline and muscimol microinjections into the vestibular nuclei on simian eye movements. Exp. Brain Res. **86**: 347–358.
76. COHEN, H., B. COHEN, T. RAPHAN & W. WAESPE. 1992. Habituation and adaptation of the vestibulo-ocular reflex: a model of differential control by the vestibulo-cerebellum. Exp. Brain Res. **90**: 526–538.
77. WAESPE, W. & V. HENN. 1977. Neuronal activity in the vestibular nuclei of the alert monkey during vestibular and optokinetic stimulation. Exp. Brain Res. **27**: 523–538.
78. WAESPE, W. & V. HENN. 1977. Vestibular nuclei activity during optokinetic after-nystagmus (OKAN) in the alert monkey. Exp. Brain Res. **30**: 323–330.
79. YOKOTA, J. I., H. REISINE & B. COHEN. 1992. Nystagmus induced by electrical microstimulation of the vestibular and prepositus hypoglossi nuclei in the monkey. Exp. Brain Res. **92**: 123–138.
80. HOLSTEIN, G. R., G. P. MARTINELLI, J. DEGEN & B. COHEN. 1996. Inhibitory neuronal circuits in the central vestibular system. Ann. N.Y. Acad. Sci. **781**: 443–457.
81. SCHNABOLK, C. & T. RAPHAN. 1992. Modelling 3-D slow phase velocity estimation during off-vertical axis rotation (OVAR). J. Vest. Res. **2**: 1–14.
82. WYLIE, D. R. W. & B. J. FROST. 1999. Complex spike activity of Purkinje cells in ventral uvula and nodulus of pigeons in response to translational optic flow. J. Neurophysiol. **81**: 256–266.

Characteristics of the VOR in Response to Linear Acceleration

GARY D. PAIGE[a] AND SCOTT H. SEIDMAN

Department of Neurobiology and Anatomy and the Center for Visual Science, University of Rochester, Rochester, New York 14642, USA

ABSTRACT: The primate linear VOR (LVOR) includes two forms. First, eye-movement responses to translation [e.g., horizontal responses to interaural (IA) motion] help maintain binocular fixation on targets, and therefore a stable bifoveal image. The translational LVOR is strongly modulated by fixation distance, and operates with high-pass dynamics (>1 Hz). Second, other LVOR responses occur that cannot be compensatory for translation and instead seem compensatory for head tilt. This reflects an otolith response ambiguity—that is, an inability to distinguish head translation from head tilt relative to gravity. Thus, ocular torsion is appropriately compensatory for head roll-tilt, but also occurs during IA translation, since both stimuli entail IA acceleration. Unlike the IA-horizontal response, IA torsion behaves with low-pass dynamics (with respect to "tilt"), and is uninfluenced by fixation distance. Interestingly, roll-tilt, like IA translation, also produces both horizontal (a translational reflex) and torsional (a tilt reflex) responses, further emphasizing the ambiguity problem. Early data from subjects following unilateral labyrinthectomy, which demonstrates a general immediate decline in translational LVOR responses, are also presented, followed by only modest recovery over several months. Interestingly, the usual high-pass dynamics of these reflexes shift to an even higher cutoff. Both eyes respond roughly equally, suggesting that unilateral otolith input generates a binocularly symmetric LVOR.

INTRODUCTION TO THE VESTIBULO-OCULAR REFLEX

The primate vestibulo-ocular reflex (VOR) stabilizes binocular fixation on visual targets during head movements by generating compensatory eye movements in response to head motion, thereby maintaining a stable foveal image. The VOR consists of a set of reflexes that respond to angular and linear head motion in any combination or direction. This is required because natural behavior routinely includes complex motion, even during such simple tasks as walking. The VOR must compensate for all aspects of head movement to accomplish its overall goal of maintaining fixation stability. The VORs are fast and operate synergistically with slower visual following mechanisms to ensure effective target fixation over the broad range of natural head movements. We limit considerations here to primates (including humans) because they have binocular foveate vision and a vergence mechanism to align the eyes precisely on targets in depth; features that are intimately tied to the VOR.

---

[a]Address for communication: Gary D. Paige, Department of Neurobiology and Anatomy, Box 603, University of Rochester, 601 Elmwood Avenue, Rochester, New York 14642. Phone: 716/275-2591; fax: 716/442-8766; e-mail: gary_paige@urmc.rochester.edu

Research over the past decade has guided us toward a relatively simple view of VOR function. Briefly, the VOR can be viewed as a set of basic reflexes whose dynamic properties reflect different labyrinthine inputs (canal-driven AVORs vs. otolith-driven LVORs) and distinct processes within central pathways (translational vs. tilt LVORs), all of which combine linearly during complex motion. Experimental evidence leading to this view is explored below, with particular emphasis on the LVOR.

FUNCTIONAL OVERVIEW OF THE LVOR

Linear acceleration is detected by the two otolith organs on each side, the utriculus, which lies close to the horizontal canal plane, and the sacculus, which is oriented roughly vertically. Each contains an array of hair cells responsive to linear acceleration that approximately covers the 2-D plane of that organ. Taken together, a 3-D representation of linear acceleration arises across a bandwidth of DC to at least 8 Hz.[13] An important problem in the LVOR is that the otoliths respond to both head tilt relative to gravity and to translational motion,[15] a physical ambiguity inherent in all linear accelerometers. Thus, a head roll-tilt toward the left shoulder and a prolonged linear acceleration to the right produce equivalent signals in utricular afferents sensitive to interaural (IA) forces. However, LVOR responses must differ for the two types of head movement if fixation stability is to be maintained. Roll tilt requires ocular torsion, while IA translation requires a horizontal ocular response. If different head movements produce the same otolith signal, do they also produce the same response? In fact, they do, though with different dynamic properties. Torsional responses are greatest during prolonged stimuli and become progressively weaker as stimulus frequency rises,[16,20,29] while horizontal responses are greatest during transient or high-frequency motion (>2 Hz) and fall off with declining frequency.[29,40] Thus, a partial resolution of the otolith ambiguity problem is applied by the CNS, in which low-frequency linear acceleration is interpreted as tilt and high-frequency stimuli as translation.[14,17,29] This requires central processing (filtering) of afferent input within separate low-pass and high-pass reflex pathways in order to account for their fundamental physiologic differences that are not apparent in afferent response properties.

Frequency parsing is parsimonious with natural behavior and perception.[14,29,34] Translations typically occur in the context of high-frequency motion (e.g., locomotion). They are correctly perceived as translations, not tilts, and produce translational-LVOR responses. Unlike translations, rapid tilts are faithfully detected by the canals and do not necessarily require otolith input. In contrast, static and low-frequency linear accelerations arise naturally during prolonged head tilts (e.g., looking down while walking, lying down). They are appropriately perceived as tilts, not translations, and produce tilt-LVOR responses. This is fortunate, because if static head tilt were interpreted as prolonged translation, horizontal nystagmus would occur, as would erroneous perceptions. In short, frequency parsing is a compromise, and perhaps even an adaptation, in the brain's utilization of inherently ambiguous sensory input.

How many specific LVORs exist in response to linear acceleration? This question was addressed by Paige and Tomko[29,30] in squirrel monkeys that were sinusoidally translated (0.5–4.0 Hz) with their heads in different orientations in order to record LVOR responses to interaural (IA), dorsoventral (DV), and naso-occipital (NO) motion. Responses under these conditions proved limited to just five types. These could be parsed into those compensatory for head translation and those compensatory for head tilt, based upon the direction and sign of eye movements with respect to the visual world. From the previous examples, horizontal responses to IA acceleration (IA-horizontal) and vertical responses to DV acceleration (DV-vertical) help maintain binocular fixation stability during IA and DV *translation*, respectively. In contrast, torsional responses during IA acceleration cannot possibly compensate for IA translation, and are instead presumably compensatory for effective roll-tilt (recall the ambiguity problem). Similarly, vertical responses to low-frequency NO acceleration (NO-vertical) primarily reflect a compensatory response to effective pitch-tilt of the head. Finally, high-frequency NO motion produces complex horizontal and vertical responses (see below). We will outline the fundamental characteristics of these different LVOR response types, and then proceed to more recent contributions that further solidify their dynamic properties.

The Translational LVOR

The properties of the translational LVOR depend upon the line of sight relative to linear motion.[29,30] When gaze is orthogonal to motion, as during IA and DV translation, horizontal and vertical responses occur, composing the IA-horizontal and DV-vertical LVORs, respectively. Both display similar properties, including high-pass dynamics. In addition, geometry dictates that these LVORs should be nearly absent when fixating a distant target, but most become progressively larger as the target nears. They indeed behave according to these kinematic requirements, though imperfectly. The primary CNS signal conveying fixation distance is linked to vergence, but is likely derived from a premotor command driving both vergence and VOR modulation,[24,30,42] since (1) changes in VOR responses precede actual changes in vergence, and (2) they occur in darkness during voluntary vergence. Accommodation (lens focus) might provide an alternative or additional signal related to fixation distance,[35] but is more sluggish, less accurate, and in any case operates together with vergence during natural viewing.

When gaze is parallel to the axis of motion (e.g., NO translation), the kinematic requirements of the LVOR are extraordinary.[30] To maintain fixation on targets during forward motion, the eye must track the complex geometry of visual object motion relative to the eye.[46] In a field of many objects, the motion profiles of each collectively form an "optic flow" field, in which the radial motion of objects appears to emanate from a stationary center called the focus of expansion (or motion "null point"). To maintain fixation on a particular element in the field during forward translation, the eye must not rotate when fixating a target along its own NO axis (the reflex null point), but must rotate rightward when gaze is to the right (i.e., when fixating a target to the right), leftward when gaze is left, upward when gaze is up, and downward when gaze is down. Responses must be governed by both fixation distance and angle from the axis of motion, and must be different (even opposing at times) for the two eyes due to their lateral separation in the head. We have demonstrated that the NO-LVOR indeed generates this remarkable behavior, primarily at high frequencies (e.g., 5 Hz). To accomplish such unique kinematic properties, the neural pathways underlying the NO-LVOR must likewise be unique. The fact that forward motion can generate responses in any direction requires that NO-responsive afferents establish connections with every extraocular muscle on each eye. Both excitatory and inhibitory connections must coexist in each muscle's control path, and these must then be dynamically modulated by the current state of the system (e.g., fixation distance and eccentricity). The complexities of this fascinating reflex are obvious, but there is little to add to its fundamental understanding from the recent literature. An exception is the recent finding that the reflex's geometry is under adaptive control, since NO motion combined with viewing through binocular parallel wedge prisms results in predictable and specific shifts in its gaze-dependent properties.[36]

The Tilt-LVOR

The tilt-LVOR[29] includes torsional responses to IA accelerations (IA-torsion) during roll-tilt or IA translation, and vertical responses to NO accelerations (NO-vertical) during pitch-tilt or NO translation. These are often termed ocular counterrolling and counterpitching, respectively. Both reflexes operate with low-pass dynamics. The tilt-LVOR is closely related to the AVOR, since both compensate for head *rotations*. The distinction is that the tilt-LVOR is naturally activated by gravity during head tilt, while the AVOR is not. The complementary dynamics of the two reflexes (high-pass AVOR and low-pass tilt-LVOR) help to provide relatively flat performance across a broad range of frequencies,[3,33,47] a property that we have also confirmed in squirrel monkeys (unpublished observations).

An interesting, if not natural, tilt-related stimulus is constant-velocity off-vertical axis rotation (OVAR),[31] of which "BBQ-spit" rotation[2,14] around an earth-horizontal axis is a simple form. This produces a linear acceleration vector of 1 g rotating opposite the head. A continuous ocular response that opposes head rotation results, called the "bias" component (effectively a form of tilt-LVOR responding to the rate of tilt), along with at least one "modulation" component that oscillates at a frequency also related to the rate of rotation. These modulations can be viewed as responses to accelerations along single head axes corresponding to known tilt and translational LVORs (e.g., IA-horizontal and NO-vertical).[2] Thus, the characterization of primate LVOR classes derived from translational studies can serve as general "building blocks" for composing overall responses to any form of linear acceleration.[29,30] The OVAR bias response belongs on the list of building blocks. It operates with low-pass dynamics, which explains why it was not observed by Paige and Tomko during high-frequency translation. How the rate of head tilt is derived from otolith afferent signals is not trivial, as addressed by others,[3,12] but a common theme is that it requires the sequential activation of 2-D otolith inputs (not just IA acceleration, for example).

RECENT CONTRIBUTIONS

The Translational LVORs

Earlier work on the LVOR demonstrated robust translational and tilt LVORs during linear motion at high frequencies,[29,30,46] but were limited. Important questions remained with regard to the precise response dynamics of the different LVORs, their linearity, and their kinematics and modulating influences (e.g., vergence influence and vision). These concerns were recently explicitly addressed.[44] The IA-horizontal and DV-vertical translational LVORs were reevaluated in squirrel monkeys at 0.5–4.0 Hz, but over a 0.1–0.7-g range of peak amplitude, and across a broad range of vergences. LVOR response sensitivities (in deg/cm) were linearly related to vergence (in m^{-1}, or meter-angles, MA), but this relationship was most robust (though still suboptimal) at 4 Hz, and declined with decreasing frequency to become negligible by 0.5 Hz. A small residual response appeared even when vergence was zero, following similar dynamics. In contrast to sensitivity, response phase was not systematically modulated by vergence. Phase leads were near zero (compensatory) at 4.0 Hz, but increased with declining frequency. Collectively, these findings quantify and confirm the high-pass dynamics of translational LVORs, together with a multiplicative interaction with fixation distance. Finally, the translational LVORs were found to behave linearly over a 7-fold range of stimulus amplitudes.

Tilt-Translation Ambiguity

Recall that IA translation generates small torsional responses simultaneously with horizontal ones. Unlike the IA-horizontal LVOR, IA-torsional responses behaved with low-pass dynamics (relative to effective head roll-tilt). Further, this reflex was not modulated by vergence, despite the simultaneous and strongly modulated horizontal (translational) response. These are characteristics expected of a tilt-LVOR.

Previous studies of the LVOR in response to translation have typically limited stimuli to high frequencies, well within the operating range of the translational LVOR, but far above that of the tilt-LVOR. Perhaps the reflexes would behave differently during low-frequency accelerations (e.g., 0.01–0.1 Hz) that emphasize the useful bandwidth of the tilt-LVOR. We have recently initiated a broad-band assessment of both LVORs, in order to address this important concern.[26] Both tilts and translations were studied over a broad frequency range.

Translational acceleration at low frequency is not readily feasible. However, we have devised a unique technique for accomplishing this task in the form of a rotating dynamically controlled sled, or sled/rotator.[26,38] The squirrel monkey version includes a superstructure with manual axes to place subjects in any orientation in 3-D space, mated to a motorized cantilevered arm, the "chair axis," which controls earth–horizontal-axis rotation—in other words, head tilt. This structure is mounted on a 1.2-m motorized linear sled, which is in turn fixed to a "base axis" that controls earth–vertical-axis rotation. Sinusoidal translations at high frequencies are produced by the sled alone, but since peak acceleration is related to peak excursion, sled motion is limited to ≥ 0.5 Hz to achieve even modest stimulus intensities. To overcome this limitation, the sled/rotator can be exploited to increase the effective sled "length" by rotating the base axis at constant velocity while oscillating the subject linearly back and forth. The rotating sled generates centripetal accelerations proportional to head eccentricity. The sled is then used to simply vary head eccentricity sinusoidally. The linear acceleration generated by the sinusoidal motion is negligible at low frequency, but combines with the changing centripetal acceleration to yield large effective amplitudes that would otherwise require enormous sled lengths. One caveat is that tangential accelerations are also induced by sled motion, but this is only problematic above 0.1 Hz, where these tangential effects exceed a third of the primary (centripetal) stimulus. Trials are performed after an initial angular acceleration (start-up) in darkness to bring the sled to typically 180 deg/s constant velocity with the subject on center, and after all AVOR activity is extinguished. Sled oscillation at 0.5-m peak excursion then yields a peak acceleration of 0.5 g. The effective peak excursion now becomes 12.4 m at 0.1 Hz and 1.24 km at 0.01 Hz!

One key concern can now be addressed directly—if otolith afferents cannot in fact distinguish tilt from translation, do the LVORs that are driven by these afferents respond equally to both translation and tilt? We had known that the translational and tilt LVORs (IA-horizontal and IA-torsional) simultaneously respond to IA translation. Do both horizontal and torsional responses also occur during head roll-tilt? Recent data suggest that the answer is yes. This is readily appreciated during low-frequency (0.025-Hz) translation and tilt trials exemplified in FIGURE 1. We display a low-frequency example because high-frequency head roll strongly activates the roll AVOR, which primarily produces torsional responses, but can also generate horizontal eye movements when gaze is displaced upward or downward relative to the head's axis of roll.[37] This can obscure concurrent LVORs. Fortunately, the roll AVOR falls off dramatically below 0.1 Hz, allowing access to the LVORs. The important surprise from FIGURE 1 is the robust horizontal response to low-frequency head roll that closely resembles the response to IA translation. Both presumably reflect the IA-horizontal (translational) LVOR operating far below its natural range. Note that what appears to be a large horizontal response actually calculates as a tiny LVOR sensitivity (in deg/cm) due to the enormous effective translational excursions entailed at low frequencies. The torsional responses seen in FIGURE 1 presumably reflect the tilt-LVOR, together with a small AVOR during the tilt trial.

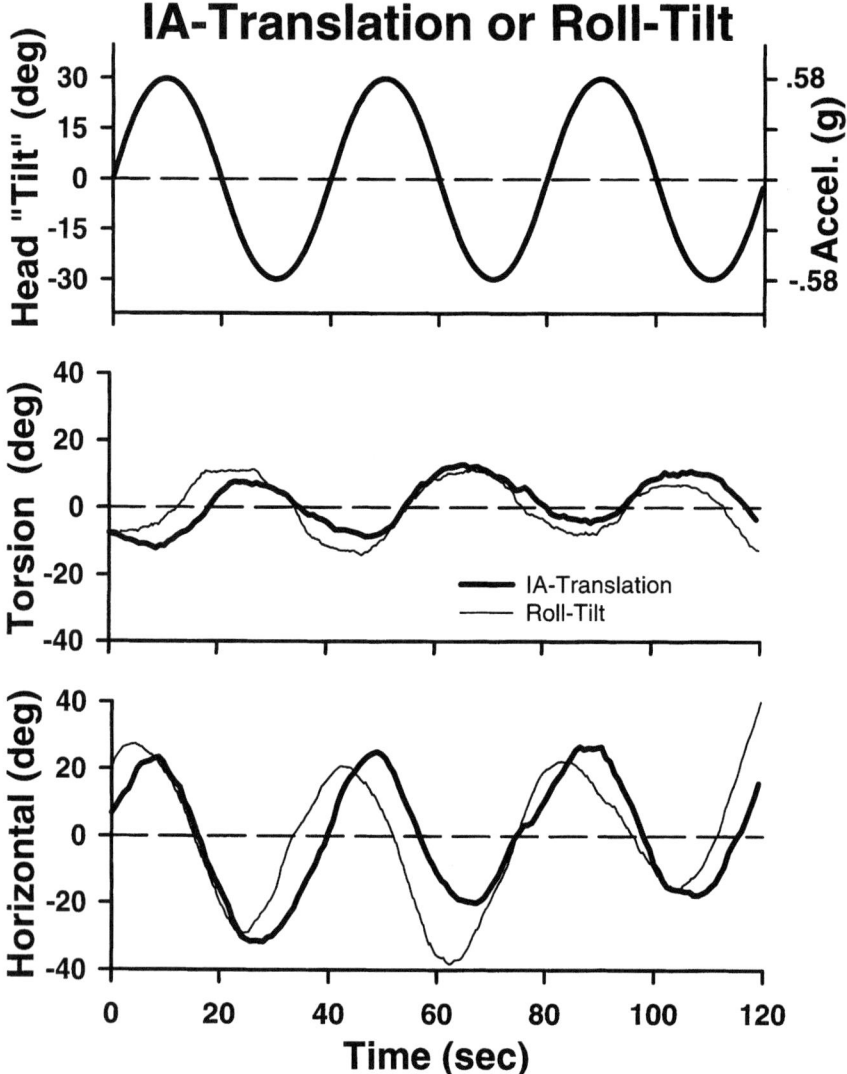

FIGURE 1. Horizontal (**bottom records**) and torsional (**middle records**) eye-movement responses to IA acceleration (**top records**) produced during IA-translational motion and by head roll-tilt (*see key*). Eye-movement traces have been desaccaded to create cumulative eye-position records.

Another key finding that we have repeatedly observed is that the IA-horizontal (translational) and IA-torsional (tilt) LVORs appear whether the head is oriented either upright or nose-up. This is particularly important in the case of the tilt-LVOR, because the determination of true head tilt requires a measurement of the resultant of translational and gravitational force [gravitoinertial force, or (GIF)]. Head orientation should be crucial in governing a true tilt response, because GIF "swings" in a plane determined by the head tilt relative to *gravity*. For example, IA acceleration with the head upright simulates head roll-tilt (in relation to GIF) and should produce torsion, but with the head nose-up the same stimulus simulates head yaw (in relation to GIF) and should produce a horizontal (not torsional) response. In fact, torsional responses persist. Thus, true tilt is not utilized in the LVOR, and instead this simplified *quasi-tilt* LVOR seems driven by IA acceleration transduced by primarily utricular afferents.[29,44] Early results suggest that the same conclusion holds for the NO-vertical "tilt" LVOR. If influences of orientation and GIF exist at all, they remain small[20] and incapable of reorienting the LVOR, and thus are of little functional importance.

To best illustrate the frequency-dependent parsing of otolith-driven reflexes, we combined results from low-frequency (rotating sled) and high-frequency IA translation trials in order to construct a broad-band depiction of LVOR response amplitudes as a function of stimulus frequency. FIGURE 2 clearly displays the high-pass translational (horizontal) and low-pass-tilt (torsional) forms of LVOR, simultaneously recorded in each animal and then averaged across three subjects.

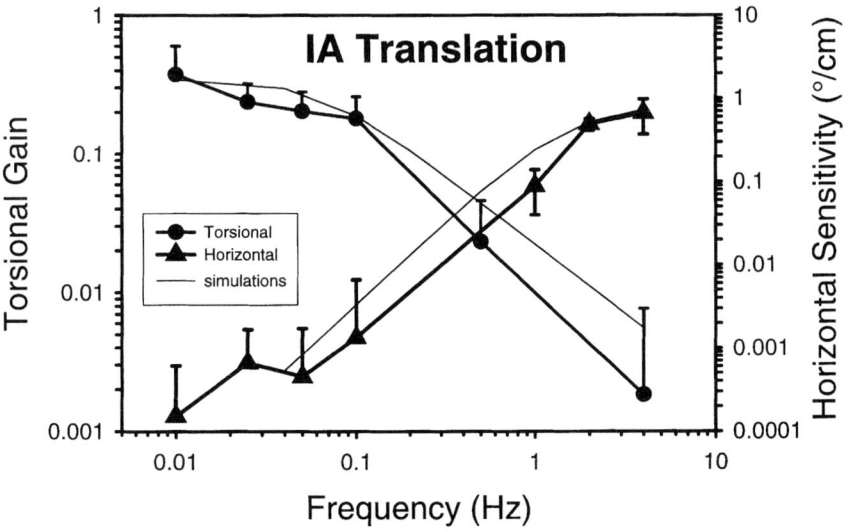

FIGURE 2. Response gain of the IA-torsional tilt-LVOR and response sensitivity of the IA-horizontal translational LVOR plotted together across a broad frequency bandwidth. The data represent the average of three monkeys. The two response types were recorded simultaneously during IA translation trials. Model simulations (see FIG. 3) appear as fine traces.

Visual–Vestibular Interactions (VVI) in the Translational LVOR

Visual–vestibular interaction (VVI) has traditionally focused on relations between vision and the AVOR. These interactions are synergistic and complementary during natural behavior.[4,11,25] The high-pass AVOR compensates for higher-frequency head perturbations, while low-pass visual systems subserve more sluggish movements. When placed in conflict (e.g., head-fixed targets), vision suppresses the AVOR, but only up to around 1 Hz, above which the AVOR is not easily altered. A fascinating form of VVI is the well-known AVOR modulation by imagined target motion in darkness.[5] That is, AVOR gain is augmented when subjects imagine earth-fixed targets and diminished for head-fixed ones. The dynamics of this effect resemble that of smooth pursuit.

We have recently characterized the *human* LVOR and its interactions with vision and fixation distance during IA oscillation at 0.5–4 Hz.[27] Horizontal responses were robust, binocular, and modulated by vergence, though the relationship between LVOR sensitivity and vergence was somewhat less than in monkeys. The same was observed by others.[8] The human LVOR displayed similar high-pass dynamics as in monkeys. The presence of earth-fixed or head-fixed targets activated direct visual influences that combined with the LVOR to improve the overall goal of target fixation at lower frequencies. When targets were extinguished, but subjects continued to imagine tracking them, this context enhanced (for earth-fixed "targets") or suppressed (for head-fixed "targets") the LVOR. However, the effect declined systematically as frequency rose, as for real targets, and at 4 Hz, little influence remained. Thus, influences of both real and imagined visual targets on the LVOR followed pursuit-like low-pass dynamics, as for the AVOR. The influence of imagined targets was limited, and accounted for roughly a third of the influence of real targets on the VORs at 0.5 Hz.

Canal–Otolith Interactions

How do the different VORs interact, and is the interaction simply a linear combination of independent subreflexes? We directly addressed this question quantitatively in squirrel monkeys. The AVOR was studied in isolation and in combination with the translational LVOR during high-frequency earth–vertical-axis rotations with the head positioned centrally or displaced radially from the axis of motion [dynamic "eccentric rotation" (ER)]. ER activates both the AVOR and LVOR simultaneously. Geometry dictates that when the head is upright and facing nose-out, the IA-horizontal LVOR (driven by tangential acceleration) should *add* to the AVOR, but when facing nose-in, the LVOR is reversed and should *subtract* from the AVOR. We indeed found this to be the case, but with dynamic properties that reflected the underlying LVOR and AVOR components.[43] All of these properties applied equally well to combinations of the vertical AVOR and DV-vertical LVOR.

Responses during ER generally showed high-pass properties directly attributable to the LVOR components as measured during translation alone, in concert with the AVOR as likewise recorded alone. The influence of vergence also appeared in ER responses.[45] Thus, at 4 Hz with monkeys facing nose-out, gain increased dramatically with increasing vergence. With the animals turned nose-in, LVOR–AVOR interactions became more complicated. When fixation distance was far, the LVOR was small and the AVOR dominated the response (phase ~180 deg). However, gain declined as fixation distance decreased, and when closer than the rotation axis, the response reversed (phase ~0 deg), presumably because the LVOR became dominant. In summary, the AVOR and LVOR interact linearly (as vector additions) during ER, as a function of both frequency and vergence. Other studies[10,41,49] support this conclusion. Recent experiments on humans in this laboratory (unpublished observations) demonstrate the same results, including the response modulation and reversal as fixation shifts from far to near during 4-Hz oscillation. This response reversal does not occur at 1 or 0.5 Hz, presumably because the LVOR is less robust at these frequencies and cannot overcome the AVOR. Clinical relevance is strong, since an LVOR-dependent response inversion is an unequivocal observation, and loss in the LVOR (otoliths) relative to the AVOR (canals) should eliminate the phenomenon.

Compensation in the LVOR for Endorgan Lesions

Unilateral labyrinthectomy renders the vestibular system unbalanced and asymmetric. Most previous studies on compensation have focused on the AVOR and tilt-LVOR. We have recently begun to study compensation in the translational LVOR in monkeys.[28] Early observations provide two fundamental conclusions relevant to the structure and function of LVOR pathways: (1) the otolith organs on one side drive both eyes conjugately and roughly symmetrically, just as in the canal-driven AVOR; and (2) after bilateral labyrinthectomy, all VORs are abolished, implying that nonvestibular inputs do not influence eye movements during angular or linear motion. After labyrinthectomy on one side, we found that translational-LVOR sensitivities declined, both across frequencies at dark vergence, and as a function of vergence at 4 Hz. Phase lead rose across the bandwidth as well, as if the high-pass characteristics of the LVOR had shifted to a higher cutoff. Interestingly, while AVOR gain also declined, it recovered over four months, while the LVOR remained nearly the same. We suspect that the LVOR cannot be easily recalibrated because the low-pass limits of vision overlap poorly with the high-pass range of the LVOR, and it is the interaction between vision and the VOR that is required for adaptive plasticity to occur.[22,23] Even the AVOR is difficult to adapt at high frequencies.[6]

MODEL OF THE LVOR AND ITS INTERACTIONS WITH THE AVOR

The key structural and functional properties of the LVORs and AVORs are expressed conveniently in the model of FIGURE 3, here limited to the horizontal AVOR and IA-LVORs for simplicity. IA head acceleration is first transduced by the utriculus to yield an afferent response that is split into translational and tilt-LVOR pathways. The basic *tilt-position pathway* simply requires a low-pass filter and scaling (G_{tilt}) to convert its input into effective head roll-tilt to drive torsional eye position. Note that IA acceleration, not GIF, drives the reflex. Therefore, actual head orientation is irrelevant, as observed experimentally. A one-pole low-pass filter suffices to replicate our squirrel monkey data (see fine line associated with torsion in FIG. 2), with a time constant of 2.5 s (0.06-Hz corner) and gain of 0.5. A novel *tilt-rate pathway* is included, utilizing velocity storage (quite weak in roll), to account for the bias component of OVAR responses.[3,31] This pathway likely also influences the tilt-LVOR, as we have often observed phase leads at 0.025 Hz and below. Inputs from the vertical canals during head roll are not drawn in FIGURE 3, but they surely exist, and are readily included as inputs to the summing junction that combines tilt rate and roll-velocity storage, in parallel construction with the horizontal-canal pathway. Note that there is no need for vergence modulation anywhere in the roll-torsion pathway.

The *translational-LVOR pathway* is quite different, and must account for the reflex's high-pass dynamics, amplitude modulation proportional to reciprocal fixation distance, and residual response when fixation distance is infinitely far. Initial processing includes an integration of head acceleration and high-pass filtering. Note that the one-pole low-pass element ("leaky integrator") is equivalent to the combination of a pure integrator and a one-pole high-pass filter. While the actual mechanism is debatable,[1,32] the outcome is

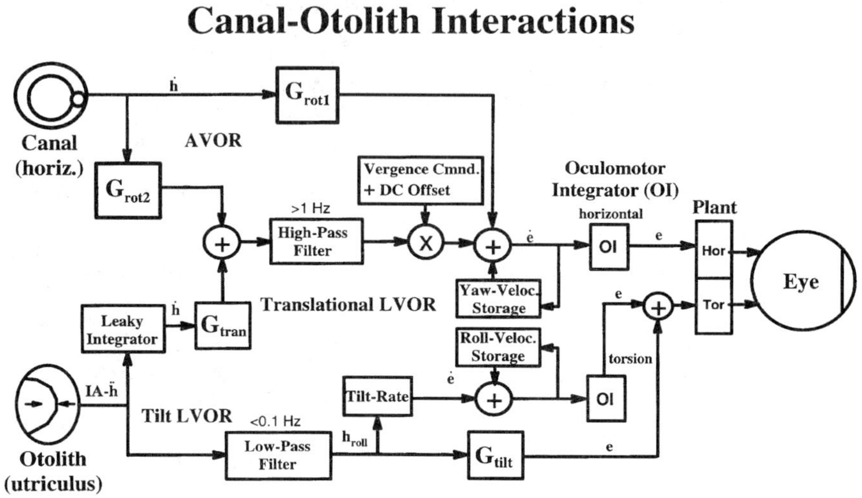

Canal-Otolith Interactions

FIGURE 3. Model of the AVOR and LVOR (both translational and tilt forms), restricted to reflect head motion in its horizontal plane and limited to head yaw and/or IA translation.

ultimately a high-pass signal proportional to head velocity. One key requirement is that the overall filtering process must be second order to prevent persistent head acceleration (head tilt) from producing continuous nystagmus.[44] Experimental support is provided by two of three animals in the rotating-sled study and by other reports;[2,7] phase lead shows a roughly 180-deg shift (not 90 deg) with decreasing frequency. The filtered signal is next multiplied by a vergence command that includes a positive DC offset. A vergence command of zero then leaves the reflex with a small response, as observed. From this model, the data are well simulated by time constants of 0.25 s and 0.05 s for the leaky integrator and high-pass filter, respectively, and a gain (G_{tran}) of 0.4 (see fine line associated with the IA-horizontal–LVOR in FIG. 2).

The AVOR is simplified in the model of FIGURE 3, missing interactions with vision and some complexities of velocity storage that are not of fundamental concern here. A key feature is the vergence-dependent pathway that our data (monkey and human) suggest must operate with high-pass dynamics, just as in the LVOR. This may explain why we and others have observed meager, if not detrimental, effects of vergence on AVOR gain at modest frequencies.[10,39] What remains unclear is whether the filter and multiplier elements in the AVOR are shared with the LVOR (as in the model) or remain separate (an alternative structure).

Note that otolith ambiguity is inherent in the model, since IA-acceleration drives both reflex pathways simultaneously, regardless of whether produced by roll-tilt or IA translation. An interesting query arises—What translational response is expected during a rapid roll-tilt? Simulation suggests that a 20-deg step in tilt yields a brief (1–2-s) horizontal response of only ~3 deg at dark vergence. This could easily be lost within the large concurrent AVOR response, and may account for why this response attribute remains elusive.[21]

Although the model's single labyrinthine input and single ocular output are oversimplifications, the extension to a bilateral input and output[30] is straightforward. Similarly, while the model shown is limited to horizontal inputs (e.g., IA or yaw head motion), generalization is simple. DV acceleration would stimulate DV-sensitive saccular afferents whose input would be processed by a DV-vertical translational-LVOR pathway, but without a tilt pathway. NO motion would activate both a tilt pathway that drives vertical eye movement and a translation pathway that includes modulation by gaze eccentricity as well as vergence, modeled as a series multiplication.[30]

Is there cellular evidence relevant to the model? The modulation of vestibular signals by vergence and gaze have now been recorded in identified second-order neurons in the vestibular nuclei, along with the exciting finding that they also carry a signal related to the IA translational LVOR during ER.[9,18,19,48] This fits nicely within our model (FIG. 3), as implied by the summing junction just before the oculomotor integrator.

REFERENCES

1. ANGELAKI, D. E. & B. J. HESS. 1996. Organizational principles of otolith- and semicircular canal-ocular reflexes in rhesus monkeys. Ann. N.Y. Acad. Sci. **781**: 332–347.
2. ANGELAKI, D. E. & B. J. M. HESS. 1996. Three-dimensional organization of otolith-ocular reflexes in rhesus monkeys. 1. Linear acceleration responses during off-vertical axis rotation. J. Neurophysiol. **75**: 2405–2424.
3. ANGELAKI, D. E. & B. J. M. HESS. 1996. Three-dimensional organization of otolith-ocular reflexes in rhesus monkeys. 2. Inertial detection of angular velocity. J. Neurophysiol. **75**: 2425–2440.
4. BARNES, G. R. 1993. Visual-vestibular interaction in the control of head and eye movement: the role of visual feedback and predictive mechanisms. Progr. Neurobiol. **41**: 435–472.
5. BARR, C. C., L. W. SCHULTHEIS & D. A. ROBINSON. 1976. Voluntary, non-visual control of the human vestibulo-ocular reflex. Acta Otolaryngol. (Stockh.) **81**: 365–375.
6. BELLO, S., G. D. PAIGE & S. M. HIGHSTEIN. 1991. The squirrel monkey vestibulo-ocular reflex and adaptive plasticity in yaw, pitch, and roll. Exp. Brain Res. **87**: 57–66.
7. BOREL, L. & M. LACOUR. 1992. Functional coupling of the stabilizing eye and head reflexes during horizontal and vertical linear motion in the cat. Exp. Brain Res. **91**: 191–206.
8. BUSETTINI, C., F. A. MILES, U. SCHWARZ & J. R. CARL. 1994. Human ocular responses to translation of the observer and of the scene: Dependence on viewing distance. Exp. Brain Res. **100**: 484–494.
9. CHEN-HUANG, C. & R. A. MCCREA. 1998. Viewing distance related sensory processing in the ascending tract of deiters vestibulo-ocular reflex pathway. J. Vest. Res. **8**: 175–184.
10. CRANE, B. T., E. S. VIIRRE & J. L. DEMER. 1997. The human horizontal vestibulo-ocular reflex during combined linear and angular acceleration. Exp. Brain Res. **114**: 304–320.
11. DEMER, J. L. 1992. Mechanisms of human vertical visual-vestibular interaction. J. Neurophysiol. **68**: 2128–2146.
12. FANELLI, R., T. RAPHAN & C. SCHNABOLK. 1990. Neural network modelling of eye compensation during off-vertical-axis rotation. Neural Networks **3**: 265–276.
13. FERNÁNDEZ, C. & J. M. GOLDBERG. 1976. Physiology of peripheral neurons innervating otolith organs of the squirrel monkey. III. Response dynamics. J. Neurophysiol. **39**: 996–1008.
14. GUEDRY, F. E. 1974. Psychophysics of vestibular sensation. *In* Handbook of Sensory Physiology. Vol. VI/2: Vestibular System, H. H. Kornhuber, Ed.: 3-154. Springer-Verlag. Berlin.
15. HOLLY, J. E. & G. MCCOLLUM. 1996. The shape of self-motion perception I. Equivalence of classification for sustained motions. Neuroscience **70**: 461–486.
16. LICHTENBERG, B. K., L. R. YOUNG & A. P. ARROTT. 1982. Human ocular counterrolling induced by varying linear accelerations. Exp. Brain Res. **48**: 127–136.
17. MAYNE, R. 1974. A systems concept of the vestibular organs. *In* Handbook of Sensory Physiology. Vol.VI/2: Vestibular System, H.H. Kornhuber, Ed.: 493-580. Springer-Verlag. Berlin.
18. MCCONVILLE, K. M. V., R. D. TOMLINSON & E. Q. NA. 1996. Behavior of eye-movement-related cells in the vestibular nuclei during combined rotational and translational stimuli. J. Neurophysiol. **76**: 3136–3148.

19. MCCREA, R. A., C. CHEN-HUANG, T. BELTON & G. T. GDOWSKI. 1996. Behavior contingent processing of vestibular sensory signals in the vestibular nuclei. Ann. N.Y. Acad. Sci. **781**: 292–303.
20. MERFELD, D. M., W. TEIWES, A. H. CLARKE, H. SCHERER & L. R. YOUNG. 1996. The dynamic contributions of the otolith organs to human ocular torsion. Exp. Brain Res. **110**: 315–321.
21. MERFELD, D. M. & L. R. YOUNG. 1995. The vestibulo-ocular reflex of the squirrel monkey during eccentric rotation and roll tilt. Exp. Brain Res. **106**: 111-122.
22. MILES, F. A. & B. B. EIGHMY. 1980. Long-term adaptive changes in primate vestibuloocular reflex. I. Behavioral observations. J. Neurophysiol. **43**: 1406–1425.
23. PAIGE, G. D. 1983. Vestibuloocular reflex and its interactions with visual following mechanisms in the squirrel monkey. II. Response characteristics and plasticity following unilateral inactivation of horizontal canal. J. Neurophysiol. **49**: 152–168.
24. PAIGE, G. D. 1991. Linear vestibulo-ocular reflex (LVOR) and modulation by vergence. Acta Otolaryngol. **481**(Suppl): 282–286.
25. PAIGE, G. D. 1994. Senescence of human visual-vestibular interactions: smooth pursuit, optokinetic, and vestibular control of eye movements with aging. Exp. Brain Res. **98**: 355–372.
26. PAIGE, G. D., P. BOULOS & S. H. SEIDMAN. 1995. Eye movement responses to low frequency tilt and translation in the squirrel monkey. Soc. Neurosci. Abstr. **21**: 138.
27. PAIGE, G. D., L. TELFORD, S.H. SEIDMAN & G.R. BARNES. 1998. Human vestibuloocular reflex and its interactions with vision and fixation distance during linear and angular head movement. J. Neurophysiol. **80**: 2391–2404.
28. PAIGE, G. D., L. TELFORD, S. H. SEIDMAN & P. BOULOS. 1996. The linear vestibulo-ocular reflex (LVOR) following labyrinthectomy. Neurosci. Abst. **22**: 661.
29. PAIGE, G. D. & D. L. TOMKO. 1991. Eye movement responses to linear head motion in the squirrel monkey. I. Basic characteristics. J. Neurophysiol. **65**: 1170–1182.
30. PAIGE, G. D. & D. L. TOMKO. 1991. Eye movement responses to linear head motion in the squirrel monkey. II. Visual-vestibular interactions and kinematic considerations. J. Neurophysiol. **65**: 1183–1196.
31. RAPHAN, T. & B. COHEN. 1985. Velocity storage and the ocular response to multidimensional vestibular stimuli. *In* Reviews of Oculomotor Research. Vol. I. Adaptive Mechanisms in Gaze Control, A. Berthoz and G. Melvill Jones, Ed.: 123-143. Elsevier. Amsterdam.
32. RAPHAN, T., S. WEARNE & B. COHEN. 1996. Modeling the organization of the linear and angular vestibulo-ocular reflexes. Ann. N.Y. Acad. Sci. **781**: 348–363.
33. RUDE, S. A. & J. F. BAKER. 1988. Dynamic otolith stimulation improves the low frequency horizontal vestibulo-ocular reflex. Exp. Brain Res. **73**: 357–363.
34. SCHÖNE, H. & H. G. MORTAG. 1968. Variation of the subjective vertical on the parallel swing at different body positions. Psychol. Forsch. **32**: 124–134.
35. SCHWARZ, U. & F. A. MILES. 1991. Ocular responses to translation and their dependence on viewing distance. I. Motion of the observer. J. Neurophysiol. **66**: 851–863.
36. SEIDMAN, S. H., G. D. PAIGE & D. L. TOMKO. 1999. Adaptive plasticity in the naso-occipital linear vestibuloocular reflex. Exp. Brain Res. In press.
37. SEIDMAN, S. H., L. TELFORD & G. D. PAIGE. 1995. Vertical, horizontal, and torsional eye movement responses to head roll in the squirrel monkey. Exp. Brain Res. **104**: 218–226.
38. SEIDMAN, S. H., L. TELFORD & G. D. PAIGE. 1998. Tilt perception during dynamic linear acceleration. Exp. Brain Res. **119**: 307–314.
39. SHELHAMER, M., D. M. MERFELD & J. C. MENDOZA. 1995. Effect of vergence on the gain of the linear vestibulo-ocular reflex. Acta Otolaryngol. **520**(Suppl): 72–76.
40. SKIPPER, J. J. & G. R. BARNES. 1989. Eye movements induced by linear acceleration are modified by visualisation of imaginatry targets. Acta Otolaryngol. **468**(Suppl): 289–293.
41. SNYDER, L. H. & W. M. KING. 1992. The effect of viewing distance and location of the axis of head rotation on the monkey's vestibulo-ocular reflex: I. Eye movement responses. J. Neurophysiol. **67**: 861–874.
42. SNYDER, L. H., D. M. LAWRENCE & W. M. KING. 1992. Changes in vestibulo-ocular reflex (VOR) response anticipate changes in vergence angle. Vision Res. **32**: 569–575.
43. TELFORD, L., S. H. SEIDMAN & G. D. PAIGE. 1996. Canal-otolith interactions driving vertical and horizontal eye movements in the squirrel monkey. Exp. Brain Res. **109**: 407–418.

44. TELFORD, L., S. H. SEIDMAN & G. D. PAIGE. 1997. Dynamics of squirrel monkey linear vestibuloocular reflex and interactions with fixations distance. J. Neurophysiol. **78**: 1775–1790.
45. TELFORD, L., S. H. SEIDMAN & G. D. PAIGE. 1998. Canal-otolith interactions in the squirrel monkey vestibulo-ocular reflex and the influence of fixation distance. Exp. Brain Res. **118**: 115–125.
46. TOMKO, D. L. & G. D. PAIGE. 1992. Linear vestibuloocular reflex during motion along axes between nasooccipital and interaural. Ann. N.Y. Acad. Sci. **656**: 233–241.
47. TOMKO, D. L., C. WALL III, F. R. ROBINSON & J. P. STAAB. 1988. Influence of gravity on cat vertical vestibulo-ocular reflex. Exp. Brain Res. **69**: 307–314.
48. TOMLINSON, R. D., K. M. V. MCCONVILLE & E.-Q. NA. 1996. Behavior of cells without eye movement sensitivity in the vestibular nuclei during combined rotational and translational stimuli. J. Vest. Res. **6**: 145–158.
49. VIIRRE, E., D. TWEED, K. MILNER & T. VILIS. 1986. A reexamination of the gain of the vestibuloocular reflex. J. Neurophysiol. **56**: 439–450.

Functional Organization of Primate Translational Vestibulo-Ocular Reflexes and Effects of Unilateral Labyrinthectomy

DORA E. ANGELAKI,[a] M. QUINN McHENRY, SHAWN D. NEWLANDS, AND J. DAVID DICKMAN

Department of Surgery (Otolaryngology) and Anatomy, University of Mississippi Medical Center, 2500 North State Street, Jackson, Mississippi 39216-4505, USA

ABSTRACT: Translational vestibulo-ocular reflexes (trVORs) are characterized by distinct spatio–temporal properties and sensitivities that are proportional to the inverse of viewing distance. Anodal (inhibitory) labyrinthine stimulation (100 µA, <2 s) during motion decreased the high-pass filtered dynamics, as well as horizontal trVOR sensitivity and its dependence on viewing distance. Cathodal (excitatory) currents had opposite effects. Translational VORs were also affected after unilateral labyrinthectomy. Animals lost their ability to modulate trVOR sensitivity as a function of viewing distance acutely after the lesion. These deficits partially recovered over time, albeit a significant reduction in trVOR sensitivity as a function of viewing distance remained in compensated animals. During fore–aft motion, the effects of unilateral labyrinthectomy were more dramatic. Both acute and compensated animals permanently lost their ability to modulate fore–aft trVOR responses as a function of target eccentricity. These results suggest that (1) the dynamics and viewing distance-dependent properties of the trVORs are very sensitive to changes in the resting firing rate of vestibular afferents and, consequently, vestibular nuclei neurons; (2) the most irregularly firing primary otolith afferents that are most sensitive to labyrinthine electrical stimulation might contribute to reflex dynamics and sensitivity; (3) inputs from both labyrinths are necessary for the generation of the translational VORs.

INTRODUCTION

Primary otolith afferent signals transduced in the central nervous system differ significantly in spatio–temporal organization from those of semicircular-canal-afferent information. Primary otolith afferents, for example, code linear acceleration rather than velocity, as is the case for primary semicircular-canal afferents. Furthermore, primary otolith afferents innervating a single macula differ in their spatial organization, allowing for possible spatio–temporal central processing.[1] In addition, the dynamics and spatial organization of the translational vestibular-ocular reflexes (trVORs) are remarkably different from those of the rotational VORs (rotVORs). TrVOR responses depend strongly on viewing distance and target eccentricity, particularly during motion parallel to gaze direction. Nevertheless, it has

---

[a]To whom correspondence may be addressed. Phone: 601/984-5090; fax: 601/984-5107.

been commonly assumed that trVORs are generated through signal processing that is generally similar to that of the rotVORs. As shown in FIGURE 1 (top), otolith afferent signals coding for linear acceleration ($\ddot{d}$) are centrally integrated to yield a signal proportional to linear velocity ($\dot{d}$). The velocity-to-position integrator subsequently performs the second integration of velocity into position signals. To account for the high-pass filtered properties of the reflex, a 'high pass filter' is often added to this signal processing.[2]

As stated previously,[3] the proposal of a double brainstem integration is problematic given the strong high-pass filter properties of the reflex. The proposal of a high-pass filter cascaded with the double integration does not eliminate the problem. To make matters worse, the traditional concept of trVOR organization totally ignores the dynamics of the eye-plant. There is ample experimental and theoretical evidence to suggest that the eye plant dynamics comprise a second- or even third-order system with low-pass filtered properties.[4–6] For a high-frequency oculomotor response like the trVORs, the low-pass filtered dynamics of the eye plant cannot be neglected. For primates in particular, the dynamics of the plant become important at frequencies higher than ~0.5 Hz. One possibility that is inherent to the traditional concept of a canal-like organization for the trVORs is that the three poles of the primate eye plant are centrally compensated through "lead" elements that consist of parallel pathways to the neural integrator.[7,8] If this were the case, the trVOR organization would consist of a series of lead–lag elements that would be redundant and largely cancel each other. Stated differently, the traditional concept that relies on a central integrator, would require extensive high-pass filtering not only to generate the high-pass filtered properties of the reflex, but also to completely compensate for the eye plant. Alternatively, it is possible that the eye plant is used constructively by the reflex to perform the "missing" integration (FIG. 1, bottom). This simple and refreshing idea has been recently formulated into a model that could account for velocity-like properties in central vestibular neurons by Green and Galiana[9]. Assuming that this simple alternative is true, trVOR and rotVOR processing would partly be segregated within the brainstem.

To further address a potentially different functional and computational organization of the trVORs and rotVORs, two different aspects of the reflex were further examined: the effects of bilateral labyrinthine electrical stimulation, and the acute and chronic properties of the reflex after unilateral labyrinthectomy.

Traditional concept

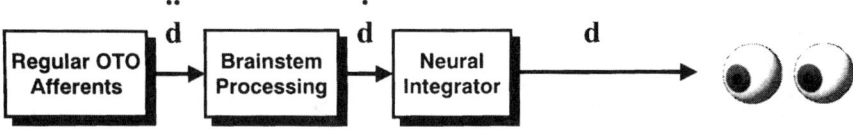

Green and Galiana (1998)

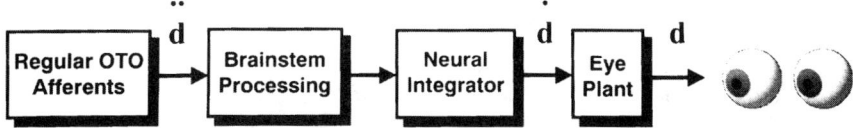

FIGURE 1. Conceptual framework for the basic processing to generate the trVORs (vergence and eye-position-dependent effects have not been illustrated).

METHODS

Five rhesus monkeys (3–4 kg) were used in the present studies. Each animal was chronically implanted for head restraint and dual eye coils in both eyes.[10] All surgical procedures were performed under sterile conditions in accordance with the NIH guidelines. After animals were satisfactorily trained to fixate and follow far and near targets for juice reward, three animals were implanted with bilateral labyrinthine electrodes.[11,12] In the remaining two animals, all semicircular canals were first inactivated as part of a different set of studies.[13] Examination of the trVORs before and after canal plugging demonstrated that the vergence–distance horizontal responses were unaffected by the plugging (e.g., FIG. 4). Approximately 3–4 months after canal plugging, animals were subjected to unilateral (left in one and right in the other) labyrinthectomy. Eye-movement responses during translation were collected one week to three months after the lesion.

Translational motion was delivered using a 2-m linear sled (Acutronics Inc). Binocular 3-D eye movements were recorded using the magnetic search coil technique (CNC Engineering). Three-dimensional eye positions were expressed as rotation vectors using straight ahead as the reference position. Torsional, vertical, and horizontal eye position and velocity were the components of the eye-position and eye-velocity vectors along the naso–occipital, interaural, and vertical head axes, respectively. Positive directions were clockwise (as viewed from the animal, i.e., rotation of the upper pole of the eye toward the right ear), downward and leftward for the torsional, vertical, and horizontal components, respectively.

Animals were sinusoidally oscillated either along their interaural axis (lateral motion) or along their naso–occipital axis (fore–aft motion) at frequencies ranging between 0.5 and 30 Hz. Sinusoidal motion took place either in complete darkness or while fixating a centered target at different distances. For behaviorally controlled experiments, each trial was initiated under computer control when the animal satisfactorily fixated the target light in a dimly illuminated environment for a random period of approximately 300–1000 ms. During motion, the target remained illuminated but the background lights were turned off. Since saccade-free runs were desired, eye velocity and eye acceleration were computed on-line such that a trial was aborted if eye acceleration exceeded a certain value (usually set to 800–1000 deg/s^2) before the end of the stimulus. Animals were given a fluid reward (contingent upon fixation within the prestimulus position window and upon compliance with the eye-acceleration threshold for both eyes) only after the sled had come to a stop.

Data were low-pass filtered (200 Hz, 6-pole Bessel), digitized at a rate of 833.33 Hz (Cambridge Electronics Design, model 1401, 16-bit). Sensitivity and phase were determined by fitting a sine function (and a DC offset) to both response and stimulus (output of the 3-D linear accelerometer) using a nonlinear least-squares algorithm based on the Levenberg–Marquardt method. Sensitivity was expressed as the ratio of response (eye velocity) peak amplitude to peak linear velocity. Phase was expressed as the difference (in degrees) between peak eye velocity and peak linear velocity. Because all data were devoid of fast-phase eye movements, the labyrinthectomy responses were not contaminated by the problem of gaze-holding deficits usually associated with unilateral labyrinthectomy.[14–16]

RESULTS

Functional Ablation of the Most Irregular Afferents

Bilateral anodal labyrinthine electrical stimulation, which has been shown to selectively silence or decrease spontaneous activity of the most irregularly firing vestibular afferents,[7,17] when delivered during translation altered the sensitivity, phase, and frequency dependence of the trVORs. As shown in FIGURE 2, constant anodal stimulation during motion decreased horizontal eye velocity. In contrast, cathodal labyrinthine stimulation increased the magnitude of

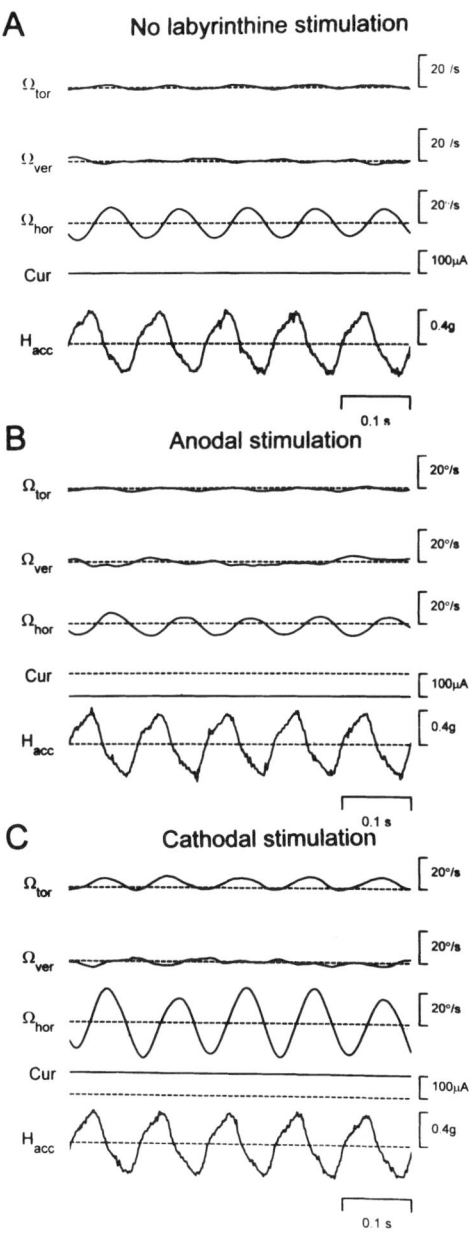

FIGURE 2. Torsional, vertical, and horizontal components of eye velocity (Ω_{tor}, Ω_{ver}, Ω_{hor}) during lateral translation at 10 Hz in the absence (**A**) and in the presence of anodal (inhibitory) (**B**) and cathodal (excitatory) (**C**) electrical labyrinthine stimulation (100 μA). For all motion stimuli, the animal fixated on a target located approximately in-between the two eyes at a distance of 10 cm. *Dotted lines* illustrate zero eye velocity. Cur: constant labyrinthine stimulation; H_{acc}: the output of a linear accelerometer mounted on the animal's head.

the eye movement generated during lateral motion at frequencies >2 Hz. Examination of labyrinthine stimulation across a wide range of frequencies and different target distances revealed the following: (a) the decrease (increase) of trVOR sensitivity in the presence of anodal (cathodal) labyrinthine electrical stimulation was viewing- distance-dependent, such that not only the zero-intercept but also the slope of the regression lines that related trVOR sensitivity with the inverse of viewing distance was affected by the currents (FIG. 3). Anodal stimulation decreased, whereas cathodal stimulation increased the dependence of sensitivity on viewing distance. (b) Not only the sensitivity but also the phase of the reflex was affected by the currents. At frequencies >2 Hz, cathodal stimulation resulted in increased phase leads. The opposite was true during anodal stimulation. (c) Cathodal stimulation increased the high-pass filtered properties of the trVOR. In contrast, anodal stimulation decreased the slope of sensitivity increase versus frequency.

Effects of Unilateral Labyrinthectomy

The trVOR dependence on viewing distance, as well as its spatio–temporal response properties were also substantially affected by unilateral labyrinthectomy. The two animals included in this study had all semicircular canals plugged approximately three to four months prior to the unilateral labyrinthectomy. A comparison of the horizontal response components before and after plugging revealed that canal inactivation induced no consistent change in the sensitivity, dynamics, and viewing-dependent properties of the trVORs (compare filled squares and circles in FIG. 4; see also reference 10). Because of the large spontaneous nystagmus right after unilateral labyrinthectomy, animals were unable to adequately fixate during the first week after the operation. Thus, animals were only tested after one week postsurgery, then again at regular intervals through 3 to 4 months after the operation.

One week after unilateral labyrinthectomy, trVOR sensitivity had lost its ability to modulate as a function of target distance (FIG. 4, left, open circles). Over time, animals partially recovered, such that 2 to 3 months later trVOR sensitivities seemed to reach and maintain approximately half the sensitivity of intact animals. This was consistent in both

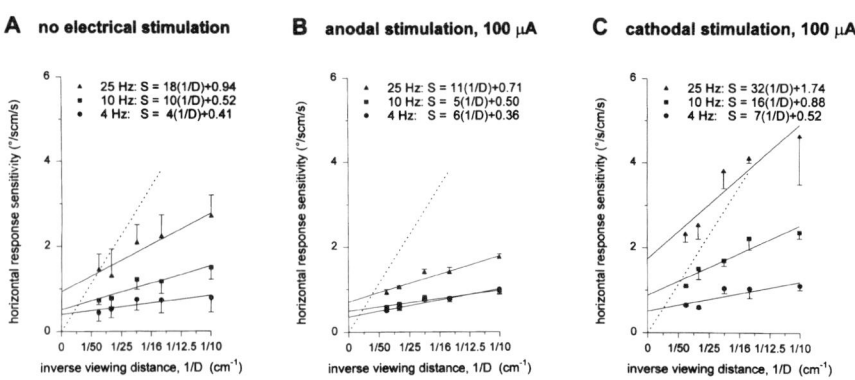

FIGURE 3. Horizontal response sensitivity during lateral motion as a function of inverse viewing distance without (**A**) and during anodal (**B**) and cathodal (**C**) bilateral labyrinthine stimulation. *Error bars* represent standard deviations. *Solid lines* are linear regressions through the data. *Dotted lines* illustrate ideal response (sensitivity, in deg/s/cm/s, should increase linearly as a function of the inverse of viewing distance with a slope of 57).

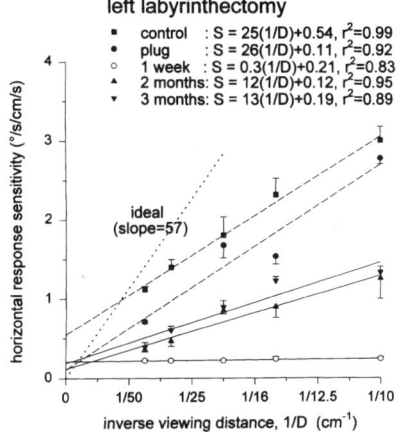

FIGURE 4. Horizontal response sensitivity during lateral motion as a function of inverse viewing distance in two animals with left and right labyrinthectomies. Data (10 Hz) acquired in the intact animal (*filled squares*), in the canal-plugged animal ~1 week before unilateral labyrinthectomy (*filled circles*), and at different times after lesion. *Error bars* represent standard deviations. *Solid lines* are linear regressions through the data. *Dotted lines* illustrate ideal responses.

animals (FIG. 4, left and right traces). It should be noted that trVOR sensitivities during fixation at near targets were consistently undercompensatory, even in animals with intact labyrinths.

The loss of the ability to modulate trVOR sensitivity as a function of viewing distance when trVOR was tested within one month after labyrinthectomy and its partial recovery (at least for high-frequency translations) is also illustrated in FIGURE 5. In addition to the loss of the dependence on viewing distance, trVOR dynamics were also significantly altered 1 week after left labyrinthectomy (FIG. 5, left). The sensitivity to translation was lower than intact animals at all stimulus frequencies tested. The decrease in sensitivity was larger at lower frequencies. This resulted in response sensitivities that increased sharply as a function of frequency. In addition, large phase leads were present at low frequencies such that eye movements were not compensatory at frequencies <10 Hz. These deficits seemed to have partially recovered 2 months after the operation.

The loss of the viewing distance-dependence and a recovery of trVOR dynamics in the animal with the right labyrinthectomy are illustrated in FIGURE 6. Even though the intact animal was characterized by significant differences in trVOR sensitivity in darkness (vergence at ~1-2 MA[10]) and during fixation of a center target at a distance of 20 cm (compare open circles with open triangles), this was no longer the case after labyrinthectomy. TrVOR sensitivities were the same whether in darkness or during near target fixation and very close to the intact trVOR sensitivities in darkness (FIG. 6, compare filled circles with filled triangles).

Despite a partial recovery of trVOR properties during lateral motion, more dramatic deficits were observed in the trVOR during fore–aft motion. Kinematic requirements dictate that the trVOR also depends on the spatial relationship between the translational displacement and the gaze direction. Specifically during fore–aft motion, fixation of a center, near target should generate negligible horizontal and vertical eye movements. Fixation to the left or right should evoke large horizontal eye movements with reversed polarity (or equivalently, phase during sinusoidal motion). As shown in FIGURE 7A for intact (and

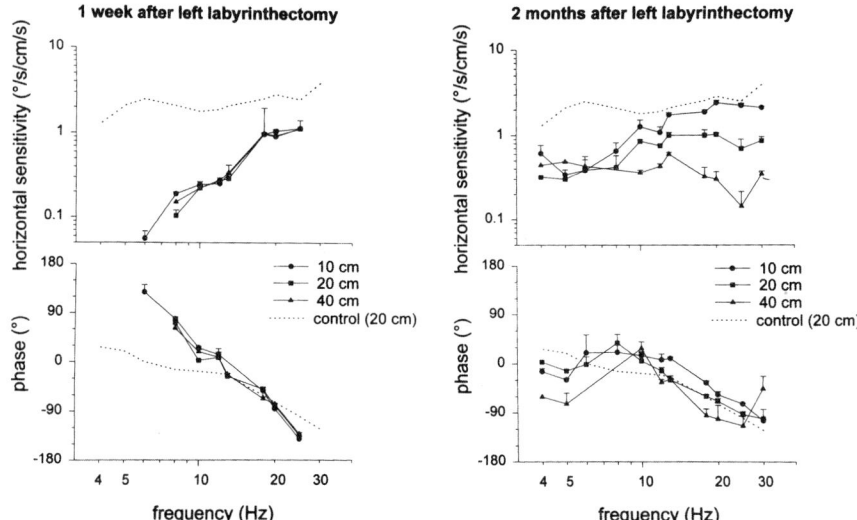

FIGURE 5. Horizontal response sensitivity and phase during lateral translation at different viewing distances 1 week and 2 months after left labyrinthectomy. Responses prior to the lesion at a distance of 20 cm have been included as *dotted lines*.

canal-plugged) animals, fore–aft oscillations while fixating an approximately centered, near target at a distance of 20 cm elicited negligible horizontal and vertical eye movements. Nevertheless, the same vestibular stimulus generated robust horizontal eye movements during fixation at a target to the right (forward motion elicits rightward and backward motion elicits leftward eye movements). Eye movements with opposite phase were elicited during fixation to the left (FIG. 7A).

When tested 1 week after left labyrinthectomy, the dependence of trVOR velocity on horizontal fixation position was no longer observed (FIG. 7B). In fact, in the animal with the left labyrinthectomy, compensatory eye movements could still be generated during fixation away from the lesion side (i.e., to the right). During fixation to the left, the evoked trVOR responses were anticompensatory. In essence, the eye movements evoked in the lesioned animal were not appropriately dependent on target position. This was also true in the animal with a right labyrinthectomy, but the trVOR responses were appropriate for fixation to the left (data not shown). These dramatic deficits seen during fore–aft motion were still present when tested 2 to 3 months after the lesion (FIG. 7C).

DISCUSSION

As part of an on-going effort to understand the neural processing in the trVORs, we have recently studied the behavioral properties of the binocular eye movements elicited during head translation. We first recorded trVOR responses in complete darkness (vergence angles of ~1 MA) such that we could quantify the dynamics of the reflex in a broad frequency range.[10] Interestingly, a low-order, lumped rational transfer function could not adequately describe the trVOR dynamics in a broad frequency range. We have taken this as evidence to suggest that a simple, feed-forward, canal-like processing might not be adequate to describe the trVORs.

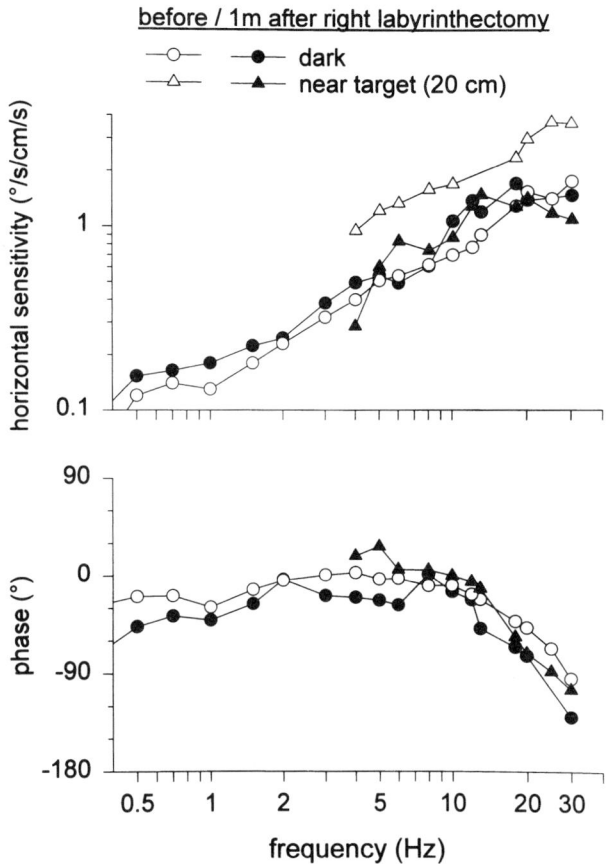

FIGURE 6. Horizontal response sensitivity and phase during lateral translation in darkness (*circles*) and during fixation at a near target at a distance of 20 cm (*triangles*) before and 1 month after right labyrinthectomy (*open* vs. *filled symbols*).

As a next step to understand the behavioral aspects that might elucidate the underlying organization of the reflex, trVORs were studied after unilateral labyrinthectomy. Currently, the bilateral organization of the trVORs is not well understood. Other than a few preliminary studies,[18–20] the effects of unilateral labyrinthectomy on the trVORs are unknown. In contrast, the acute and compensated effects of unilateral labyrinthectomy on rotVORs have been extensively studied in several species, including primates.[15,21,22] Static deficits (e.g., spontaneous nystagmus) are the first to recover, whereas dynamic reflex properties slowly follow. Acutely after unilateral labyrinthectomy, rotVOR gain is reduced, primarily during rotation toward the injured side, resulting in a relatively large asymmetry that is exaggerated during high-frequency or high-acceleration rotations.[23–27]

In the otolith system, each utricular macula is sensitive to accelerations in many directions because of the differing morphological polarizations of the receptor cells (e.g., reference 28). Thus, a single labyrinth might be sufficient for the trVORs. We found that this is not the case. The major effects of unilateral labyrinthectomy were a dramatic loss of the reflex dependence on target distance and eccentricity. In fact, a significant reflex impairment remains in compensated animals. In addition, trVOR dynamics during lateral motion

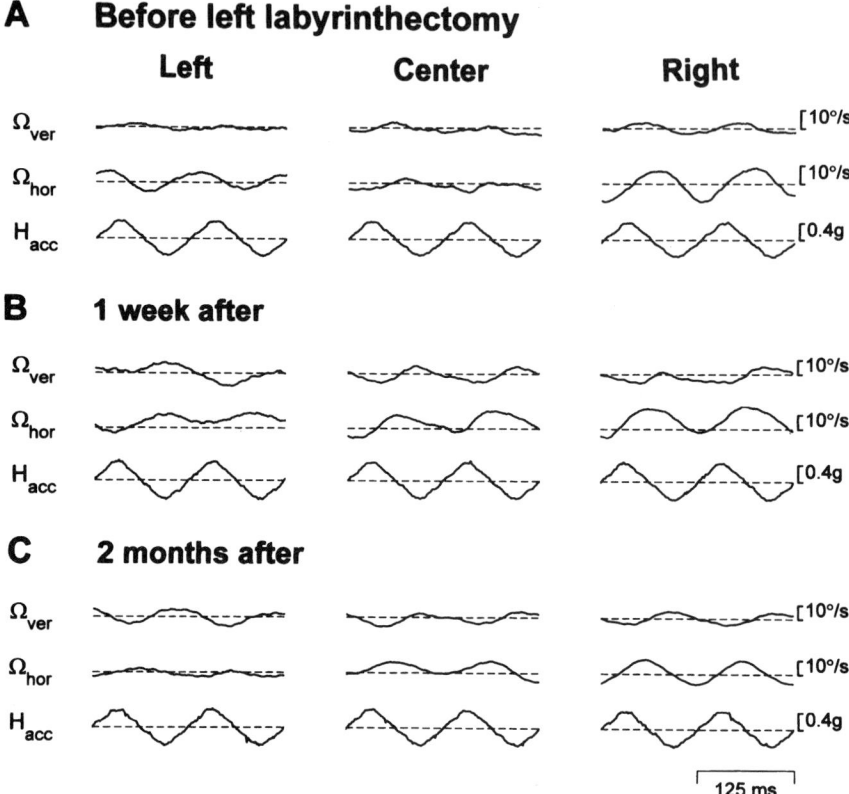

FIGURE 7. Vertical and horizontal components of eye velocity during fore–aft motion before (**A**), 1 week (**B**), and 2 months (**C**) after left labyrinthectomy (8 Hz). From left to right, data during fixation of a left, center, or right target (6-cm eccentric at a distance of 20 cm). *Dotted lines* illustrate zero eye velocity. H_{acc}: the output of a linear accelerometer mounted on the animal's head. Leftward and downward eye movements are positive. Backward acceleration is positive.

also exhibited more high-pass filtered properties acutely after unilateral labyrinthectomy. During fore–aft motion, the effects of unilateral labyrinthectomy were even more dramatic. Both acute and compensated animals permanently lost their ability to modulate fore–aft trVOR responses as a function of target eccentricity. These results suggest that it is convergent signals from the two labyrinths that are responsible for scaling the response proportional to viewing distance and eye position.

In addition to unilateral labyrinthectomy, bilateral electrical stimulation has also been used as a means of decreasing the spontaneous activity of the most sensitive, irregularly firing vestibular afferents.[7,17,29] When anodal (inhibitory) labyrinthine currents are delivered, a selective, reversible ablation (silencing) of neural firing is produced in the most irregular afferents that lasts for the duration of the electrical stimulation. Regularly firing afferents, on the other hand, are little affected by even the largest current levels used.[17,30] Even though constant anodal or cathodal currents elicit horizontal and torsional nystagmus in both eyes when only a single labyrinth is stimulated, bilateral anodal stimulation results in a decrease of primary afferent discharge with no nystagmus being generated. Bilateral constant anodal stimulation has then become a useful tool to study the contribution of the most irregular vestibular afferents to the production of the VORs. Using this technique, it was demonstrated that the most irregular vestibular afferents do not contribute to the rotVOR in the dark at frequencies between 0.5 and 4 Hz.[7] There has been some evidence, however, that irregular vestibular afferents might contribute to the inertial vestibular system[11,12] and the viewing distance-dependent changes in the rotVOR during near-target fixation.[31] Despite these modest effects of irregular vestibular afferent ablation on the rotVOR, vestibular nuclei neurons have been shown to receive inputs from the whole continuum of afferents.[29,32–34]

In the present studies, the effects of functional, reversible ablation and recruitment of the most irregular otolith afferents on the dynamics and sensitivity of the trVORs were investigated using short-duration (<2 s) bilateral, anodal (inhibitory), and cathodal (excitatory) currents (100 µA) during translational motion. The presence of anodal labyrinthine stimulation decreased horizontal trVOR sensitivity and its dependence on viewing distance for both lateral and fore–aft motion stimuli. In addition, anodal currents during sinusoidal oscillations decreased the slope of trVOR sensitivity as a function of frequency. Cathodal currents had opposite effects. That is, the presence of cathodal stimulation during translational motion increased trVOR sensitivity and its dependence on viewing distance, as well as resulted in more high-frequency trVOR dynamics during both sinusoidal oscillations and transient motion stimuli. Changes in resting firing rates and/or reversible ablation and recruitment of part of the afferent continuum could be responsible for these observations. Accordingly, it is possible that the most irregularly firing otolith afferents might participate in the high-frequency-response dynamics and the viewing-distance-dependent changes of the translational VORs.

Taken together, our present understanding of the anatomical, neural, and functional organization of the trVORs still remains rather rudimental. Even though considerable behavioral data have been gathered during the past decade, even the simplest conceptual framework, much more neural implementation, is still highly debatable. Despite our limited knowledge, we believe that available evidence suggests that trVORs may not be organized similarly to the rotVORs. Alternatively, partly segregated central pathways, bilateral spatio–temporal processing, as well as different computational principles might characterize otolith versus canal–ocular pathways (FIG. 8).

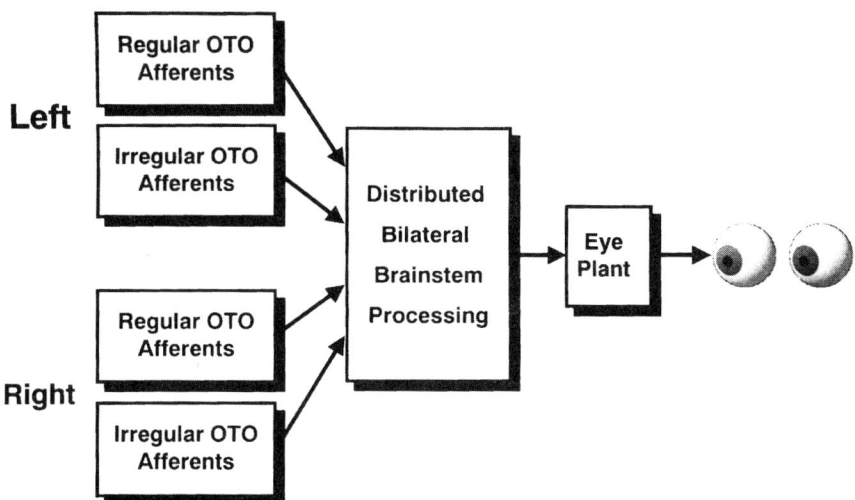

FIGURE 8. Proposed framework for addressing the basic processing underlying the trVORs.

ACKNOWLEDGMENT

The work was supported by grants from NIH (EY10851), NASA (NAGW-4377) and the Air Force Office of Scientific Research (F-49620).

REFERENCES

1. ANGELAKI, D. E. 1992. Spatiotemporal convergence (STC) in otolith neurons. Biol. Cybern. **67**: 83–96.
2. TELFORD, L., S. H. SEIDMAN & G. D. PAIGE. 1997. Dynamics of squirrel monkey linear vestibuloocular reflex and interactions with fixation distance. J. Neurophysiol. **78**: 1775–1790.
3. ANGELAKI, D. E. & B. J. HESS. 1996. Organizational principles of otolith and semicircular canal-ocular reflexes in rhesus monkeys. Ann. N.Y. Acad. Sci. **781**: 332–347.
4. FUCHS, A. F., C. A. SCUDDER & C. R. S. KANEKO. 1988. Discharge patterns and recruitment order of identified motoneurons and internuclear neurons in the monkey abducens nucleus. J. Neurophysiol. **60**: 1874–1895.
5. ROBINSON, D. A. 1962. The mechanics of human saccadic eye movement. J. Physiol. (Lond.) **174**: 245–264.
6. STAHL, J. S. & J. I. SIMPSON. 1995. Dynamics of abducens nucleus neurons in the awake rabbit. J. Neurophysiol. **73**(4): 1383–1395.
7. MINOR, L. B. & J. M. GOLDBERG. 1991. Vestibular-nerve inputs to the vestibular-ocular reflex: a functional-ablation study in the squirrel monkey. J. Neurosci. **11**: 1636–1648.
8. SKAVENSKI, A. A. & D. A. ROBINSON. 1973. Role of abducens neurons in vestibuloocular reflex. J. Neurophysiol. **36**: 724–738.
9. GREEN, A. M. & H. L. GALIANA. 1998. A hypothesis for shared central processing of canal and otolith signals. J. Neurophysiol. **80**: 2222–2228.
10. ANGELAKI, D. E. 1998. Three-dimensional organization of otolith-ocular reflexes in rhesus monkeys. III. Responses to translation. J. Neurophysiol. **80**: 680–695.
11. ANGELAKI, D. E. & A. A. PERACHIO. 1993. Contribution of irregular semicircular canal afferents to the horizontal vestibuloocular response during constant velocity rotation. J. Neurophysiol. **69**(3): 996–999.
12. ANGELAKI, D. E., A. A. PERACHIO, M. MUSTARI & C. L. STRUNK. 1992. Role of irregular otolith afferents in the steady-state nystagmus during off-vertical axis rotation. J. Neurophysiol. **68**: 1895–1900.
13. ANGELAKI, D. E., M. Q. MCHENRY, J. D. DICKMAN, S. D. NEWLANDS & B. J. M. HESS. 1999. Computation of inertial motion: neural strategies to resolve ambiguous otolith information. J. Neurosci. **19**(1): 316–327.
14. CANNON, S. C. & D. A. ROBINSON. 1987. Loss of the neural integrator of the oculomotor system from brain stem lesions in monkey. J. Neurophysiol. **57**(5): 1383–1409.
15. FETTER, M. & D. S. ZEE. 1988. Recovery from unilateral labyrinthectomy in rhesus monkey. J. Neurophysiol. **59**: 370–393.
16. REY, C. G. & H. L. GALIANA. 1993. Transient analysis of vestibular nystagmus. Biol. Cybern. **69**: 395–405.
17. GOLDBERG, J. M., C. E. SMITH & C. FERNÁNDEZ. 1984. Relation between discharge regularity and responses to externally applied galvanic currents in vestibular nerve afferents of the squirrel monkey. J. Neurophysiol. **51**: 1236–1256.
18. BRONSTEIN, A. M., M. A. GRESTY & G. B. BROOKES. 1991. Compensatory otolithic slow phase eye movement responses to abrupt linear head motion in the lateral direction: findings in patients with labyrinthine and neurological lesions. Acta Otolaryngol. Suppl. **481**: 42–46.
19. LEMPERT, T., C. C. GIANNA, M. A. GRESTY & A. M. BRONSTEIN. 1997. Effect of otolith dysfunction. Impairment of visual acuity during linear head motion in labyrinthine defective subjects. Brain **120**: 1005–1013.
20. PAIGE, G. D., G. R. BARNES, L. TELFORD & S. H. SEIDMAN. 1996. Influence of sensorimotor context on the linear vestibulo-ocular reflex. Ann. N.Y. Acad. Sci. **781**: 322–331.
21. TAKAHASHI, M., M. IGARASHI & J. L. HOMICK. 1977. Effect of otolith end organ ablation on horizontal optokinetic nystagmus and optokinetic afternystagmus in the squirrel monkey. ORL **39**: 74–81.

22. WOLFE, J. W. & C. M. KOS. 1977. Nystagmic responses of the rhesus monkey to rotational stimulation following unilateral labyrinthectomy: final report. Trans. Am. Acad. Ophthalmol. Otolaryngol. **84**: 38–45.
23. ALLUM, J. H. J., M. YAMANE & C. R. PFLATZ. 1988. Long-term modifications of vertical and horizontal vestibulo-ocular reflex dynamics in man. Acta Oto-laryngol. (Stockholm) **105**: 328–337.
24. FETTER, M. & J. DICHGANS. 1990. Adaptive mechanisms of VOR compensation after unilateral peripheral vestibular lesions in humans. J. Vestibular Res. **1**: 9–22.
25. MAIOLI, C. & W. PRECHT. 1984. The horizontal optokinetic nystagmus in the cat. Exp. Brain Res. **55**: 494–506.
26. MAIOLI, C., W. PRECHT & S. RIED. 1983. Short- and long-term modifications of vestibulo-ocular response dynamics following unilateral vestibular nerve lesions in the cat. Exp. Brain Res. **50**: 259–274.
27. VIBERT, N., C. DE WAELE, M. ESCUDERO & P. P. VIDAL. 1993. The horizontal vestibulo-ocular reflex in the hemilabyrinthectomized guinea-pig. Exp. Brain Res. **60**: 263–273.
28. LINDEMAN H. H. 1969. Studies on the Morphology of the Sensory Regions of the Vestibular Apparatus: 1–113. Springer-Verlag. Berlin.
29. CHEN-HUANG, C., R. A. MCCREA & J. M. GOLDBERG. 1997. Contributions of regularly and irregularly discharging vestibular-nerve inputs to the discharge of central vestibular neurons in the alert squirrel monkey. Exp. Brain Res. **114**: 405–422.
30. DICKMAN, J. D. & D. E. ANGELAKI. 1993. Changes in response of canal-related vestibular neurons via galvanic ablation of irregular afferent fibers in pigeons. Soc. Neurosci. Abstr. **19**:139.
31. CHEN-HUANG, C. & R. A. MCCREA. 1998. Contribution of vestibular nerve irregular afferents to viewing distance-related changes in the vestibulo-ocular reflex. Exp. Brain Res. **119**(1): 116–130.
32. BOYLE, R., J. M. GOLDBERG & S. M. HIGHSTEIN. 1992. Inputs from regularly and irregularly discharging vestibular nerve afferents to secondary neurons in squirrel monkey vestibular nuclei: II. Correlation with vestibulospinal and vestibuloocular output pathways. J. Neurophysiol. **68**: 471–484.
33. GOLDBERG, J. M., S. M. HIGHSTEIN, A. K. MOSCHOVAKIS & C. FERNÁNDEZ. 1987. Inputs from regularly and irregularly discharging vestibular nerve afferents to secondary neurons in the vestibular nuclei of the squirrel monkey. I. An electrophysiological analysis. J. Neurophysiol. **58**: 700–718.
34. HIGHSTEIN, S. M., J. M. GOLDBERG, A. K. MOSCHOVAKIS & C. FERNÁNDEZ. 1987. Inputs from regularly and irregularly discharging vestibular nerve afferents to secondary neurons in the vestibular nuclei of the squirrel monkey. II. Correlation with output pathways of secondary neurons. J. Neurophysiol. **58**: 719–738.

Inertial Processing of Vestibulo-Ocular Signals

BERNHARD J. M. HESS[a,c] AND DORA E. ANGELAKI[b]

[a]*Department of Neurology, University Hospital Zürich, Frauenklinikstrasse 26, CH-8091, Zürich, Switzerland*

[b]*Department of Surgery (Otolaryngology), University of Mississippi Medical Center, 2500 North State Street, Jackson, Mississippi 39216-4505, USA*

ABSTRACT: New evidence for a central resolution of gravito-inertial signals has been recently obtained by analyzing the properties of the vestibulo-ocular reflex (VOR) in response to combined lateral translations and roll tilts of the head. It is found that the VOR generates robust compensatory horizontal eye movements independent of whether or not the interaural translatory acceleration component is canceled out by a gravitational acceleration component due to simultaneous roll-tilt. This response property of the VOR depends on functional semicircular canals, suggesting that the brain uses both otolith and semicircular canal signals to estimate head motion relative to inertial space. Vestibular information about dynamic head attitude relative to gravity is the basis for computing head (and body) angular velocity relative to inertial space. Available evidence suggests that the inertial vestibular system controls both head attitude and velocity with respect to a gravity-centered reference frame. The basic computational principles underlying the inertial processing of otolith and semicircular canal afferent signals are outlined.

INTRODUCTION

Control of self-motion and orientation in space depends on vestibular information that is supplemented by visual and somatosensory motion cues. The vestibular sense organs, which are located in the base of the skull, are fast and precise detectors of head rotation and translation. Their output is processed in the vestibular nuclei and the cerebellum. Since the peripheral sense organs move together with the physical reference, that is, the head, relative to inertial space, the central nervous system faces computational problems in processing the sensor signals that are analogous to problems known from man-made inertial navigation systems.[1,2,3] As in inertial navigation, estimation of self-motion in space requires representation of position, speed, and heading relative to an acceleration-free reference system. In a biological context, any coordinate system in which gravity is always directed along one of the coordinate axes may serve this purpose. We call such reference systems *inertial reference* systems.[e] For a realization, the central nervous system needs to solve two interrelated issues: (1) estimate head (and body) attitude relative to gravity by distinguishing translational from gravitational acceleration at any time, and (2) estimate

---

[c]To whom correspondence may be addressed. Phone: 411/255-5500; fax 411/255-4507; e-mail: bhess@neurol.unizh.ch

[e]As long as inertial forces due to rotation of the earth can be ignored, any gravity-centered coordinate frame can be considered to be equivalent to an acceleration-free reference-frame modulo subtraction of the constant acceleration of gravity along the respective axes.

head angular velocity in space. Available evidence suggests that there are at least two networks in the brainstem that are involved in establishing an inertial reference system.

One of these brainstem networks processes head (and body) attitude relative to gravity. There are two principal ways to accomplish this goal. In the first way, head angular velocity, measured by the semicircular canals, could be integrated to estimate head (and body) attitude relative to space, and thus relative to gravity. This solution is realized in modern inertial navigation systems in which the direction cosine matrix (also called attitude or rotation matrix), relating the vehicle (body) reference frame to a ground-based reference coordinate system, is computed based on the turn rate measured by high-precision three-dimensional rate sensors. The accuracy of this approach depends essentially on the accuracy with which the rotation can be determined and on the computation rate. Gravity plays no role in this approach other than for a correction of linear accelerometer measurements at a later stage, which is based on a internal model of gravity (see, e.g., reference 4). From a biological standpoint, it is the head (and body) attitude relative to gravity that is of major concern for posture and balance control rather than the precise three-dimensional orientation with respect to a fixed ground-based coordinate reference. Thus, one rotational degree of freedom, namely rotation about the earth-vertical, might not need to be controlled. This simplification relaxes the computational burden of signal processing. The following schema realizes such a strategy: Here, velocity signals from the semicircular canals (and possibly other sources) would be processed together with the resultant gravito-inertial signals from the otolith organs (network N_i in FIG. 1). The output of this network N_i encodes translational (inertial) and gravitational head acceleration. Algorithms that discriminate gravity from inertial acceleration, based on angular velocity, gravito-inertial acceleration and time rate of change of gravito-inertial acceleration have been proposed by Viéville and Faugeras[5] and others.[6] The implicit evaluation of head attitude of these algorithms relative to gravity suffices for establishing and maintaining a gravity-centered (and thus inertial) coordinate reference frame.

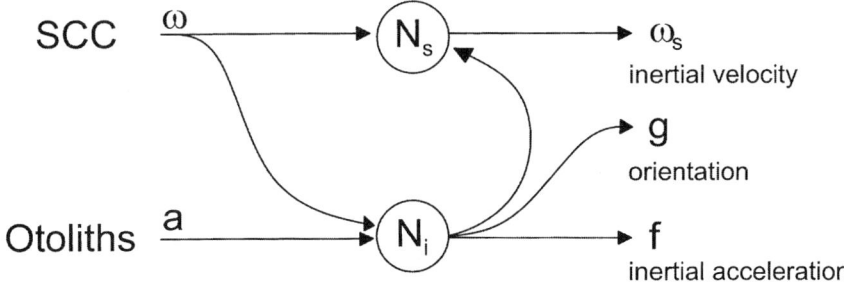

FIGURE 1. Schematic diagram of inertial vestibular signal processing: Semicircular-canal (ω) and otolith afferent signals (**a**) are processed in the vestibular system to compute head orientation and motion in space: The gravito-inertial signals coded by otolith afferents are processed by a nonlinear network (N_i) that discriminates the inertial acceleration components (**f**) due to head translation from the gravitational components associated with current head orientation relative to gravity (**g**). The extracted information about head orientation relative to gravity (**g**) is used to compute head velocity in space (ω_s).

The second network is related to the estimation of inertial velocity. Not until recently has it become clear that one of the fundamental functions of the so-called velocity storage integrator[7,8] is related to central estimation of head angular velocity with respect to a gravity-centered reference system.[9,10] Transformation of semicircular canal signals from intrinsic head-centered coordinates to a gravity-centered reference frame relies on veridical information about instantaneous head orientation relative to gravity (see N_s in FIG. 1). The output of this network that operates in the low-frequency range is inertial angular velocity. Since estimation of inertial head velocity, that is, velocity of the head (and body) relative to a space, is mathematically correlated with a simultaneous estimation of head attitude (relative to the same reference),[4,11] the two networks N_i and N_s presumably operate in close register. If the vestibular system does not care about the full knowledge of head (and body) orientation relative to a fixed ground-based reference in all three rotational degrees of freedom, but would essentially keep track only of head attitude relative to gravity, one would predict the same properties for central estimation of inertial head velocity. At present, there is no pertinent experimental evidence available.

Since self-motion in space must be based on a gravity-centered rather than a head-centered reference frame, it is not surprising that new insights in inertial signal processing have come from studies on the three-dimensional spatial tuning properties of the vestibulo-ocular reflex (VOR). In the following paragraphs we first describe some recent experiments that address the problem of discrimination of translational and gravitational accelerations, a problem that is central in inertial vestibular signal processing. Second, we discuss mechanisms by which gravity information can be used to represent head velocity in space.

METHODS

All data were obtained from juvenile rhesus monkeys (*Macaca mulatta*) that were chronically prepared with a head holder device for restraining the head during the experiment and with scleral dual-search coils for recording three-dimensional eye movements. Details of fabrication and implantation of the dual-search coils have been reported elsewhere.[12] Three-dimensional eye position was measured with a two-field search-coil system. The search-coil signals were calibrated as described in Hess *et al.*[13] Horizontal, vertical, and torsional eye positions were digitized at a sampling rate of 833 Hz and stored on a computer for off-line data analysis. All eye positions are expressed as rotation vectors, $\mathbf{E} = \tan(\rho/2)\ \mathbf{u}$, where $\mathbf{u}$ is a unit vector pointing along the rotation axis of the eye, and ρ is the angle of rotation about this axis. Rotation vectors were expressed relative to a right-handed coordinate system, where the y-axis was aligned with the interaural axis (positive direction leftward as seen from the monkey), and the x-y plane rotated upward by about 15–18 deg relative to stereotactic horizontal. A positive torsional (E_{tor}), vertical (E_{ver}) or horizontal (E_{hor}) eye-position component corresponded to a clockwise, downward, or leftward rotation of the eye (from the subject's point of view). The eye angular velocity vector, $\Omega = (\Omega_{tor}, \Omega_{ver}, \Omega_{hor})$, was computed from the eye position vector, $\mathbf{E}$, according to the equation: $\Omega = 2\ (\ \dot{\mathbf{E}} + \mathbf{E} \wedge \dot{\mathbf{E}}\ ) / (1 + \|\mathbf{E}\|^2)$. In addition to intact animals, data from two animals will be presented in which all semicircular canals had been plugged as described elsewhere.[14,15]

EXPERIMENTAL PARADIGMS

Eye movements were induced with the animals seated in a primate chair and secured to a motor-driven multiaxis rotator (Acutronic, Inc). In most of the experiments, this rotator was riding on top of a linear sled that was powered by a servocontrolled linear motor (track of 2 m length; Acutronics Inc). The rotator was used to deliver rotatory (roll) movements, whereas the linear sled was used to translate the animals laterally. All eye movements were recorded in complete darkness. Results from two sets of experiments were as presented and discussed in the following sections.

Roll-Tilt Translation Paradigms to Test Tilt-Translation Discrimination

In these experiments, the stimuli consisted of (1) sinusoidal lateral translations on a linear sled, (2) in roll-tilt oscillations about the naso–occipital axis, and (3) in combined linear and roll-tilt oscillations, all at a frequency of 0.5 Hz. The lateral translations and roll-tilts were matched in amplitude in order to generate the same interaural peak shear acceleration of 0.37 G. Thus, lateral translations of the animals combined with out-of phase roll-tilt oscillation resulted in zero interaural acceleration ("roll-tilt minus translation" paradigm), while translations combined with in-phase roll-tilt generated a peak interaural acceleration of 0.74 G ("roll-tilt plus translation" paradigm). The acceleration in the "roll-tilt minus translation" paradigm is illustrated in FIGURE 2 with respect to an

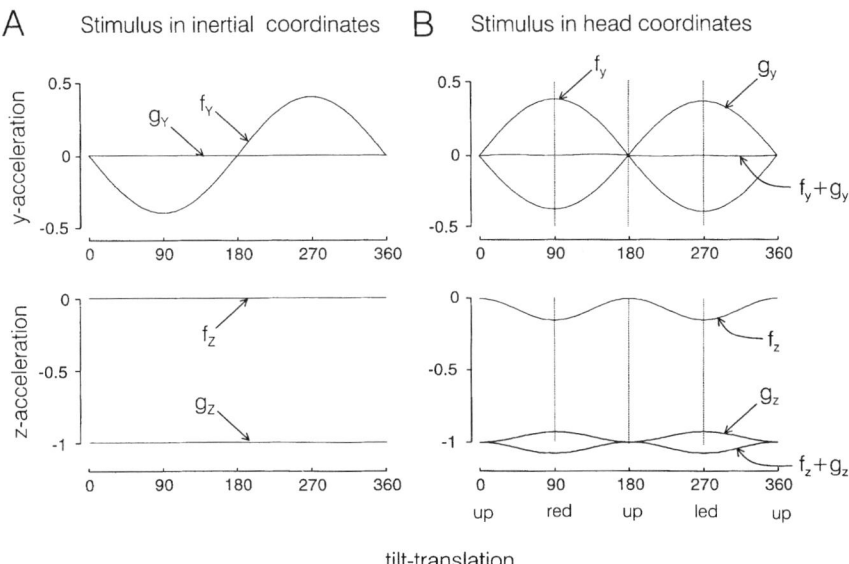

FIGURE 2. (**A**) Linear acceleration stimuli during combined roll-tilt and translation. Head acceleration (**f**) and gravity (**g**) measured relative to inertial coordinates (Y-axis parallel to the direction of travel, Z-axis aligned with gravity, and X-axis orthogonal to Y and Z) during sinusoidal translation with peak amplitude of 0.4 g combined with roll oscillations with peak tilt angle of 21.8 deg. Note constant acceleration of gravity at the −1-G level. (**B**) Same motion measured in noninertial, head-centered coordinates (x-axis: naso–occipital, y-axis: interaural and z-axis: perpendicular to x and y). Note modulation of both the translational (**f**) and gravitational (**g**) acceleration along the y- and z-axes due to the sinusoidal roll oscillation. The z-components oscillate at twice the frequency of roll oscillation. Also shown are the resultant accelerations, $a_y = f_y + g_y$ and $a_z = f_z + g_z$, along the y- and z-axes. Accelerations along the x-axis are zero (not shown).

inertial and a head-centered reference frame. In the inertial reference frame (X, Y, Z), translational acceleration is along the Y-axis and gravitational acceleration along the Z-axis, whereas in head-centered coordinates (x, y, z) both translational and gravitational acceleration exhibit nonzero y- and z-components. Due to matched amplitudes in the "roll-tilt minus translation" paradigm, the y-components cancel each other, whereas there remains a small resultant z-component. However, this component modulates at twice the frequency of translation. If the VOR can distinguish translational from gravitational acceleration, one would expect modulation of horizontal eye velocity at the frequency of translation and a small vertical component at twice that frequency (to compensate for the roll motion). In contrast, if the VOR could not distinguish the two accelerations, no horizontal eye velocity modulation would be expected.

Rotation-Tilt Paradigm to Test Coordinate Reference Frame of VOR

These experiments were designed to test the coordinate reference frame in which central head-velocity signals are encoded. During an earth-vertical-axis rotation, which is the standard procedure for testing per- and postrotatory VOR, there is no difference in representation of head velocity with respect to head-centered and gravity-centered coordinates. To induce a physical dissociation of these two reference systems, fast short-lasting passive head (and body) tilts (tilt velocity: 180 deg/s, tilt acceleration: 180 deg/s^2) were delivered immediately after stop of a constant velocity earth-vertical-axis rotation at 90 deg/s. Three-dimensional postrotatory eye velocity was measured and analyzed in spatial phase plots to test the reference frame in which head-velocity signals were centrally coded. To fully characterize the spatial characteristics of postrotatory VOR, animals were tilted in different combinations from upright, supine, or 90 deg right (left) ear-down in pitch, roll, and yaw. For a more complete account of experiments based on these vestibular paradigms, see references 10 and 16.

VESTIBULAR STRATEGIES ASSESSING HEAD ATTITUDE RELATIVE TO GRAVITY

Without discrimination of translational from gravitational acceleration, head attitude relative to gravity cannot be determined from vestibular signals, since afferents do not distinguish between these two forms of accelerations. To investigate possible vestibular discrimination strategies, context-specific responses of the VOR have long been used as a model system. For example, during lateral translation, the VOR generates horizontal eye movements.[17–22] A typical response from a rhesus monkey is illustrated in FIGURE 3A, showing the horizontal VOR to a translatory oscillation at 0.5 Hz ± 0.37 G. The controversial question is, what is the specific force driving this reflex? Is it lateral head (translational) acceleration, an acceleration that is not directly detected by the otolith afferents, or is it the total gravito-inertial acceleration that is encoded by otolith afferents (see vector **f** and **g** in monkey sketch, FIG. 3A, bottom)? To address this question, let us compare the responses to the four stimulus conditions illustrated in FIGURE 3. In contrast to lateral translation, sinusoidal roll tilts produce a torsional VOR with little or no modulation of horizontal eye velocity (FIG. 3B). These observations may suggest that the VOR responds selectively to linear accelerations (i.e., translational or gravitational), which in the two examples were matched with regard to the amplitude along the interaural axis.

In order to further test whether the animals distinguish head roll from lateral translation we delivered combined roll-tilt—translation stimuli such that the inertial and gravitational accelerations along the interaural axis were either in or out of phase, and therefore

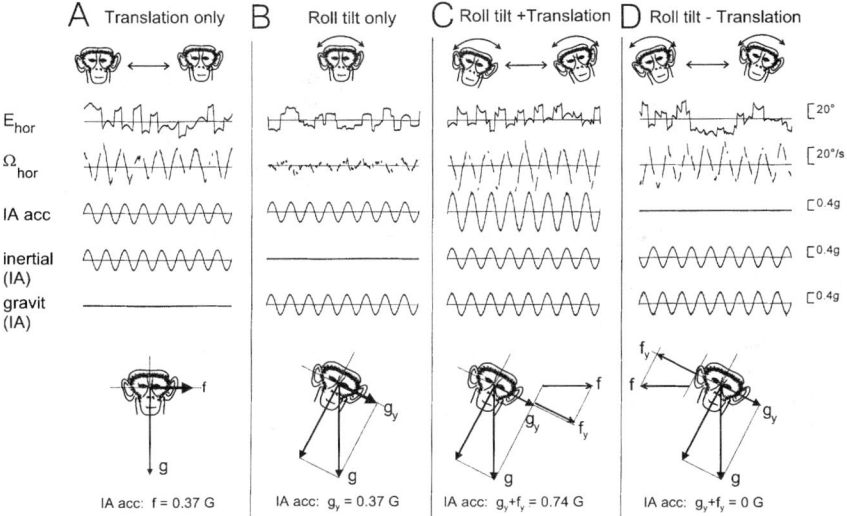

FIGURE 3. Translational vestibulo-ocular response to sinusoidal roll-tilt translation in rhesus monkey. (**A**)–(**D**) Modulation of horizontal eye position (E_{hor}) and slow phase eye velocity (Ω_{hor}) to lateral translation only, roll-tilt only, roll-tilt + translation and roll-tilt − translation at 0.5 Hz in complete darkness. Thin lines indicate zero eye position (gaze straight ahead) and zero velocity. Stimulus traces show lateral head acceleration (IA accel), translational (inertial), and graviational (gravit) acceleration components along the y-axis. The sketches of the monkey heads on the bottom row show the peak of the gravitational (**g**) and inertial (**f**) acceleration acting on the otoliths.

added to or subtracted from each other (see stimulus traces in FIG. 3C, 3D). The in-phase stimulus generated a modulation of horizontal eye velocity that was similar in magnitude to what it was during translation only (FIG. 3C). When the animal was submitted to an out-of-phase tilt-translation stimulus such that the translational and gravitational acceleration components subtracted from each other, there was still a horizontal modulation that was phase locked to lateral head translation (FIG. 3D). Thus, it appears that the horizontal translational VOR responds specifically to head translation in space rather than to the component of acceleration along the interaural axis. This was further corroborated by testing the VOR responses to tilt-translation stimuli in which peak roll-tilt changed while the translational acceleration remained constant. In each of these cases, horizontal eye velocity was found to be independent of the gravitational acceleration component along the y-axis (FIG. 4A). Responses always reflected the translational acceleration component rather than the resultant acceleration (FIG. 4; compare peak horizontal velocity curves in A with dashed curve f_y versus solid curve $f_y + g_y$ in B). Taken together, these results suggest that the translational VOR responds specifically to the translational acceleration (f_y in FIG. 2A). It further implies the existence of an efficient brainstem mechanism that identifies the inertial acceleration component from otolith afferent signals.

Two hypothesis have been proposed as to how the brainstem could filter the gravito-inertial signals carried by otolith afferents to obtain gravitational acceleration (indicating head orientation) and inertial acceleration information (indicating linear or curvilinear head translation). The "multisensory" integration hypothesis states that information from many different sensory sources, including the otolith organs and semicircular canals, are used to differentiate between head tilt and translation.[3,23,24] Alternatively, the "frequency-

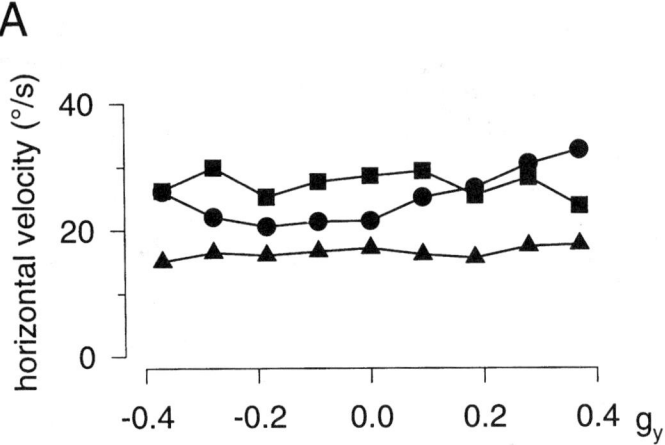

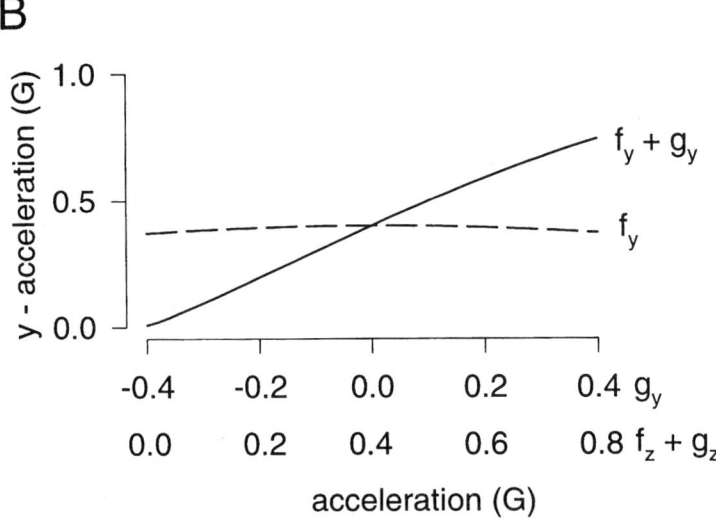

FIGURE 4. Responses to combined roll-tilt translation with different peak roll amplitudes. (**A**) Peak horizontal slow phase velocity as a function of interaural acceleration. Note the linear correlation of the response amplitude with stimulation (*circles*: $r^2 = 0.97$; *squares*: $r^2 = 0.99$; *triangles*: $r^2 = 0.99$; symbols designate different animals). (**B**) Interaural component of translational (f_y, *dashed lines*) and resultant acceleration ($f_y + g_y$, *solid lines*) as a function of stimulus acceleration.

segregation" hypothesis states that the frequency content of otolith signals alone determines the source of acceleration such that high-frequency accelerations are interpreted as translations, whereas low-frequency accelerations are interpreted as head tilts.[20,22] To investigate whether semicircular canal signals are necessary for discriminating head tilt from translation, we tested the just described tilt-translation paradigms in two monkeys in which all semicircular canals had been inactivated by plugging. From these experiments two immediate observations could be made: First, these animals exhibited a horizontal VOR in response to pure roll-tilt that was similar to the horizontal response (i.e., translational VOR) to translation (FIG. 5A and 5B). Second, these animals exhibited no horizontal response in the "roll-tilt minus translation" paradigm in which the interaural acceleration was nullified by oppositely directed gravitational and translational acceleration components of equal magnitude (FIG. 5D).

From these findings we conclude that semicircular canal input is essential in centrally resolving the tilt-translation ambiguity in otolith afferent signals, at least for frequencies > 0.1 Hz where the VOR is important for gaze stabilization.[16] At lower frequencies when canal dynamics becomes prevalent, other sensory cues must provide veridical velocity information for correct tilt-translation discrimination of otolith afferent signals.

INERTIAL REPRESENTATION OF HEAD ANGULAR VELOCITY IN THE VESTIBULO-OCULAR REFLEX

An important parameter in the control of self-motion in space is the accurate estimation of head velocity. Since semicircular-canal afferent signals are coded in head-centered coordinates, these signals must be centrally transformed relative to an inertial reference frame. Inertial representation of head velocity depends on access to information about head attitude relative to gravity. As previously outlined, there is evidence for central vestibular mechanisms that separate gravito-inertial information in VOR into its constituent components of inertial and gravitational acceleration. Motion in space could thus be represented at the brainstem level by encoding head velocity (and position) in gravity-centered (inertial) coordinates.

Evidence for an inertial representation of head angular velocity has been provided by a number of studies of the VOR. One of the early key observation has been that head tilts away from an upright position decrease the time constant of a horizontal postrotatory nystagmus.[25,26] This phenomenon, usually described as "tilt suppression" or "tilt dumping,"[27]

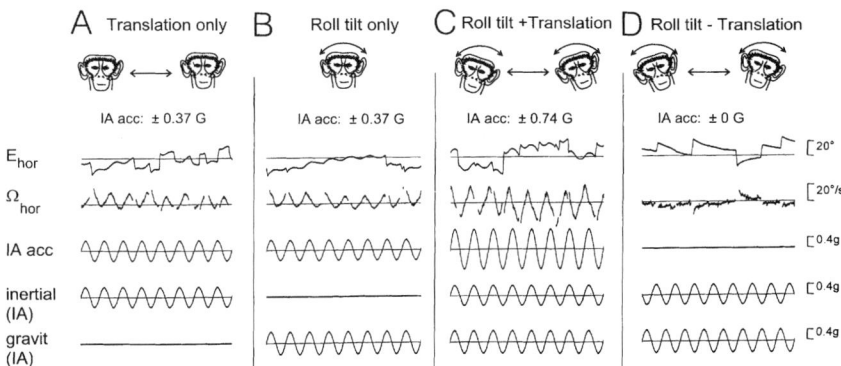

FIGURE 5. Translational vestibulo-ocular response to sinusoidal roll-tilt translation in rhesus monkey after inactivation of all semicircular canals. Horizontal eye position (E_{hor}) and slow phase eye velocity (Ω_{hor}) during roll-tilt ± lateral translation at 0.5 Hz in complete darkness. Same stimulus protocols as in FIGURE 3. Note that there is no horizontal response during out-of-phase roll-tilt translation motion ("roll-tilt minus translation").

has been found to be associated with the generation of an eye-velocity component along an axis in the plane of the tilt, orthogonal to the principal response axis.[10,28–32] FIGURE 6A and 6B illustrate the spatial transformation of a horizontal postrotatory VOR following a head tilt in the pitch plane. Note the strong amplitude dumping of the horizontal velocity component (principal response) and the rapid rise of a torsional response component

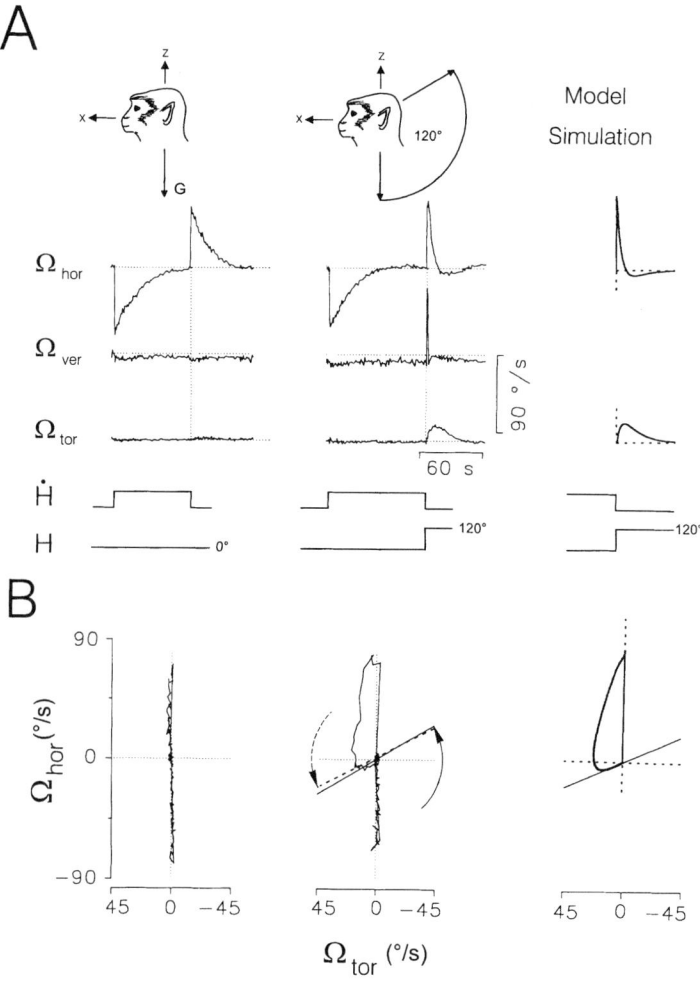

FIGURE 6. Spatial reorientation of the horizontal VOR after a pitch tilt. (**A**) Horizontal, vertical, and torsional eye-velocity components (Ω_{hor}, Ω_{ver}, Ω_{tor}) plotted versus time. *Horizontal* and *vertical dotted lines* indicate the stop of head rotation and the zero velocity line. *First column*: no tilt; *second column*: pitch-tilt through 120 deg (see sketch of monkey head); *third column*: model simulation (for details see text). (**B**) Phase plots of eye-velocity vectors in the pitch plane. *Solid lines* (*solid arrows*): new direction of gravity after the tilt. *Dashed lines* (*dashed arrows*): fit to eye-velocity trajectory yielding a tilt angle of 116.5 deg.

(orthogonal component) (FIG. 6A). The resultant change in orientation of head angular velocity is best illustrated in a phase plot (FIG. 6B). Similar shifts of the postrotatory responses toward alignment with gravity have also been observed in the vertical system following head (and body) tilts in the yaw plane. Interestingly, experimental analysis of the spatial response characteristics revealed that fundamentally different geometric mechanisms underlay the spatial transformations in the horizontal and vertical system.[10,33,34]

In the horizontal system, the transformation of eye velocity can best be described with a leaky integration in cascade with a gravity-dependent rotatory transformation (T_g in FIG. 7A). In this parametric approach, the afferent velocity signals from the horizontal canals, conceived as push–pull velocity signals from the right and left lateral semicircular canals (see vector $\mathbf{v}_{rll}$ in FIG. 7A), are integrated by a simple first-order differential equation (i.e., $\mathbf{v}_{rlh} = \dot{\mathbf{u}} + A\mathbf{u}$). This integration is followed by a mechanism that rotates the integrated lateral canal signals (**u**) in the pitch, roll, or an intermediate pitch-roll plane toward alignment with gravity (i.e., $\omega = T_g \mathbf{u}$, where T_g represents a gravity-dependent unitary transformation). A model simulation of the observed spatial transformation of horizontal postrotatory VOR during a head tilt in the pitch plane is illustrated in FIGURE 6 (rightmost

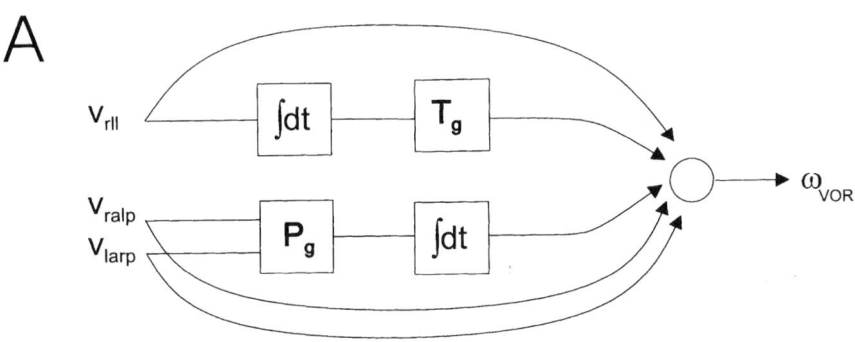

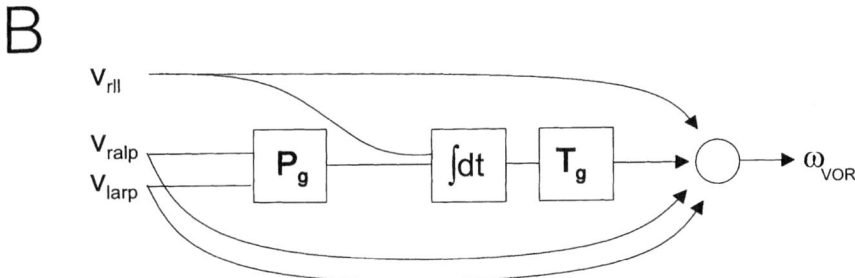

FIGURE 7. Flow diagram of spatial transformation mechanisms of semicircular-canal signals. (**A**) Separate spatial transformation channels for processing lateral and vertical semicircular-canal signals. (**B**) Combined spatial transformation channels for processing lateral and vertical semicircular canal-signals. *Abbreviations*: T_g: gravity-dependent rotatory transformation; $\int dt$: leaky integrator; $\mathbf{v}_{rll}$: vectorial input signal from activation of coplanar lateral semicircular-canals (subscript "rll" for right–left lateral); $\mathbf{v}_{ralp}$, $\mathbf{v}_{larp}$ vectorial input signal from activation of coplanar vertical semicircular canals (subscript "ralp" for right-anterior left-posterior, subscript "larp" for left-anterior right-posterior); ω_{VOR}: resultant vectorial angular velocity signal.

column). For this simulation, the semicircular canal afferent response was modeled with the torsion pendulum model of cupular motion.[35] The time constant of central eye-velocity integration was chosen to be equal along each of the principal axes (diagonal matrix $A = [a_{ik}]$ with $a_{ii} = 1/T_c = 1/15\ \text{s}^{-1}$). Note that the simulation predicts both the response wave form and the time constants of the principal and orthogonal components.

In contrast to the horizontal system, the spatial transformation in the vertical system can best be characterized by a gravity-dependent projection mechanism P_g (FIG. 7A) that is followed by a leaky integrator. Only a projection mechanism can account for the spatial frequency of the orthogonal response component in the vertical system that varies in contrast to the horizontal system with double the spatial frequency of head tilt.[10] The rotation and projection mechanisms provide an adequate description of the experimentally observed transformation characteristics of postrotatory eye velocity from horizontal VOR toward vertical or torsional VOR and between torsional and vertical VOR, respectively. A slightly different mechanism, however, seems to underlie the spatial shifts observed in postrotatory torsional or vertical VOR during head tilts from supine or 90-deg right or left ear-down position toward upright. The experimental data show a ratio of spatial frequency of the orthogonal response component and head tilt that lies between 1 and 2.[10] A possible explanation for this "mixed" transformation behavior could be that vertical semicircular canal signals share, after an initial projection transformation, part of the horizontal inertial network (FIG. 7B).

One of the important characteristics associated with the spatial representation of head velocity in the VOR concerns the marked attenuation of the principal response during a head tilt that can be described as a reduction of the apparent time constant (see FIG. 8). A

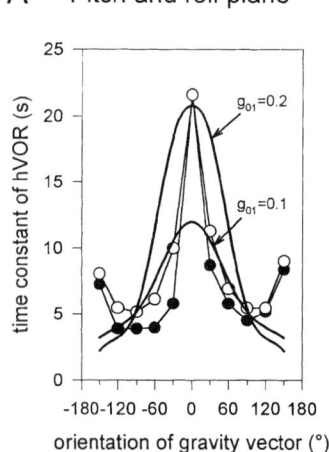
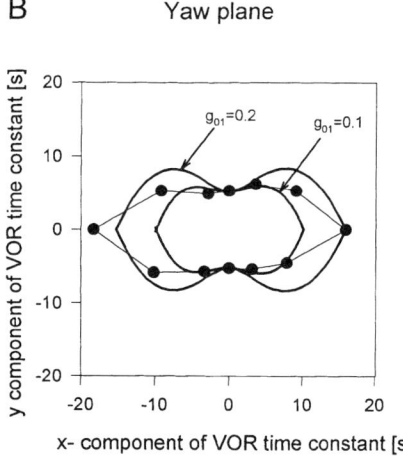

FIGURE 8. Vestibulo-ocular reflex dumping characteristics. (**A**) Experimental time constants of horizontal postrotatory response as a function of tilt in the pitch (*solid circles*) and roll (*open circles*) plane. (**B**) Time constants of torsional postrotatory response as a function of head tilt in the yaw plane. All experimental data from reference 10 (figs. 5 and 14). *Thick curves* represent model predictions for the horizontal VOR based on a downstream rotation model (A) and for the torsional VOR based on the projection model (B). *Model parameters*: Torsion pendulum model for afferent canal response with lower and upper cutoff time constant of $t_1 = 5$ s and $t_2 = 0.003$s, respectively; central leaky integration with $T = 15$ s; coupling ratio between inertial and direct response $g_{01} = 0.1$ and $g_{02} = 0.2$ (see *arrows*).

simple scaling mechanisms can explain only part of the experimentally observed dumping characteristics. Model simulations predict that most of this dumping can be obtained by choosing a convergence ratio of the direct and the inertially transformed response of the order of magnitude of 10:1. Such down-scaling of the inertial response could occur on convergence neurons at the output of the system (ω_{VOR} in FIG. 7A). As illustrated in FIGURE 8, this assumption provides an approximate description of the experimentally observed dumping characteristics. In the horizontal system it can account for the dumping in roll/pitch tilt responses for up to about 120 deg. For tilt angles larger than 120 deg, the VOR starts to exhibit underdamped response properties characterized by a secondary increase of the time constant. In the vertical system, the dumping characteristics can be described with a down-scaling of similar magnitude (solid curves marked with $g_{oi}=0.1$ in FIG. 8A, 8B). However, in both the horizontal and the vertical system, zero-tilt responses along the principal axis exhibited much larger time constants than those predicted by the assumed convergence ratio given earlier. On the other hand, the dumping expected on the basis of a larger convergence ratio of 5:1 that would account for the time constant observed with no tilt rather purely predicts the actually observed response dumping (solid curves marked with $g_{oi}=0.2$ in FIG. 8). This rather rough sensitivity analysis of the influence of the central weighing between direct and inertial response suggests that tilt dumping cannot be described with a linear model, but may rather require a jumplike nonlinear change in system behavior when a sudden head (and body) tilt occurs.

The VOR shares the inertial vestibular system with the optokinetic reflex that exhibits similar spatial reorientation of postrotatory eye velocity when tested in statically tilted head positions.[36–38] Similar to the VOR, the torsional and vertical optokinetic afterresponses exhibit quite different spatial response characteristics from the horizontal optokinetic afterresponse. For example, in an upright position, torsional or vertical velocity of optokinetic afternystagmus is dumped to zero without any appreciable reorientation toward gravity, a behavior that could be described with a projection mechanism. Interestingly, when tested with dynamic head tilts, the torsional and vertical optokinetic afterresponses exhibit similar, even though more variable, spatial response characteristics to the torsional and vertical VOR.[39] Taken together, the underlying spatial transformation mechanisms are likely to be analogous to those hypothesized for the VOR (FIG. 7B).

ACKNOWLEDGMENT

This study was supported by grants from the Swiss National Science Foundation (#31-47287.96), NIH (EY10851), NASA (NAGW-4377), and the Air Force Office of Scientific Research (F49620).

REFERENCES

1. BARLOW, J. S. 1964. Inertial navigation as a basis for animal navigation. J. Theor. Biol. **6**: 76–117.
2. BARLOW, J. S. 1966. Inertial navigation in relation to animal navigation. J. Inst. Navigation **19**: 3.
3. MAYNE, R. A. 1974. A systems concept of the vestibular organs. *In* Handbook of Sensory Physiology—The Vestibular System, H. H. Kornhuber, Ed.: 493–580. Springer-Verlag. New York.
4. TITTERTON, D. H. & J. L. WESTON. 1997. Strapdown Inertial Navigation Technology. Peter Peregrinus. Stevenage.
5. VIÉVILLE, T. & O. D. FAUGERAS. 1990. Cooperation of the inertial and visual systems. NATO ASI Series f **63**: 339–350.
6. GLASAUER, S. & D. M. MERFELD. 1997. Modelling three-dimensional vestibular responses during complex motion stimulation. *In* Three-dimensional Kinematics of Eye, Head, and Limb

Movements, M. Fetter, T. Haslwanter, H. Misslisch, and D. Tweed, Eds.: 387–398. Harwood. Amsterdam.
7. RAPHAN, T., V. MATSUO & B. COHEN. 1977. A velocity storage mechanism responsible for optokinetic nystagmus (OKN), optokinetic afternystagmus (OKAN) and vestibular nystagmus. *In* Control of Gaze by Brain Stem Neurons, R. Baker and A. Berthoz, Eds. Elsevier/North-Holland. Amsterdam.
8. RAPHAN, T., V. MATSUO & B. COHEN. 1979. Velocity storage in the vestibulo-ocular reflex arc (VOR). Exp. Brain Res. **35**: 229–248.
9. RAPHAN, T. & B. COHEN. 1988. Organizational principles of velocity storage in three dimensions: the effect of gravity on cross-coupling of optokinetic afternystagmus. Ann. N.Y. Acad. Sci. **545**: 74–92.
10. ANGELAKI, D. E. & B. J. M. HESS. 1994. Inertial representation of angular motion in the vestibular system of rhesus monkeys. I. Vestibuloocular reflex. J. Neurophysiol. **71**: 1222–1249.
11. HESS, B. J. M. & D. E. ANGELAKI. 1997. Inertial vestibular coding of motion: concepts and evidence. Curr. Opin. Neurobiol. **7**: 860–866.
12. HESS, B. J. M. 1990. Dual-search coil for measuring 3-dimensional eye movements in experimental animals. Vision Res. **30**: 597–602.
13. HESS, B. J. M., A. J. VAN OPSTAL, D. STRAUMANN & K. HEPP. 1992. Calibration of 3-dimensional eye position using search coil signals in the rhesus monkey. Vision Res. **32**: 1647–1654.
14. ANGELAKI, D. E., B. J. M. HESS, Y. ARAI & J.-I. SUZUKI. 1996. Adaptive modification of primate vestibulo-ocular reflex to altered peripheral vestibular inputs. I. Frequency-specific recovery of horizontal VOR after inactivation of the lateral semicircular canals. J. Neurophysiol. **76**: 2941–2953.
15. EWALD, J. R. 1892. Physiologische Untersuchungen über das Endorgan des Nervus octavus. Bergmann. Wiesbaden, Germany.
16. ANGELAKI, D. E., M. Q. MCHENRY, J. D. DICKMAN, S. D. NEWLANDS & B. J. M. HESS. 1999. Computation of inertial motion: neural strategies to resolve ambiguous otolith information. J. Neurosci. **19**: 319–327.
17. ANGELAKI, D. E. 1998. Three-dimensional organization of otolith-ocular reflexes in rhesus monkeys. III. Responses to translation. J. Neurophysiol. **80**: 680–695.
18. BALOH, R. W., K. BEYKIRCH, V. HONRUBIA & R.D. YEE. 1988. Eye movements induced by linear acceleration on a parallel swing. J. Neurophysiol. **60**: 2000–2013.
19. NIVEN, J. I., W.C. HIXSON & M. J. CORREIA. 1966. Elicitation of horizontal nystagmus by periodic linear acceleration. Acta Otolaryngol. **62**: 429–441.
20. PAIGE, G. D. & D. L. TOMKO. 1991. Eye movement responses to linear head motion in the squirrel monkey. I. Basic characteristics. J. Neurophysiol. **65**: 1170–1182.
21. SCHWARZ, U. & F. A. MILES. 1991. Ocular responses to translation and their dependence on viewing distance. I. Motion of the observer. J. Neurophysiol. **66**: 852–864.
22. TELFORD, L., S. H. SEIDMAN & G. D. PAIGE. 1997. Dynamics of squirrel monkey linear vestibuloocular reflex and interactions with fixation distance. J. Neurophysiol. **78**: 1775–1790.
23. GUEDRY, F. E. 1974. Psychophysics of vestibular sensation. *In* Handbook of Sensory Physiology—The Vestibular System, Part 2—Psychophysics, Applied Aspects and General Interpretations, H. H. Kornhuber, Ed.: 3–154. Springer-Verlag. Berlin.
24. YOUNG, L. R. 1974. Perception of the body in space: mechanisms. *In* Handbook of Physiology—The Nervous System, Chap. 22; I. Darian-Smith, J. M. Brookhart, and V. B. Mountcastle, Eds.: 1023–1066. Springer-Verlag. Berlin.
25. BENSON, A. J. 1974. Modification of the response to angular acceleration by linear acceleration. *In* Handbook of Sensory Physiology, Vol. VI/2. H. H. Kornhuber, Ed.: 281–320. Springer-Verlag. Berlin.
26. BENSON, A. J. & M. A. BODIN. 1966. Comparision of the effect of the direction of the gravitational acceleration on post-rotational responses in yaw, pitch and roll. Aerosp. Med. **37**: 889–897.
27. WAESPE, W., B. COHEN & T. RAPHAN. 1985. Dynamic modification of the vestibuloocular reflex by the nodulus and uvula. Science. **228**: 199–202.
28. ANGELAKI, D. E. & J. H. ANDERSON. 1991. The horizontal vestibulo-ocular reflex during linear acceleration in the frontal plane of the cat. Exp. Brain Res. **86**: 40–46.

29. HARRIS, L. R. 1987. Vestibular and optokinetic eye movements evoked in cat by rotation about a tilted axis. Exp. Brain Res. **66**: 522–532.
30. HARRIS, L. R. & G. R. BARNES. 1987. Orientation of vestibular nystagmus is modified by head tilt. *In* The Vestibular System: Neurophysiology and Clinical Research, M. D. Graham and J. L. Keminink, Eds.: 539–549. Raven. New York.
31. MERFELD, D. M., L. R. YOUNG, D. L. TOMKO & G. D. PAIGE. 1991. Spatial orientation of VOR to combined vestibular stimuli in squirrel monkeys. Acta Otolaryngol. Suppl. **481**: 287–292.
32. MERFELD, D. M., L. R. YOUNG, G. D. PAIGE & D. L. TOMKO. 1993. Three-dimensional eye movements of squirrel monkeys following postrotatory tilt. J.Vest. Res. **3**:123–139.
33. ANGELAKI, D. E. & B. J. M. HESS. 1995. Inertial representation of angular motion in the vestibular system of rhesus monkeys. II. Otolith-controlled transformation that depends on an intact cerebellar nodulus. J. Neurophysiol. **73**: 1729–1751.
34. ANGELAKI, D. E., J.-I. SUZUKI & B. J. M. HESS. 1995. Differential processing of semicircular canal signals in the vestibulo-ocular reflex. J. Neurosci. **15**: 7201–7216.
35. FERNÁNDEZ, C. & J. M. GOLDBERG. 1971. Physiology of peripheral neurons innervating semicircular canals in the squirrel monkey. II. Response to sinusoidal stimulation and dynamics of peripheral vestibular system. J. Neurophysiol. **34**: 661–675.
36. DAI, M., T. RAPHAN & B. COHEN. 1991. Spatial orientation of the vestibular system: dependence of optokinetic afternystagmus on gravity. J. Neurophysiol. **66**: 1422–1438.
37. RAPHAN, T. & D. STURM. 1991. Modelling the spatiotemporal organization of velocity storage in the vestibuloocular reflex by optokinetic studies. J. Neurophysiol. **66**: 1410–1420.
38. GIZZI, M., T. RAPHAN, S. RUDOLF & B. COHEN. 1994. Orientation of human optokinetic nystagmus to gravity: a model based approach. Exp. Brain Res. **99**: 347–360.
39. HESS B. J. M. & D. E. ANGELAKI. 1994. Inertial representation of visual and vestibular self-motion signals. *In* Multisensory Control of Posture. T. Mergner and F. Hlavacka, Eds.: 183–190. Plenum Press. New York.

Cross-Striolar and Commissural Inhibition in the Otolith System

Y. UCHINO,[a,e] H. SATO,[a] K. KUSHIRO,[a] M. ZAKIR,[a] M. IMAGAWA,[a] Y. OGAWA,[b] M. KATSUTA,[c] AND N. ISU[d]

[a]*Department of Physiology, Tokyo Medical University, 6-1-1 Shinjuku, Shinjuku-ku, Tokyo 160-0022, Japan*

[b]*Department of Otolaryngology, Tokyo Medical University, 6-1-1 Shinjuku, Shinjuku-ku, Tokyo 160-0022, Japan*

[c]*Department of Orthopedic-Surgery, Tokyo Medical University, 6-1-1 Shinjuku, Shinjuku-ku, Tokyo 160-022, Japan*

[d]*Department of Information and Knowledge Engineering, Faculty of Engineering, Tottori University, Tottori 680-0945, Japan*

ABSTRACT: Neural connections from the saccular and utricular nerves to the ipsilateral vestibular neurons and the commissural effects were studied by using intracellular recordings of excitatory (E) and inhibitory (I) postsynaptic potentials (PSPs) in vestibular neurons of cats after focal stimulation of the saccular and the utricular maculae. Neural circuits from the maculae to vestibular neurons, termed cross-striolar inhibition, may provide a mechanism for increasing the sensitivity to linear acceleration and tilt of the head. It was examined whether secondary vestibular neurons activated by an ipsilateral otolith organ received a commissural inhibition from a contralateral otolith organ that occupied the same geometric plane. Results suggest that utricular-activated vestibular neurons receiving commissural inhibition may provide a mechanism for increasing the sensitivity to horizontal linear acceleration and tilt of the head. The commissural inhibition of the saccular system was much weaker than that of the utricular system.

INTRODUCTION

In mammals, the otolith organs of the inner ear sense linear acceleration. These organs consist of two receptors, the saccular macula and the utricular macula, oriented in planes perpendicular to each other. The saccular macula senses mainly vertical, and the utricular macula senses mainly horizontal linear acceleration. The mechanisms of excitation and inhibition of these receptors are fundamentally identical. Hair cells in the receptors are geometrically arranged with morphological polarity and a directional kinocilium.[1–3] Two groups of oppositely polarized hair cells are divided by the striola in the middle of the receptor. The striola is shaped sigmoidally in the saccular macula and semicircularly in the utricular macula, so that by any given direction of linear acceleration, some hair cells will

---

[e]To whom correspondence may be addressed. Phone 81-3-3351-6141; fax: 81-3-3351-6544; e-mail: y-uchino@tokyo-med.ac.jp

be depolarized.[4,5] At the same time, some hair cells on the opposite side of the striola will be hyperpolarized.

During a series of experiments we often recorded monosynaptic EPSPs followed by IPSPs with large amplitude in vestibular neurons after selective stimulation of the saccular nerve. We hypothesized that primary afferents innervating hair cells on one side of the striola terminate monosynaptically on vestibular neurons, and that primary afferents innervating hair cells on the other side disynaptically project on the same neurons via inhibitory interneurons. The sensitivity of the neurons to linear acceleration should then increase by a "disinhibitory mechanism." We recently found this disinhibitory mechanism in the saccular system, and named it "cross-striolar inhibition."[6] If the same mechanism occurs in the utricular system, the cross-striolar inhibition would be generalized in the otolith system. We studied input convergence from both sides of the striola in the utricular macula to single vestibular neurons in order to determine whether the cross-striolar inhibition exists in the utricular system.

It is well known that the commissural inhibition in the semicircular-canal system functions to increase the sensitivity of secondary vestibular neurons during rotatory movements of the head[7] and selective stimulation of the ampullary nerves,[8,9] but little is known about the pathways that link the bilateral otolith organs that occupied the same geometric plane.[10,11] In the present study, we investigated whether otolith-activated secondary vestibular neurons also receive commissural inhibition from a contralateral macula in the same geometric plane. The intracellular recording from secondary vestibular neurons was performed following electrical stimulation of bilateral nerves innervating the utricular and the saccular maculae.

METHODS

Experiments were performed on 42 cats in conformity with the "Guiding Principles for the Care and Use of Animals in the Field of Physiological Sciences; The Physiological Society of Japan, 1988." Each cat was initially anesthetized with an intramusclar injection of ketamine hydrochloride (Ketalar, Parke-Davis; 15–20 mg/kg) followed by halothane (Fluothane, Zeneca) -nitrous oxide by inhalation. The cat was then decerebrated at the precollicular level, paralyzed with pancuronium bromide, and artificially ventilated. Pairs of fine silver electrodes (acupuncture needles insulated except for usually 500 µm of the tips and glued together at an interelectrode distance of approximately 0.8 mm) were inserted for the focal stimulation of the bilateral saccular and utricular maculae. The inner ear was drained of liquid by using a small piece of twisted cotton. To prevent the spread of stimulus current, the nerves and electrodes were covered with a warm semisolid paraffin–Vaseline mixture. Cathodal or anodal current pulses of 150- or 200-µs duration were applied to the saccular and the utricular maculae at a rate of 2–3 Hz.

For the focal stimulation of the saccular or the utricular macula, either the utricular or the saccular nerve and three ampullary nerves were transected in the left inner ear.[12,13] Two groups of animals were used when the saccular macula was stimulated. The group A animals had electrode placement across the striola, such that one electrode was in the rostroventral part of the saccular macula and the other was in the caudo-dorsal part (FIG. 1A, B and FIG. 2A$_S$). A small constant current (usually below 30 µA) was applied to activate a small area of the macula; the activated area could be changed by reversing the polarity of the stimulus current. In the group B, both electrodes were placed on the dorsal edge of the saccular macula, not across the striola (FIG. 2B$_S$). When the focal stimulation was given to the utricular macula, one electrode was placed beside the medial edge and the other beside the lateral edge of the utricular macula (FIG. 3C). For the study of commissural effects, a pair of electrodes was placed in the middle part of either the saccular or the utricular macula after the transection of the other vestibular nerves.

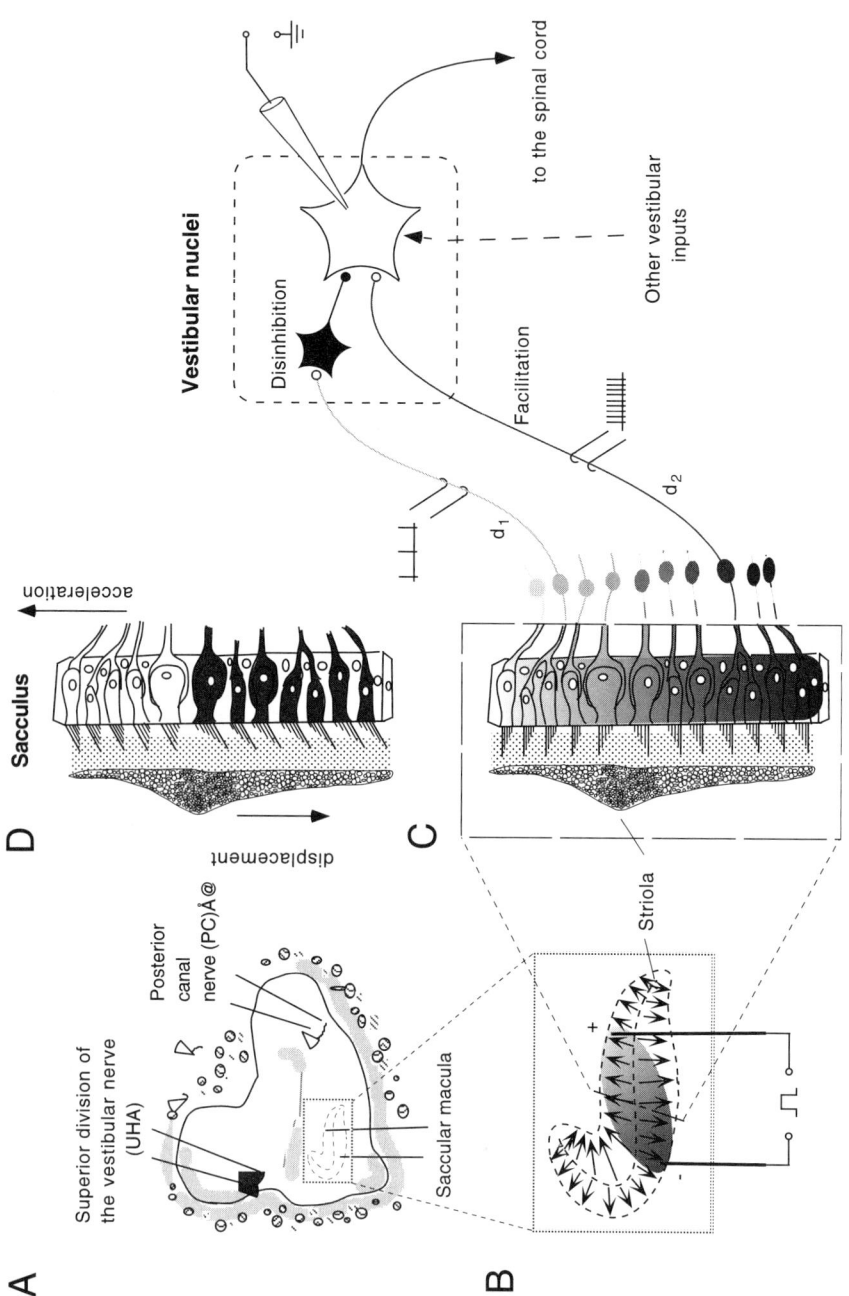

FIGURE 1. (**A**) Ventrolateral view of the cat inner ear showing the electrode position for focal stimulation of the saccular macula and the other vestibular nerves. Bipolar silver electrodes were inserted into the rostro-ventral and caudodorsal parts of the saccular macula (group A, see text). (**B**) Schematic diagram of the saccular macula showing the morphological polarity of hair cells. *Arrows* show the direction of kinocilia (the thickest and longest hair in each hair cell shown in D and C). Stimulus current was applied so that the rostro-ventral electrode "a" was negative. The size of the saccular macula is around 2.5 mm rostro-caudally and around 1.0 mm ventrodorsally. (**C**) Cross-section of the saccular receptor at the *dashed line* in B and afferent connections, and the cross-striolar inhibition predicted from the results (see text). Saccular afferents were depolarized around the cathode electrode and hyperpolarized around the anode electrode. Therefore, the firing rate of afferent d_1 decreased, while that of afferent d_2 increased. (**D**) Cross-section of the saccular receptor showing the functional pseudoequivalence of the focal stimulation in C. If the acceleration is upward, the displacement of hairs is downward; hair cells under the striola then depolarize (*dark*) and those above it hyperpolarize (*light*). Superior division of the vestibular nerve (U, utricular; H, horizontal; A, anterior) and the posterior canal nerve (PC) were stimulated during different series of experiments. (Modified from Uchino *et al.*,[6] by permission.)

The animal was suspended by hip pins and a clamp on the spinal process of the T_1 vertebra. The caudal part of the cerebellum was aspirated to expose the floor of the fourth ventricle. The amplitudes of N1 field potentials, which are due to the monosynaptic activation of secondary vestibular neurons evoked by provoked fibers of the otolith nerve, were measured from the caudal part of the lateral vestibular nucleus or the rostral part of the descending

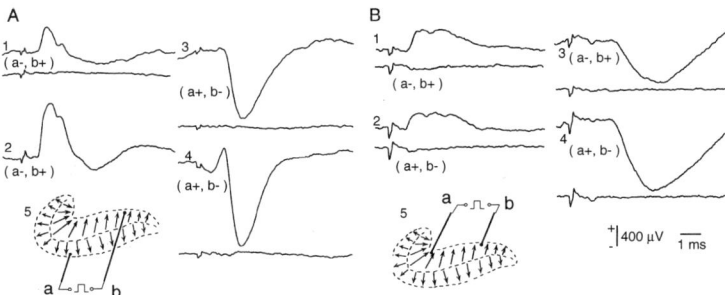

Focal Stimulation of the Saccular Macula

FIGURE 2. (**A**) Intracellular responses in a vestibular neuron after stimulation of the saccular macula in group A (see text). A_1 and A_2: When an 8-µA (1.3 × N1 threshold) or a 10-µA (1.7 × N1 threshold) stimulus current was applied to the saccular macula with the polarity shown at panel A_5, a monosynaptic EPSP followed by an IPSP was recorded in a vestibular neuron. A_3 and A_4 : When a 3-µA (0.5 × N1 threshold) or a 6-µA (1.0 × N1 threshold) stimulus current was applied with the opposite polarity to that shown in panel A_5, a disynaptic IPSP (A_3) or an EPSP-IPSP sequence (A_4) was recorded in the same neuron. The *lower trace* in each record indicates extracellular potential. (**B**) Intracellular responses in two vestibular neurons in the same cat after stimulation of the dorsol edge of the saccular macula in group B (see text). B_1: When a 35-µA (2.3 × N1 threshold) stimulus current was applied to the saccular macula with the electrode position shown in panel B_5, a monosynaptic EPSP followed by a small EPSP was recorded in a vestibular neuron. B_2: When a 100-µA (3.3 × N1 threshold) stimulus current was applied with opposite polarity, a monosynaptic EPSP similar to that of B_1 was recorded in the same neuron. B_3: When a 40-µA (2.7 × N1 threshold) stimulus current was applied as shown in panel B_5, a disynaptic IPSP was recorded in a vestibular neuron. B_4: When an 80-µA (2.7 × N1 threshold) stimulus current was applied with the opposite polarity, an IPSP similar to those of B_3 was recorded in the same neuron. The *lower trace* in each record indicate extracellular potential. (Modified from Uchino *et al.*,[6] by permission.)

Focal stimulation of the utricular macula

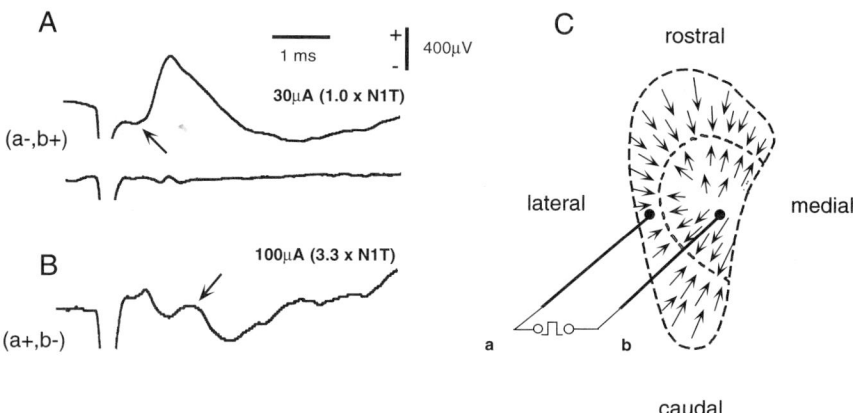

FIGURE 3. (**A, B**) Intracellular responses in a vestibular neuron after focal stimulation of the utricular macula. (**C**) Sechematic diagram of the utricular macula showing the morphological polarity (*arrows*) of hair cells. The striola and the margin of the macula are indicated by *broken lines*. (**A**) When a 30-µA (1.0 x N1 threshold) stimulus current was applied to the utricular macula with the polarity shown in C (lateral "a," negative, and medial "b," positive), a monosynaptic EPSP followed by an IPSP was recorded in a vestibular neuron. (**B**) When a 100-µA (3.3 x N1 threshold) stimulus current was applied with opposite polarity to C ("a," lateral positive, and "b," medial negative), a disynaptic IPSP was recorded in the same neuron. *Oblique arrows* in A and B indicate diverging points of the intracellular and extracellular recordings. The *lower trace* in each record indicates extracellular potential.

nucleus with glass micropipettes containing a 2-M NaCl solution saturated with Fast Green dye to check a possible current spread to the other vestibular nerves.[13] The intracellular recording from lateral and descending vestibular nucleus neurons was done by using micropipettes filled with 2-M K citrate and having a resistance of 3–10 MΩ. Postsynaptic potentials and field potentials were routinely averaged 10–30 times. The thresholds of the N1 field potential were comparable with those reported in previous papers.[6,13–15]

RESULTS

Cross-striolar Inhibition in the Saccular System

EPSPs and IPSPs were recorded in lateral and descending vestibular nucleus neurons after focal stimulation of the saccular macula. In group A, 22 (61%) of 36 neurons showed the opposite pattern of PSPs when the stimulus polarity was reversed; EPSPs with one stimulus polarity and IPSPs with the other polarity, or vice versa. In 11 of these 22 neurons, the EPSPs were followed by smaller IPSPs as stimulus intensity was increased up to around 0.7–2.0 × N1 threshold. Similarly, the IPSPs were preceded by smaller EPSPs in 3 of the 22 neurons when stimuli of reversed polarity were increased in intensity up to around 0.5–1.0 × N1 threshold. The typical patterns of PSPs evoked by focal stimulation of the saccular macula are shown in FIGURE 2A. After focal stimulation of the macula (FIG. 2A$_5$), an EPSP followed by an IPSP was recorded in a vestibular neuron (FIG. 2A$_1$, electrode "a" negative and electrode "b" positive polarity). When the stimulus current was increased, the amplitudes of both components became larger (FIG. 2A$_2$). The EPSP had a latency of 0.9 ms at an intensity of around 2 × N1 threshold. Focal stimulation of the opposite polarity (electrode "a" positive and electrode "b" negative polarity) evoked an IPSP with a latency of 1.5 ms in the same neuron (FIG. 2A$_3$). When the stimulus current was increased, an EPSP was evoked prior to the IPSP (FIG. 2A$_4$). The late IPSP evoked by the former polarity of stimulation and the early EPSP evoked by the latter polarity are presumably due to the spread of the excited area beyond the striola. In the remaining 14 (39%) neurons, however, the patterns of PSPs were not changed irrespective of stimulus polarity: EPSPs in 12 neurons (including EPSP-IPSP sequences in 5 neurons) and IPSPs in 2 neurons. The majority of latencies of the EPSPs were monosynaptic (≤1.2 ms), while those of the IPSPs were disynaptic (≥1.5 ms).[6] Thus, it appears that a group of saccular afferents had monosynaptic excitatory contacts with vestibular neurons, and another group of afferents that originated from hair cells with an opposite morphological polarity had disynaptic connections with the same vestibular neurons through single inhibitory interneurons (FIG. 1C).

In group B, monosynaptic EPSPs or disynaptic IPSPs were evoked in most of the vestibular neurons by focal stimulation of the dorsal edge of the saccular macula (FIG. 2B). In these cells, the response patterns were always the same regardless of the polarity of stimulus current at an intensity of ≤ 2.0 × N1 threshold. FIGURE 2B$_{1,2}$ illustrates responses of a cell, in which a monosynaptic EPSP (latency of 0.9 ms) was evoked by both stimulus polarities with a threshold approximately the same as the N1 threshold. Another cell had the other response that a disynaptic IPSP (latency of 2.3 ms) was evoked by both stimulus polarities with a threshold of ≤ 2.0 × N1 threshold (Fig. 2B$_{3,4}$). In 31 neurons located in the ventral part of the lateral nucleus and the rostral part of the descending nucleus, the response patterns kept the same when the stimulus polarity was changed: EPSPs in 25 neurons (including EPSP-IPSP sequences in 5 neurons) and IPSPs in 6 neurons. The majority of latencies of the EPSPs were monosynaptic (≤1.2 ms), while those of the IPSPs were disynaptic (≥1.5 ms).

Cross-striolar Inhibition in the Utricular System

Ten (40%) of 25 neurons showed opposite patterns of PSPs when the polarity of stimulus current to the ipsulateral utricular macula was reversed; EPSPs with one stimulus polarity and IPSPs with the other polarity, or vice versa. In 3 of the 10 neurons, an EPSP was followed by a small IPSP after a stimulus at an intensity of approximately 1.0 × N1 threshold. Similarly, an IPSP was preceded by a small EPSP after a stimulus of reversed polarity at an intensity of approximately 1.0 × N1 threshold in 1 of the 10 neurons. The typical pattern of PSPs evoked by focal stimulation of the utricular macula is shown in FIGURE 3. After focal stimulation of the macula (lateral electrode "a" negative and medial electrode "b" positive polarity, FIG. 3C), an EPSP followed by an IPSP was recorded in a vestibular neuron (FIG. 3A). The EPSP had a latency of 0.8 ms at an intensity of around 1.0 × N1 threshold. Focal stimulation of the opposite polarity (medial electrode "b" negative and lateral electrode "a" positive polarity) evoked an IPSP with a latency of 1.8 ms in the same neuron (FIG. 3B). In 5 of the other 15 neurons, however, the pattern of PSPs was not changed irrespective of stimulus polarity; EPSPs in 3 neurons (including a EPSP-IPSP sequence in 1 neuron) and IPSPs in 2 neurons were evoked by the both polarity. In the remaining 10 neurons, EPSPs were evoked only when the lateral electrode (the medial electrode for one neuron) was a cathode. The majority of the latencies of the EPSPs were monosynaptic (≤1.2 ms), while latencies of the IPSPs were disynaptic or polysynaptic (≥1.5 ms).[8,14,16–18] Thus, it appears that a group of utricular afferents had monosynaptic excitatory contacts with some vestibular neurons, and another group of afferents that originated from hair cells with an opposite morphological polarity had disynaptic or polysynaptic connections with the same vestibular neurons through inhibitory interneurons.

Commissural Effects of the Saccular and the Utricular Systems

The commissural inhibition on 49 saccular-activated secondary vestibular neurons was studied by the selective electrical stimulation of the contralateral saccular nerve. The secondary vestibular neurons were identified by their orthodromic latencies (≤ 1.2 ms, i.e., monosynaptic) to stimulation of the saccular nerve ipsilateral to the recording side.[19] The majority of saccular-activated secondary vestibular neurons (40/49, 82%) showed no visible potentials in response to stimulation of the contralateral saccular nerve (FIG. 4B). Their latencies of responses to ipsilateral saccular nerve stimulation are shown in FIGURE 4B. Seven (14%) of 49 neurons received inhibition and two (4%) of 49 neurons received facilitation with polysynaptic latencies (≥ 1.8 ms) from the contralteral saccular nerve. FIGURE 4 (insets) shows PSPs evoked by bilateral saccular nerve stimulation in a same neuron. The neuron generated a monosynaptic EPSP following stimulation of the ipsilateral saccular nerve at an intensity of 40 µA (2.7 × N1T) (left inset), and hyperpolarized in response to

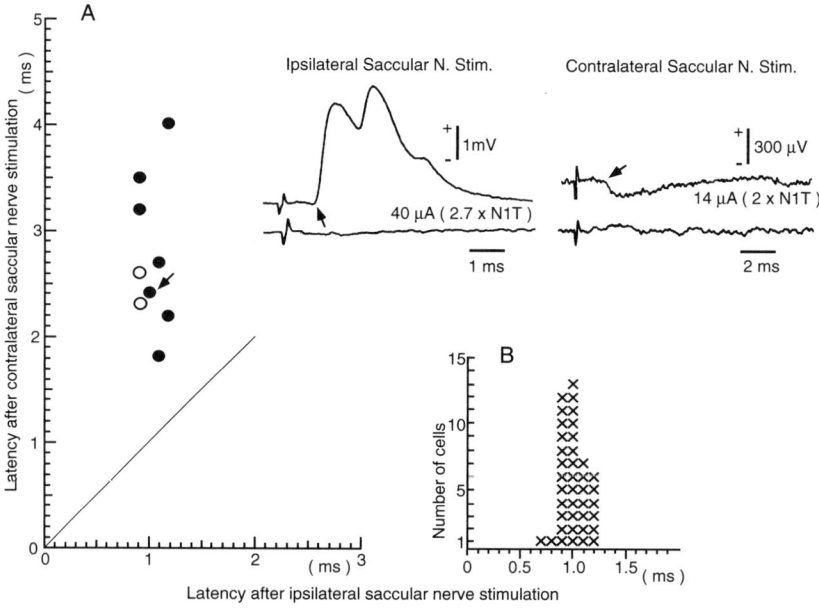

FIGURE 4. Latencies of intracellular responses of vestibular neurons to bilateral saccular nerve stimulation. Ipsilateral and contralateral saccular nerve stimulation evoked a monosynaptic (1.0 ms) double-peaked EPSP (*left inset*) and a disynaptic (2.4 ms) inhibitory potential (*right inset*), respectively. *Lower trace* in each record indicates juxtacellular record. *Arrows* indicate diverging points of the intracellular and the juxtacellular records. (**A**) Latencies of monosynaptic EPSPs following stimulation of the ipsilateral saccular nerve (*absissa*) versus latencies of polysynaptic inhibitory (*closed circles*) or facilitatory (*open circles*) potentials after stimulation of the contralateral saccular nerve (*ordinate*) is shown. Arrow shows the cell of the inset. (**B**) Latency histogram of monosynaptic EPSPs recorded from ipsilaterally saccular-activated cells that generated no visible potentials by contralateral saccular nerve stimulation.

stimulation of the contralateral saccular nerve at an intensity of 14 µA (2 × N1T)(right inset). FIGURE 4A illustrates latencies of commissural inhibition (closed circles) and facilitation (open circle) versus those of responses to ipsilateral stimulation. Amplitude of commissural inhibitory potentials in response to stimulation of the contralateral saccular nerve was 216±62 µV (mean ± SD, $n=7$).

On the other hand, about half (14/32, 44%) of the utricular-activated secondary vestibular neurons received commissural inhibition from the contralateral utricular nerve with a latency of disynaptic or more ($\geqq 1.8$ ms) (FIG. 5, inset). One (3%) of the 32 neurons received facilitation, instead. FIGURE 5A illustrates latencies of commissural inhibition (closed circles) and facilitation (open circle) versus those of responses to ipsilateral stimulation. The remaining 17 (53%) neurons made no visible responses following stimulation of the contralateral utricular nerve. Their latencies of responses to ipsilateral utricular nerve stimulation are shown in FIGURE 5B. The amplitudes of commissural inhibitory potentials evoked by stimulation of the utricular nerve were larger [1400 ± 1300 µV(mean ± SD), $n=13$] than those of commissural inhibitory potentials evoked by the saccular nerve stimulation (FIG. 6).

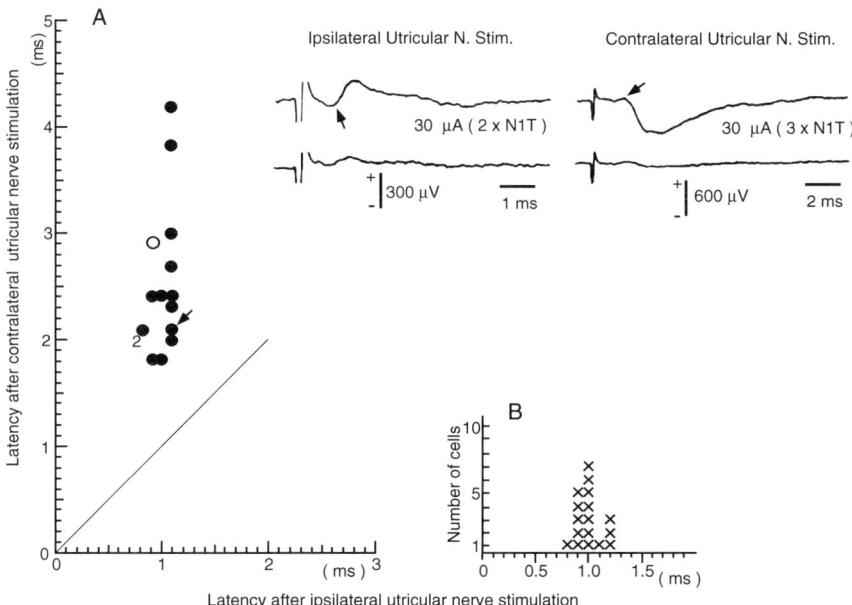

FIGURE 5. Latencies of intracellular responses in vestibular neurons to bilateral utricular nerve stimulation. Ipsilateral and contralateral utricular nerve stimulation evoked a monosynaptic (1.1 ms) EPSP (*left inset*) and a disynaptic (2.2 ms) inhibitory potential (*right inset*), respectively. *Lower trace* in each record indicates juxtacellular record. *Arrows* indicate diverging points of the intracellular and the juxtacellular records. (**A**) Latencies of monosynaptic EPSPs following stimulation of the ipsilateral utricular nerve (*abscissa*) versus latencies of polysynaptic inhibitory (*closed circles*) or facilitatory (*open circles*) potentials after stimulation of the contralateral utricular nerve. *Arrow* shows the cell of the inset. (**B**) Latency histogram of monosynaptic EPSPs recorded from ipsilaterally utricular-activaterd cells that generated no visible potentials by contralateral utricular nerve stimulation.

DISCUSSION

In the mammalian semicircular canal system, hair cells in the crista of individual canals have the same polarity in direction of activation. When hair cells in one semicircular canal depolarize during angular acceleration, those in the contralateral coplanar canal hyperpolarize. This circuitry is in keeping with the morphology of the ampullary crista. Sensitivity is then increased by the convergence of facilitation from one side and disinhibition from the other side, or by disfacilitation and inhibition, the latter in each case via commissural fibers.[7,8,16,20–22]

In the otolith system, two groups of hair cells are divided by the striola in the middle of each macula. Hair cells across the striola in each organ have opposite polarities of sensitive direction. When hair cells on one side of the striola depolarize during linear acceleration, those on the opposite side hyperpolarize. We hypothesized that facilitation–disinhibition or disfacilitation–inhibition circuits could exist unilaterally. In this study and a previous investigation,[6] we showed that stimulation of afferents from the both sides of the striola in saccular macula frequently evoked monosynaptic EPSPs and

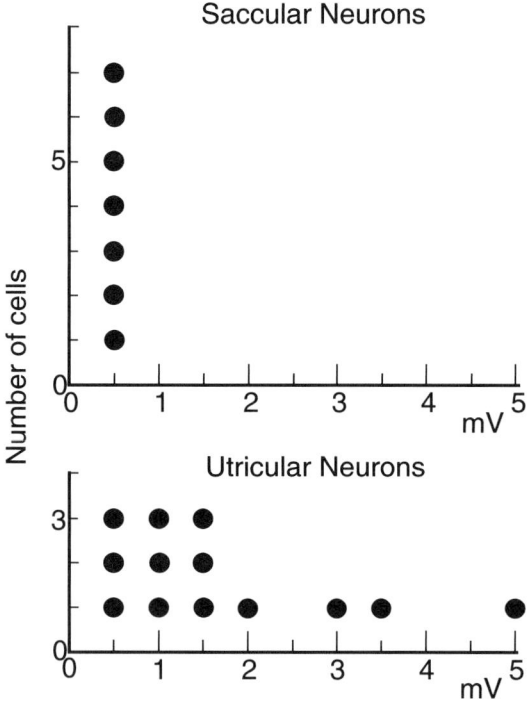

FIGURE 6. Amplitude histograms of commissural inhibitory potentials after saccular nerve and utricular nerve stimulations. Otolith-activated secondary vestibular neurons partly received commisural inhibition from the contralateral coplanar macula. The greater part of utricular-activated neurons showed larger amplitude of commissural inhibition compared with saccular-activated neurons.

disynaptic IPSPs in the same vestibular neurons. This push–pull circuit, which we named "cross-striolar inhibition," may explain the increasing sensitivity to vertical linear acceleration (FIG. 1C). This circuit not only has high sensitivity but also noise-resistant input characteristics; common-mode noise is rejected.

In the utricular system, focal stimulation of the medial and the lateral edges of the utricular macula also evoked monosynaptic EPSPs and disynaptic IPSPs in the same vestibular neurons (FIG. 3). The percentage of neurons that showed "cross-striolar inhibition" however, was not as large as that in the saccular system. The "cross-striolar inhibition," seems to function more impotantly in the saccular system than the utricular system. Commissural inhibition might function primarily in the utricular system instead. We reexamined the existence of commissural connections in the utricular system, and hypothesized that vestibular neurons that received primary afferents innervating hair cells on the medial part of the striola inhibit contralateral vestibular nucleus neurons that received primary afferents innervating hair cells on the medial part of the striola in the contralateral macula. This hypothesis is based on the morphological polarity of hair-cell arrangement; when hair cells on the medial part of the striola depolarize during head translation to the contralateral side, those on the contralatelar side hyperpolarize. The same neuronal mechanism is also applicable to hair cells on the lateral part of the striola.

In the present study, approximately half of the vestibular neurons activated by stimulation of the utricular nerve received inhibition from the contralateral utricular nerve. The amplitudes of the commissural inhibitory potentials evoked by the stimulation of the contralateral utricular nerve were larger (1400 ± 1300 μV) than those of the commissural inhibitory potentials evoked by the saccular nerve stimulation. In the saccular system, about 85% of vestibular neurons showed no visible potentials following stimulation of the contralateral saccular nerve, and the amplitudes of the commissural inhibitory potentials were small (216 ± 62 μV), if any (see the Results section). Our findings suggest that the commissral inhibitory mechanism may more strongly act in the utricular system compared with the saccular system. This mechanism may play a major role in increasing sensitivity to horizontal linear accelerations in the utricular system.

In the present study on commissural effects, we stimulated the whole utricular macula in each side. However, we need to stimulate the two different groups of hair cells separately to examine our hypothesis mentioned earlier. If the commissural inhibition is involved in the mechanism to increase the sensitivity to horizontal linear accelerations, the commissural inhibition should originate from the contralateral hair cells with opposite polarity to that of the ipsilateral hair cells. If a commissural effect stems from contralateral hair cells with the same polarity, the effect should be facilitatory to increase the sensitivity. In our preliminary experiments, commissural facilitation was rarely seen in the utricular system. This scarcity of commissural facilitation may result from improper positioning of the stimulus electrodes. The present configuration of the stimulus electrodes in the utricular macula might not have been across the striola. We cannot completely rule out this possibility, because it is difficult to asertain the striola, because it is covered by the utricular nerve in the utricular macula.

ACKNOWLEDGMENTS

This study was supported by a research grant from the Japan Space Forum promoted by NASDA (National Space Development Agency of Japan) and by a grant from the Japan Ministry of Education, Science, and Culture, grant-in-aid for scientific research 09671760. We thank Miss K. Takayama for secretarial assistance.

REFERENCES

1. FLOCK, Å. 1964. Structure of the macula utriculi with special reference to directional interplay of sensor responses as revealed by morphological polarisation. J. Cell Biol. **22**: 413–431.
2. LINDEMAN, H. H. 1973. Anatomy of the otolith organs. Adv. Oto-Rhino-Laryngol. **20**: 405–433.
3. WERSÄLL, J. 1956. Studies on the structure and innervation of the sensory epithelium of the cristae ampullaries in the guinea pig: a light and electronmicroscopic investigation. Acta Oto-Laryngol. Suppl. **126:** 1–85.
4. GOLDBERG, J. M., et al. 1990.The vestibular nerve of the chinchilla. V. Relation between afferent discharge properties and peripheral innervation patterns in the utricular macula. J. Neurophysiol. **63**: 791–804.
5. SHOTWELL, S. L., R. JACOBS & A. J. HUDSPETH. 1981. Directional sensitivity of individual vertebrate hair cells to controlled deflection of their hair bundles. Ann. N.Y. Acad. Sci. **374**: 1–10.
6. UCHINO, Y., H. SATO & H. SUWA. 1997. Excitatory and inhibitory inputs from saccular afferents to single vestibular neurons in the cat. J. Neurophysiol. **78**: 2186–2192.
7. SHIMAZU, H. & W. PRECHT. 1966. Inhibition of central vestibular neurons from the contralateral labyrinth and its mediating pathway. J. Neurophsyiol. **29**: 467–492.
8. KASAHARA, M. & Y. UCHINO. 1974. Bilateral semicircular canal inputs to neurons in cat vestibular nuclei. Exp. Brain Res. **20**: 285–296.
9. SANS, A., J. RAYMOND & R. MARTY. 1972. Projections des crêtes ampullaires et de l' utricle dans les noyaux vestibulaires primaires. Étude microphysiologique et corrélations anatomo-fonctionelles. Brain Res. **44**: 337–355.
10. SHIMAZU, H. & C. SMITH. M.1971. Cerebellar and labyrinthine influence on single vestibular neurons identified by natural stimuli. J. Neurophysiol. **34**: 493–508.
11. WILSON, V., et al. 1978. Properties of central vestibualr neurons fired by stimulation of the saccular nerve. Brain Res. **143**: 251–261.
12. SASAKI, M., et al. 1991. Is there a three neuron arc in the cat utriculo-trochlear pathway? Exp. Brain Res. **86**: 421–425.
13. UCHINO, Y., et al. 1997. Sacculocollic reflex arcs in cats. J. Neurophysiol. **77**: 3003–3012.
14. UCHINO, Y., et al. 1994. Monosynaptic and disynaptic connections in the utriculo-ocular reflex arc of the cat. J. Neurophysiol. **71**: 950–958.
15. UCHINO, Y., et al. 1996. Utriculoocular reflex arc of the cat. J. Neurophysiol. **76**: 1896–1903.
16. GOLDBERG, J. M., et al. 1987. Inputs from regularly and irregularly discharging vestibular nerve afferents to secondary neurons in the vestibular nuclei of the squirrel monkey. I. An electrophysiological analysis. J. Neurophysiol. **58**: 700–718.
17. PRECHT, W. & H. SHIMAZU. 1965. Functional connections of tonic and kinetic vestibular neurons with primary vestibular afferents. J. Neurophysiol. **28**: 1014–1028.
18. WILSON, V. J. & G. MELVILL JONES. 1979. Mammalian Vestibular Physiology. Plenum Press. New York and London.
19. SATO, H., et al. 1997. Properties of saccular nerve-activated vestibulospinal neurons in cats. Exp. Brain Res. **116**: 381–388.
20. MARKHAM, C. II. 1968. Midbrain and contralateral labyrinth influences on brainstem vestibular neurons in the cat. Brain Res. **9**: 312–333.
21. MANO, N., T. OSHIMA & H. SHIMAZU. 1968. Inhibitory commissural fibers interconnecting the bilateral vestibular nuclei. Brain Res. **8**: 378–382.
22. UCHINO, Y., et al. 1986. The commissural inhibition on secondary vestibulo-ocular neurons in the vertical semicircular canal systems in the cat. Neurosci. Lett. **70**: 210–216.

Human Ocular Counterrolling During Roll-Tilt and Centrifugation

H.G. MACDOUGALL,[a] I. S. CURTHOYS,[a,c] G. A. BETTS,[a] A. M. BURGESS,[a] AND G. M. HALMAGYI[b]

[a]*Vestibular Research Laboratory, Department of Psychology, The University of Sydney, Sydney, New South Wales, Australia*

[b]*Eye and Ear Research Unit, The Department of Neurology, Royal Prince Alfred Hospital, Sydney, New South Wales, Australia*

ABSTRACT: To test a hypothesis about how otoliths resolve roll-tilts from translations, we measured human ocular torsion position [ocular counterrolling (OCR)] to maintained linear acceleration stimuli. All subjects ($n=8$) were tested in two conditions where the same magnitude of shear along an interaural axis was generated in one of two ways: either by roll-tilt on a tilt-chair in a 1-g environment, or by centripetal linear acceleration during constant velocity rotation 1 m from the axis of rotation on a fixed-chair human centrifuge. The interaural shear to the otoliths was the same for these two conditions, but the dorsoventral shear was different and for all eight subjects the OCR on the centrifuge was significantly greater than the torsion on the tilt-chair, although the resultant angle was in fact smaller on the centrifuge than on the tilt-chair. The results confirm that dorsoventral shear is important for determining OCR. The otoliths may resolve potential stimulus ambiguities between tilts and translations by virtue of the different patterns of interaural and dorsoventral shear that these stimuli generate.

INTRODUCTION

Linear accelerations stimulate the otolith receptors in ways that seem to be ambiguous.[1,2] For example, a roll-tilt of the head around a naso–occipital axis toward a subject's right shoulder generates a shear force across the subject's interaural axis, as does a passive lateral translation of head to the subject's left, and the otoliths must resolve this ambiguity.[3] We propose that the ambiguity between tilts and translations only occurs if interaural shear force is considered in isolation. When both interaural and dorsoventral shear forces are considered in combination the apparent ambiguity of tilt and translation stimuli disappears since there are differences in the dorsoventral shear force between tilts and translations, and we suggest that tilts and translations are resolved in part by the different patterns of interaural and dorsoventral shear forces between tilts and translations. That suggestion was tested in the present experiment by testing subjects in conditions where interaural shear was constant but dorsoventral shear differed in order to identify whether an otolith-dependent response—static ocular torsion (ocular counterrolling or

[c]To whom correspondence may be addressed. Phone: 61 2 9351 3570; fax: 61 2 9351 2603; e-mail: ianc@psych.usyd.edu.au

OCR)—shows a difference between the two conditions. If interaural shear alone were important for generating ocular torsion, then the same torsion magnitude should be produced under both conditions. Significant differences in torsion between the two conditions would suggest that dorsoventral shear (probably predominantly saccular stimulation; see Curthoys *et al.*, this volume) has a role in the generation of torsion.

If a subject on a tilt-chair is given a roll-tilt around a naso–occipital axis, there will be a systematic increase in the magnitude of the force along the subject's interaural axis, while there will be simultaneously a decrease in the force along the subject's dorsoventral axis (FIGS. 1 and 2). On the other hand, systematically increasing the constant velocity for a seated subject, facing along a tangent in a fixed-chair human centrifuge, will systematically increase the shear force acting along the subject's interaural axis, whereas the shear force acting along the subject's dorsoventral axis will remain constant at 1 g (FIGS. 1 and 2). We have conducted experiments delivering such linear accelerations to the utricular macula by roll-tilt and by centrifugation and measuring the magnitude of the ocular torsion in the two situations using our own calibrated and validated video image-processing procedure called VTM[4,5] to record ocular torsional position.

Centrifuge : 172 °/ s   Tilt-chair : 67 °

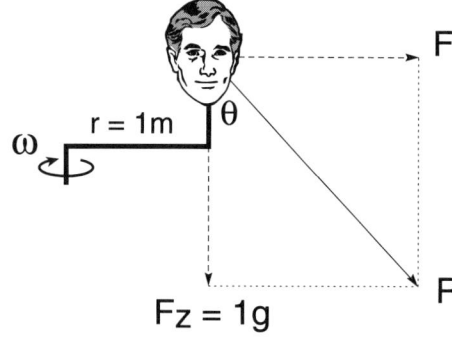 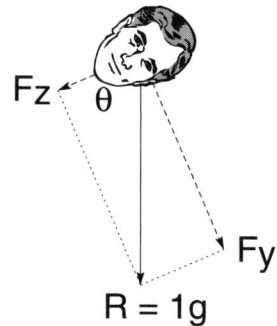

$F_y = \omega^2 r$ $F_y = 0.92$ g $F_y = R \sin\theta$ $F_y = 0.92$ g
$R = \sqrt{F_y^2 + F_z^2}$ $F_z = 1.00$ g $F_z = R \cos\theta$ $F_z = 0.39$ g
     $R = 1.36$ g $R = 1.00$ g
$\theta = \tan^{-1}\left(\frac{F_y}{F_z}\right)$ $\theta = 42.6°$ $\theta = 67.0°$

Where:
F_y = interaural shear force
F_z = force along the midline of the body
R = resultant gravitoinertial force.

g = gravitational force
θ = angle between head z-axis and the resultant force
ω = angular velocity in radians per second

FIGURE 1. A comparison of the static forces on the tilt-chair and centrifuge during roll-tilt stimulation.

A. Tilt-chair   Increasing roll-tilt angle: dorsoventral shear decreases

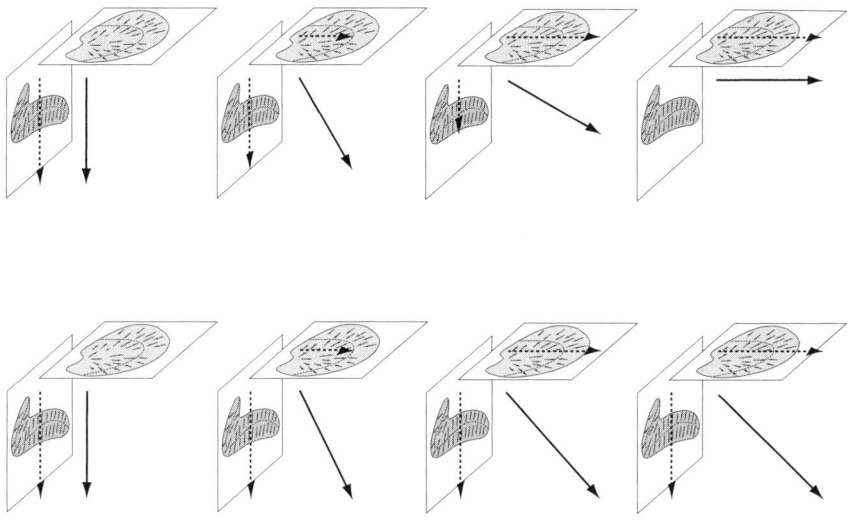

B. Centrifuge   Increasing centrifugal force: dorsoventral shear constant

FIGURE 2. To contrast the manner in which successively larger roll-tilt stimuli on the tilt-chair and centrifuge (*rows*) generate very different patterns of interaural and dorsoventral shear stimulation, each figure represents the head-fixed planes of one labyrinth during linear acceleration stimulation. In each column the magnitude of the interaural shear stimulus (shown by the *dashed arrow* across the hypothetical projection of the utricular macula into the horizontal plane) is identical, but the saccular shear (*dashed arrow* across the hypothetical projection of the saccular macula into the sagittal plane) is different. The maculae have been sketched onto the planes in order to remind one of the different relative extents of the activation patterns of the two maculae. (**A**) Tilt-chair. Increasing role-tilt angle: dorsoventral shear decreases. (**B**) Centrifuge. Increasing centrifugal force: dorsoventral shear constant.

MATERIALS AND METHODS

Eight subjects aged between 21 and 54 were tested. No subject reported any history of vestibular or auditory dysfunction. The ocular torsion of each subject was measured (monocular—left eye) in both tilt-chair and fixed-chair centrifuge with the order of testing randomized between subjects. All procedures were in accordance with the Declaration of Helsinki and were approved by the Human Ethics Committee of the University of Sydney.

The two experimental conditions are shown schematically in FIGURES 1 and 2. A roll-tilt of 67 deg and a centripetal linear acceleration arising from constant-velocity centrifugation (172 deg/s) of a subject 1 m from the axis of rotation on a fixed-chair human centrifuge and positioned so that the linear acceleration is directed along the interaural axis, both generate a shear force of 0.92-g units directed along an interaural line. However, the dorsoventral shear is very different in the two conditions, being 0.42-g units on the tilt-chair and 1 g on the centrifuge.

The roll-tilt device was a standard motor-driven tilting chair. Subjects were seated in this chair and supported by head, trunk, and hip supports, strapped in, and roll-tilted (at 2.5 deg/s) to an angle of 67 deg from vertical and held at that angle for 4 minutes and then returned to upright at the same speed. Only right-ear down tilts were measured. The centrifuge was a Servo-Med fixed-chair human centrifuge. Subjects were seated 1 m from the axis of rotation, facing along a tangent, right ear out, so that during constant velocity rotation the centripetal linear acceleration was directed along the subject's interaural (Y) axis. Subjects were held in place by means of comparable head, trunk, and hip supports, as well as seat belts. For each subject the head was held in the same standard position as used on the tilt-chair.

The centrifuge was accelerated at 5 deg/s/s to 172 deg/s, and subjects were maintained at that constant velocity for a total of 4 minutes in order to allow the effect of the angular acceleration stimulation of the semicircular canals (and hence any horizontal or vertical nystagmus and its concomitant effects) to dissipate completely. The average of the ocular torsion measure over the final 2 minutes of the 4 minutes was measured, and the difference between this value and the baseline measure was taken as the magnitude of ocular torsion. The subject was then returned to rest to ensure that the torsion value returned to zero.

Ocular torsion position [ocular counterrolling (OCR)] was measured by an objective image-processing method (called VTM) from images provided by a small head-mounted video camera.[4,5] All measures on both tilt-chair and centrifuge were in darkness with the exception of a single small dim fixation light attached to the chair at eye height, 80 cm from the subject's eye. During centrifugation this fixation light acted to suppress vestibular nystagmus.

The subject's head was positioned using a bubble vial from a carpenter's level that was glued to a tightly fitting, individually moulded, thermoplastic mask made of Sansplint (Smith and Nephew) that the subject wore throughout the experiment. This vial and mask allowed a very simple check at the end of the test of whether the head had shifted, but also allowed accurate comparable positioning of the person's head on both the tilt-chair and centrifuge. The head was positioned so that the horizontal canals were earth horizontal. This is a comfortable head position, and placing each subject's head in this standard position allowed us to specify the stimuli in the two conditions with respect to each head axis, a condition that we consider to be crucial to the interpretation of this experiment and the understanding of otolith function.

The right-hand rule was used so that clockwise ocular torsional rotation (where the upper pole of the eye rotated toward the subject's right shoulder) is positive and counter-clockwise ocular torsion is negative.

RESULTS

In order to attain the target stimulus values on the tilt-chair and centrifuge, there were differences in the pattern of semicircular canal stimulation: on the tilt-chair the vertical canals (predominantly) were stimulated by the angular acceleration and deceleration during roll-tilt. On the centrifuge the horizontal canals (predominantly) were stimulated during yaw angular acceleration. To minimize the influence of the differing canal stimuli in our experiment, torsion measures were not taken until at least 2 minutes after the end of the angular acceleration, by which time any effect of the canal stimuli should have dissipated since it is more than three times the human semicircular-canal long-time constant (about 20 s).

FIGURE 3 shows the time series of ocular torsion position during stimulation on the centrifuge for subject HM. Each point is a single measure of ocular torsion at a sampling rate

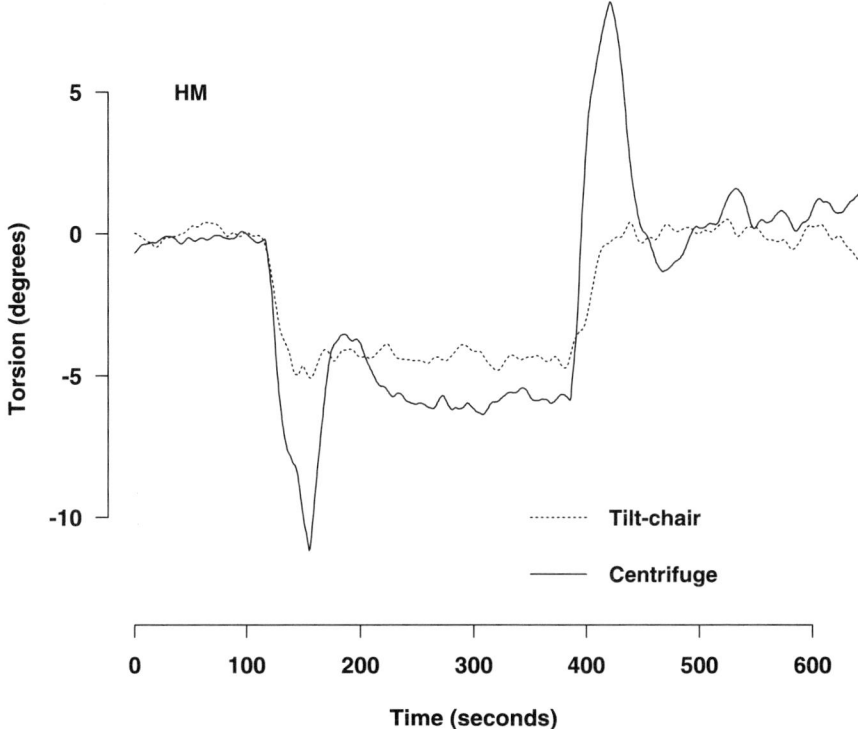

FIGURE 3. Time series of ocular torsion position of one subject (HM) before, during, and after linear acceleration stimulation on a centrifuge. At the onset of angular acceleration there is an increase in torsion (to more than −12 deg), which then decays at the end of the angular acceleration to an average value of −5.5 deg. At deceleration there is again a large angular-acceleration-dependent torsion that overshoots the baseline and then slowly returns to the baseline value. The constant velocity stimulation extends from 150 s to 400 s. The *dashed lines* show the torsion of this same subject during roll-tilt to 67 deg on a tilt-chair.

of 1.5 Hz. The solid line is the best-fitting line to the raw data using the lowess procedure.[6] The initial portion is at rest (0 deg/s for 2 minutes) to establish the baseline.

During the angular acceleration, there is a large change in torsional position that declines to a steady value once the chair has attained constant velocity. This ocular torsion position change during yaw angular acceleration was reported by us previously,[7] and is known to be due to the activation of the semicircular canals, since it has been shown to occur when the subject is given an on-center angular acceleration with a chair-fixed fixation light (which suppresses vestibular nystagmus), and our recent unpublished experiments varying head pitch show that this torsion is due to horizontal and not posterior canal activation. At the end of the angular acceleration the angular-acceleration-dependent torsion declines and approaches an asymptotic value. The torsion during roll-tilts on the tilt-chair did not show the acceleration-dependent torsion, but was maintained at a steady value on reaching the roll-tilted position (FIG. 3). Even so, we measured for 4 minutes in both conditions and only used the torsion values during the final 2 minutes to ensure that the torsion measures were free of any canal contribution.

We sought to minimize the effect of the variability between subjects[8] by comparing each subject's measured torsion on the tilt-chair with their own measured torsion on the centrifuge and conducting statistical tests on the difference measures within subjects. Each subject showed a statistically significantly larger torsion to the centrifuge stimulus as compared to the tilt-chair stimulus (TABLE 1). The average difference across the eight subjects at 0.92-g interaural shear was 2.5 deg ± 1.6 deg (standard deviation). This average difference was significantly greater than zero ($t = 4.29$; $df=7$; $p< 0.01$).

DISCUSSION

For stimuli whose interaural utricular shear was equated, every one of the eight subjects showed significantly larger ocular torsion on the centrifuge than on the tilt-chair. That result shows that dorsoventral shear is important for generating this otolith-dependent response because, whereas the interaural shear was the same for both tilt-chair and centrifuge, the dorsoventral shear differed between the two conditions. Objective measures of the forces using a linear accelerometer confirmed that our calculations were correct: the same interaural shear was generated on both centrifuge and tilt-chair. It should be noted that whereas the resultant angle (42.5 deg) of the gravitoinertial force on the centrifuge was smaller than the resultant angle on the tilt-chair (67 deg), the measured ocular torsion was larger on the centrifuge than on the tilt-chair.

These results show for the first time using direct comparison of the same subjects under two different maintained otolith stimulation conditions with equated interaural shear that there is a larger ocular torsion during centrifugation as compared to roll-tilt. Our results confirm and extend other recent reports that the otolith-ocular response during dynamic otolith stimulation is a weighted sum of interaural and dorsoventral shear.[9-11] Merfeld et al.'s study[10] concerned dynamic linear acceleration stimulation of subjects with their head in different head positions on a linear sled, but in that study maintained (static) linear acceleration stimuli were not presented. Also scleral search coils were used, but because of coil slip in torsion, scleral search coils are not reliable tools for measuring ocular torsional position. In the study by de Graaf et al.[9] torsion was measured in light and linear accelerations were maintained, but there were no direct comparisons of the same subjects on tilt-chair and centrifuge with the same interaural shear, and torsion was measured in light.

Such a combined effect of interaural and dorsoventral stimulation may also explain how other apparent otolithic stimulus ambiguities are resolved. Many head positions would be ambiguous without dorsoventral input; for example, maintained head positions of 60 deg and 120 deg roll-tilted around an X-axis from vertical would be indistinguishable, since the

TABLE 1. Mean ocular torsion position in degrees for equivalent interaural shear on the centrifuge and tilt-chair.

Subject	Centrifuge	Tilt-Chair
AB	9.9 ± 0.4	6.5 ± 0.4
AC	10.5 ± 0.7	4.9 ± 0.3
DG	7.0 ± 1.0	5.0 ± 0.8
EN	5.4 ± 0.5	4.6 ± 0.6
HM	5.7 ± 0.3	4.4 ± 0.3
IC	8.3 ± 0.4	6.7 ± 0.7
JS	8.8 ± 0.8	5.0 ± 0.7
SE	7.6 ± 0.5	6.2 ± 0.5

Note: means ± standard deviation.

interaural shear in both of those maintained roll-tilt positions is identical. In fact, these two roll-tilts are readily distinguishable perceptually, and the present results suggest that it is likely that the dorsoventral shear assists that resolution. On the basis of our results we suggest that dorsoventral shear also assists in the resolution of tilts from translations.

We have commenced comparable experiments using visual perception to index roll-tilt perception.[12] However using a visual indicator to measure perceived roll-tilt must be done with care, since otolithic stimulation influences both the person's perceived orientation and also the ocular torsional position of the eyes.[13] If a visual stimulus is used to indicate perceived roll-tilt, then errors may occur because we have shown that the brain does not "take into account" the torsional position of the eye, with the result that an image of a gravitationally vertical line falling on a torted eye will appear to be tilted.[13] The flaw in using visual measures of roll-tilt perception is that the otolith stimulus affects both perceived roll-tilt and also the torsional position of the eye, so that if a visual indicator (such as a visible line) is used to indicate perceived roll-tilt,[14] then the otolith-determined change in torsional eye position alone will cause the perceived orientation of the line to change irrespective of the subject's felt position. So using a visual stimulus to index roll-tilt is essentially "double counting," since the roll-tilt affects the sensation of roll-tilt and also generates erroneous visual indication of tilt using a visual indicator. This unanticipated effect of ocular torsion has confounded some perceptual studies of otolithic function.

ACKNOWLEDGMENTS

We gratefully acknowledge the support of NHMRC of Australia and for partial support by Research Grant 5 R01 DC 02372-02 from the National Institute on Deafness and Other Communication Disorders, National Institutes of Health. The assistance of Warren Davies is gratefully acknowledged.

REFERENCES

1. SHOTWELL, S. L., R. JACOBS & A. J. HUDSPETH. 1981. Directional sensitivity of individual vertebrate hair cells to controlled deflection of their hair bundles. Ann. N.Y. Acad. Sci. **374**: 1–10.
2. FERNANDEZ, C. & J. M. GOLDBERG. 1976. Physiology of peripheral neurons innervating otolith organs of the squirrel monkey. I. Response to static tilts and to long-duration centrifugal force. J. Neurophysiol. **39**: 970–984.
3. MAYNE, R. 1974. A systems concept of the vestibular organs. In Handbook of Sensory Physiology. Vestibular System, Vol. 6, Pt. 2, H. H. Kornhuber, Ed.: 493–580. Springer-Verlag. Berlin.
4. MOORE, S. T., I. S. CURTHOYS & S. G. MCCOY. 1991. VTM–an image-processing system for measuring ocular torsion. Comput. Methods Programs Biomed. **86**: 219–230.
5. MOORE, S. T., T. HASLWANTER, I. S. CURTHOYS & S. T. SMITH. 1996. A geometric basis for measurement of three-dimensional eye position during image processing. Vision Res. **36**: 445–459.
6. BECKER, R. A., J. M. CHAMBERS & A. R. WILKES. 1988. The new S language: a programming environment for data analysis and graphics. Wadsworth (Brooks/Cole Computer Science Series). Pacific Grove, Calif.
7. SMITH, S. T., I. S. CURTHOYS & S. T. MOORE. 1995. The human ocular torsion position response during yaw angular acceleration. Vision Res. **35**: 2045–2055.
8. DIAMOND, S. G., C. H. MARKHAM, N. E. SIMPSON & I. S. CURTHOYS. 1979. Binocular counterrolling in humans during dynamic rotation. Acta Otolaryngol. **87**: 490–498.
9. DE GRAAF, B., J. E. BOS & E. GROEN. 1996. Saccular impact on ocular torsion. Brain Res. Bull. **40**: 321–330.
10. MERFELD, D. M., T. TEIWES, A. H. CLARKE, H. SCHERER & L. R. YOUNG. 1996. The dynamic contributions of the otolith organs to human ocular torsion. Exp. Brain Res. **110**: 315–321.

11. MILLER, E. F., II. 1962. Counterrolling of the human eyes produced by head tilt with respect to gravity. Acta Otolaryngol. **54**: 479–501.
12. CURTHOYS, I. S. & G. A. BETTS. 1997. The role of utricular stimulation in determining perceived postural roll-tilt. Aust. J. Psychol. **49**: 134–138.
13. WADE, S. W. & I. S. CURTHOYS. 1997. The effect of ocular torsional position on the perception of the roll-tilt of visual stimuli. Vision Res. **37**: 1071–1078.
14. DAI, M. J., I. S. CURTHOYS & G. M. HALMAGYI. 1989. Linear acceleration perception in the roll plane before and after unilateral vestibular neurectomy. Exp. Brain Res. **77**: 315–328.

Canal and Otolith Afferent Activity Underlying Eye Velocity Responses to Pitching While Rotating

T. RAPHAN,[a,e] M. DAI,[b] J. MARUTA,[b] W. WAESPE,[b] V. HENN,[c] J.-I. SUZUKI,[d] AND B. COHEN[b]

[a]*Institute of Neural & Intelligent Systems, Department of Computer and Information Science, Brooklyn College of the City University of New York, 2900 Bedford Avenue, Brooklyn, New York 11210, USA*

[b]*Departments of Neurology and Physiology and Biophysics, Mount Sinai School of Medicine, New York, New York 10029, USA*

[c]*Neurology Department, Zurich University Hospital, CH-8091, Zurich, Switzerland*

[d]*Department of Otolaryngology, Teikyo University, Tokyo 117-0003, Japan*

ABSTRACT: Pitching the head while rotating (PWR) combines periodic activation of the semicircular canals and the otoliths to generate pitch and roll eye deviations and continuous horizontal nystagmus. Monkeys were tested after individual pairs of semicircular canals were plugged and single units were recorded in the vestibular nerve while the animals were sinusoidally pitched 20–40 deg about a spatial horizontal axis with 5- and 16-s periods and simultaneously rotated about a spatial vertical axis at 30–120 deg/s. As previously shown, the steady-state horizontal response disappeared after plugging the vertical semicircular canals, but was maintained when the lateral canals were plugged. When the left anterior and right posterior canal (LARP) pair was left intact, the steady-state response depended on the axis about which the pitching took place. When the axis was normal to the LARP plane, there was no steady-state response. When the pitching axis was perpendicular to the LARP normal, the response was maximal. Firing rates of otolith units were approximately in phase with pitch position, and the addition of rotation about a vertical axis did not change the response. Lateral canal units did not have a steady-state modulation during pitch or constant velocity rotation. During PWR, they oscillated at twice the pitch frequency. This corresponded to the frequency at which the canal was maximally activated as it aligned with the plane of rotation. The amplitude of modulation increased proportionally to rotational velocity, but the phase remained the same. These characteristics were unchanged during roll while rotating (RWR), which induces little continuous nystagmus. Anterior and posterior canal units were maximally excited near pitch-velocity maxima and minima, respectively, during pure pitching. During PWR, however, the phases of both components simultaneously

[e]To whom correspondence may be addressed. Phone: 718-951-4193; Fax: 718-951-4489; e-mail: raphan@nsi.brooklyn.cuny.edu

shifted toward each other and toward being in phase with otolith units. The peak excitation tended toward a forward-pitch position when the rotation was to the ipsilateral side, and toward a backward pitch position when the rotation was to the contralateral side. With 120-deg/s rotation during a 16-s pitch period, the phase difference between anterior and posterior canal units was as small as 17 deg. These data support the postulate that the correlation between vertical canal and otolith units is the critical factor in generating continuous unidirectional horizontal nystagmus during PWR.

INTRODUCTION

Maintenance of gaze during complex stimuli that activate the visual, vestibular, and somatosensory systems demands resolution of a wide range of input from the semicircular canals and otoliths. Pitching the head while rotating (PWR) combines periodic activation of the semicircular canals and the otoliths. All semicircular canals are activated as their planes come into and out of the plane of rotation. The otoliths are activated by gravity induced by the pitching stimulus. The slow phase velocity of the nystagmus during PWR is composed of oscillating horizontal, vertical, and roll components and a continuous steady-state horizontal eye velocity.[1,2] The steady-state horizontal response disappeared after plugging the vertical semicircular canals, but was maintained when the lateral canals were plugged.[1,3] This suggests that the response arises in the lateral, not the lateral canals. Pitching about axes above and below the interaural line did not alter the direction, magnitude, or modulation characteristics of the horizontal nystagmus, indicating that Coriolis forces probably are not significant in generating the response.[1] Cutting the lateral canal nerve afferents inactivated velocity storage and caused a loss of the steady-state component of the horizontal nystagmus.[1]

One hypothesis that has been put forward to explain the ongoing horizontal nystagmus during PWR is that a signal approximately proportional to the phase between vertical canal and otolith afferents is used to activate the velocity storage integrator to generate continuous horizontal nystagmus.[1] The purpose of this study was to test this hypothesis by recording horizontal eye velocity in animals with selected canal plugs and by recording activity of canal and otolith afferents during PWR.

METHODS

The methods of animal preparation and eye-movement recordings are detailed elsewhere.[4-6] Briefly, under anesthesia, scleral search coils were implanted in the frontal plane to record horizontal and vertical eye position, and a roll coil was implanted on top of the same eye of each animal to record torsional eye movements.[7,8] Eye movements were recorded with the animal's head fixed to square Helmholtz field coils that were 12.5 cms on a side. Bolts for holding the head painlessly were positioned stereotaxically, and analgesics (morphine sulfate 2 mg, intramuscular × 2) and antibiotics (cephalothin 100 mg, intramuscular, daily × 5) were given after surgery to reduce postoperative pain and infection. Eye movements were calibrated by rotating the animal about an earth vertical axis in light in yaw, pitch, or roll at 30 deg/s. The individual voltages were differentiated and were used as references. For horizontal and vertical calibrations, the gain was considered to be close to unity,[9-11] whereas the gain for roll was taken as 0.66.[11] Canal plugging was performed by grinding across the canal with a diamond burr until the membranous canal was interrupted.[5,12,13] The region of the canal was packed with bone and covered with a small piece of muscle.

Neuronal activity was recorded using varnished tungsten microelectrodes having impedances of 1–5 MΩ at 1 kHz. They were introduced normal to the stereotaxic horizontal plane. The recorded signals were amplified, filtered using a bandpass filter of 0.2–5 kHz, and spikes were identified using a window discriminator. The output of the discriminator was a standard pulse synchronized with the firing of the unit. Canal afferents were identified by oscillating the animal by hand about axes approximately normal to the canal planes. Modulation of their activity was in phase with head velocity at frequencies above 0.1 Hz. Otolith afferents were identified by their modulation in phase with head position with regard to gravity. Eye velocity and unit activity was recorded on FM magnetic tape and read into a PC using a DAOS data-acquisition system for off-line analysis.

The apparatus used to induce rotations for behavioral studies was a vestibular and optokinetic stimulator that provided four axes of rotation (Contraves Goerz, Neurokinetics) (see references 4 and 14 for a description). To generate pitch while rotating, animals were rotated continuously about a spatial vertical axis at 60 deg/s and simulataneously pitched with an amplitude of 20 deg about a spatial horizontal axis. For single-unit studies, another apparatus (Toennies) that provided essentially the same set of movements (see reference 1 for description), was used from the laboratory of Volker Henn.

Pitch position was used as the reference for determining the phases of pitch and roll eye velocity. The animal's head could also be positioned such that its interaural axis was at an angle relative to the pitching axis. We defined this as the pitching angle. It was 0 deg when the animal was pitched and 90 deg when it was rolled. Steady-state horizontal eye velocity and the changes in the phase of pitch and roll eye velocity relative to pitch position were determined as a function of the pitching angle.

RESULTS

Behavioral Eye Velocity Responses During PWR

Rotation about a spatial vertical axis at 60 deg/s with the monkey upright induced perrotatory nystagmus, whose horizontal eye velocity decayed to zero (FIG. 1A, left). When the monkey was then pitched around its interaural axis with an amplitude of ± 20 deg and a frequency of 0.1 Hz (FIG. 1A, middle), there were roll, pitch, and yaw oscillations in eye velocity. The roll eye velocity (R VEL) was close to being in phase with pitch position, while the phase of pitch eye velocity (V VEL) was between head position and head velocity. A yaw eye velocity (H VEL) was also induced. It had a pronounced steady-state component, which built up with a time constant of approximately 10–15 s (FIG. 1A, middle). When pitching of the head was stopped during continuous rotation about the spatial vertical, yaw eye velocity decayed back to zero with a time constant of approximately 10–15 s, the dominant time constant of the aVOR and velocity storage[10] (FIG. 1A, right). This indicates that PWR generates a steady-state estimate of head velocity about a yaw axis resulting in a continuous compensatory yaw eye velocity. The dominant time constant of the response decline is consistent with the postulate that the steady-state estimate of head velocity is stored in velocity storage to generate the continuous unidirectional yaw eye velocity.[1]

All canals and otoliths are activated by this stimulus paradigm. Therefore, we plugged individual canal pairs to determine how the intact canals might contribute to the steady-state response. When the lateral canals were plugged bilaterally, leaving the vertical canals intact (VC animal), there was a negligible yaw component of eye velocity in response to a step of rotation about the spatial vertical axis (FIG. 1B, left).[6] During PWR, however, yaw eye velocity increased with a time constant similar to the

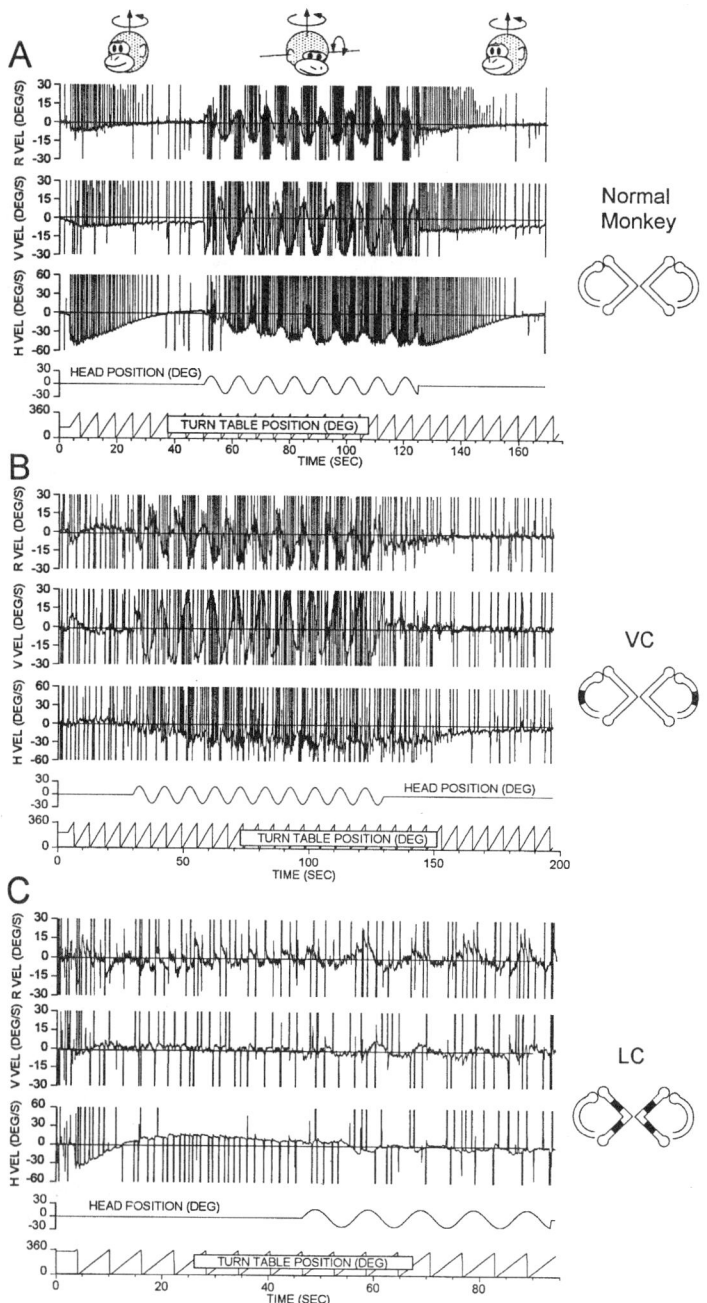

normal animal (FIG. 1B, middle). There were also modulations in eye velocity (FIG. 1B, middle) and a decay in slow phase eye velocity when the pitching was stopped that were similar to those in the normal animal (FIG. 1B, right). This shows that the vertical canals, together with the otoliths, are sufficient to activate velocity storage to generate the steady-state yaw eye velocity.

In contrast, when all four vertical canals were plugged, leaving the lateral canals intact (LC animal), the yaw component of eye velocity in response to a step of rotation was close to normal (FIG. 1C, left), but the response to PWR was abolished (FIG. 1C, right). This indicates that the lateral canals and the otoliths are not sufficient and that the vertical canals are essential for maintaining steady-state yaw eye velocity during PWR. These findings are consistent with our previous hypothesis that the vertical canals in conjunction with the otoliths combine to generate a signal that maintains steady-state yaw eye velocity through velocity storage.[1]

We next considered the mechanism by which the vertical canals and otoliths might contribute to the steady-state eye velocity. When only the right anterior and left posterior canals were left intact (RALP animal), the steady-state yaw eye velocity during PWR was strongly dependent on the plane in which the head was oscillated (FIG. 2). When the animal was pitched in the intact canal plane (RALP plane) and rotated, no steady-state compensatory yaw eye velocity was produced (FIG. 2A). In this case, pitch and roll eye velocities were close to being 90 deg out of phase relative to pitch position. In contrast, when the animal was pitched in the plane orthogonal to the RALP plane and rotated, it produced the maximal steady-state yaw eye velocity (FIG. 2B). The phases of both pitch and roll eye velocity were close to that of pitch position, indicating that the phase of vertical canal activation was close to that of the otolith activation. Other angles of head pitching generated steady-state responses between these two extremes.

Primary Afferent Unit Activity During PWR

From the data on canal plugging, we postulated that the phase of vertical canal activation relative to that of the otoliths, which encode pitch position, determines the steady-state horizontal eye velocity during PWR. FIGURE 3 shows a schematic comparing the the theoretically expected behavior of primary afferents during one cycle of pitching of the head relative to gravity and their behavior during PWR. During pure pitching of the head relative to gravity, otolith activity would essentially be in phase with the head position signal

---

FIGURE 1. Pitch while rotating (PWR) in a normal monkey. Eye-velocity components are in accordance with a right-hand rule: counterclockwise looking at the animal was positive for roll, pitching down is positive for vertical, and yaw to the left is positive for horizontal. Animals were rotated in darkness to the left at 60 deg/s as noted by insert at the top of eye-velocity traces. After the nystagmus had decayed to zero, pitching about an interaural axis was begun at 0.167 Hz, with an amplitude of 20 deg. (**A**) Roll, vertical, and horizontal eye velocity oscillated with the pitching frequency of the stimulus. Yaw (horizontal) eye velocity had a steady-state value in addition to the oscillations, which rose with a time constant of approximately 10 s. It then decayed with the same time constant when the pitching was stopped. (**B**) When the lateral semicircular canals were plugged, leaving the vertical canals intact (VC), characteristics of all three components of eye velocity were similar to that of the normal animal. (**C**) When the vertical canals were plugged, leaving the lateral canals intact (LC), a vestibular-induced yaw (horizontal) response was elicited. During PWR, only weak oscillatory components were present, while the steady-state horizontal nystagmus was absent.

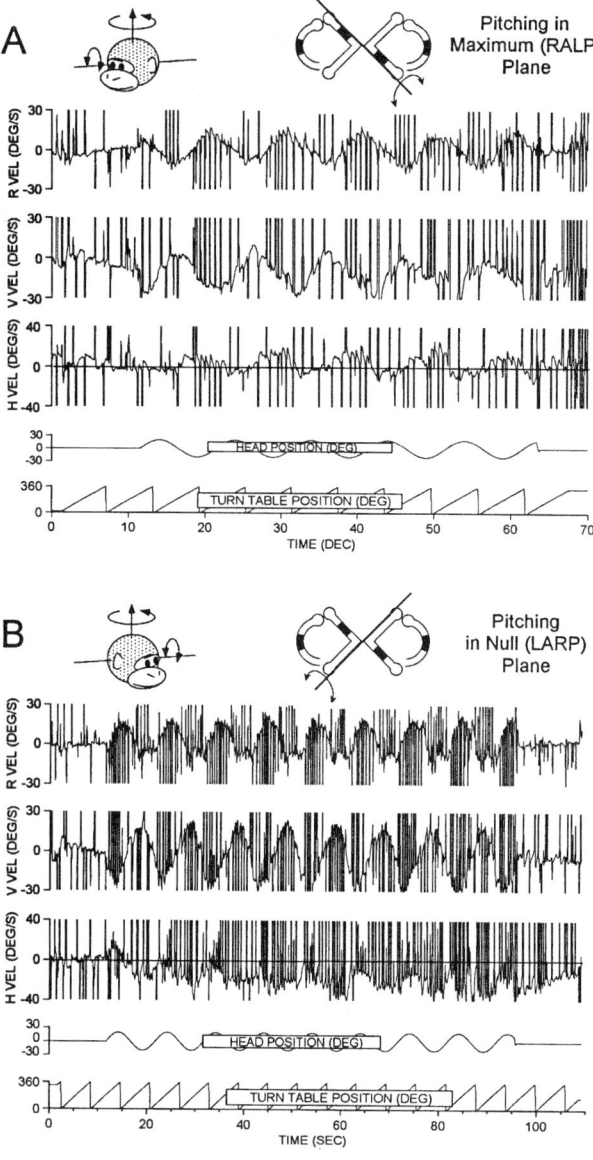

FIGURE 2. Pitch while rotating (PWR) in a monkey with only the right anterior and left posterior (RALP) intact. Positive eye-velocity components and stimulus parameters are as in FIGURE 1. (**A**) When the pitching was done about an axis normal to the RALP plane, the resulting eye-velocity components are all oscillatory and in phase with each other. There is no steady-state horizontal component. (**B**) When the animal was pitched about an axis, which is spatially horizontal and in the LARP plane, it caused the relative alignment of the normal to the RALP plane with the spatial vertical to be a sinusoidal function of pitching angle. This produced oscillating eye-velocity components as well as steady-state horizontal component similar to the normal animal.

(solid trace, FIG. 3A; dash-dot trace, FIG. 3B). The anterior and posterior canal afferent signals on one side would be 180 deg out of phase with each other (black and gray solid traces; FIG. 3B) and 90 deg or 270 deg out of phase with the otolith signal (FIG. 3B). PWR would combine the activation of the canal afferents to the pitching as well as the movement of the canal into and out of the plane of rotation. At any given rotation velocity, a relatively low-frequency pitching motion would shift the anterior and posterior canal afferent signals close to the otolith signal as the plane of the associated canal moved into the plane of rotation (black and gray solid traces, FIG. 3C). In addition, for symmetrical pitching about the position where the vertical canals are spatially vertical, anterior and posterior canal afferents would modulate their activity close to being in phase with each other. Lateral canal afferents would modulate their activity at twice the frequency of the head pitching (dotted trace, FIG. 3C).

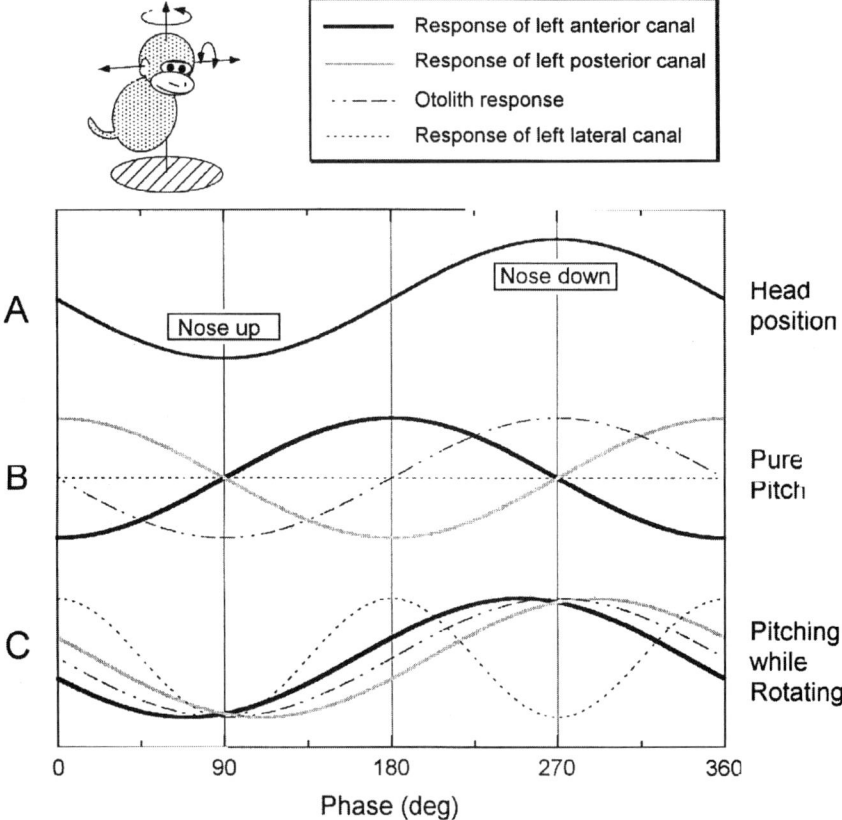

FIGURE 3. Predicted dominant variation and phase relationships among left lateral, anterior, and posterior canal afferents and otolith afferents during pitching with and without rotation to the left. (**A**) One cycle of head pitching. (**B**) Predicted activity of otolith afferent (*dash-dot trace, gray solid*), left anterior (*black solid trace*), left posterior (*gray solid trace*); left lateral (*dotted trace*) during pitching the head about an interaural axis. (**C**) Predicted activity of otolith afferent (*dash-dot trace*), left anterior (*black solid trace*); line left posterior (*gray solid trace*); left lateral (*dotted trace*) during pitching the head about an interaural axis during PWR.

Recordings of canal afferents during pure pitching and PWR at approximately 0.06 Hz support this hypothesis (FIG. 4). When the head was pitched, there was little or no modulation in lateral canal activity (light trace, LC; FIG. 4). There was modulation in anterior and posterior canal activity in phase with head velocity (light trace, AC and PC; FIG. 4). The anterior and posterior canal activity were 180 deg out of phase with each other as expected. Also, as expected, there was a significant modulation of otolith afferent activity that was in phase with head position in response to the pitching (light trace, Oto). During PWR, otolith modulation (dark trace, Oto; FIG.4) was similar to that during pure pitching (light trace, Oto; FIG. 4). However, anterior and posterior canal activity shifted so that they were modulated in phase closer to each other and to otolith modulation (dark trace, AC, PC; FIG. 4).

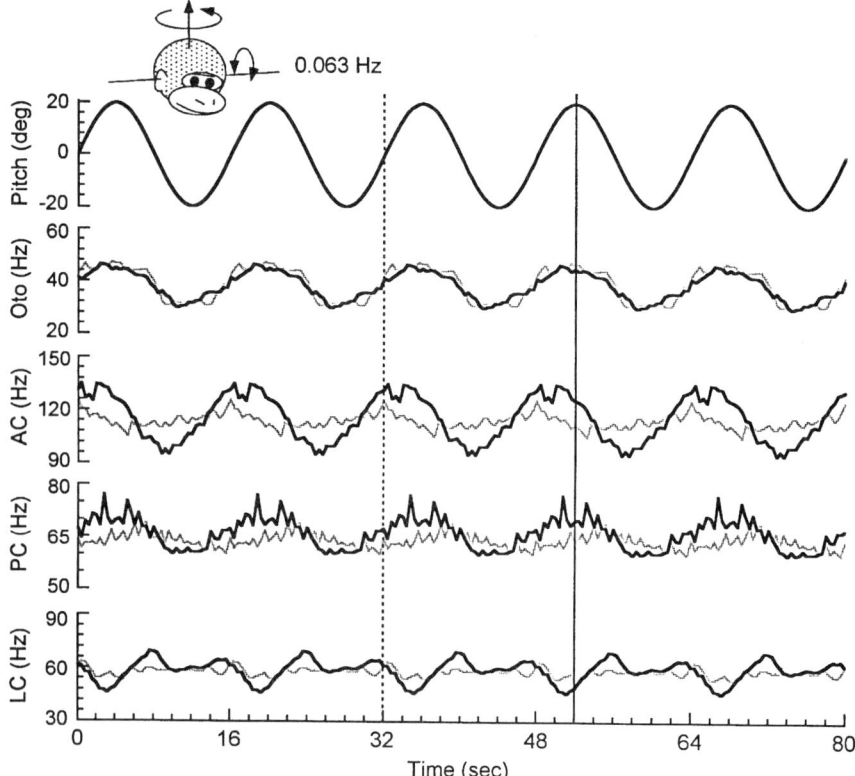

FIGURE 4. Comparison of anterior, posterior, and lateral canal and otolith unit activity during pitching with (*dark trace*) and without (*light trace*) rotation. *Vertical solid line* shows the phase when head is pitched forward with zero velocity. *Dotted vertical line* is the phase of maximum velocity when the head is upright. Rotation was at 60 deg/s to the left with unit activity recorded from the left side. Otolith activity (Oto) is in phase with head position during pitching with and without rotation. Anterior (AC) and posterior (PC) canal afferents are modulated with small amplitudes in approximately in phase and 180 deg out of phase with head velocity, respectively, during pitching alone. During PWR, both the AC and PC afferents have greater modulation amplitude and the phases move closer to being in phase with each other and the Oto modulation. The lateral canal afferent activity is essentially unmodulated during pitching alone, but has a frequency of modulation approximately twice that of the dominant frequency of pitching during PWR.

The stimulus to the vertical canals during PWR comprises two components. The canals are activated by pitch, approximately in phase with pitch velocity, but they are also activated as they are brought into and out of the plane of rotation, approximately in phase with pitch position. For a given amplitude, the pitch at high frequencies will contribute more to activation of the canals than pitch at low frequencies, relative to the activation produced by bringing the canal into and out of the plane of rotation. This would shift the phase away from head position toward head velocity. These shifts in phase are seen in the activity of a left anterior canal afferent during rotation to the right at 30 deg/s (left AC; FIG. 5). For pure oscillation in pitch at 0.063 Hz (FIG. 5A), the response phase of the unit, shown by the solid vertical line, was close to that of head velocity in the forward direction (dark solid trace, left AC; FIG. 5A). During PWR, the phase shifted back (gray trace, left AC; FIG. 5A) and was close to being in phase with pitch position, shown by the vertical dotted line.

When PWR occurred at 0.2 Hz, the phase of modulation of the canal activity during pitching was approximately the same as at 0.063 Hz, shown by the alignment of the two vertical solid lines (dark solid trace, left AC; FIG. 5B). During PWR, however, the phase

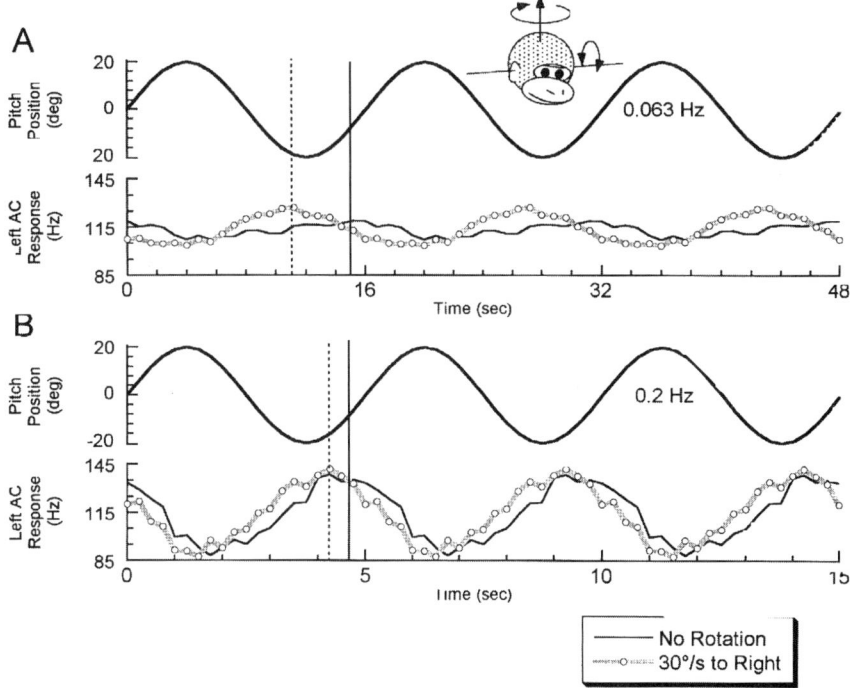

FIGURE 5. Comparison of anterior canal unit activity during pitching with and without rotation at different frequencies. Rotation was at 30 deg/s to the right with unit activity recorded from the left side. The *vertical solid line* shows the approximate peak of the modulation when there was pitching and no rotation. The *vertical dotted line* shows the approximate peak during PWR. (**A**) Pitching was at 0.067 Hz with an amplitude of 20 deg. When there was pitching without rotation, the unit was modulated in phase with head velocity (*solid vertical line*). When there was rotation to the right, the phase shifted so that the maximum of the modulation corresponded to the head-back position (*dotted vertical line*). (**B**) At a pitching frequency of 0.2 Hz, the phase of the response during PWR was closer to the response during pitching alone.

of the modulation in the left AC response shifted only a fraction of that at the lower frequency (compare gray trace and vertical dotted lines in FIG. 5A and 5B). The frequency of the lateral canal afferent had a dominant component, which was approximately twice the pitching frequency. The characteristics of the modulation in lateral canal and otolith afferent modulation were not affected by the increased pitching frequency (not shown). The otolith activity responded at the frequency of pitching and always had the same phase relative to pitch position. We hypothesize that it is the reduction in phase between the otolith and vertical canal activity at the higher frequency of pitching that resulted in a reduced steady-state response of horizontal eye velocity.[3]

Unit activity of the left anterior canal afferents during pitching about axes coincident with and perpendicular to the normal left anterior canal plane demonstrated the changes in phase during PWR as compared to pitching alone (FIG. 6). When pitched about an axis normal to the left anterior canal plane, which was the approximate maximal response plane for this unit, the frequency of firing was modulated with a phase close to head

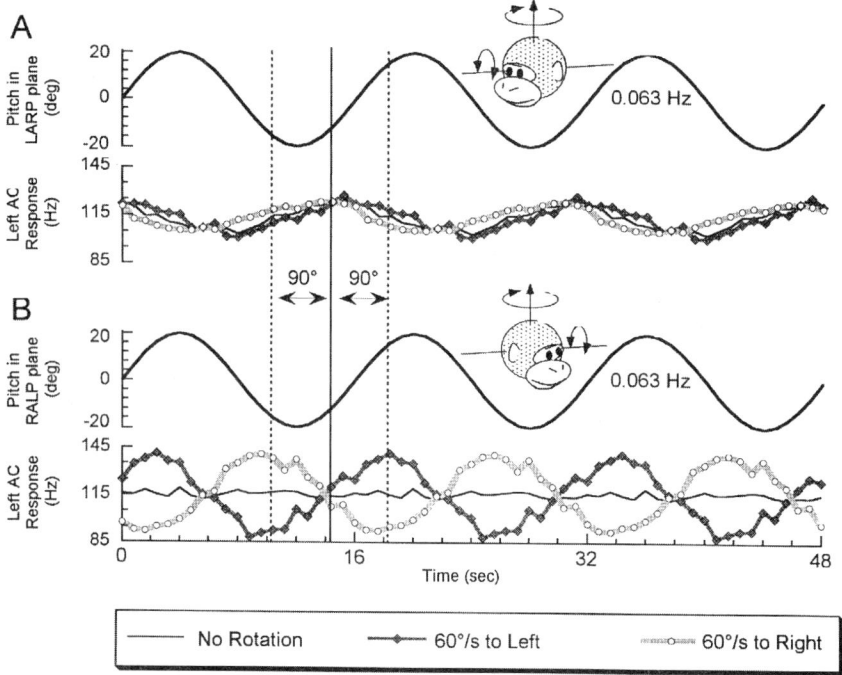

FIGURE 6. The response patterns of a left-anterior canal-related unit during pitching alone and PWR about an axis normal to its maximal plane of activation (LARP) (A) and an orthogonal plane (RALP) (B). Rotation was at 60 deg/s to the left and right. (**A**) For pure pitching about an axis normal to the LARP plane, unit activity was modulated with a phase slightly leading pitching head velocity. Rotation to the right or to the left had no effect on the phase of the response. This modulation is approximately 90 deg out of phase with ipsilateral otolith modulation (see FIG. 4, Oto). (**B**) Pure pitching and no rotation about an axis normal to the RALP plane generated little or no modulation in unit activity. For 60 deg/s rotation to the left, the phase of the modulation advanced approximately 90 deg from the pitching in the maximal plane toward being in phase with pitch position. For 60 deg/s rotation to the right, the phase lagged approximately 90 deg, being approximately 180 deg out of phase with pitch position and with the afferent modulation during rotation to the left.

velocity (dark solid trace, FIG. 6A), shown by the solid vertical line. The modulation in activity was similar during rotation to the right or to the left (dark-diamond and gray-circle traces, FIG. 6A). When the animal was pitched about an axis orthogonal to the normal of the left anterior canal, which was its approximate null response plane, there were shifts in phase dependent on the direction of rotation (FIG. 6B). During pitching without rotation, there was no modulation in unit activity (dark solid trace, FIG. 6B). When the animal was rotated to the left at 60 deg/s and pitched, the phase shifted so that it lagged the phase during pure pitching by ≈90 deg (dark diamond trace, FIG. 6A and 6B), and peak unit activation corresponded to the nose-down position (vertical dotted line to right of solid vertical line, FIG. 6B). For rotation to the right at 60 deg/s, the phase shifted ≈90 deg, leading pitch head velocity (vertical dotted line to left of solid vertical line, FIG. 6B), so that peak activation occurred close to when the head was tilted back (gray circle trace, FIG. 6B).

Roll While Rotating

Roll while rotating (RWR) also activates lateral, vertical, and otolith units, but generates unidirectional horizontal nystagmus with a greatly reduced gain.[1,3] Modulation of lateral canal afferents during RWR (LC, FIG. 7) was the same as during PWR (**LC**, FIG. 4). Otolith modulation was also the same as during PWR, with no differences between pure rolling (light trace, Oto; FIG. 7) and roll while rotating (dark trace, Oto; FIG. 7). However, anterior canal afferent modulation shifted phase (dark trace, AC; FIG. 7) relative to the pure rolling condition (light trace, AC; FIG. 7), lagging by ≈270 deg, relative to the PWR condition (dark trace, AC; FIG. 4) for the same direction of rotation. Posterior canal modulation was unaltered from the PWR condition. Thus, during PWR, anterior and posterior canal afferent modulations were in phase with each other and in phase with otolith afferent modulation. During RWR, anterior and posterior canal afferent modulations were 180 deg out of phase with each other.

DISCUSSION

The results of this study show that semicircular canals, usually associated with producing eye velocity in a particular plane, can interact with the otoliths to generate and maintain eye velocity in an orthogonal plane. This is manifest as a continuous horizontal nystagmus when the head is pitched while there is an ongoing rotation. The mechanism that produces the steady-state response is related to a correlation of signals generated by the vertical canals and the otoliths, which extracts phase information and utilizes velocity storage to maintain the continuous horizontal eye velocity.[1,3,15,16]

The semicircular canals form the sensory basis for detecting angular motion of the head. Yet they are angular acceleration transducers that can only detect angular velocity motion over a short time governed by their dominant time constants, which are approximately 5 s. Continuous rotation goes undetected unless central mechanisms can estimate these continuous velocities. Velocity storage[10,17,18] appears to be a mechanism that prolongs motion estimation, but it too has a dominant time constant of about 10–20 s. Consequently, angular motions of longer duration at constant velocity would not be accurately detected. Therefore, pitching the head while rotating is a means for the central vestibular system to combine canal and otolith signals to detect and estimate velocities of the head during long-term rotations. The fact that it utilizes velocity storage as an intermediary step in oculomotor control may smooth oscillations in the yaw eye velocity, but limits the compensatory velocities to the range of saturation of the velocity storage integrator.

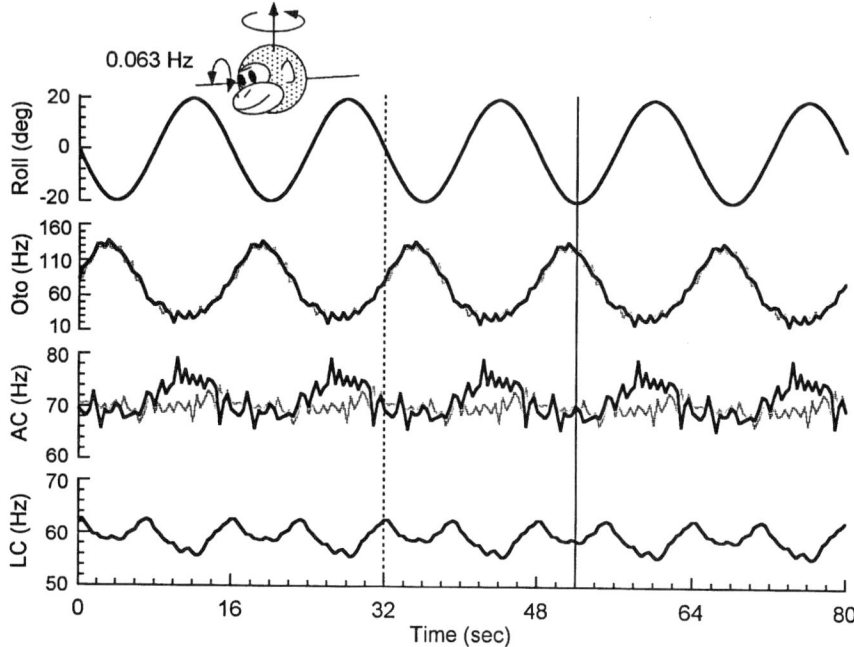

FIGURE 7. The response patterns of left otolith, anterior, and lateral canal related units during pure roll stimulation and rolling the head about a naso-occipital axis while rotating about a spatial vertical axis (RWR). Roll position of the head was positive for right side down in accordance with a right-hand rule convention. RWR was given at 30 deg/s and a rolling frequency of 0.063 Hz. Rolling alone (*light trace*) generated otolith (Oto) and anterior canal (AC) afferent modulation, which were approximately 90 deg out of phase with each other. Oto modulation was closer to being in phase with left side down head position (*solid vertical line*), while anterior canal afferent modulation (AC, *light trace*) was closer to being in phase with head angular velocity toward left side down (*dotted vertical line*). There was no lateral canal afferent modulation during rolling of the head similar to that obtained during pitching (not shown). During RWR, otolith modulation was unaltered (Oto, *dark solid trace*). The phase of the left anterior canal AC shifted toward being in phase with left side down. Posterior canal afferent activity would be 180 deg out of phase with the anterior canal afferent behavior. Lateral canal afferent modulation oscillated at approximately twice the frequency of the roll stimulus as during PWR.

An interesting observation is that for the same amplitude of pitch, the phase of the anterior and posterior canal afferents are close to head position for relatively low frequencies of head pitching. For higher frequencies, the phase is closer to head velocity (FIG. 6). This can be explained by the fact that the trajectory of the head in space during PWR is different from the trajectory during pure pitching. As a result of this trajectory, there are two components of head velocity that are responsible for vertical canal activation during PWR and RWR. One component is due to the pitching or rolling alone, which would predominantly activate the canal in phase with vertical head velocity. Another component is activated when the vertical canals periodically move into the plane of continuous rotation. This component would be in phase with head position relative to gravity. Therefore, the phase of the modulation of activity of particular afferent is dependent on the relative contribution of these components. For a given pitching amplitude, a high pitching frequency induces a large pitch head velocity and generates a large modulation of vertical canal activity close to the phase of pitch velocity. This, in turn, would produce a small steady-state horizontal eye velocity. At lower pitch frequencies, the vertical canal afferent activity is mainly activated by the movement of the canal plane in and out of the plane of rotation and very little by the pitching velocity. Under these circumstances, the phase during PWR is more closely aligned with head position and otolith activity, resulting in a higher yaw eye velocity. If, however, the frequency is too low, then the phase of the vertical canal afferents are shifted by the dynamics of the canal, which has a dominant time constant of 3–5 s, and would reduce the magnitude estimate of continuous head velocity. For higher rotational velocities and/or larger pitch amplitudes, there would be a greater contribution to the afferent activity from the canal coming into the plane of rotation, even at higher pitch frequencies. Thus, the hypothesis that the estimate of yaw head velocity is obtained from the phase of anterior and posterior unit activity relative to otolith activity[1] predicts that PWR velocity estimation is tuned to a pitching frequency that is dependent on rotational velocity and pitching amplitude. For rotational velocities of about 60 deg/s, and pitch amplitudes of 20 deg, the tuning is over a range between 0.05 and 0.5 Hz.

One question that arises is why PWR induces strong steady-state horizontal eye velocity while RWR does not.[3] One possible reason for this might be that the phase of the anterior and posterior canal afferent signals relative to otolith signals are utilized in estimating steady-state head velocity in space. During PWR, ipsilateral anterior and posterior canal afferents are in phase with each other. This kind of phase relationship normally occurs when rotating about a head yaw axis. When these signals are closely correlated with otolith afferents, the CNS interprets this as continuous rotation about a yaw axis. During RWR, however, the activation of the vertical canals when they are brought into the plane of rotation is as when pitching the head. That is, anterior canal afferents are in phase with each other, but they are 180 deg out of phase with posterior canal afferents. Therefore, out-of-phase modulation of the anterior and posterior canal afferents on one side could cancel the effects of having one of the canal afferents modulating in phase with the otolith afferents. This would produce a greatly reduced continuous nystagmus. It would be predicted that if there were no cancellation of vertical canal afferent activity, as when only an anterior or posterior canal is present on one side, there would be continuous nystagmus during PWR as during RWR. Data on the LARP animal, which had continuous yaw nystagmus during both PWR and RWR, are consistent with this prediction.

Analogous to PWR and RWR, there is steady-state roll nystagmus when animals are pitched or yawed while rotating about a spatial vertical in a supine position.[2] The gain of the steady-state roll eye velocity during pitching while rotating was considerably larger during pitching than for yawing. Similarly, roll and yaw oscillations during continuous rotation about a spatial vertical with the animal on its side, generated continuous vertical nystagmus.[2] The steady-state pitch eye velocity for yaw oscillation was considerably higher than during roll oscillations. These data, together with our behavioral and unit data on PWR and RWR with the animal upright, suggest that the central vestibular system does a three-dimensional cross-correlation between otolith and canal afferent information to obtain an estimate of three-dimensional continuous rotation about a spatial vertical.

Thus, we have shown that the semicircular canal and otolith afferents already have the raw information to estimate head velocity about a spatial yaw axis if the head is pitched during the rotation. We have recently confirmed the findings of Pozzo *et al*.[19] that there is substantial head pitch when walking at normal velocities. We also have indications that there are high rotational head velocities when turning corners during circular locomotion.[20] The combination of these two head motions potentially could contribute toward generating compensatory horizontal eye velocities when there is prolonged circular locomotion.

ACKNOWLEDGMENT

This work was supported by grants from the National Institutes of Health (NIH) EY04148, DC03284, NS00294, and core center grant EY01867. We thank Victor Rodriguez for help with making the figures.

REFERENCES

1. RAPHAN, T., B. COHEN & V. HENN. 1983. Nystagmus generated by sinusoidal pitch while rotating. Brain Res. **276**: 165–172.
2. HESS, B. J. M. & D. E. ANGELAKI. 1993. Angular velocity detection by head movements orthogonal to the rotation plane. Exp. Brain Res. **95**: 77–83.
3. DAI, M., T. RAPHAN, J.-I. SUZUKI, Y. ARAI & B. COHEN. 1992. Contribution of individual vertical semicircular canal pairs to estimation of yaw eye velocity during pitch while rotating. Proc. Soc. Neurosci. **216**(4).
4. DAI, M., T. RAPHAN & B. COHEN. 1991. Spatial orientation of the vestibular system: dependence of optokinetic after nystagmus on gravity. J. Neurophysiol. **66**: 1422–1438.
5. YAKUSHIN, S. B., M. DAI, J.-I. SUZUKI, T. RAPHAN & B. COHEN. 1995. Semicircular canal contribution to the three-dimensional vestibulo-ocular reflex: a model-based approach. J. Neurophysiol. **74**: 2722–2738.
6. YAKUSHIN, S. B., T. RAPHAN, J.-I. SUZUKI, Y. ARAI & B. COHEN. 1998. Dynamics and kinematics of the angular vestibuloocular reflex in monkey: effects of canal plugging. J. Neurophysiol. **80**: 3077–3099.
7. ROBINSON, D. A. 1963. A method of measuring eye movement using a scleral search coil in a magnetic field. IEEE Trans. Biomed. Eng. **BME-10**: 137–145.
8. JUDGE, S. J., B. J. RICHMOND & F. C. CHU. 1980. Implantation of magnetic search coils for measurement of eye position: an improved method. Vision Res. **20**: 535–538.
9. SKAVENSKI, A. A. & D. A. ROBINSON. 1973. Role of abducens neurons in vestibuloocular reflex. J. Neurophysiol. **36**: 724–738.
10. RAPHAN, T., V. MATSUO & B. COHEN. 1979 Velocity storage in the vestibulo-ocular reflex arc (VOR). Exp. Brain Res. **35**: 229–248.
11. CRAWFORD, J. D. & T. VILIS. 1991. Axes of eye rotation and Listing's law during rotations of the head. J. Neurophysiol. **65**: 407–423.
12. EWALD, J. R. 1892. Physiologische Untersuchungen über das Endorgan des Nervus Octavus. Bergmann. Wiesbaden, Germany.
13. MONEY, L. B. & J. W. SCOTT. 1962. Functions of separate sensory receptors of non-auditory labyrinth of the cat. Am. J. Physiol. **202**: 1211–1220.
14. REISINE, H. & T. RAPHAN. 1992. Neural basis for eye velocity generation in the vestibular nuclei during off-vertical axis rotation. Exp. Brain Res. **92**: 209–226.
15. RAPHAN, T. & B. COHEN. 1985. Velocity storage and the ocular response to multidimensional vestibular stimuli. *In* Adaptive Mechanisms in Gaze Control; Facts and Theories, A. Berthoz and G. Melvill Jones, Eds.: 123–143. Elsevier. Amsterdam.
16. RAPHAN, T. & B. COHEN. 1986. Multidimensional organization of the vestibulo-ocular reflex (VOR). *In* Adaptive Processes in Visual and Oculomotor System, E. L. Keller and D. S. Zee, Eds.: 285–292. Pergamon Press. New York.
17. RAPHAN, T., V. MATSUO & B. COHEN. 1977. A velocity storage mechanism responsible for optokinetic nystagmus (OKN), optokinetic after-nystagmus (OKAN) and vestibular nystagmus. *In* Control of Gaze by Brain Stem Neurons, R. Baker and A. Berthoz, Eds.: 37–47. Elsevier/North-Holland. Amsterdam.
18. WAESPE, W., B. COHEN & T. RAPHAN. 1983. Role of the flocculus in optokinetic nystagmus and visual-vestibular interactions: effects of flocculectomy. Exp. Brain Res. **50**: 9–33.
19. POZZO, T., A. BERTHOZ & L. LEFORT. Head stabilization during various locomotor tasks in humans. I. Normal subjects. Exp. Brain Res. **82**: 97–106.
20. IMAI, T., E. HIRASAKI, S. M. MOORE, T. RAPHAN & B. COHEN. 1998. Stabilization of gaze when turning corners during overground walking. Soc. Neurosci. **24**: 162.5.

Clinical Testing of Otolith Function

G. M. HALMAGYI[a] AND I. S. CURTHOYS[b,c]

[a]*Eye and Ear Research Unit, The Department of Neurology, Royal Prince Alfred Hospital, Sydney, New South Wales, Australia*

[b]*Vestibular Research Laboratory, Department of Psychology, The University of Sydney, Sydney, New South Wales, Australia*

ABSTRACT: The subjective visual horizontal and vestibular-evoked myogenic potentials are simple, robust, and reproducible tests of otolith dysfunction that can prove clinically useful diagnostic information in patients with vertigo and other balance disorders. While they appear to have high specificity for unilateral otolith dysfunction, further clinical research will be required to establish their sensitivity.

INTRODUCTION

The peripheral vestibular system is sensitive to both linear and angular acceleration: the semicircular canals sense angular acceleration, while the otoliths, the saccule, and the utricle, sense linear acceleration. Many different ways of testing otolith function have been proposed, including measurement of horizontal, vertical, and torsional eye movements as well as of psychophysical settings, in response to the linear accelerations produced by swings,[1] sleds,[2] centrifuges,[3] tilt-chairs,[4] and barbeque-spits.[5] However, in order for a test to be clinically useful it must be safe, practical, robust, and reproducible. In this case it also needs to be specific for and sensitive to otolith dysfunction, in particular unilateral otolith hypofunction. In our view only two tests—the subjective visual horizontal and vestibular evoked myogenic potentials—fulfill these requirements.

THE SUBJECTIVE VISUAL HORIZONTAL (OR VERTICAL): A TEST OF UTRICULAR FUNCTION

Physiological Background

Sitting upright in a totally darkened room normal subjects can align, that is, *set,* a dimly illuminated bar to within 2 deg of the gravitational horizontal[3] or vertical.[6] Studying the ability of subjects to do this after total, unilateral, surgical vestibular deafferentation (uVD) yields important results.[7-10] Whereas before uVD the subjects' settings of the subjective visual horizontal (SVH) are generally within the normal range after uVD, they invariably set the bar tilted toward the deafferented side, sometimes by 15 deg or more. They *set* the bar toward the deafferented side because they *see* the bar, which is actually gravitationally horizontal, as if it were tilted toward the intact side. Although their settings of the bar

---

[c]To whom correspondence may be addressed. Phone: 61 2 9351 3570; fax: 61 2 9351 2603; e-mail: ianc@psych.usyd.edu.au

return toward the gravitational horizontal over about 6 weeks, the settings are still offset by a mean of 4 deg six months after uVD. It appears therefore that a slight ipsilesional offset of the SVH is a permanent legacy of uVD.

What could cause such a perceptual error? Is it an offset of the internal representation of the gravitational vertical as a result of the profound asymmetry in otolithic input to the vestibular nuclei that must occur after uVD? Arguing against this mechanism is the observation that despite the uVD, the patients do not feel that their own bodies are tilted; on the contrary, they feel themselves to be normally upright, even in the dark. In other words, although they set the bar toward the uVD side, it is not in order to null a perceived tilt of the bar with the body, toward the intact side.

Another possible cause of the deviation of the SVH is a torsional deviation of eye position as a part of the ocular tilt reaction. The ocular tilt reaction is a postural synkinesis consisting of head tilt, conjugate eye torsion, and hypotropia, all toward the same side. Some patients temporarily develop a florid ipsilesional tonic ocular tilt reaction—including conjugate ocular torsion—after a unilateral peripheral vestibular lesion.[11] Following surgical uVD there is invariably a tonic ipsilesional deviation of torsional eye position: the 12 o'clock meridians of both eyes are tilted toward the uVD side.[7] One week after uVD there is up to 15 deg ipsilesional eye torsion, and the magnitude of the eye torsion correlates closely with the magnitude of the offset of the SVH ($r=0.95$). Furthermore, the eye torsion gradually resolves in tandem with the SVH. One month after uVD both the eye torsion and the offset of the SVH are about half the one-week value. A slight but statistically significant eye torsion and offset of the SVH (4–5 deg) appears to be a permanent legacy of uVD. Further support for the tight linkage between ocular torsional position and the SVH has come from related experiments in normals that show that eye torsion produces matching changes in perceived visual orientation.[12]

What is the mechanism of the eye torsion after uVD? It could be similar to that of the spontaneous nystagmus that always occurs after uVD and is due to the decreased resting activity in ipsilesional vestibular nucleus neurons, which itself is due to loss of input from primary vestibular neurons, in this case from primary utricular neurons. The concept that tonic eye torsion is utricular in origin comes from the argument that tonic eye torsion represents an offset of the ocular counterrolling mechanism, which is generally thought to be under utricular control.[4]

Vestibular Lesions and Settings of the Subjective Visual Horizontal or Vertical

Clinical studies show that patients with acute spontaneous unilateral peripheral vestibular lesions due to diseases such as vestibular neuritis, also set the bar so that it is no longer aligned with gravity but is consistently tilted toward the side of the lesion.[13,14]

Patients with acute focal brainstem lesions also offset the SVH or the subjective visual vertical (SVV).[15,16] Patients with lower brainstem lesions involving the vestibular nucleus set the SVV or SVH toward the side of the lesion, whereas patients with unilateral upper brainstem lesions involving the interstitial nucleus (of Cajal), and patients with unilateral cerebellar lesions involving the nodulus[17] set the SVV or SVH away from the side of the lesion. In most patients, there is also a tonic deviation of the torsional eye position (also called, mainly in the ophthalmic literature, *cyclotorsion*) in the same direction as the offset of the SVV. The relationship between the magnitude of the offset of the SVV and of the eye torsion is not as tight as with peripheral lesions, and there can be large differences in the magnitude of the torsional deviation and in the offset of the SVV, between the two eyes. For example, with lateral medullary infarcts involving as they usually do, the vestibular nucleus, the excyclotorsion of the ipsilesional eye is much larger than the incyclotorsion of the contralesional eye.[16]

Clinical Significance

The clinical significance of these findings is that careful standardized measurement of the SVH or SVV, using a dim light-bar in an otherwise totally darkened room, can give valuable diagnostic information. An offset of the SVV or the SVH indicates acute unilateral otolithic hypofunction from a lesion either at the level of the end-organ, the vestibular nerve, or the lower brainstem on the side to which the patient sets the bar, or at the level of the upper brainstem or caudal cerebellum on the side opposite to which the patient sets the bar. The greater the deviation of the SVH or SVV, the more acute or more extensive is the lesion. A small deviation of the SVH or SVV is a permanent legacy of many central and peripheral vestibular lesions. Finally, it needs to be remembered that the SVH and SVV tests are insensitive to bilaterally symmetrical impairment of otolith function, since they seem to depend on bilateral asymmetry in tonic resting activity of otolithic neurons in the vestibular nuclei.

VESTIBULAR-EVOKED MYOGENIC POTENTIALS: A TEST OF SACCULAR FUNCTION

Physiological Background

A brief (0.1 ms) loud (>95 dB NHL) monaural click produces a large (60–300 mV) short latency (8 ms) inhibitory potential in the tonically contracting ipsilateral sternocleidomastoid muscle.[18,19] The initial positive–negative potential that has peaks at 13 ms (*p13*) and at 23 ms (*n23*) is abolished by selective vestibular neurectomy, but not by profound sensorineural hearing loss. In other words even if the patient cannot hear the clicks there can be nonetheless normal *p13–n23* responses. Later components of the evoked response do not share the properties of the *p13–n23* potential and probably do not depend on vestibular afferents. Failure to distinguish between these early and late components could explain why earlier work along similar lines was inconclusive (e.g., reference 20).

For the reasons just given we have called the *p13–n23* response the *vestibular evoked myogenic potential* (*VEMP*).[18] Unlike neural-evoked potentials, such as the brainstem auditory-evoked potentials, which are generated by the synchronous discharge of nerve cells, the VEMP is generated by synchronous discharges of muscle cells, or rather of motor units. Being a myogenic potential the VEMP can be 500–1000 times larger than a brainstem potential, for example, 200 µV versus less than 1µV. Single motor-unit recordings in the tonically contracting sternocleidomastoid muscle show a decreased firing rate synchronous with the surface VEMP (see reference 21, figure 8).

When recording VEMPs it is essential to remember that the amplitude of the VEMP is linearly related to the intensity of the click, and to the intensity of sternomastoid activation during the period of averaging, as measured by the mean rectified EMG.[22] A conductive hearing loss abolishes the response simply by attenuating the intensity of the stimulus (see reference 21, figure 9); in such cases the VEMP can be elicited by a brisk tap to the forehead.[23] Inadequate contraction of the sternomastoid muscles will reduce the amplitude of the VEMP and failure to control the intensity of muscle activation produces spurious results (e.g., reference 24).

There are many reasons to suppose that the VEMP arises from stimulation of the saccule. First the saccule is the most sound-sensitive of the vestibular end-organs,[25,26] possibly because it lies just under the stapes footplate,[27] in an ideal position to receive the full impact of a loud click delivered to the tympanic membrane. Second, not only do click-sensitive neurons in the vestibular nerve respond to tilts,[28,29] most originate in the saccular macula[30,31] and project to the lateral and descending vestibular nuclei, as well as to other structures.[32]

The VEMP measures vestibular function indirectly through a vestibulocollic reflex transmitted via the medial and lateral vestibulospinal tracts. The short latency and exclusively ipsilateral distribution of the VEMP suggests that it is mediated by a disynaptic pathway, most likely in the ipsilateral medial vestibulospinal tract.[33]

METHOD

The technique for VEMP testing is simple and any equipment suitable for recording brainstem auditory potentials will also be capable of recording VEMPs. Since the amplitude of the VEMP is linearly related both to the intensity of the click and to the intensity of sternomastoid activation during the period of averaging, it is essential to ensure that the sound source is correctly calibrated and that the background level of rectified sternomastoid EMG activation is measured. Two reasons why the VEMPs could be absent or less than 50 µV in amplitude are a conductive hearing loss and inadequate contraction of the sternomastoid muscles.

The VEMP test does not cause dizziness and individual averages of 128 clicks can be completed in less than 3 minutes. Three superimposed averages usually give a clear result. The patient lies down for the test and activates the sternomastoid muscles for the averaging period by keeping her or his head raised from the pillow. The test cannot be done on uncooperative or unconscious patients and is difficult to do on patients with painful neck problems.

The function of one ear is best evaluated by comparing the amplitude of its VEMP with the amplitude of the VEMP from the other ear. Minor differences in latency commonly occur and might reflect differences in electrode placement over the muscle, or differing muscle anatomy. For simple clinical screening 3 runs of 128 averages with clicks of 100-dB intensity are sufficient. The peak-to-peak amplitudes can then be expressed relative to the level of background mean rectified EMG to create a ratio that largely removes the effect of differences in muscle activity. More accurate, but more time-consuming correction can be made by making repeated observations with differing levels of tonic activation.[18] We take asymmetries of greater than 2.5:1 as evidence of vestibular hypofunction on the side with the smaller potential.

Clinical Applications

Tullio Effect

Nystagmus produced by sound was first described, in pigeons, by Tullio. Loud sounds can sometimes produce nystagmus in humans. Patients who have nystagmus produced by loud sounds, that is, the Tullio effect, complain of sound-induced oscillopsia more than of sound-induced vertigo. The oscillopsia is due to a vertical–torsional nystagmus, which appears to be of superior semicircular-canal origin.[34,35] These patients can have different underlying ear problems. Some appear to have a hypermobile stapes,[36,37] while others appear to have a dehiscent superior semicircular canal.[35] A characteristic abnormality in patients with the Tullio effect is an abnormally large VEMP and an abnormally low threshold of the VEMP.[38] In normal subjects the VEMP, just like the acoustic reflex, has a threshold, usually 90–95 dB. In patients with the Tullio effect the VEMP threshold is >20 dB lower than in normals, and the VEMP amplitude at the usual 100–105-dB NHL stimulus level is large (> 500 µV). If a VEMP can be consistently elicited at 70-dB NHL, this suggests that the patient has the Tullio effect.

Vestibular Neuritis-Neurolabyrinthitis and BPPV

After an attack of vestibular neuritis about one patient in three will develop posterior semicircular-canal-type benign paroxysmal positioning vertigo (BPPV), usually within 3 months.[39] It is of some interest that patients who develop BPPV after vestibular neuritis have intact VEMPs, whereas those who do not have absent VEMPs.[39] In other words, an intact VEMP seems to be a prerequisite for the development of postvestibular neuritis BPPV. The reason for this could be that in those patients who develop postvestibular neuritis BPPV, only the superior vestibular nerve, which innervates the anterior SCC, lateral SCC, and the utricle, is involved. Since the inferior vestibular nerve innervates the posterior SCC and the saccule, the presence of posterior-canal BPPV and the preservation of the VEMP imply that the inferior vestibular nerve must have been spared. Support for such an explanation comes from data that show preservation of posterior SCC vestibulo-ocular reflexes in some patients with vestibular neuritis, patients who presumably have only involvement of the superior vestibular nerve.[40]

Acoustic Neuroma

Although most patients with acoustic neuromas (vestibular schwannomas) present with unilateral hearing loss, some present with vestibular ataxia. This is not entirely surprising, since most "acoustic" neuromas in fact arise not from the acoustic nerve but from one of the vestibular nerves, usually the inferior vestibular nerve.[41] The VEMP, which is transmitted via the inferior vestibular nerve, is abnormal—of low amplitude or absent—in four out of five patients with acoustic neuromas.[42] Since the VEMP does not depend on cochlear or on lateral semicircular-canal function, it can be diagnostically valuable in a patient suspected of having an acoustic neuroma. This is because the VEMP can be abnormal even if a brainstem auditory-evoked potential cannot be measured because cochlear function is absent, and even if the caloric test of lateral semicircular-canal function is normal.

Other Conditions

The VEMP can provide valuable information additional to that obtained by tests of lateral semicircular canal function, such as the caloric or rotational test, or by tests of utricular function, such as the SVH test described earlier. We see patients with symptoms of unilateral or bilateral vestibulopathy (e.g., vertigo, ataxia), who have normal tests of lateral semicircular-canal function at a time when the VEMP test was unequivocally abnormal. A typical example follows.

CASE HISTORY

ZG, a 43-year-old woman presented in March 1996 with a history of more than 20 attacks of acute spontaneous vertigo and vomiting lasting several hours, over a period of 18 months, without any hearing disturbance either during the attack or at any other time. Since her last attack 3 months previously she had noticed persisting mild imbalance and vertical oscillopsia when walking.

On examination there was no spontaneous, or positioning nystagmus, but she did have impairment of the left lateral and vertical semicircular canal vestibulo-ocular reflexes on

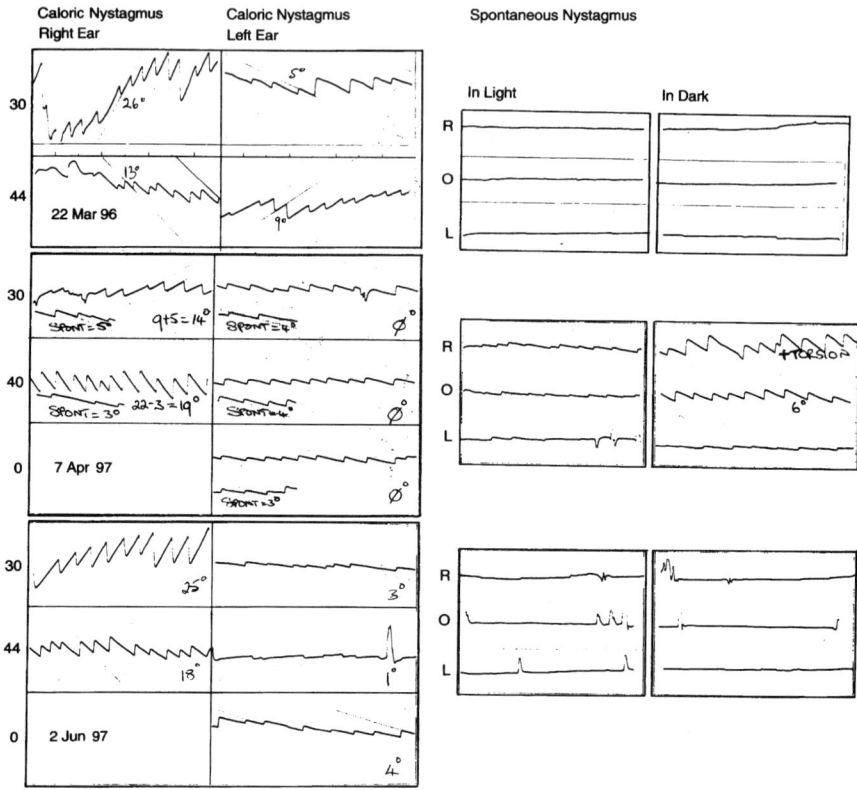

FIGURE 1. Spontaneous, and gaze-evoked nystagmus at 20 deg right and 20 deg left (**right column**), and bithermal caloric tests (**left column**) on patient ZG, before (22 Mar. 1996), during (7 Apr. 1997), and after (2 June 1997) a severe attack of acute spontaneous vertigo due to a left peripheral vestibular lesion causing repeated attacks of acute left vestibular hypofunction. The first test (22 Mar.) shows a 50% left-canal paresis (normal <25%), but no spontaneous or gaze-evoked nystagmus in light or in dark (i.e., with or without visual fixation), indicating a moderately severe compensated loss of left-lateral semicircular-canal function. The second test (7 Apr.) during an acute attack of vertigo, shows marked spontaneous nystagmus, effectively suppressed by visual fixation, with rightward and clockwise quick phases (both from the patient's point of view). This is the type of nystagmus that is produced by acute hypofunction of all 3 semicircular canals on the left. Caloric stimulation of the left ear with water at 30, 44, and 0 deg C now produces no modulation of this spontaneous nystagmus, indicating total, or subtotal insensitivity of the left-lateral semicircular canal. The third test (2 June) shows that the spontaneous nystagmus has now disappeared, and although some caloric responsiveness of the left-lateral semicircular-canal has returned, there is now a more severe loss of left-lateral semicircular-canal function than before the attack; there is now a 90% left-canal paresis. These results indicate a stepwise, acute episodic loss of left-lateral semicircular-canal function, with partial recovery of function between the attacks.

impulsive testing.[43,44] Bithermal caloric tests confirmed severe impairment of left-lateral semicircular-canal function (FIG. 1, left column, top panel). Click VEMPs were normal (FIG. 2, left column, top panel). Pure-tone audiogram, tone-burst transtympanic electrocochleogram, and brain MRI were all also normal.

She was thought to have recurrent vestibular neuritis.[45] A left-vestibular neurectomy was considered, but the patient was reluctant to undergo intracranial surgery and so no specific treatment was recommended in the hope that the attacks would eventually stop. She

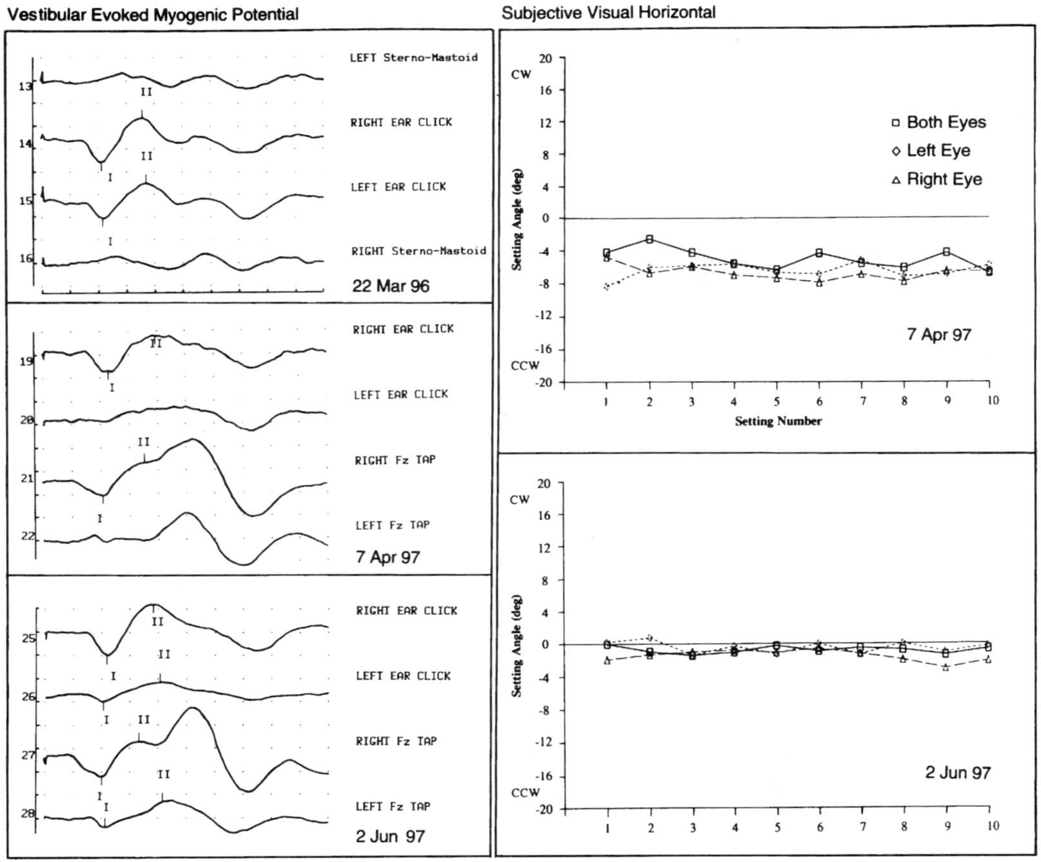

FIGURE 2. Vestibular evoked myogenic potentials (**left column**) and the subjective visual horizontal (**right column**) from ZG. These tests were done at the same time as the caloric tests shown in FIGURE 1. The first test (left column, top panel—22 Mar.) shows normal click-VEMPs: p13 (I) has a latency of 10.5 ms on the right and 10.7 ms on left; n23 (II) has a latency of 17.8 ms on the right and 18.5 ms on the left; the p13–n23 complex has an amplitude of 96 µV on the right and 77 µV on the left. Note that the VEMP is normally exclusively ipsilateral—there is no response from the left sternomastoid muscle to a right-ear click and no response from the right sternomastoid muscle to a left-ear click. The SVH was not measured at this time. During the acute attack (left column, middle panel—7 Apr.), the VEMP from the left ear is absent in response to a tap as well as in response to a click stimulus, indicating that its absence is not due to an intercurrent conductive problem in the left middle ear. The SVH now deviates by about 6 deg to the left (right column, top panel) viewing either with left, right, or both eyes (normal < 2 deg). The third test (2 June), about 2 months after the acute attack, shows that the VEMP from the left ear (to both tap and click stimuli) has reappeared, but the p13–n23 amplitude (I–II) is now reduced (40 µV), compared with the opposite side (110 µV) and with the same side before the attack (77 uV). The SVH has now returned to normal (right column, bottom panel). Note that the p13–n23 amplitudes from the right, and the p13 and the n23 latencies from left and right remain about the same on all 3 tests.

returned for review during an attack of vertigo about a year later. On examination at that time she was distressed, pale, sweaty, and retching. There was third-degree right-beating horizontal-torsional nystagmus present even with visual fixation on the light (FIG. 1, right column, middle panel). Impulsive testing still showed impaired responses from all three left semicircular canals; on the 30-step Unterberger test, she rotated to the left by 90 deg and she was unable to stand on a foam mat with her eyes closed (positive-matted Romberg test). While she admitted on specific questioning that she has some left-sided tinnitus and fulness, the pure-tone audiogram remained normal. Caloric testing now showed no responses at all from the left-lateral semicircular canal (FIG. 1, left column, middle panel). Click and tap VEMPs from the left were now absent (FIG. 2 left column, middle panel). The SVH was deviated by about 5 degrees to the left (FIG. 2, right column, top panel).

She was treated symptomatically and when reviewed 2 days later felt and looked better. She remarked that she had developed a severe left occipital headache lasting several hours, soon after the vertigo began to resolve. On examination the spontaneous nystagmus was no longer present and she no longer deviated on the Unterberger test, but the results of the head impulse test, and the matted Romberg test were unchanged.

When reviewed four months later, she remarked that she had had no further attacks. Caloric tests showed that some responsiveness had reappeared in the left-lateral semicircular canal (FIG. 1, left column, bottom panel); the SVH was now set in the normal range (FIG. 2, right column, bottom panel). The VEMP from the left ear had reappeared, at normal latency, but at the lower limit of the normal range in amplitude (FIG. 2, left column, bottom panel).

Comment

The attacks of acute spontaneous vertigo in this patient were clearly due to repeated, partly reversible, episodes of impaired vestibular function on the left. The caloric tests and the VEMP tests showed improvement between the second and the third examinations, indicating that both the left-lateral semicircular canal and the left saccular function were fluctuating. The return of the SVH settings to normal between the second and third examination does not necessarily indicate improvement in left utricular function; it could simply indicate brainstem compensation for a fixed impairment in left utricular function.[46] While the patient did develop left tinnitus and fulness during the last vertigo attack, her pure-tone thresholds and electrocochleogram were consistently normal. Could this be an example of "*vestibular* Ménière's disease"? Will left cochlear function be eventually impaired? The vertical oscillopsia and positive matted Romberg indicate chronic vestibular insufficiency, which sometimes occurs in not only patients with bilateral loss of vestibular function but also in patients with severe unilateral loss of vestibular function.[47,48]

ACKNOWLEDGMENTS

This work was supported by the National Health and Medical Research Council, the Garnet Passe and Rodney Williams Memorial Foundation, and by the RPA Neurology Department Trustees.

REFERENCES

1. BALOH, R. W., K. BEYKIRCH, V. HONRUBIA & R. D. YEE. 1988. Eye movements induced by linear acceleration on a parallel swing. J. Neurophysiol. **60**: 2000–2013.

2. GIANNA, C. C., M. A. GRESTY & A. M. BRONSTEIN. 1997. Eye movements induced by lateral acceleration steps. Exp. Brain Res. **114**: 124–129.
3. DAI, M. J., I. S. CURTHOYS & G. M. HALMAGYI. 1989. Perception of linear acceleration before and after unilateral vestibular neurectomy. Exp. Brain Res. **77**: 315–328.
4. DIAMOND, S. G. & C. H. MARKHAM. 1983. Ocular counter-rolling as an indicator of vestibular otolith function. Neurology **33**: 1460–1469.
5. FURMAN, J., R. H. SCHOR & D. B. KAMERER. 1993. Off vertical axis rotational responses in patients with peripheral vestibular lesions. Ann. Otol. Rhinol. Laryngol. **102**: 137–143.
6. BETTS, G. A. & I. S. CURTHOYS. 1998. Visually perceived vertical and visually perceived horizontal are not orthogonal. Vision Res. **38**: 1989–1999.
7. CURTHOYS, I. S., M. J. DAI & G. M. HALMAGYI. 1991. Human torsional ocular position before and after unilateral vestibular neurectomy. Exp. Brain Res. **85**: 218–225.
8. BÖHMER, A. & J. RICKENMANN. 1995. The subjective visual vertical as a clinical parameter of vestibular function in peripheral vestibular disease. J. Vestibular Res. **5**: 35–45.
9. BERGENIUS, J., A. TRIBUKAIT & K. BRANTBERG. 1996. The subjective horizontal at different angles of roll-tilt in patients with unilateral vestibular impairment. Brain Res. Bull. **40**: 385–391.
10. TABAK, S., H. COLLEWIJN & L. J. J. M. BOUMANS. 1997. Deviation of the subjective vertical in long-standing unilateral vestibular loss. Acta Otolaryngol. (Stockh.) **117**: 1–6.
11. HALMAGYI, G. M., M. A. GRESTY & W. P. R. GIBSON. 1979. Ocular tilt reaction due to peripheral vestibular lesion. Ann. Neurology **6**: 80–83.
12. WADE, S. & I. S. CURTHOYS. 1997. The effect of ocular torsional position on perception of the roll-tilt of visual stimuli. Vision Res. **37**: 1071–1078.
13. FRIEDMANN, G. 1971. The influence of unilateral labyrinthectomy on orientation in space. Acta Otolaryngol. **71**: 289–298.
14. VIBERT, D, A. B. SAFRAN & R. HAUSLER. 1993. Evaluation clinique de la fonction otolithique par measure de la cyclotorsion oculaire et de la "skew deviation." Ann. Oto-laryngol (Paris) **110**: 87–91.
15. HALMAGYI, G. M., T. BRANDT, M. DIETERICH, I. S. CURTHOYS, R. STARK & W. F. HOYT. 1990. Tonic contraversive ocular tilt reaction with unilateral meso-diencephalic lesion. Neurology **40**: 1503–1509.
16. DIETERICH, M. & T. BRANDT. 1993. Ocular torsion and tilt of the subjective visual vertical are sensitive brainstem signs. Ann. Neurology **33**: 292–299.
17. MOSSMAN, S. & G. M. HALMAGYI. 1997. Partial tonic ocular tilt reaction due to unilateral cerebellar lesion. Neurology **49**: 491–3.
18. COLEBATCH, J. G., G. M. HALMAGYI & N. F. SKUSE. 1994. Myogenic potentials generated by a click-evoked vestibulocollic reflex. J. Neurol. Neurosurg. Psychiatry **57**: 190–197.
19. ROBERTSON, D. D. & D. J. IRELAND. 1995. Vestibular evoked myogenic potentials. J. Otolaryngol. **24**: 3–8.
20. BICKFORD, R. G., J. L. JACOBSON & D. T. R. CODY. 1964. Nature of averaged evoked potentials to sound and other stimuli in man. Ann. N.Y. Acad. Sci. **112**: 204–223.
21. HALMAGYI, G. M., J. G. COLEBATCH & I. S. CURTHOYS. 1994. New tests of vestibular function. Balliere's Neurol. **3**: 485–500.
22. LIM, C. L., P. CLOUSTON, G. SHEEAN & C. YIANNIKAS. 1995. The influence of voluntary EMG activity and click intensity on the vestibular evoked myogenic potential. Muscle Nerve **18**: 1210–1213.
23. HALMAGYI, G. M., R. A. YAVOR & J. G. COLEBATCH. 1995. Tapping the head activates the vestibular system: a new use for the clinical reflex hammer. Neurology **45**: 1927–1929.
24. FERBER-VIART, C., N. SOULIER, D. DUCLAUX & C. DUBREUIL. 1998. Cochleovestibular afferent pathways of trapezius muscle responses to clicks in human. Acta Otolaryngol. (Stockh.) **118**: 6–10.
25. YOUNG, E. D., C. FERNANDEZ & J. M. GOLDBERG. 1977. Responses of squirrel monkey vestibular neurons to audio-frequency sound and head vibration. Acta Otolaryngol. (Stockh.) **84**: 352–360.
26. DIDIER, A. & Y. CAZALS. 1989. Acoustic responses recorded from the saccular bundle on the eighth nerve of the guinea pig. Hearing Res. **37**: 123–128.

27. ANSON, B. J. & J. A. DONALDSON. 1973. Surgical Anatomy of the Temporal Bone and Ear: 285. Saunders. Philadelphia.
28. MUROFUSHI, T., I. S. CURTHOYS, A. N. TOPPLE, J. G. COLEBATCH & G. M. HALMAGYI. 1995. Responses of guinea pig primary vestibular neurons to clicks. Exp. Brain Res. **103**: 174–178.
29. MUROFUSHI, T., I. S. CURTHOYS & D. P. GILCHRIST. 1996. Response of guinea pig vestibular nucleus neurons to clicks. Exp. Brain Res. **111**:149–152.
30. MUROFUSHI, T. & I. S. CURTHOYS. 1997. Physiological and anatomical study of click-sensitive primary vestibular afferents in the guinea-pig. Acta Otolaryngol. (Stockh.) **117**: 66–72.
31. MCCUE, M. P. & J. J. GUINAN. 1997. Sound-evoked activity in primary afferent neurons of the mammalian vestibular system. Am. J. Otol. **18**: 355–360.
32. KEVETTER, G. A. & A. A. PERACHIO. 1986. Distribution of vestibular afferents that innervate the sacculus and posterior canal in the gerbil. J. Comp. Neurol. **254**: 410–424.
33. UCHINO, Y., H. SATO, M. SASAKI, H. IKEGAMI, N. USI & W. GRAF. 1997. The sacculocollic reflex arc in cats. J. Neurophysiol. **77**: 3003-3012.
34. ROTTACH, K. G., R. D. VON MAYDELL, A. O. DISCENNA, A. Z. ZIVOTOFSKY, L. AVERBUCH-HELLER & R. J. LEIGH. 1996. Quantitative measurements of eye movements in a patient with Tullio phenomenon. J. Vestib. Res. **6**: 255–259.
35. MINOR, L. B., D. SOLOMON, J. S. ZINREICH & D. S. ZEE. 1998. Tullio's phenomenon due to dehiscence of the superior semicircular canal. Arch. Otolaryngol. Head Neck Surg. **124**: 249–258.
36. DEECKE, L., T. MERGNER & D. PLESTER. 1981. Tullio phenomenon with torsion of the eyes and subjective tilt of the visual surround. Ann. N.Y. Acad. Sci. **374**: 650–655.
37. DIETERICH, M., T. BRANDT & W. FRIES. 1989. Otolith function in man. Results from a case of otolith Tullio phenomenon. Brain **112**: 1377–1392.
38. COLEBATCH, J. G., J. C. ROTHWELL, A. BRONSTEIN & H. LUDMAN. 1994. Click-envoked vestibular activation in the Tullio phenomenon. J. Neurol. Neurosurg. Psychiatry **57**: 1538–1540.
39. MUROFUSHI, T., G. M. HALMAGYI, R. A. YAVOR & J. G. COLEBATCH. 1996. Vestibular evoked myogenic potentials in vestibular neuritis: an indicator of inferior vestibular nerve involvement. Arch. Otolaryngol. Head Neck Surg. **122**: 845–848.
40. FETTER, M. & J. DICHGANS. 1996. Vestibular neuritis spares the inferior division of the vestibular nerve. Brain **119**: 755–763.
41. CLEMIS, J., W. J. BALLAND, P. J. BAGGOT & S. T. LYON. 1986. Relative frequency of inferior vestibular schwannoma. Arch. Otolaryngol. **112**: 190–194.
42. MUROFUSHI, T., M. MATSUZAKI & M. MIZUNO. 1998. Vestibular evoked myogenic potentials in patients with acoustic neuromas. Arch. Otolaryngol. Head Neck Surg. **124**: 509–512.
43. HALMAGYI, G. M. & I. S. CURTHOYS. 1988. A clinical sign of canal paresis. Arch. Neurol. **45**: 737–739.
44. CREMER, P. D., G. M. HALMAGYI, S. T. AW, I. S. CURTHOYS, L. A. MCGARVIE, M. J. TODD, R. A. BLACK & I. P. HANNIGAN. 1998. Semicircular canal plane impulses detect absent function of individual semicircular canals. Brain **121**: 699–716.
45. SCHUKNECHT, H. F. & R. L. WITT. 1985. Acute bilateral sequential vestibular neuritis. Am. J. Otolaryngol. **6**: 255–257.
46. CURTHOYS, I. S. & G. M. HALMAGYI. 1995. Vestibular compensation. A review of the oculomotor, neural and clinical consequences of unilateral vestibular loss. J. Vestibular Res. **5**: 67–107.
47. HALMAGYI, G. M. 1994. Vestibular insufficiency following unilateral vestibular deafferentation. Aust. J. Otolaryngol. **1**: 510–512.
48. REID, C. B., R. EISENBERG, G. M. HALMAGYI & P. A. FAGAN. 1996. The outcome of vestibular nerve section for intractable vertigo: the patients' point of view. Laryngoscope **106**: 1553–1556.

Directional Abnormalities of Vestibular and Optokinetic Responses in Cerebellar Disease

MARK F. WALKER[a,c] AND DAVID S. ZEE[b]

[a]Departments of Neurology, The Johns Hopkins University School of Medicine, Pathology 2-210, 600 North Wolfe Street, Baltimore, Maryland 21287, USA

[b]Departments of Neurology, Ophthalmology, and Otolaryngology-Head and Neck Surgery, The Johns Hopkins University School of Medicine, Baltimore, Maryland 21287, USA

ABSTRACT: Directional abnormalities of vestibular and optokinetic responses in patients with cerebellar degeneration are reported. Three-axis magnetic search-coil recordings of the eye and head were performed in eight cerebellar patients. Among these patients, examples of directional cross-coupling were found during (1) high-frequency, high-acceleration head thrusts; (2) constant-velocity chair rotations with the head fixed; (3) constant-velocity optokinetic stimulation; and (4) following repetitive head shaking. Cross-coupling during horizontal head thrusts consisted of an inappropriate upward eye-velocity component. In some patients, sustained constant-velocity yaw-axis chair rotations produced a mixed horizontal-torsional nystagmus and/or an increase in the baseline vertical slow-phase velocity. Following horizontal head shaking, some patients showed an increase in the slow-phase velocity of their downbeat nystagmus. These various forms of cross-coupling did not necessarily occur to the same degree in a given patient; this suggests that different mechanisms may be responsible. It is suggested that cross-coupling during head thrusts may reflect a loss of calibration of brainstem connections involved in the direct vestibular pathways, perhaps due to dysfunction of the flocculus. Cross-coupling during constant-velocity rotations and following head shaking may result from a misorientation of the angular eye-velocity vector in the velocity-storage system. Finally, responses to horizontal optokinetic stimulation included an inappropriate torsional component in some patients. This suggests that the underlying organization of horizontal optokinetic tracking is in labyrinthine coordinates. The findings are also consistent with prior animal-lesion studies that have shown a role for the vestibulocerebellum in the control of the direction of the VOR.

INTRODUCTION AND BACKGROUND

The Ideal Vestibuloocular Reflex

The purpose of the vestibulo-ocular reflex (VOR) is to ensure that the line of sight of each eye continues to point to the object of interest when the head is moved. For the rotational

---

[c]To whom correspondence may be addressed. Phone: 410/614-1575; fax: 410/614-1746; e-mail: mwalker@dizzy.med.jhu.edu

VOR, an ideal response must meet several requirements. Consider the situation where the object of interest is distant from the head, such that no adjustment is needed for translation of the orbits. In this case, an ideal VOR has unity gain; that is, the absolute magnitude of eye velocity relative to the head must equal the magnitude of head velocity. Second, the axes of eye and head velocity must be parallel, but the eye must rotate in the direction opposite to head rotation. Third, head and eye velocity must be synchronized; there must be a minimal time delay or phase shift of eye velocity relative to the head.

Processing of Labyrinthine Signals to Create an Ideal VOR

There are inherent anatomic features of the labyrinths and the orbits that add further complexity to the transformation of labyrinthine inputs into the appropriate eye movement signals. The three pairs of semicircular canals are neither exactly orthogonal nor are the two canals within any single-pair coplanar. Moreover, canal orientations vary among individuals.[1] The pulling directions of the extraocular muscles are only approximately aligned with their corresponding canal planes, and pulling directions vary with orbital position, particularly when the effects of the orbital smooth-muscle pulleys are taken into account.[2]

An additional consideration relates to differences in processing of labyrinthine information depending on the frequency composition of head motion. First, when the speed, acceleration, or frequency of head rotation is high, nonlinearities appear in the VOR. Excitation of a semicircular canal becomes a much more effective stimulus than inhibition, and there is a velocity-dependent inhibitory cutoff. These nonlinearities, which form the basis for Ewald's Second Law, are manifest in the setting of unilateral vestibular loss, where there is a marked asymmetry between the responses to the intact and to the lesioned side.[3] Second, during sustained, low-frequency rotations, a velocity-storage mechanism comes into play. This mechanism extends the bandwidth over which the VOR can effectively stabilize the line of sight. In addition, in conditions such as head tilts and centrifugation, in which the axis of head rotation is not parallel to the gravito-inertial axis, velocity storage, in response to otolith inputs, realigns eye velocity away from the head-velocity axis and toward the gravito-inertial axis.[4,5]

Thus, the VOR is a far more complicated reflex than its basic underlying three-neuron arc suggests. What then are the mechanisms by which the VOR remains accurate, adjusting for anatomical variations, making immediate on-line corrections for changes in target location, and, over the long term, maintaining an appropriate calibration?

The Cerebellum and the VOR

About twenty-five years ago, Ito suggested that the vestibulocerebellum was a structure involved in the long-term adaptive control of the amplitude (gain) of the VOR.[6] Lesion studies supported this idea; adaptive changes in the amplitude of the VOR no longer occur in the absence of the cerebellar flocculus.[7,8] Lesion experiments have also implicated the vestibulocerebellum in the control of the direction of the VOR. Schultheis and Robinson,[9] in cats, used a cross-axis adaptation training paradigm in which the visual scene was oscillated around an axis orthogonal to that of the head oscillation. Several hours of exposure to this stimulus served to shift the axis of eye rotation toward the axis of visual motion; this was true even when the VOR was measured in the dark. Combined floccular and nodular lesions abolished the ability of these cats to undergo cross-axis adaptation. Additional evidence for cerebellar control of VOR direction comes from monkeys that have undergone nodulo-uvulectomy; in these animals, eye velocity no longer realigns with a shifted gravito-inertial axis.[5,10]

Based upon the results of animal experimentation we predicted that patients with cerebellar dysfunction would show abnormalities of the control of the direction of the VOR during vestibular stimulation. Here we present preliminary data indicating that this indeed is the case. We found examples in our cerebellar patients of abnormal cross-coupling during high-frequency, high-acceleration head thrusts; during sustained low-velocity head rotations; and following head shaking. We also observed abnormalities of the axis of eye rotation during optokinetic stimulation. For a given patient, there was no clear correlation of the directional abnormalities of slow-phase eye velocity elicited by these various types of vestibular and optokinetic stimulation. This suggests that different parts of the cerebellum might determine the axis of eye rotation for different types of stimuli.

METHODS

Patients

We recorded eye movements in eight patients with cerebellar degeneration and four normal control subjects. Four patients were men and four women. The mean age was 57 (range 33 to 75). Patients did not have significant extracerebellar neurologic abnormalities, except for one patient who also had bilateral vestibular hypofunction and one with extrapyramidal findings, suggesting the diagnosis of multiple system atrophy. All patients had a spontaneous downbeat nystagmus with slow-phase velocities ranging from 0.7 to 7.8 deg/s in darkness (mean 3.6 ± 2.7).

Four normal subjects had a mean age of 33 (range 23–53). Head thrusts were performed in all normal subjects; chair rotations were performed in two and head shaking in one.

Subjects were recorded using a three-field magnetic search coil system. During chair rotations, the head was fixed relative to the field coils using a bite-bar restraint. In other paradigms, the head was free and rotated manually by one of the authors. A single three-dimensional surface eye coil (usually on the left eye) and a three-dimensional head coil (attached to a bite bar or taped to the forehead) were used. Eye and head coil signals were sampled at 500 Hz.

Stimuli

1. Head thrusts: High acceleration head rotations were manually applied in the horizontal plane with the head upright (head pitched about 10 to 20 deg nose up from Reid's stereotaxic plane). Typical peak head velocities were about 100 to 200 deg/s and accelerations about 5000 deg/s.[2] For most subjects, head thrusts were performed with the subject facing the examiner and with fixation of a near target (the examiner's nose). For one control subject and one patient, head thrusts were applied from behind during fixation of a central LED target at 126 cm. All recordings were made in full room illumination. The timing and direction of head thrusts were unpredictable to the subject.

2. Chair rotations: Constant-velocity chair rotations in the yaw plane with the head restrained in the upright position were applied at 30 or 60 deg/s in complete darkness with recording of per- and postrotatory nystagmus. During continued rotation, after the per-rotatory nystagmus had decayed, lights were turned on to record optokinetic responses. Before the chair was stopped, lights were extinguished, and optokinetic after-nystagmus (OKAN) was recorded.

3. Head shaking: High-frequency (typically about 2 Hz) manual head shaking was performed in the horizontal and vertical planes in complete darkness.

Analysis

Angular velocities of the head, eye-in-space (eye-in-field), and eye-in-head were derived using standard three-dimensional eye analysis techniques.[11,12] For chair rotations (during which the head was restrained) and for post-head-shaking nystagmus (when the head was held still), angular eye velocity was referenced to a field-fixed coordinate system. For head thrusts, eye-in-head velocity was compared to head velocity. This comparison requires that both head and eye velocities be referenced to the same coordinate frame;[13] therefore, it was necessary to perform an additional transformation to obtain head velocity relative to head position. This was accomplished through a matrix multiplication of the inverse (transpose) of the rotation matrix representing head position and the angular head-velocity vector in space-fixed coordinates.

The onset of a head thrust was defined as the time at which the absolute value of the head angular velocity along the principal axis of movement exceeded 5 deg/s. For the purposes of the present paper, which focuses on directional abnormalities rather than the exact timing of responses, this criterion proved to be adequate and gave consistent results. In addition, all results were interactively verified.

For analysis of nystagmus associated with chair rotations and head shaking, an interactive technique was used to mark slow phases. Then the median value during the central 50% of each slow phase was calculated for each of the three components of angular eye velocity.

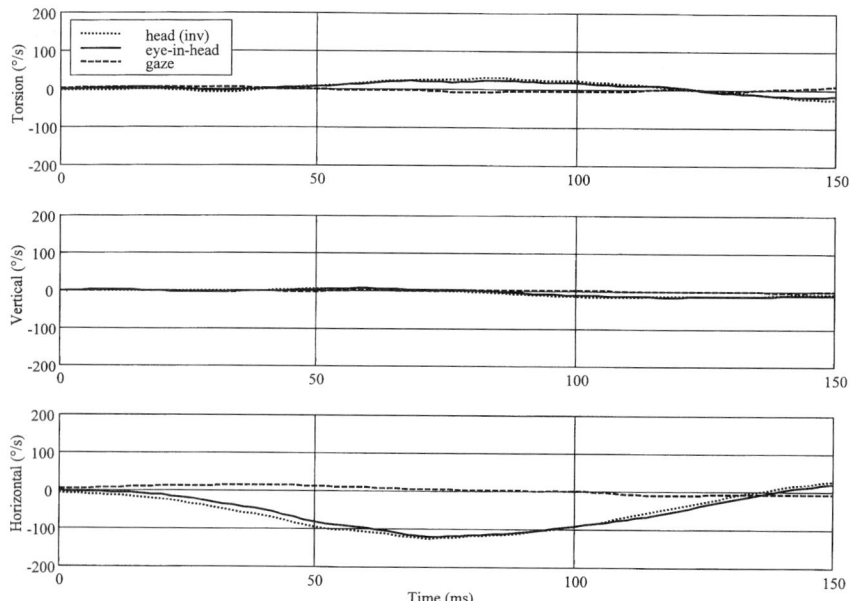

FIGURE 1. Rightward yaw head thrust in a normal subject during fixation of an LED target (126 cm) in the light. Plots show each of the three components of angular velocity for eye-in-head (*solid line*), head-in-head (*dotted line*), and eye-in-space (gaze error; *dashed line*). By convention, positive velocities are clockwise (from the subject's perspective), upward, and rightward. For comparison to eye velocity, each component of head velocity is inverted.

RESULTS

In cerebellar patients, directional errors (cross-coupling) were seen in eye movement responses to both vestibular and optokinetic stimuli. We will illustrate each of these with specific examples from high-frequency head thrusts, constant-velocity chair rotations, head-shaking, and optokinetic stimulation.

High-Frequency VOR: Head Thrusts

The typical eye movement response to a horizontal head thrust in a normal subject during fixation of a distant target is shown in FIGURE 1. When a rapid rightward head rotation is applied, there is a compensatory eye velocity to the left, which closely follows the head velocity, thus minimizing gaze error. Orthogonal (vertical and torsional) components are negligible.

FIGURE 2 compares head and eye velocities in a two-dimensional representation; the vertical component of head or eye velocity is plotted against the corresponding horizontal

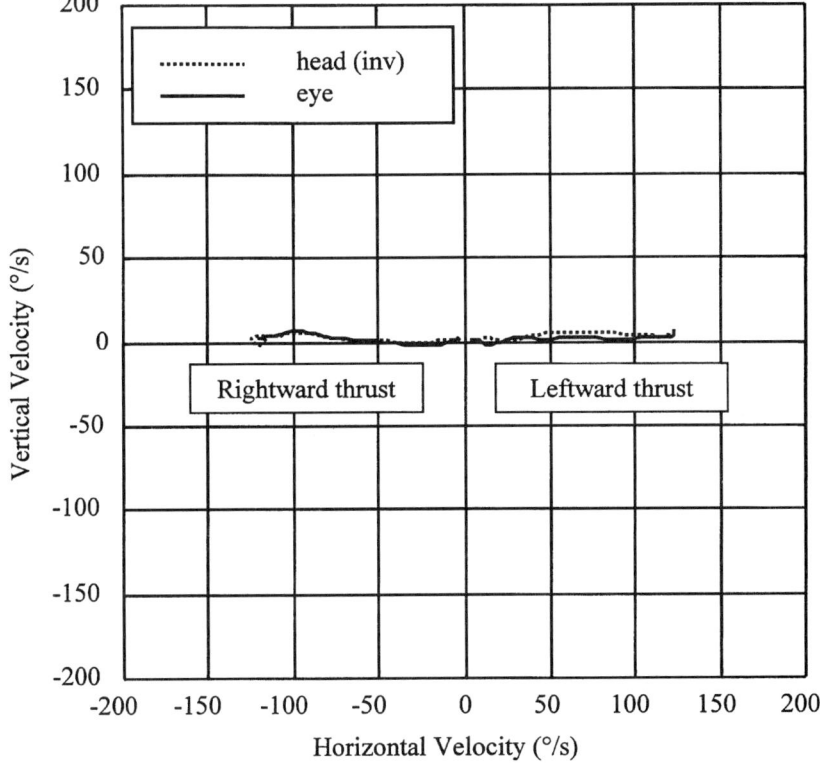

FIGURE 2. 2-D representation of yaw head thrusts (first 80 ms after onset of head movement) in a normal subject. The vertical component of the angular velocity is plotted against the horizontal component for the head (*dotted line*) and eye (*solid line*). Head velocity is inverted. Representative rightward and leftward head thrusts are shown.

component for the first 80 ms after the onset of the head movement. Representative rightward and leftward head thrusts are shown. Again, note that the axes of head and eye velocities are closely matched.

FIGURE 3 shows the response to a leftward horizontal head thrust (with distant fixation) in a cerebellar patient. We focus on the initial vestibular slow phase (approximately the first 60 ms), before the onset of the corrective quick phase. There are two main findings. First, there is a large upward eye velocity inappropriate for the head movement. Second, the gain (eye velocity / head velocity) of the horizontal slow-phase component is much greater than unity. The resultant oblique eye movement creates a large gaze error, which requires a corrective quick phase to return gaze to the desired point of fixation. The upward eye velocity cannot be attributed simply to a coincident upward slow phase of the patient's spontaneous downbeat nystagmus. Note that the vertical slow-phase velocity was not constant, but rose nearly linearly with increasing horizontal eye velocity. At the peak, it was much greater (more than 30-fold) than the velocity of the spontaneous nystagmus (about 3 deg/s). Moreover, the timing of the vertical component relative to the head movement was consistent from trial to trial. Note also that the vertical response also cannot be explained by a vertical component to the applied head thrust, which was minimal.

Responses to leftward and rightward head thrusts are depicted in a two-dimensional plot in FIGURE 4A for the same patient. Upward vertical cross-coupling is observed in the responses to head thrusts in both directions, although there is some degree of asymmetry in the magnitude and timing of these responses. A two-dimensional plot of horizontal head thrust responses from another patient is given in FIGURE 4B, showing similar upward cross-coupling.

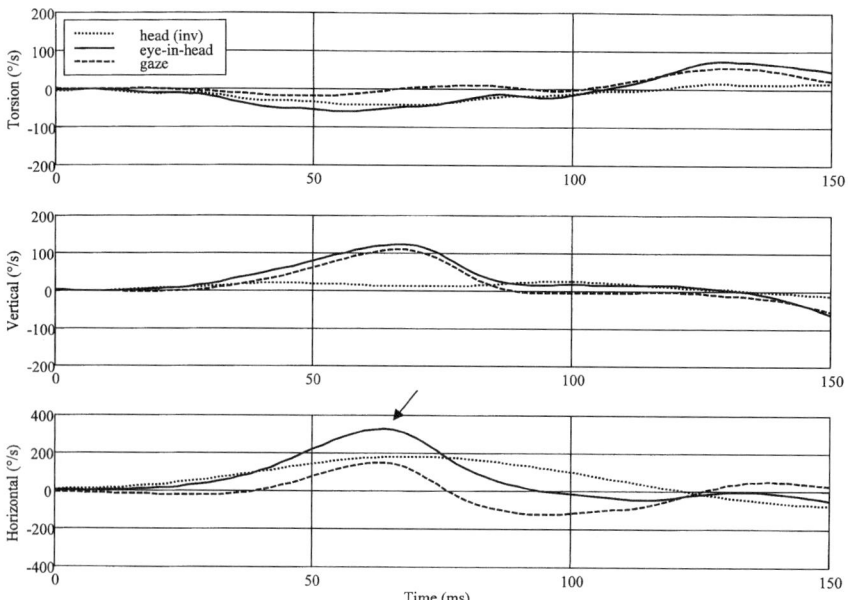

FIGURE 3. Leftward yaw head thrust in a cerebellar patient. The response is characterized by an inappropriate upward eye velocity and a horizontal component with increased gain, resulting in an oblique gaze error. The *arrow* marks the onset of an apparent corrective eye movement.

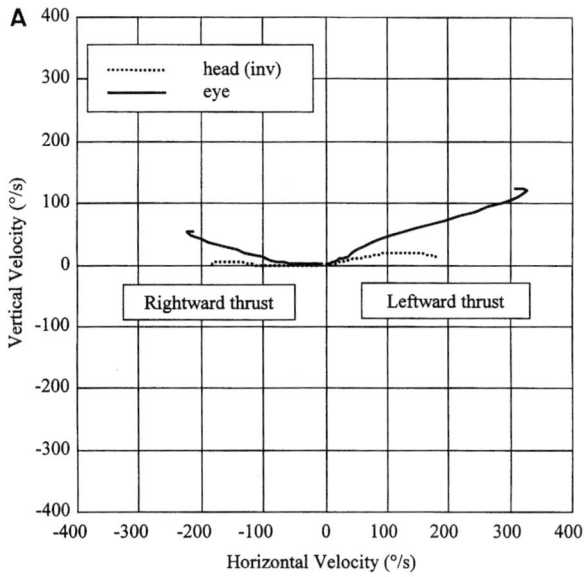

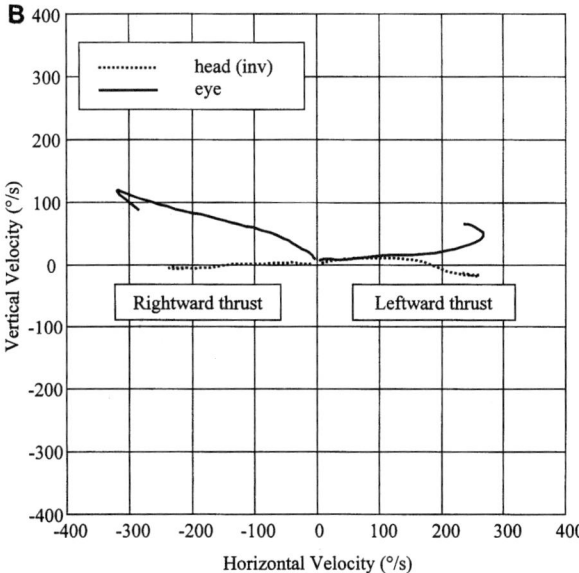

FIGURE 4. 2-D representation of head thrust responses in cerebellar patients. (**A**) Same patient as in FIGURE 3 (126-cm fixation), (**B**) another patient (near fixation). In both cases, there is an inappropriate upward eye velocity with head thrusts in either direction, although magnitude and timing differ.

Low-Frequency VOR: Velocity Step Rotations

Some cerebellar patients exhibited cross-coupling in their responses to constant-velocity rotations in the yaw plane. An example is seen in FIGURE 5, the response to a leftward velocity step of 60 deg/s. Both onset and offset of rotation produced an increase in the upward slow-phase velocity. The vertical responses were prolonged, and, at least for the postrotatory response, vertical velocity was not maximal at onset but built up over several seconds. Responses were asymmetric; in this case, a greater vertical component was seen in the postrotatory than in the pre-rotatory nystagmus. This corresponded to a greater increase in upward slow-phase velocity during the per-rotatory phase with rightward chair rotation (not shown). As with head thrusts, vertical cross-coupling always occurred in the upward direction.

Cross-coupling into torsion is also evident in this case and was observed similarly in several other patients. The direction typically corresponded to the direction of the horizontal slow phase; with rightward slow phases, the torsional component was clockwise (relative to the subject), and with leftward slow phases, the torsion was counterclockwise. The torsional component decayed more slowly than the horizontal component.

The same data are presented in a three-dimensional plot in FIGURE 6 and compared to the results in a normal subject. In the normal subject, the nystagmus was essentially confined to the horizontal plane, with minimal vertical and torsional components.

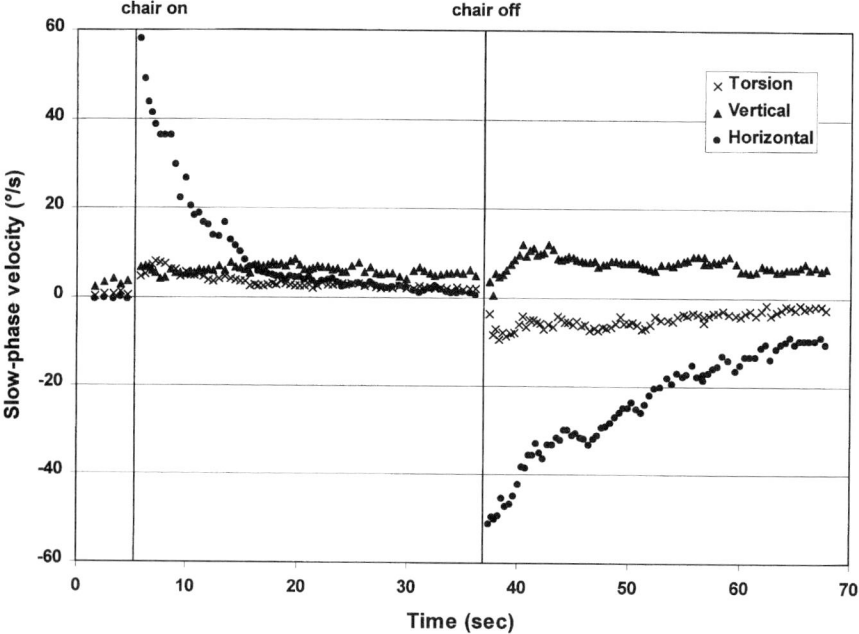

FIGURE 5. Leftward velocity step (60 deg/s) in a cerebellar patient. In this case, lights were out for the duration of the recording, and there was no optokinetic stimulation. Note in addition to the expected horizontal pre- and postrotatory nystagmus that there are inappropriate vertical and torsional components. The vertical component consists of an increase in upward slow-phase velocity above the baseline of the spontaneous downbeat nystagmus. The torsional component changes direction and corresponds to the side of the stimulated labyrinth. Cross-coupled components appear to decay more slowly (i.e., longer time constants) than the horizontal component.

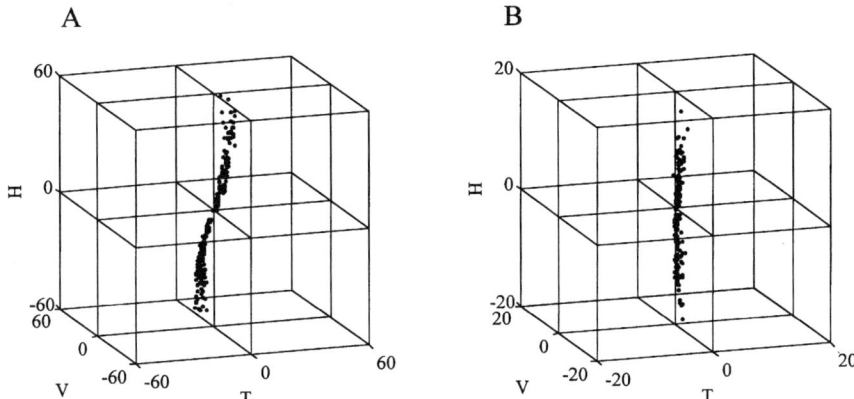

FIGURE 6. (**A**) 3-D eye velocity axis (H: horizontal; V: vertical; T: torsion) during yaw velocity steps (60 deg/s) for the same patient as in FIGURE 5. Responses to leftward and rightward rotation are shown, including both pre- and post-rotatory phases. (**B**) Eye velocity axis in a normal subject in responses to 30 deg/s yaw velocity steps. All velocities are given in deg/s.

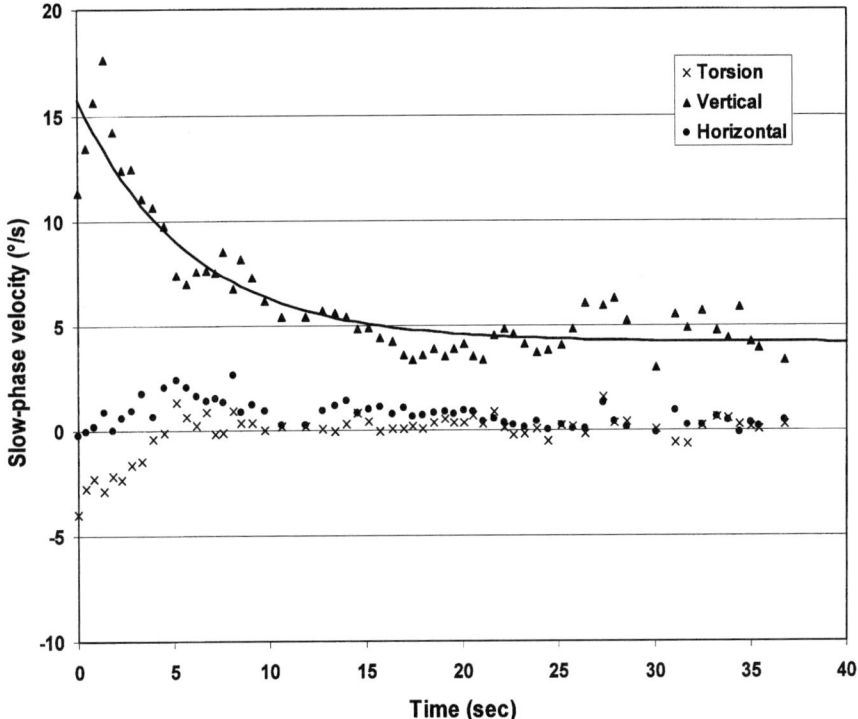

FIGURE 7. Post-head-shaking nystagmus in a cerebellar patient. The 40-ms interval immediately following cessation of horizontal head shaking (about 1.8 Hz for 19 s) is shown. An exponential fit to the decaying vertical slow-phase velocity is shown with a *solid line* (time constant 5.9 s).

Head Shaking

High-frequency horizontal and vertical head shaking were followed by an exponentially decaying downbeat nystagmus in some patients. This is illustrated in FIGURE 7 for a patient with a spontaneous upward slow-phase velocity of about 4 deg/s. Following 19 s of horizontal head shaking at a mean frequency of 1.8 Hz, the upward slow-phase velocity had increased to about 15 deg/s. This decayed with a time constant of 5.9 s back to the baseline. A similar nystagmus was elicited by vertical head shaking (not shown). Vertical nystagmus was not seen after head shaking in the control subject. Unlike responses to chair rotations, when there was an inappropriate torsional component, after head shaking the eye velocity was almost purely vertical.

Optokinetic Nystagmus

Optokinetic responses were tested by turning on the room lights during a prolonged constant-velocity horizontal chair rotation. In normal subjects, this elicited the expected horizontal nystagmus with little response in the torsional or vertical planes. However, in several patients, horizontal optokinetic nystagmus included a considerable torsional component, the direction of which was related to the direction of the horizontal slow phase. This is illustrated for one patient in FIGURE 8. During a 30 deg/s chair rotation to the right with lights on, there was a mixed horizontal-torsional nystagmus with mean slow-phase

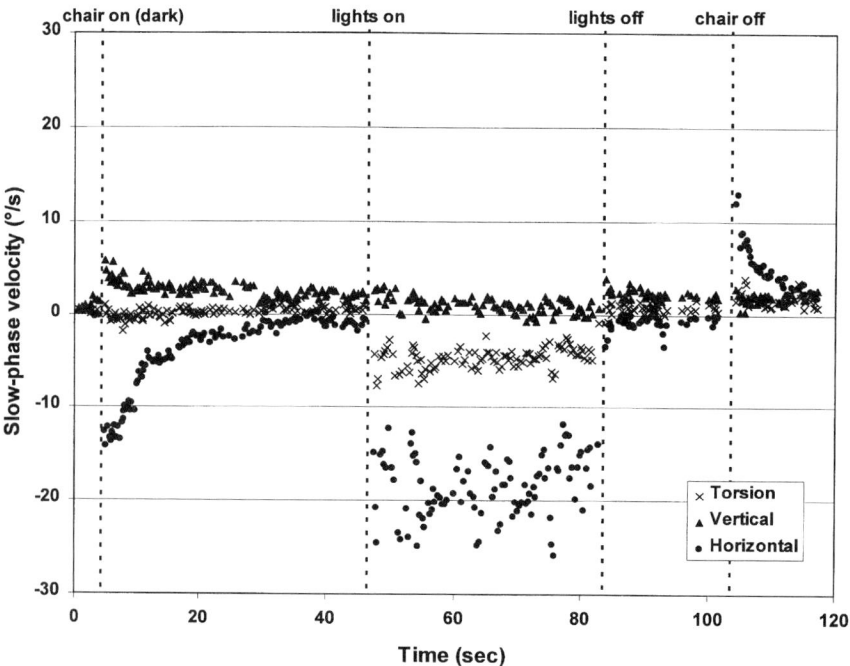

FIGURE 8. Rightward velocity step (30 deg/s) in another cerebellar patient. Note the torsional cross-coupling during optokinetic stimulation when the lights are turned on during rotation.

velocity of 18.5 deg/s to the left and 4.7 deg/s counterclockwise. With leftward chair rotation (not shown), the direction of torsion reversed. Note in this patient the absence of correlation between the cross-coupled responses to the vestibular and to the optokinetic stimuli. Pre- and post-rotatory nystagmus exhibited a small vertical but no torsional slow-phase component. In contrast, horizontal optokinetic stimulation was associated with torsional, but not vertical, cross-coupling.

DISCUSSION

The main finding in this study is that in response to vestibular and optokinetic stimuli, patients with cerebellar lesions frequently show cross-coupling abnormalities in generating compensatory eye movements of the correct direction. During earth-vertical-axis rotation, patterns of cross-coupling included inappropriate upward and torsional eye velocity. Cross-coupled responses were observed with both high-frequency/high-acceleration (head thrusts) and low-frequency (velocity steps) stimulation, although not necessarily correlated in an individual patient. In some patients, there was torsional cross-coupling during horizontal optokinetic stimulation when the head was upright.

Cross-coupling and the Cerebellum

Our results implicate the cerebellum in the control of the direction of the VOR. This finding is not surprising, given that lesions of the vestibulocerebellum in experimental animals impair cross-axis VOR adaptation[9] and the ability to reorient eye velocity with shifts in the gravito-inertial axis.[5,10] That our patients showed directional abnormalities in response to both high- and low-frequency stimuli implies that both semicircular canal pathways and the velocity storage system are affected. This is likely mediated in different substructures of the cerebellum, as the pattern of abnormalities differed among patients.

A common pattern of VOR cross-coupling was the generation of an inappropriate upward eye-velocity component with horizontal head rotation. While all of our patients had a spontaneous nystagmus with slow phases upwards, we do not think the spontaneous nystagmus explains the dynamic cross-coupling during head rotation. First, not all patients with downbeat nystagmus had inappropriate cross-coupling during the VOR. Second, the simple superposition of a spontaneous vertical nystagmus upon the horizontal VOR does not explain the finding that the VOR could have vertical velocities much larger than those of the spontaneous nystagmus, nor does it explain the close relationship between the primary and cross-coupled components of eye velocity (see FIG. 4). What then is the mechanism of this upward bias in both static (spontaneous nystagmus) and dynamic (VOR) conditions?

Prior studies have pointed to asymmetries in the pathways involved in slow vertical eye movements. These asymmetries have been invoked to explain the emergence of downbeat nystagmus when cerebellar function is disrupted. Floccular inhibitory output related to the vertical VOR appears to be predominantly, if not exclusively, directed toward the anterior canal pathway.[14,15] Floccular inactivation selectively impairs the downward VOR.[16] This asymmetry also holds for vertical pursuit. Floccular Purkinje cells discharge during downward but not upward pursuit; the optimal direction suggests alignment with the ipsilateral anterior canal.[17] This correlates with the finding that in cerebellar patients, downward pursuit is impaired more severely than upward pursuit.[18] The overall hypothesis is that the flocculus inhibits anterior canal pathways to facilitate smooth downward eye movements, whether of vestibular or visual origin.

A loss of tonic floccular inhibition might also explain the common finding of downbeat nystagmus in cerebellar patients, if this serves to uncover an underlying upward eye-velocity bias. It has been suggested that such a bias might originate from the pursuit[18] or central vestibular[19] systems, or from the vestibular periphery.[20]

As described in the Introduction, there are a number of factors, including canal alignment, VOR nonlinearities, and pulling directions of eye muscles, that must be taken into account to generate a VOR of the correct direction. Compensation for some of these factors could be achieved through a convergence of connections from different semicircular canals upon individual cells within the vestibular nuclei; these brainstem connections of the VOR have been described mathematically as a 3×3 matrix.[21,22] The cerebellum might calibrate the VOR direction by adjusting the relative synaptic strengths of these convergent canal inputs and thus their relative contributions to the response of a given eye-muscle pair. Disruption of this calibration might produce eye movements of the wrong direction. For example, when yaw-axis stimulation produces a vertical eye-movement response (as in our patients or in cross-axis adaptation), there might be an excessive horizontal-canal contribution to the pathways mediating the vertical response.

Cross-Coupling During Head Thrusts

In some patients, yaw-axis head thrusts were associated with an upward eye-velocity component. Again, this is somewhat analogous to cross-axis adaptation experiments, in which a horizontal head rotation is coupled with a vertical visual stimulus to produce vertical cross-coupling of the horizontal VOR. This suggests that the mechanism of cross-coupling in our patients may involve an inappropriate calibration of the brainstem connections (or parallel side pathways) between the semicircular canals and the extraocular muscle pairs.

An important difference between these results and those of cross-adaptation experiments is the nonlinearity of these head-thrust responses. Horizontal head thrusts in either direction produced an upward slow phase, whereas cross-axis adaptation serves to rotate the eye-velocity axis, such that horizontal rotation in one direction is associated with an upward eye movement and rotation in the other direction with a downward eye movement. Thus, in our patients, a simple adjustment of the strengths of brainstem connections (or, equivalently, of the values of the elements of the Robinson brainstem matrix) will not suffice. Additional nonlinear factors must come into play. In the case of head thrusts this nonlinearity may be derived from the known asymmetry between canal excitation and inhibition at these high frequencies, which forms the basis for the asymmetry of head-thrust responses in patients with unilateral vestibular lesions.[23] Given this asymmetry, what might be the source of upward cross-coupling?

We offer two hypotheses for the origin of vertical cross-coupling in response to yaw head thrusts in our patients; these are not necessarily mutually exclusive, nor do we imply that there are no other possibilities. As mentioned earlier, prior studies have found that the inhibitory output of floccular Purkinje cells to the VOR is largely directed toward cells receiving input from the horizontal and anterior canals.[14,15] Our hypotheses are based on those findings and on the assumption that there are secondary connections from the horizontal canals to the vertical canal pathways, as predicted by cross-axis adaptation experiments.[9,21] The latter is supported by single-unit recordings; vertical canal flocculus target neurons have a greater sensitivity to horizontal head rotation than do neurons that do not receive floccular inhibition.[24]

The first hypothesis is that upward cross-coupling during head thrusts results from an increased gain of anterior canal pathways with the loss of floccular inhibition or from inappropriate synaptic modification due to floccular dysfunction. Although the anterior canals are usually considered to be only minimally sensitive to yaw head rotations, head thrusts

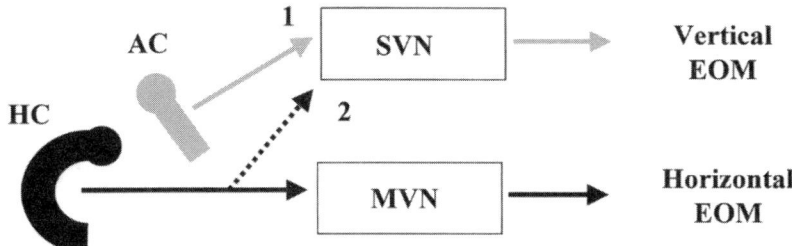

FIGURE 9. Schematic representation of proposed mechanisms for vertical cross-coupling during head thrusts (see text). Sites of potentially increased gain are: (**1**) Direct input of the (contralateral) anterior canal, and (**2**) convergent input from the (ipsilateral) horizontal canal to the anterior canal pathways. AC: anterior canal; HC: horizontal canal; SVN: superior vestibular nucleus; MVN: medial vestibular nucleus; EOM: extraocular muscles.

in our study were typically performed with the head upright in a position that is about 10 to 20 deg nose-up from Reid's stereotaxic plane. In this position, earth-vertical-axis rotations would be expected to stimulate the contralateral anterior canal.[1,25] That, in combination with an increased gain of anterior-canal pathways, might result in an upward component to the slow phase evoked by horizontal head rotation.

A second possibility is that the upward eye velocity originates not from the anterior canal but from the ipsilateral horizontal canal through a projection onto the anterior-canal pathway. The gain of this projection might also be increased with lesions of the flocculus. This hypothesis might be distinguished from the first by performing earth-vertical head thrusts in various pitch positions. A mechanism involving stimulation of the anterior canal would be expected to be more sensitive to changes in pitch, due to changes in the pattern of vertical-canal activation.[25] The two hypotheses are illustrated in schematic form in FIGURE 9.

An additional finding of our experiments is that horizontal head thrust responses may have an increased gain of the primary horizontal component (see FIGS. 3 and 4). This extends the results of prior studies that have reported increased gains in cerebellar patients in response to velocity steps and sinusoidal stimulation up to 1.6 Hz.[26,27]

Cross-Coupling During Velocity Steps

Cross-coupling in response to velocity steps included a torsional eye velocity whose direction was matched to that of the horizontal component (clockwise with rightward and counterclockwise with leftward slow phases). Some patients also showed an increase in upward slow-phase velocity. The mechanism is likely to be different from that mediating cross-coupling during head thrusts. The nonlinearities present in head-thrust responses should not play a role during low-frequency stimulation. Perhaps the directional abnormalities in this case relate to the orientation of eye velocity by the velocity-storage system. As described earlier, lesions to the cerebellar nodulus and uvula abolish the reorientation of eye velocity to the gravito-inertial axis in experimental animals. Perhaps, in some cases of cerebellar dysfunction, the default orientation in the upright position is no longer correctly oriented with gravity.

Another possibility is that at least some of the torsional "cross-coupling" relates to a different relationship to Listing's Law during the VOR in cerebellar patients. In normal subjects, the horizontal VOR includes a small torsional component defined by the "1/4-angle rule" or "half-Listing strategy": at any orbital position, the plane of angular eye-velocity vectors is tilted 25% of the angle between the primary position and the current eye position. Perhaps in these patients the tilt of eye velocity is greater, producing more torsion. Alternatively or additionally, the primary position could be much more eccentric relative to straight-ahead gaze.

Cross-Coupling after Head Shaking

Following 2-Hz horizontal or vertical head shaking, some patients developed an increase in the slow-phase velocity of their downbeat nystagmus (FIG. 5). The resulting nystagmus was nearly purely vertical. Head-shaking nystagmus is usually attributed to the accumulation of asymmetrical activity from the vestibular periphery within the velocity-storage system. This is most commonly observed in patients with a unilateral vestibular deficiency, who typically show a nystagmus after horizontal head shaking in which slow phases are directed to the deficient side. In our patients, the vertical nystagmus might represent an accumulation of an upward bias associated with stimulation of either labyrinth; alternatively, it might be another reflection of an upward bias within the velocity-storage system itself.

Cross-Coupling during Optokinetic Stimulation

During visual tracking there must be a transformation of information encoded in retinal coordinates into the appropriate motor signals for eye movements. We have previously described patients with focal cerebellar lesions in the region of the middle cerebellar peduncle who showed cross-coupling of vertical into torsional eye movements during attempted vertical smooth pursuit.[28] We hypothesized that this finding could be explained if the signals used for vertical smooth pursuit are, at some stage, perhaps in the vestibular nuclei or vestibulocerebellum, encoded in a (vertical) semicircular-canal coordinate framework. If this is the case, the cerebellar lesion could then interfere with the transformation of visual information into ocular motor signals of the correct direction.

In some of our patients here we have also found inappropriate cross-coupling, in this case an inappropriate torsional response, during horizontal optokinetic stimulation. We do not yet have data on these patients during attempted tracking of a small target, but since in humans, full-field "optokinetic" tracking and smooth pursuit of small targets probably share much circuitry in the vestibular nuclei and vestibulocerebellum, it is likely that we would have found the same pattern of abnormality during horizontal smooth pursuit. Even so, there are differences between smooth pursuit of small targets and full-field stimulation with respect to how well they obey Listing's Law,[29] so a comparison between these two types of tracking stimuli in the same patients is needed.

How do we explain the pattern of cross-coupling during optokinetic stimulation in our patients using the idea that visual tracking signals are encoded in a labyrinthine frame of reference? One possibility is that the (horizontal) visual inputs were being transformed directly into a lateral semicircular-canal frame of reference, that is, being passed directly to the same secondary vestibular neurons within the brainstem that mediate afferent stimulation from the lateral semicircular canals. In humans, stimulation restricted to a lateral semicircular canal (the plane of which is about 30 deg above earth horizontal when the head is in its natural upright position) gives a mixed horizontal-torsional movement with rightward slow phases associated with clockwise torsion and leftward slow phases associated with the counterclockwise torsion. This is the same pattern of coupling of the directions of horizontal and torsional eye movements as shown by our patients during optokinetic nystagmus. Thus, one might predict inappropriate cross-coupling of torsion during horizontal optokinetic stimulation, unless there were a mechanism to correct for the misalignment between the plane of the lateral semicircular canals in the head and the earth-horizontal plane when the head is in its natural upright position. We speculate that the cerebellum normally superimposes a correction on the ocular motor (vestibular) response to horizontal retinal image motion during OKN, and so ensures that the axis of eye rotation during visual tracking is correctly oriented.

As with the VOR, an additional possibility is that the apparent torsional cross-coupling is somehow related to Listing's Law. In this case, that would seem unlikely to be the only explanation, given the large magnitude of the torsional relative to the horizontal velocity.

CONCLUSION

In conclusion, cerebellar patients frequently show abnormalities of the *direction* of the compensatory slow phase during vestibular and optokinetic stimulation. Likely, these abnormalities reflect involvement of the same structures that have been shown to control the amplitude of the high-frequency VOR (the flocculus), and the direction of the slow-phase generated by the velocity-storage mechanism during low-frequency stimulation (the nodulus).

ACKNOWLEDGMENTS

This work was supported by NIH Grant EY01849 (D.S.Z.), by NIH Grant 5-P60-DC00979 to the Research and Training Center for Hearing and Balance (D.S.Z., M.F.W.), and by a grant from the Arnold Chiavi Foundation, Inc. (D.S.Z., M.F.W.). The National Aeronautics and Space Administration supported this work through NASA Cooperative Agreement NCC 9-58 with the National Space Biomedical Research Institute. A. Lasker and D. Roberts provided valuable technical assistance.

REFERENCES

1. BLANKS, R. H. I., I. S. CURTHOYS & C. H. MARKHAM. 1975. Planar relationships of the semicircular canals in man. Acta Otolaryngol. (Stockh.) **80**: 185–196.
2. MILLER, J. M. & J. L. DEMER. 1996. New orbital constraints on eye rotation. *In* Three-dimensional Kinematics of Eye, Head and Limb Movement, M. Fetter, T. Haslwanter, H. Misslisch, and D. Tweed, Eds.: 349–357. Harwood. Amsterdam.
3. HALMAGYI, G. M., *et al.* 1990. The human horizontal vestibulo-ocular reflex in response to high-acceleration stimulation before and after unilateral vestibular neurectomy. Exp. Brain Res. **81**: 479–490.
4. ANGELAKI, D. E. & B. J. M. HESS. 1994. Inertial representation of angular motion in the vestibular system of rhesus monkeys. I. Vestibuloocular reflex. J. Neurophysiol. **71**: 1222–1249.
5. WEARNE, S., T. RAPHAN & B. COHEN. 1998. Control of spatial orientation of the angular vestibuloocular reflex by the nodulus and uvula. J. Neurophysiol. **79**: 2690–2715.
6. ITO, M. 1972. Neural design of the cerebellar motor control system. Brain Res. **40**: 81–84.
7. ROBINSON, D. A. 1976. Adaptive gain control of vestibuloocular reflex by the cerebellum. J. Neurophysiol. **39**: 954–969.
8. LISBERGER, S. G., F. A. MILES & D. S. ZEE. 1984. Signals used to computer errors in the monkey vestibuloocular reflex: possible role of the flocculus. J. Neurophysiol. **52**: 1140–1153.
9. SCHULTHEIS, L. W. & D. A. ROBINSON. 1981. Directional plasticity of the vestibulo-ocular reflex in the cat. Ann. N.Y. Acad. Sci. **374**: 504–512.

10. ANGELAKI, D. E. & B. J. M. HESS. 1995. Inertial representation of angular motion in the vestibular system of rhesus monkeys. II. Otolith-controlled transformation that depends on an intact cerebellar nodulus. J. Neurophysiol. **73**: 1729–1751.
11. STRAUMANN, D., *et al.* 1995. Transient torsion during and after saccades. Vision Res. **35**: 3321–3335.
12. HASLWANTER, T. 1995. Mathematics of three-dimensional eye rotations. Vision Res. **35**: 1727–1739.
13. AW, S. T., *et al.* 1996. Three-dimensional vector analysis of the human vestibuloocular reflex in response to high-acceleration head rotations. I. Responses in normal subjects. J. Neurophysiol. **76**: 4009–4020.
14. ZHANG, Y., A. M. PARTSALIS & S. M. HIGHSTEIN. 1995. Properties of superior vestibular nucleus flocculus target neurons in the squirrel monkey. I. General properties in comparison with flocculus projecting neurons. J. Neurophysiol. **73**: 2261–2278.
15. SATO, Y. & T. KAWASAKI. 1990. Operational unit responsible for plane-specific control of eye movement by cerebellar flocculus in cat. J. Neurophysiol. **64**: 551–564.
16. FUKUSHIMA, K., E. V. BUHARIN & J. FUKUSHIMA. 1993. Responses of floccular Purkinje cells to sinusoidal vertical rotation and effects of muscimol infusion into the flocculus in alert cats. Neurosci. Res. **17**: 297–305.
17. KRAUZLIS, R. J. & S. G. LISBERGER. 1996. Directional organization of eye movement and visual signals in the floccular lobe of the monkey cerebellum. Exp. Brain Res. **109**: 289–302.
18. ZEE, D. S., A. R. FRIENDLICH & D. A. ROBINSON. 1974. The mechanism of downbeat nystagmus. Arch. Ophthalmol. **30**: 227–337.
19. BALOH, R. W. & J. W. SPOONER. 1981. Downbeat nystagmus: a type of central vestibular nystagmus. Neurology **31**: 304–310.
20. BÖHMER, A. & D. STRAUMANN. 1998. Pathomechanism of mammalian downbeat nystagmus due to cerebellar lesion: a simple hypothesis. Neurosci. Let. **249**: 1–4.
21. ROBINSON, D. A. 1982. The use of matrices in analyzing the three-dimensional behavior of the vestibulo-ocular reflex. Biol. Cybern. **46**: 53–66.
22. ROBINSON, D. A. 1985. The coordinates of neurons in the vestibulo-ocular reflex. *In* Adaptive Mechanisms in Gaze Control: Facts and Theories, A. Berthoz & G. Melvill Jones, Eds.: 297–311. Elsevier. Amsterdam.
23. AW, S. T., *et al.* 1996. Three-dimensional vector analysis of the human vestibuloocular reflex in response to high-acceleration head rotations. II. Responses in subjects with unilateral vestibular loss and selective semicircular canal occlusion. J. Neurophysiol. **76**: 4021–4030.
24. QUINN, K. J. & J. F. BAKER. 1998. Processing of spatial information by floccular and nonfloccular target neurons in the alert cat. Brain Res. **780**: 143–149.
25. CURTHOYS, I. S., R. H. BLANKS & C. H. MARKHAM. 1977. Semicircular canal functional anatomy in cat, guinea pig and man. Acta Otolaryngol. **83**: 258–265.
26. THURSTON, S. E., *et al.* 1987. Hyperactive vestibulo-ocular reflex in cerebellar degeneration: pathogenesis and treatment. Neurology **37**: 53–57.
27. BALOH, R. W. & J. L. DEMER. 1993. Optokinetic-vestibular interaction in patients with increased gain of the vestibulo-ocular reflex. Exp. Brain Res. **97**: 334–342.
28. FITZGIBBON, E. J., *et al.* 1996. Torsional nystagmus during vertical pursuit. J. Neuroophthalmol. **16**: 79–90.
29. FETTER, M., *et al.* 1994. Three-dimensional human eye movements are organized differently for the different oculomotor subsystems. Neuro-ophthalmology **14**: 147–152.

Assessing Otolith Function by the Subjective Visual Vertical

ANDREAS BÖHMER[a,d] AND FRED MAST[b,c]

[a]*Department of Otorhinolaryngology, University Hospital, CH-8091 Zürich, Switzerland*

[b]*Department of Psychology, University Zürich, CH-8032 Zürich, Switzerland*

ABSTRACT: The effects of peripheral vestibular diseases on the subjective visual vertical (SVV) are resumed and provide the basis for some insights into the otolith pathophysiology. With a normal range of 0 ± 2 deg (when measured in an upright body position), the SVV was shifted by 11 ± 6 deg toward the ipsilateral ear in 40 patients following an acute unilateral vestibular deafferentiation (UVD), but in the opposite direction in 9 of 52 patients after stapes surgery. These opposite effects suggest a push–pull mechanism of the pairs of otolith organs with respect to the SVV. The dissociation between the SVV and the perception of body position indicates influences by unconscious reflexive mechanisms such as ocular cyclotorsion on the SVV. In chronic UVD patients, lateral shifts of the subjects during constant angular velocity rotation into various eccentric positions (± 16 cm) revealed a shift of the "center of graviception" close to the remaining intact contralateral inner ear. To date, this seems to be the most consistent test for clinical identification of a chronic compensated unilateral loss of otolith function. The findings regarding asymmetries in otolithic sensitivity to medially and laterally directed roll-tilts remain controversial, probably mainly because of influences of extravestibular cues.

INTRODUCTION

The otolith organs as sensors for the direction of the gravito-inertial force vector provide not only subconscious postural reflexes but also contribute to the perception of spatial orientation. The perception of the gravitational vertical may be assessed in various ways: by asking the subjects to adjust either their own body position to the vertical or horizontal (postural vertical or horizontal[1–3]), or a light bar to the vertical (subjective visual vertical), or to perform earth vertical saccades in darkness (subjective vertical saccades[4]). The subjective visual vertical (SVV) has been extensively investigated by psychologists (review, e.g., reference 5), but was ignored for a long time by the clinical neuro-otologists. Friedmann in 1971[6] reported changes of the SVV induced by unilateral ablative inner ear surgery, but his findings fell nearly into oblivion until, in the last 10 years, clinicians became interested in the potential diagnostic value of the SVV. Similar to other spatial orientation tasks, the estimation of verticality depends on multimodal sensory input. Visual information as one of the most important cues can easily be excluded by performing the

[c]Address for communicataion: F. Mast, Harvard University, Department of Psychology, William James Hall 840, 33 Kirkland Street, Cambridge, Massachusetts 02138: phone: 617/495-5921; fax: 617/496-3122; e-mail: fmast@wjh.harvard.edu

[d]Deceased October 1998.

tests in darkness, but other cues, mainly of proprioceptive origin, are more difficult to control and remain a probable source for conflicting results as outlined below. Nevertheless, the SVV has the potential of extending the clinical diagnosis of otolithic disorders.

The aim of this paper is not confined to a phenomenological description of the effects of various peripheral vestibular diseases on the SVV. Rather, we try to deduce from this phenomenology some more general insights in the black box, which provides the perception of verticality and thus hopefully in the (patho-)physiology of the otolithic organs with respect to their function as one of the important sensory cues for the SVV.

MATERIAL AND METHODS

Data presented here was taken from published papers[7–12] and from studies by the author and colleagues.[13–16] Some of the author's experimental series were updated with more recent, unpublished data. Since 1991 the SVV is routinely assessed at our institution in all neuro-otological patients by having them sitting upright with the head unrestrained[d] in darkness 1.5 m in front of a dim light bar. The bar can be rotated in the patient's frontal plane by the examiner. Starting from an oblique position, the patient verbally instructs the examiner how to rotate the bar until it appears exactly vertical ("human remote control"). Additional experimental series were performed with the same procedure while the subjects were lying horizontally on their side[13] or seated on a 26 deg slanted chair with the head, shoulder, knee, and foot of the undermost side supported but otherwise unrestrained.[14] Both in the 90 deg and 26-deg side positions, the room illumination was turned off only after positioning, thus giving the subjects strong information of their position relative to gravity. This is in contrast to the studies by Bergenius and Tribukait,[9,11,12] where the subjects were firmly fixed on a tilt chair and rolled into various oblique positions in darkness. Similarly, subjects tested in darkness on a fixed chair centrifuge,[7,14,16] where the gravito-inertial force vector was tilted by means of centrifugation, did not have visual information as regards their body position relative to the resultant force vector.

In some studies,[1,7–9,11,12] the subjective visual horizontal was assessed instead of the SVV. Measurements of both the subjective vertical and horizontal in the same patients did not reveal significantly different settings when the adjustments were taken in an upright body position.[6] During static roll-tilts, however, normal subjects showed slightly different errors (A- and E-phenomena) for the subjective horizontal and vertical.[18] In the following paragraphs, we focus on the perceptual effects of inner ear lesions relative to premorbid or normal data within the same test condition. Presuming that these lesions do not affect the subjective visual vertical differently than the subjective horizontal, both parameters will be treated as clinically equivalent and for simplicity will be referred to as the SVV.

The SVV is indicated as the angle in the subject's roll plane (frontal plane) between the subjectively earth vertical light bar and the direction of the gravito-inertial force vector (= γ) or between the light bar and the body longitudinal axis (= β). Positive values indicate deviations of the upper pole of the light bar to the right (subject's view). The lower case cipher indicates the subject's roll position in degrees relative to the gravity: $_0$ = upright, $_{+90}$ = 90 deg right ear down side position. All patients are treated as if their right inner ears were lesioned. In the present series, five settings of the SVV of each subject were averaged.

[d]In normals and patients with acute unilateral vestibular deafferentiation assessed in an upright body position, head tilts in a physiological range of spontaneous head tilts were found not to have significant effects on the SVV, that is, the SVV assessed with the head unrestrained did not differ from the SVV settings with the head firmly fixed in a vertical position.[17]

RESULTS / DISCUSSION

Push–Pull Between the Two Ears

An unilateral surgical elimination of the otolithic organs is probably the best controlled condition that is available in human beings to provide some insight in the black box of the inner ear generators of the SVV. The following limitations, however, have to be considered: there is no surgical selective ablation of the otolithic organs, and the semicircular canals are also involved in all these operations. Ablation of unilateral vestibular function is rarely performed in a normally functioning inner ear; usually there is a preexisting disease (Ménière's disease, acoustic neurinoma etc.) to which the central nervous system has already adapted. At least some branches of the inferior vestibular nerve (with afferents of parts of the saccule) are often spared in selective vestibular neurectomies in order not to compromise the blood supply to the cochlear nerve. In spite of these limitations, all studies on SVV in the acute stage of unilateral vestibular deafferentiation reported an astonishingly consistent deviation of the SVV (γ_0) of +10 to +12 deg : 9.8 ± 4.7 deg after labyrinthectomy for Ménière's disease, $n = 6$;[6] 12.4 ± 3.8° after vestibular neurectomy, $n = 21$;[7] 10.6 ± 6.5 deg after unilateral intratympanic gentamycin instillation for Ménière's disease, $n =14$;[12] and 10.7 ± 5.5 deg after vestibular neurectomy, $n = 40$ (FIG. 1A). In all instances, the SVV was shifted with the upper pole of the light bar (subjective zenith) toward the lesioned side. In spontaneously occurring vestibular diseases, such as acute peripheral vestibulopathy, the deviation of the SVV was smaller [$\gamma_0 = 6.8 \pm 7.1$, $n =20$;[13] and 8.9 ± 3.7 deg, $n =9$ (reference 9)]. In this disorder the extent of the lesions remained unknown (probably often incomplete). Consistently normal SVV settings were found in patients suffering from benign paroxysmal positional vertigo [($\gamma_0 = 0.2 \pm 0.8$, $n =19$ (reference 13)], suggesting that there is no major otolithic disorder in this disease.

In contrast to the consistent deviation of the SVV toward the affected ear in acute vestibular lesions such as neurectomy or acute peripheral vestibulopathy, a deviation in the opposite direction may occur after stapes surgery. In a study by the authors that included 52 patients (FIG. 1B), stapedotomy did not affect the SVV in 75% of the patients (i.e., the SVV was in the normal range before and after surgery), while 17% of the patients showed significant postoperative deviations of the SVV toward their healthy ear (negative γ_0). More than half of the patients that had the SVV tilted toward the healthy ear suffered from a moderate postoperative dizziness compared to only 17% in the first group. Tribukait and Bergenius[11] found significant deviations of the SVV in the same direction after stapedotomy, even in 10 of 12 patients with a group mean (γ_0) of −4.0 ± 2.5 deg. Stapes surgery with opening of the vestibule may be considered to cause a slight local labyrinthitis and an irritation of the adjacent sensory structures, especially of the otolith organs that are situated just beneath the oval window.

From this evidence one may conclude that the otolith organs in both inner ears act–similarly to the semicircular canals—as an antagonistic push–pull mechanism when determining the SVV. With a unilateral loss, the otolith organs of the contralateral ear push the SVV by some 10 deg to the opposite, diseased side. A unilateral irritation causing an increase of the neuronal resting discharge pushes the SVV toward the opposite, unaffected ear. This demonstrates that the ear with the higher resting acitivity exerts a *push* on the SVV. In UVD subjects, the remaining contraleral otolith organs, however, are also able to *pull* the SVV to their side when stimulated adequately, as will be shown below.

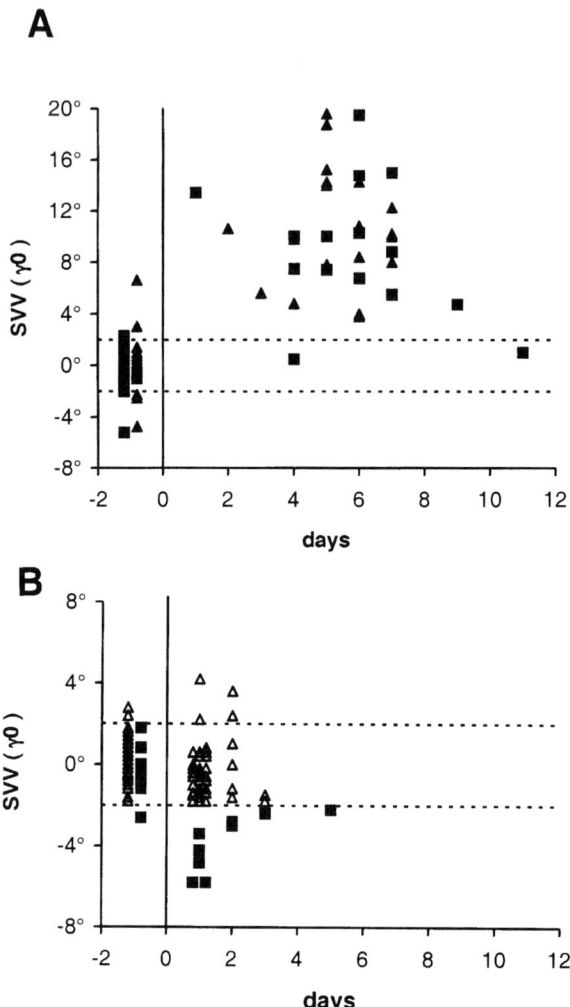

FIGURE 1. Effects of unilateral vestibular deafferentiation ($n = 40$; A) and stapedotomy ($n = 52$; B) on the subjective visual vertical as measured in an upright body position. (**A**) Each patient is represented by two symbols, one preoperatively (left) and at one given day after surgery (right); *squares* indicate patients submitted to a total eighth nerve resection, *triangles* indicate selective vestibular neurectomies. Before surgery only a few patients (mainly Ménière's disease) set their SVV beyond the normal range of ± 2 deg (*broken horizontal lines*); after surgery all but two deviated significantly toward the affected ear. The latter two patients had both an acoustic neurinoma excised with probably already preoperatively abolished otolithic function. Data from Böhmer and Rickenmann,[13] updated with more recent measurements. (**B**) Thirty-nine and 4 out of 52 patients with otosclerosis set the SVV in the normal range or slightly toward the operated ear in the first days after stapedotomy (*triangles*), while 9 patients had significantly negative γ_0 after surgery (*squares*). All but one of the latter group were in the normal range preoperatively.[15]

Is the SVV a Perceptual or a Reflexive Phenomenon?

In spite of these important deviations of the SVV following acute unilateral vestibular lesions, patients do not feel themselves tilted, that is, there is no concomitant illusionary body tilt. Patients suffering from an acute peripheral vestibulopathy could adjust themselves very accurately in a vertical position in darkness on a tilt chair (-0.3 ± 2.7 deg), although they set their SVV at the same time at 12.2 ± 11.9 deg.[10] Other examples of dissociations between SVV and body position are seen in normal subjects in horizontal side positions. When asked to adjust their body position to the physical horizontal, normal subjects showed a variably large range of settings depending on the method,[1,19] but the group means were around 90 deg indicating no systematic error. At the same time, these subjects adjusted the SVV (γ) at >20° (A-phenomenon). Proprioceptive cues do not seem to add much to the more acurate perception of body orientation compared to the SVV, because the same dissociation also occurs under water.[21,22]

Occasionally, a full triad of head tilt, skew deviation, and tonic ocular torsion occurs following an acute unilateral deviation,[23,24] but usually the head tilt is slight (4.6 ± 4.7 deg; $n = 5$,) and within the normal range of static head positions while sitting upright.[17] Static ocular torsion (OT) with the upper pole of both eye bulbs rolled toward the lesioned side, on the other hand, is a very consistent finding in these patients [9.5 ± 3.3 deg, $n = 23$ (reference 8)] and may be a contributing factor for the tilt of the SVV and the dissociation between perception of body vertical and visual vertical. OT and SVV, however, are not firmly correlated in all occasions: patients with chronic unilateral vestibular hypofunction showed a persistent pathological ocular torsion, but fully compensated SVV in the normal range.[25] A similar conjugate torsional deviation of the eyebulbs can be elicited by roll-tilts of the head in normal subjects.[26,27] This ocular counterrolling (OCR) is always compensatory, that is, in the opposite direction to the head roll tilt, but never fully compensates the amount of the lateral head tilt. SVV and OCR as a function of roll-tilt differ remarkably. In moderate roll-tilts up to some 50 deg, the SVV is occasionally set too far away from the body longitudinal axis (overcompensation; E-phenomenon), while for increasing roll tilts the SVV is set toward the body axis (undercompensation; A-phenomenon). The OCR, on the other hand, never overcompensates and saturates between 60 and 90 deg.

The dissociation between the perception of body position and the SVV, as well as the high correlation between the angle of OT and of the SVV following acute unilateral vestibular deafferentiation, suggest an important contribution of the unconscious reflexive-otolith-induced static torsion of the eye bulb and the retinal coordinates on the visual perception of the vertical. The dissociation between the body roll-tilt-induced OCR and the SVV, on the other hand, indicates additional otolith-based (perceptual) mechanisms, independent of retinal coordinates, for the perception of vertical. Implementation of information about the direction of gravity, independent from OCR was, for example, found also in 40 % of neurons in the prestriate visual cortex of rhesus monkeys.[28]

Cortico-oculomotor pathways generating voluntary saccades seem to be under similar otolith influences: when normal subjects in a horizontal side position are asked to perform earth-vertical saccades in darkness, the angle of the line connecting the upper and the lower imaginary fixation points relative to gravity correlated well with the subject's SVV in the same position (reference 4 and FIG. 2). Due to an identical frame of reference, both the SVV and the vertical saccades in darkness may thus be considered as equivalent in their diagnostic properties.

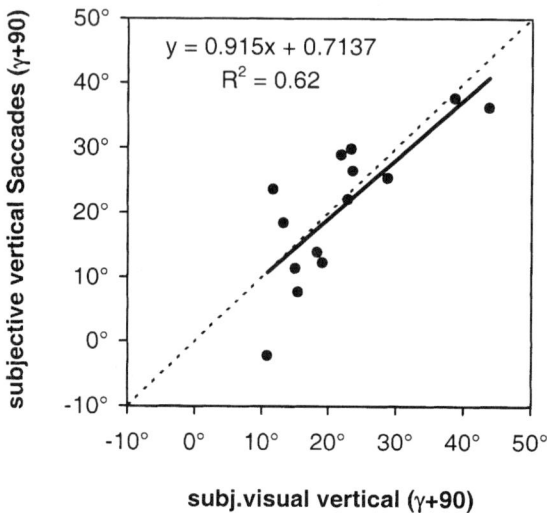

FIGURE 2. Correlation between subjective visual saccades and SVV. Fourteen normal subjects were lying in darkness on their right side and performed first earth vertical saccades and then a SVV test. Eye movements were recorded with an infrared-video system (ULMER, Synapsis, inomed) and were calibrated relative to gravity by performing saccades between two true vertical LEDs at the end of the test. All settings, except one subject's saccades, have a positive γ indicating an A phenomenon. In respect to the intraindividual scatter of repeated settings of SVV in side position, the correlation is quite high.

In Chronic UVD Subjects, the Direction of GIF Vector is Mainly Perceived by the Intact Contralateral Inner Ear

During the weeks and months following an acute unilateral vestibular deafferentiation, the SVV returns toward the normal range.[15] In the same way, slowly developing unilateral disorders usually do not induce significant deviations of the SVV. In a series of 27 patients with an acoustic neurinoma, for example, the mean γ_0 was 1.1 ± 1.5 deg, only four patients had a SVV > 2.5 deg with a maximal value of 4.0 deg (the author's unpublished data). In chronic stages, UVD subjects thus can no longer be distinguished by their SVV from normals, unless sophisticated unphysiological stimulation techniques are used. Centrifugation of the subjects in various slightly eccentric positions provides such a technique, where the GIF vectors at various sites in the head and body differ considerably in direction, and furthermore are systematically changed during the experiments. When chronic UVD subjects

FIGURE 3. SVV during off-center rotation in 21 normals (**A**) and 9 chronic UVD subjects. (Data from Böhmer and Mast.[16]) The SVV is assessed first as βstat, with the subject sitting upright on the stationary centrifuge. The GIF vector is 1 g and directed parallel to the body axis at each site in the head (*sketch at left*). The centrifuge is then accelerated and rotated with 240 deg/s. After 1 minute adaptation 5 settings of the SVV are made and the subject is then moved in pseudorandomized order in an eccentric position. In each position the SVV is measured 5 times. Each symbol is the mean of the settings in each subject and each position relative to the settings during standstill. Note that in normals the regression line crosses zero exactly in a position with the rotation axis through the center of the head. In UVD subjects (**B**), the regression crosses zero at 5.9 cm paramedian on the intact side, which is slightly lateral to the presumed position of the intact inner ear.

A

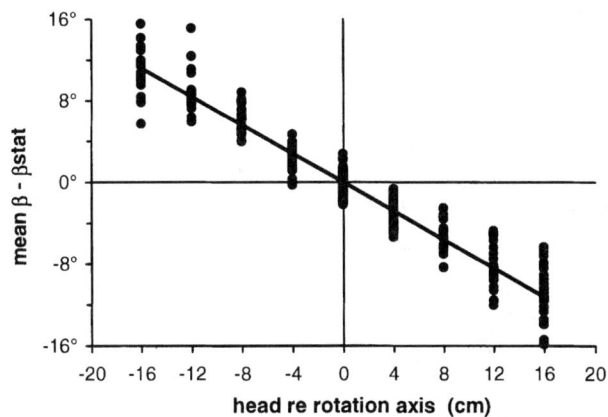
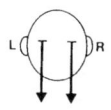
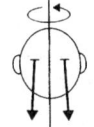
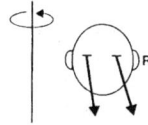

B

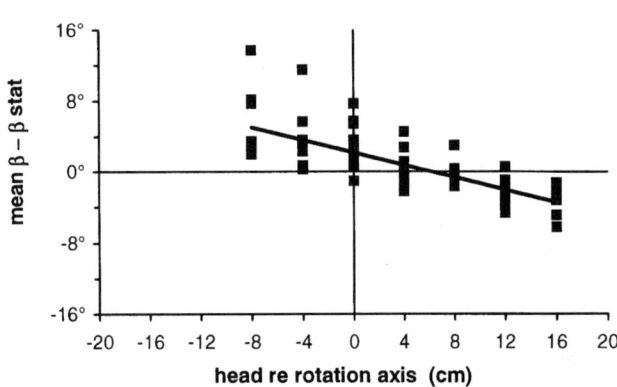

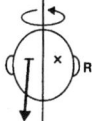
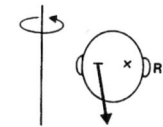

were asked to continuously adjust the SVV during such a centrifugation (± 16 cm eccentricity), their settings revealed a function of eccentricity. The angle of the SVV set during standstill before centrifugation was reached during centrifugation in eccentric positions with the axis of rotations 5.9 ± 2.5 cm paramedian on the side of the intact ear (FIG. 3 and reference 16). In this position, the GIF vector acting on the intact inner ear was close to earth vertical. Normal subjects, on the other hand, indicated the SVV in the same direction as during standstill in positions with the axis of rotation running through the center of the head (0.0 ± 1.4 cm). The observation that the "center of graviception" in chronic UVD patients was found even a little lateral to the intact inner ear suggests that the inner ear is the major sensory source for the SVV. If proprioceptive cues would have been important, they supposedly would have kept the center of graviception closer to the midline.

Asymmetric Medially and Laterally Directed Roll-Tilt Perception of the Otoliths?

The hair cells in the semicircular-canal cupulae are all unidirectionally polarized; nevertheless, (chronic) UVD subjects do not show significant asymmetries in the vestibulo-ocular reflex elicited by bidirectional rotation except in the very high-frequency range.[29] The hair cells in the otolith organs are polarized in all directions. The utriclar macula shows a moderate preponderance of hairs cells that are polarized for excitation by laterally directed roll-tilts over hair cells excited by medially directed tilts.[30] From this it is unlikely that chronic UVD subjects would reveal clinically useful asymmetries in the settings of the SVV. The corresponding clinical data, summarized in TABLE 1, however, is somewhat controversial. Unequal methods that allowed for differently strong extravestibular influences might be a cause of the differences. Another complicating factor could be the shift of the SVV (β_0, γ_0 or "bias") in acute vestibular deafferentiation. Most data on asymmetries have been obtained in acute and subacute stages after unilateral vestibular deafferentiation, where patients in an upright position have a β_0 of some 10 deg. All these studies imply a linear superimposition of β_0 to the SVV in tilt positions, but it is very possible that this interaction is more complex and nonlinear.

TABLE 1 indicates the effects of roll tilt stimulation on the SVV as roll gains, which were calculated as (β tilt–β_0) / (–angle of roll tilt) in order to allow for an easier comparison between different studies. When subjects in the acute stage after vestibular neurectomy were positioned in horizontal side positions or seated in 26 deg oblique positions, both with the affected ear up and down, the SVV (β_{90} and β_{26}) shifted relative to the settings in upright position by some 75 and 20 deg, respectively. No differences in the amount of the SVV shifts between the two sides emerged. For all four positions, the roll gain was around 0.8. When the roll-tilts were performed on a tilt chair in darkness (subjects firmly fixed to the chair), patients suffering from an acute vestibular neuritis were more sensitive to roll-tilts toward the unaffected ear. As in the previous experiments, the gains for tilts in both directions, however, were decreased compared to the corresponding roll-gain in normal subjects. In contrast, an *increased* roll gain (relative to preoperative and relative to normal) was reported in acute UVD subjects when the GIF vector acting on the remaining intact inner ear was tilted laterally by eccentric centrifugation.[7] A simultaneous important roll-gain *decrease* for medially directed roll-tilts resulted in a very prominent asymmetry of the roll-tilt sensitivity in that study. The only other roll-gain increase known to us was observed in patients acutely after stapedotomy, which showed a gain of 1.30 (compared to 1.08 preoperative) for roll-tilts toward the operated ear on a tilt chair.[11] In an upright position, these patients had also a tilt of the SVV away from the operated ear, suggesting that a surgically elicited labyrinthine "irritation'" induced not only an increased resting activity but also an increased sensitivity to laterally directed roll-tilts. In chronic stages after unilateral vestibular deafferentiation, when the SVV in an upright static position had compensated to normal values in most patients, symmetrical gain decreases for lateral and

TABLE 1. Roll gains for laterally and medially GIF in UVD subjects

Stimulated condition	Roll-tilt	Patients	n	Roll gain				Reference	
				Lateral[a] (Mean)	Tilts (S.D.)	Medial[a] (Mean)	Tilts (S.D.)	Asymmetry[b] (%)	
Side position on bed	Body; 90 deg	Vest. Neurect. Acute	14	0.85	0.10	0.84	0.12	1	13
Slanted chair	Body; 90 deg	Vest. Neurect. Acute	13	0.80	0.24	0.76	0.17	3	14
Tilt chair	Body 10, 20, 30 deg	Vest. Neuritis, Acute	11	0.90	0.31	0.56	0.36	24	9
Centrifuge, 1-m eccentric	GIF vector, 26 deg	Vest. Neurect. Acute	21	1.19		0.40		50	7
Centrifuge ± 16-cm eccentric	GIF vector, ± 16 deg	Vest. Neurect. Chron.	10	0.35	0.19	0.34	0.12	1	16

[a]Relative intact inner ear.
[b](lateral − medial) (lateral + medial).

medial roll-tilts of the GIF vector were found during minimal eccentric centrifugation. In this experimental setup, extravestibular cues probably were of minor importance because the proprioceptive stimulation was multidirectional and thus mutually canceling.

In contrast to this equal or higher sensitivity to laterally directed roll tilts for the generation of the SVV, the utricular sensitivity for the generation of linear vestibulo-ocular reflexes was lower for lateral directed than for medially directed tilts of the GIF vector.[31] This might suggest a functional separation of various parts of the utricular maculae.

CONCLUSIONS

Clinically, the SVV test performed in an upright body position is a simple and quick otoneurological examination that provides information on the tonic afferent balance or imbalance between the otolith organs in the two inner ears, and thus easily reveals acute unilateral lesions. It may be considered as an otolith analogon of the spontaneous nystagmus reflecting the semicircular canal afferent balance. In chronic or slowly developing unilateral vestibular lesions, the SVV compensates similarly to the spontaneous nystagmus. While a chronic horizontal canal loss can be assessed by caloric irrigation, more sophisticated techniques are necessary to reveal chronic unilateral otolithic lesions. Assessing the SVV in several slightly eccentric positions during constant yaw-axis rotation allows for this detection by revealing a shift of the "center of graviception" toward the remaining intact inner ear. It remains to be investigated whether the controversial medial–lateral asymmetric sensitivity of the otolithic organs may provide a technically easier test for chronic otolithic lesions.

In regard to *otolith physiology*, oppositely directed deviations after vestibular lesions and "irritations" suggest a push–pull mechanism similar to the semicircular-canal–oculomotor reflexes. The dissociation between the SVV and the perception of body position suggests contributions of the reflexive otolith-induced ocular torsion on the SVV. Cortico-oculomotor functions (saccades performed in darkness) seem to be based on the same otolith reference as the SVV. Whether the otolithic organs are differently sensitive to medially and laterally directed roll tilts in regard to the SVV remains undetermined.

REFERENCES

1. MAST, F. & T. JARCHOW. 1996. Perceived body position and the visual horizontal. Brain Res. Bull. **40**: 393–398.
2. BRONSTEIN, A. 1999. Otolith-proprioceptive interaction in the perception of the postural vertical: the effects of labyrinthine and CNS disease. This volume
3. MITTELSTAEDT, H. 1999. The role of inertial information in the control of posture, perception of the vertical and path intergration. This volume
4. VAN BEUZEKOM, A. 1999. Errors in tilt estimates based on line settings, saccadic pointing and verbal reports. This volume
5. HOWARD, I. P. 1982. Human Visual Orientation. Wiley. Chicester, England.
6. FRIEDMANN, G. 1971. The influence of unitalteral labyrinthectomy on orientation in space. Acta Otolaryngol. (Stockh.) **71**: 289–298.
7. DAI, M. J., I. S. CURTHOYS & G. M. HALMAGYI. 1989. Linear acceleration perception in the roll plane before and after unilateral vestibular neurectomy. Exp. Brain. Res. **77**: 315–328.
8. CURTHOYS, I. S., M. J. DAI & G. M. HALMAGYI. 1991. Human ocular torsion position before and after unilateral vestibular neurectomy. Exp. Brain. Res. **85**: 218–225.
9. BERGENIUS, J., A. TRIBUKAIT & K. BRANTBERG. 1996. The subjective visual horizontal at different angles of roll-tilt in patients with unilateral vestibular impairment. Brain Res. Bull. **40**: 385–391.

10. ANASTASOPOULOS, D., T. HASLWANTER, A. BRONSTEIN, M. FETTER & J. DICHGANS. 1997. Dissociation between the perception of body verticality and the visual vertical in acute peripheral vestibular disorders in humans. Neurosci. Lett. **233**: 151–153.
11. TRIBUKAIT, A. & J.BERGENIUS. 1998. The subjective visual horizontal after stapedotomy: evidence for the increased resting activity in otolithic afferents. Acta Otolaryngol. (Stockh.) **118**: 299–306.
12. TRIBUKAIT, A., J. BERGENIUS & K. BRANTBERG, 1998. The subjective visual horizontal during follow-up after uniltateral vestibular deafferentiation with gentamycin. Acta Otolaryngol. (Stockh.) **118**: 479–487.
13. BÖHMER, A. & J. RICKENMANN. 1995. The subjective visual vertical as a clinical parameter of vestibular function in peripheral vestibular diseases. J. Vest. Res. **5**: 35–45.
14. BÖHMER, A., F. MAST & T. JARCHOW. 1996. Can a unilateral loss of otolithic function be clinically detected by assessment of the subjective visual vertical? Brain Res. Bull. **40**: 423–429.
15. BÖHMER, A. 1997. Zur Beurteilung der Otolithenfunktion mit der subjektiven Visuellen Vertikalen. HNO **45**: 533–537.
16. BÖHMER, A. & F. MAST 1998. Chronic unilateral loss of otolith function revealed by the subjective visual vertical during off center yaw rotation. J. Vest. Res. In press
17. RADIVOJEVIC, V. 1998. Ist die Subjektive Visuelle Vertikale von der Kopfposition und der Augenrollung abhängig? M.D. Thesis, University of Zürich, Zürich, Switzerland.
18. BETTS, G. A. & I. S. CURTHOYS. 1996. Settings to visually perceived vertical and to visually perceived horizontal during static roll-tilt are not orthogonal. J. Vest. Res. **6**(4S): S25.
19. MITTELSTAEDT, H. 1991. The role of the otoliths in the perception of the orientation of self and world to the vertical. Zool. Jahrb. Physiol. **95**: 419–425.
20. JARCHOW, T. & F. MAST. 1999. The effect of water immersion on postural and visual orientation. Aviat. Space Enviorn. Med. In press.
21. JARCHOW, T. 1995. Der Einfluss somatosensorischer Reize auf die erlebte Körperschräglage und die subjektive Senkrechte. Unpublished Diploma Thesis, University of Zürich, Zürich, Switzerland.
22. MAST, F. & T. JARCHOW. 1996. On the relationship between subjective body position and the visual vertical. J. Vest. Res. **6**(4S): S22.
23. HALMAGY, G. M., M. A. GRESTY & W. P. R. GIBSON. 1978. Ocular tilt reaction with peripheral vestibular lesion. Ann. Neurol. **6**: 80–83.
24. WOLFE, G. I., *et al.* 1993. Ocular tilt reaction resulting from vestibuloacoustic nerve surgery. Neurosurgery **32**: 417–421.
25. SCHMID, A. Static ocular counterroll after unilateral vestibular neuritis. This volume
26. DIAMOND, S. G. & C. H. MARKHAM. 1983. Ocular counterrolling as an indicator of vestibular otolithic function. Neurology **33**: 1460–1469.
27. BUCHER, U., F. MAST & N. BISCHOF. 1992. An analysis of ocular counterrolling in response to body positions in three-dimensional space. J. Vest. Res. **2**: 213–220.
28. SAUVAN, X. M. & E. PETERHANS. 1999. Orientation constancy in neurons of monkey visual cortex. Visual Cognition. **6**: 43–54.
29. HALMAGY, G. M. & I. S. CURTHOYS. 1988. A clinical sign of canal paresis. Arch. Neurol. **45**: 737–739.
30. FERNANDEZ, C. & J. M. GOLDBERG. 1976. Physiology of peripheral neurons innervating otolith organs of the squirrel monkey. I. Response to static tilts and to long-duration centrifugal force. J. Neurophysiol. **39**: 970–984.
31. LEMPERT, T., *et al.* 1996. Transaural linear vestibulor-ocular reflexes from a single utricle. Brain Res. Bull. **40**: 311–313.

Horizontal Linear Vestibulo-Ocular Reflex Testing in Patients with Peripheral Vestibular Disorders

THOMAS LEMPERT,[a,b,c] MICHAEL A. GRESTY,[a] AND ADOLFO M. BRONSTEIN[a]

[a]*MRC Human Movement and Balance Unit, National Hospital for Neurology, London WC1N 3BG, UK*

[b]*Neurologische Klinik, Charité - Virchow-Klinikum, Humboldt-Universität, Augustenburger Platz 1, 13353 Berlin, Germany*

ABSTRACT: Horizontal eye movements in response to lateral head translation [linear vestibulo-ocular reflex (LVOR)] in normal subjects and in patients with bilateral vestibular failure ($n = 14$), unilateral vestibular nerve section ($n = 9$), and benign positional vertigo ($n = 14$), were studied.[2] LVORs were elicited in darkness by step acceleration (0.24 g) of the whole body along the interaural axis. Results and conclusions: (1) in patients with bilateral vestibular failure, LVORs were either absent or abnormal with asymmetries, diminished velocities, and prolonged latencies. Measurements of dynamic visual acuity during linear self-motion showed decreased performance in patients at 1.0 and 1.5 Hz, which correlated with absent or delayed LVORs. These findings demonstrate the functional role of LVORs for dynamic visual acuity. (2) Early after vestibular nerve section, LVORs were diminished or absent with head acceleration toward the operated ear and normal in the opposite direction. After 6–10 weeks, responses were symmetrical again. Thus, a single utricle appears to be polarized with respect to the LVOR early after unilateral vestibular loss generating mostly contraversive responses. (3) Patients with benign positional vertigo showed mostly normal LVORs, which can be explained by minor utricular damage or central compensation of a chronic unilateral deficit.

INTRODUCTION

Natural head movements comprise both angular and linear head accelerations. The vestibulo-ocular reflexes (VOR) provide oculomotor compensation for both components of movement: the angular VOR (AVOR) for head rotations and the linear VOR (LVOR) for head translations. The horizontal LVOR produces compensatory horizontal eye movements during sideward head accelerations and is a response to utricular stimulation.[1] While physiological aspects of the horizontal LVOR have been extensively investigated in animals and human subjects,[2,3] the study of LVOR pathology in patients with vestibular disorders is only at its beginning.[4] This paper summarizes our recent findings on LVOR abnormalities in patients with bilateral vestibular failure,[5] acute unilateral vestibular loss,[6] and benign positional vertigo.[7]

[c]To whom correspondence may be addressed. Phone: 030 4506 0011; fax: 030 4506 0901.

LVOR TESTING PROCEDURE

Linear acceleration along the interaural axis was provided by a car running smoothly on a track and powered by two linear motors. Patients were seated upright, facing sideways on the car with their head and body firmly restrained. The drive input to the motor was a velocity triangle generating an acceleration step of 0.24 g that lasted 650 ms and was followed by a similar period of deceleration. To determine the onset of acceleration a precision piezoresistive linear accelerometer was mounted to the patient's forehead. Velocity feedback from the car was provided by a tachowheel running on the track.

The LVOR was elicited in the dark with onset of chair movement one second after the room light had been extinguished. Between five and eight stimuli to the left and to the right were applied in a random sequence. If artifacts were observed, additional stimuli were provided. Patients were instructed to gaze passively straight ahead without fixating any particular point. To minimize the effect of convergence which may modulate the LVOR[8] they were asked to look at a gray screen straight ahead at 150-cm distance before lights went off. Horizontal eye movements were recorded by bitemporal electrooculography (EOG), filtered at 80 Hz and digitally sampled at 250 Hz. Recordings were analyzed off-line using an interactive computer programme. Saccades were identified visually and replaced by straight lines whose slopes averaged the pre- and postsaccadic slow-phase velocities. To reduce noise levels, at least five responses were averaged for each subject and direction. The LVOR was quantified from the averaged data by the slow-phase eye velocity over the interval from 300 to 500 ms after the onset of chair movement. At this time the eye velocity was steady and could be reliably measured by fitting the eye-position signal with a straight line, using a least-square error-estimation procedure (FIG. 1). The velocity asymmetry of the LVOR was expressed as a percentage using the directional preponderance formula for vestibular responses: $|R-L|/|R+L| \times 100$. The latency of the LVOR was measured from the averaged trace. Response onset was identified when the eye-position signal departed from the envelope of the baseline EOG noise present during the 200 ms preceding motion (FIG. 1).

NORMAL CONTROL DATA

Normative data for the LVOR were obtained from 21 healthy subjects between 22 and 49 years of age (mean = 34 years). Slow-phase velocities of the LVOR ranged from 4.7 to 21 deg/s (mean = 10.3 deg/s). Asymmetries between leftward and rightward responses did not exceed 13% in any subject (mean directional preponderance 6.4%) (FIG. 2). Latencies varied from 32 to 132 ms (mean = 79 ms). Values outside this range were regarded as abnormal.

LVOR IN PATIENTS WITH BILATERAL VESTIBULAR FAILURE

In this study, we tested fourteen patients (mean age 46 years; range 31–69 years) with bilaterally absent caloric responses (30 and 44°C) due to peripheral vestibular disorders. In addition to caloric irrigation, vestibular investigations included rotational testing with velocity steps of 60 deg/s and sinusoidal oscillations (0.2 Hz, peak velocity 60 deg/s). Seven patients had residual horizontal canal function with velocity gains between 0.1 and 0.4 in at least one of these two rotational tests. Therefore, the patient sample deliberately represents a spectrum at low levels of vestibular function rather than a homogeneous group with complete vestibular loss.

All patients had abnormal eye-movement responses. LVORs were absent in two patients while the remaining 12 had evidence of partial but abnormal LVOR function. Compared with normal subjects, the patient group had significantly reduced LVOR

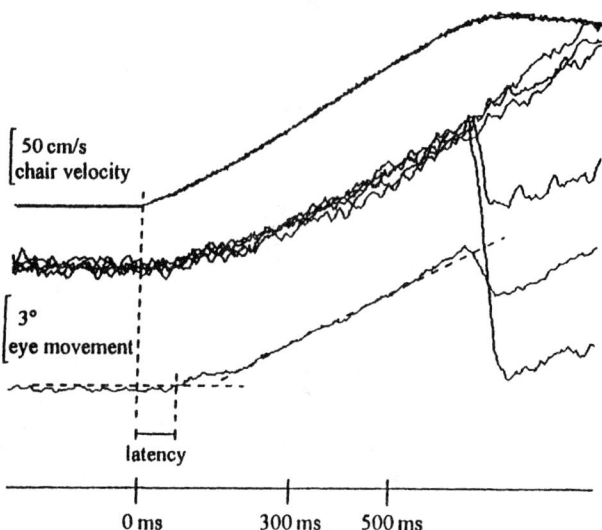

FIGURE 1. Linear vestibulo-ocular reflex (LVOR) in response to step acceleration along the interaural axis. **Top traces**: Chair velocity feedbacks for motion to the left. **Middle traces**: overlay of five raw eye movement recordings obtained in one subject for one direction of motion. Two of these recordings contain nystagmic fast phases. **Lower traces**: Average of the five raw recordings and indications of latency and velocity measurements.

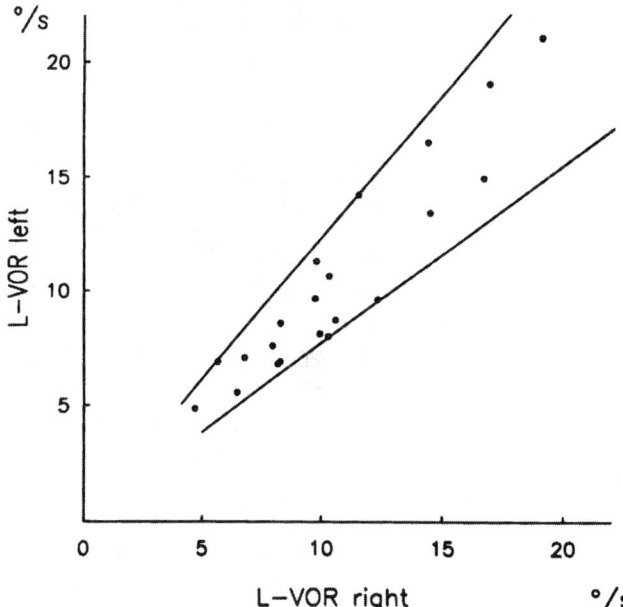

FIGURE 2. Slow-phase velocities of the linear vestibulo-ocular reflex in 21 normal subjects. *LVOR left* represents a compensatory eye movement to the left during rightward translation. *Straight lines* indicate maximum asymmetry of $\pm 13\%$.

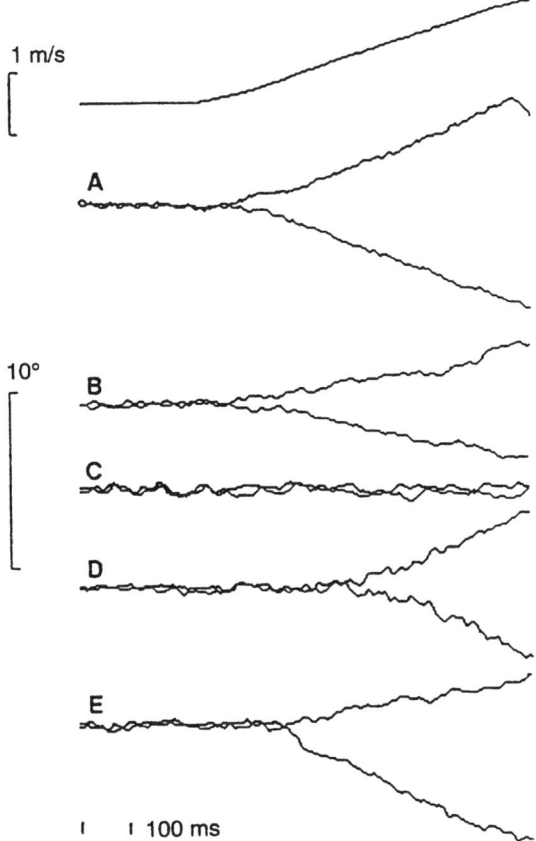

FIGURE 3. Individual averages of LVOR response in a normal subject (**A**) and in patients with bilateral vestibular failure showing abnormal LVORs: (**B**) low velocity; (**C**) absent LVOR; (**D**) prolonged latency; (**E**) asymmetry. The *top curve* is the chair velocity feedback.

velocities (mean = 6.2°/s), larger asymmetries (mean = 19.6 %), and longer latencies (mean = 144 ms) (FIG. 3). Six patients had preserved LVORs at short latency (< 130 ms), five of whom also had residual responses to rotational testing. Conversely, prolonged latencies or absence of the LVOR corresponded to absent rotational responses in six out of eight patients indicating that otolith involvement roughly paralleled the severity of horizontal canal dysfunction.

VISUAL CONSEQUENCES OF DEFECTIVE OTOLITH-OCULAR REFLEXES

The efficiency of the canal AVOR for visual stability has been demonstrated not only by eye-movement recordings but also by visual psychophysical studies.[9] In contrast, studies on the LVOR have been restricted to eye-movement recordings, which yield only indirect evidence of its visual consequences. Thus, we investigated to what extent the LVOR contributes to dynamic visual acuity during linear acceleration along the interaural axis in 14 normal subjects and in the 14 patients with bilateral vestibular failure.

Testing of Dynamic Visual Acuity

For this experiment, the car oscillated sinusoidally with the subject seated as described before at 0.5, 1.0, and 1.5 Hz, generating peak accelerations of 0.14, 0.27, and 0.39 g, and a uniform peak velocity of 42 cm/s. The target was placed at a 40-cm distance, resulting in a relative peak target velocity of 61 deg/s. For the "subject-motion" condition, the subject was moved in the car while the target was stationary (FIG. 4). For the "target-motion" condition, the target was mounted to the car while the subject remained stationary. In the "concomitant-motion" condition both subject and target were moved with the car at 40-cm target–eye separation. We presumed that during target motion pursuit was tested in isolation, whereas both the LVOR and pursuit were activated during subject motion. In the concomitant-motion condition any LVOR would have to be suppressed.

The target was a three-digit number produced by a red LED display subtending 1deg of arc in width. Each test consisted of a sequence of 40 numbers that were presented at 1-s intervals for 55 ms at the time of maximum chair velocity. Twenty of these numbers contained the digit 5. Patients were asked to maintain ocular fixation on the target and to press a button whenever they recognized a 5. To avoid artefacts related to task difficulty, the same numbers were presented for each test. Their order, however, was different for each frequency of oscillation. Two types of incorrect responses occurred: "misses" and "false recognitions." A miss was defined as a failure to signal a presented 5, and a false recognition as an indication of a 5 when the display did not contain one. For each test, performance was quantified by the total number of correct recognitions. An "otolith score" was calculated by subtracting the number of correct recognitions during target motion from the number of correct recognitions during subject motion. Thus, the otolith score quantified the contribution of the LVOR to visual stability.

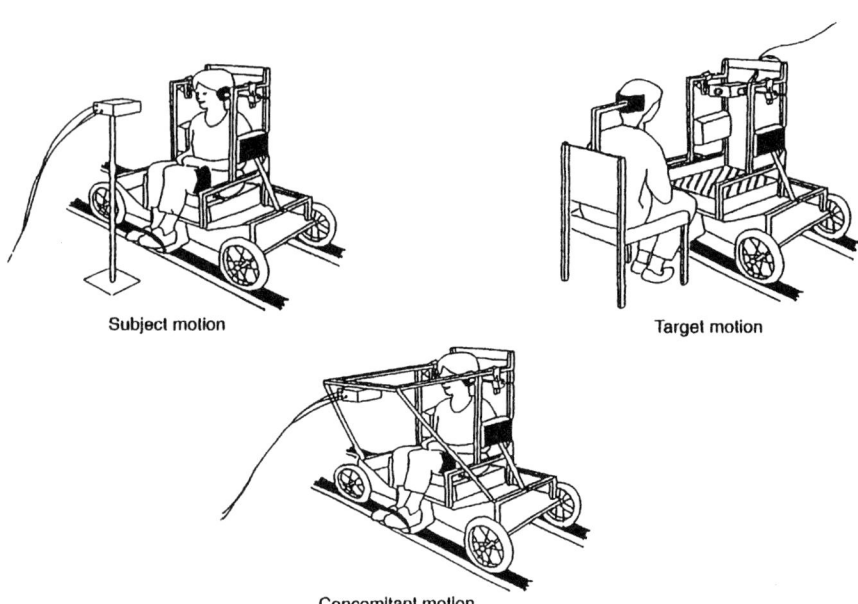

FIGURE 4. Experimental conditions. **Subject motion**: Translation along the interaural axis with target earth-fixed. **Target motion**: Stationary subject and target moves horizontally. **Concomitant motion**: Subject and target moving together on the car.

Results

At 0.5 Hz, normal subjects had similar dynamic visual acuity in all three conditions with a number of correct recognitions close to the maximum of 40 (FIG. 5). At 1 Hz, there was a significant decline in performance for target motion ($p < 0.01$), but not for subject motion. From 1 to 1.5 Hz, each subject's performance declined for both conditions ($p < 0.001$), but was invariably better for subject motion. Normal subjects reached a mean otolith score of 5.5 at 1 Hz (range, 0–13) and 7 at 1.5 Hz (range, 3–13). Otolith scores below these ranges were regarded as abnormal. During concomitant motion, a significant decline of visual acuity occurred between 1 and 1.5 Hz ($p < 0.001$) indicating that the LVOR is often incompletely suppressed at high frequencies.

In patients, performances during target motion were not different from those of normals at any frequency reflecting normal pursuit function. Unlike normal subjects, however, most patients could not enhance their dynamic visual acuity by additional otolith stimulation in the subject-motion condition: their otolith score averaged –0.6 compared with 5.5 in normals at 1 Hz ($p < 0.001$) and -0.2 compared with 7 in normals at 1.5 Hz ($p < 0.005$) (FIG. 6). At 1.5 Hz, only four patients had otolith scores in the normal range (≥ 3). All of them had preserved short latency LVORs, whereas eight out of the ten patients with abnormal scores had either prolonged latencies or absent LVORs ($p < 0.05$) (FIG. 7).

In the concomitant-motion condition, which requires suppression of the LVOR, dynamic visual acuity was not significantly different between patients and normal subjects at any frequency. Three patients with absent or low velocity LVORs (< 4.7 deg/s) had nearly complete suppression of the LVOR during concomitant motion (number of correct recognitions ≥ 37) (FIG. 8).

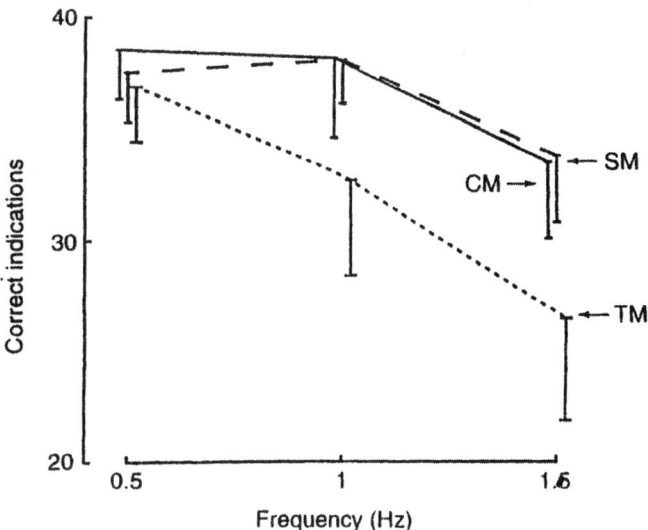

FIGURE 5. Dynamic visual acuity in 14 normal subjects: means and standard deviations of correct indications for each stimulus frequency. At 1 and 1.5 Hz, there was a statistically significant difference ($p < 0.001$) between SM and TM, reflecting improved visual acuity due to the LVOR. SM = subject motion; TM = target motion; CM = concomitant motion.

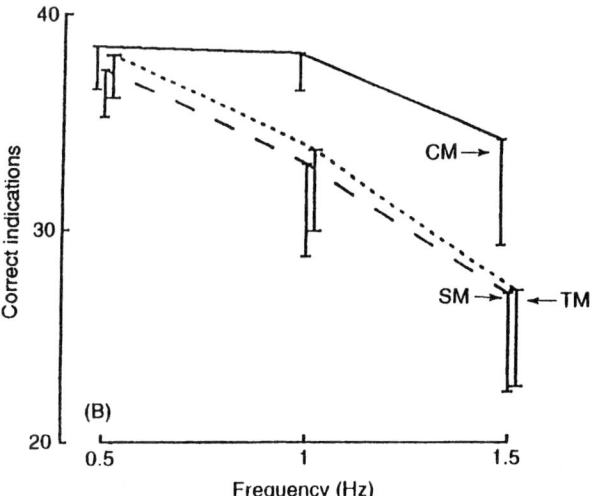

FIGURE 6. Dynamic visual acuity in 14 patients: means and standard deviations of correct recognitions for each stimulus frequency. The main difference with normals (FIG. 5) is the SM condition at 1 and 1.5 Hz: patients cannot enhance their visual acuity due to LVOR failure.

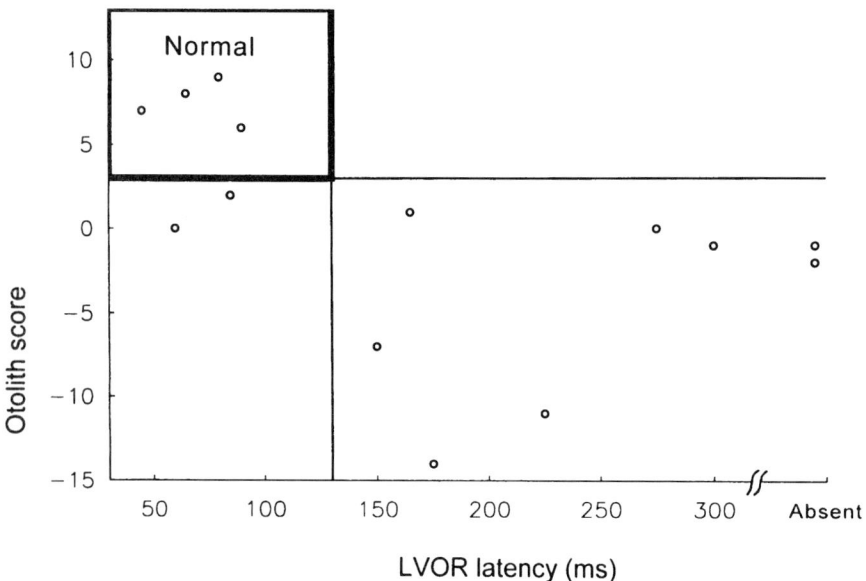

FIGURE 7. Relationship of LVOR latency and dynamic visual acuity during linear self-motion at 1.5 Hz in 14 patients with bilateral vestibular failure. The otolith score (subject motion score minus target motion score) indicates the otolithic contribution to dynamic visual acuity. Only four patients' values fell within the normal range, indicated by the top left box.

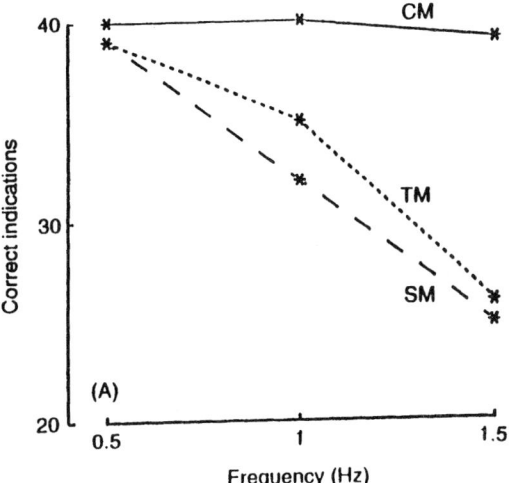

FIGURE 8. Dynamic visual acuity in one patient with complete loss of LVORs. Note a parallel decline of TM and SM as well as preserved acuity during CM, features that never occurred in normal subjects.

Conclusions

In this study, we have introduced a psychophysical approach to demonstrate that the LVOR, in fact, stabilizes vision during linear head motion beyond the frequency range of the pursuit system. The LVOR contribution to visual stability was apparent already at 1 Hz, but even more prominent at 1.5 Hz. Eye movement recordings in animals[10,11] and humans[12,13] have shown that the combination of LVOR and pursuit during linear self-motion produces enhanced compensatory eye movements when compared with pursuit alone, particulary at frequencies above 0.5 Hz. The neuronal basis of otolith–visual interaction has been identified by single-cell recordings in the vestibular nucleus of the cat.[14,15] Most of the neurons excited by linear motion were found to also respond synergistically to visual linear motion. Visual influences dominated the neuronal activity at low frequencies up to 0.25 Hz, whereas the otolith contribution prevailed in the higher frequency range.

Clinical evidence for the visual effects of the LVOR comes from our patients with bilateral vestibular failure and decreased visual acuity during high-frequency self-motion. Otolith function was abnormal in all patients and severity of otolith and canal involvement appeared to be related. Previous studies in individual patients with bilateral vestibular failure have found both sparing[16,17] and impairment or absence of the LVOR.[1,4,12,18,19]

Poor visual stability during linear motion was correlated with delay of LVORs in our patients. Delay of LVORs has been noted before both with unilateral and bilateral peripheral vestibular lesions.[16] We now show that such an abnormality has detrimental consequences for dynamic visual acuity: during sinusoidal linear motion any response delay would introduce a phase lag into the otolithic component of the eye movement that would interfere with accurate target fixation. Incongruous phase relationships between otolith and visual inputs have been shown to decrease the gain of the compensatory eye movements.[20]

LVOR IN PATIENTS WITH ACUTE UNILATERAL VESIBULAR LESIONS

The directional organization of the angular vestibulo-ocular reflex was established more than 100 years ago.[21] In contrast, the directional organization of the horizontal linear vestibulo-ocular reflex is still not well understood. Each utricular macula is sensitive to acceleration in both directions along the interaural axis due to the opposing orientations of its hair cells. The midlateral area of the utricle is activated by ipsilaterally directed head-acceleration, whereas the midmedial region is excited by medially directed acceleration (FIG. 9). The medial area is larger than the lateral one both in monkeys[22] and humans.[23] Correspondingly, 75% of peripheral utricular neurons respond to ipsilateral tilt, which corresponds to a medially directed acceleration and only 25% to contralateral tilt.[24] To determine the directional sensitivity of the horizontal LVOR we investigated nine patients with acute surgical lesions of one labyrinth. We specifically wanted to know if the remaining utricle responds bidirectionally (possibly biased toward medially directed acceleration), as is to be expected if both the medial and lateral area of the utricle provide afference for the LVOR. Alternatively, if only one of these regions feeds into the LVOR pathways, this should result in a unilateral loss of the response.

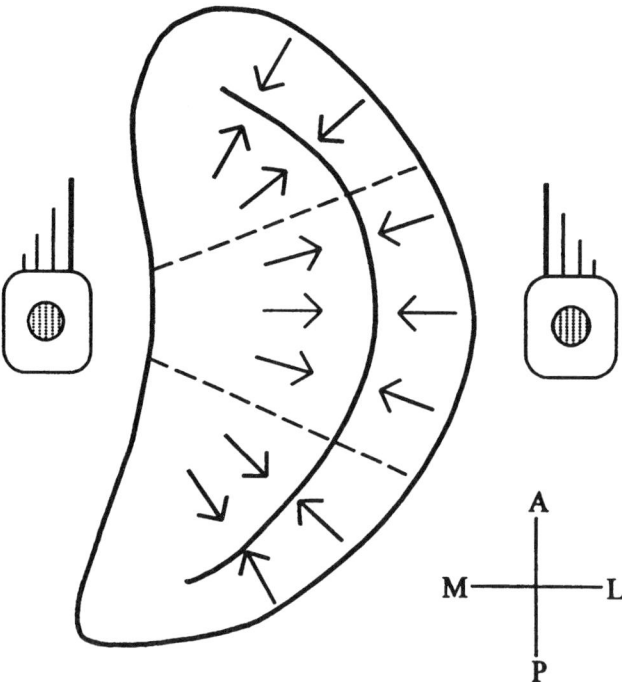

FIGURE 9. Hair-cell orientations of the right utricular macula. *Arrows* point toward kinocilia and indicate excitatory deflections of hair bundles. (A = anterior; P = posterior; L = lateral; M = medial.)

PATIENTS

We investigated nine patients who underwent selective unilateral vestibular nerve section for treatment of Ménière's disease ($n = 7$) or during removal of a small acoustic neuroma ($n = 2$). They were between 23 and 54 years of age (mean = 39 years). Before surgery, they had normal or only moderately reduced caloric function on the affected side except for one patient with an acoustic neuroma who had a minimal response only on the side of the tumor. Caloric responses were normal in the opposite ear. All patients were tested on the day before surgery and after 1 week. Five of them were tested again after 6–10 weeks.

Results

Before surgery, LVOR responses were within the normal range of velocity in eight patients averaging 10.6 deg/s with acceleration toward the affected ear (i.e., with the affected ear leading) and 11.9 deg/s in the opposite direction. Two of these patients showed abnormal asymmetries of 27 and 33%, with weaker responses when accelerated toward the affected ear. The last patient had low eye velocity, averaging 3.1 deg/s and a slight asymmetry of 18%.

When patients were tested on the seventh postoperative day, the spontaneous nystagmus was usually suppressed in the light but still present in the dark. Therefore, the individual LVOR responses were corrected by subtraction of the slow-phase velocity of the spontaneous nystagmus in the dark (mean = 3.5 deg/s; range = 0.7-8.4 deg/s). At this

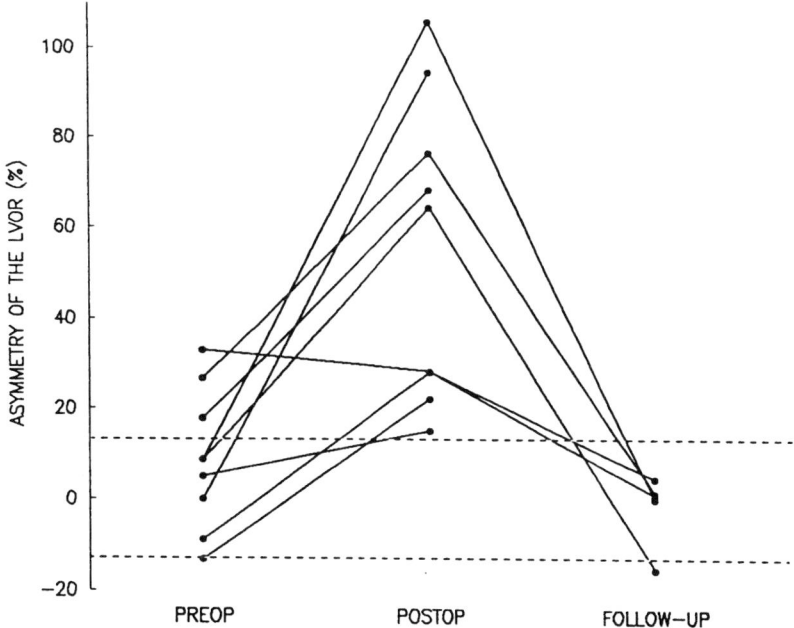

FIGURE 10. Asymmetry of the LVOR in nine patients before and after vestibular nerve section. *Positive values* indicate stronger responses with acceleration toward the intact ear. *Dashed lines* delineate normal range.

stage, the horizontal LVOR was asymmetrical in all patients (FIG. 10). After correction for spontaneous nystagmus, responses were markedly diminished or absent (mean = 4.0 deg/s) when head acceleration was directed toward the operated ear, which corresponded to a medially directed translation of the remaining utricle. In contrast, responses were almost unchanged (mean 12.5 deg/s) with head acceleration toward the intact side, which represented an ipsilaterally directed translation of the intact utricle. The difference in eye velocity measured before and after surgery was significant for acceleration toward the operated ear ($p = 0.015$) and not significant for acceleration toward the intact ear. Thus, when the velocity of of the postsurgical nystagmus was subtracted from response magnitude, postoperative LVORs for motion to the intact side were similar to preoperative responses, whereas LVOR responses to motion to the lesioned side were effectively abolished (FIG. 11). Response asymmetries ranged from 15 and 105%, and thus invariably exceeded the normal limit of 13%. The only asymmetry above 100% was obtained in one patient with negligible eye velocity (0.5 deg/s) in the same direction of head acceleration during stimuli toward the lesioned side.

After six to ten weeks the LVOR was bilaterally symmetric (< 13% asymmetry) in four out of five patients. One patient had a slight asymmetry of 15% with smaller responses when accelerated away from the lesion. Eye velocities were comparable to the preoperative values (mean = 10.2 deg/s).

Conclusions

Our findings indicate that the utricle is polarized with respect to the horizontal LVOR, producing a contraversive eye movement in response to ipsilateral head translation. The observed directional polarization of the single utricle was unexpected considering the predominance of peripheral utricular neurons responding to ipsilateral tilts,[24] which represents a medially directed acceleration. Therefore we speculated that the response bias might reflect the asymmetric resting activity of canal-related neurons in the vestibular nuclei causing spontaneous nystagmus. The asymmetries, however, persisted even after subtracting the slow-phase velocity of the spontaneous nystagmus.

The concept that each utricle provides afference only for the contraversive horizontal LVOR is supported by the effects of selective stimulation or lesion of peripheral utricular structures. Local electrical stimulation of the utricular macula was first accomplished by Fluur and Mellström in cats;[25] they observed horizontal eye movements only when activating the midlateral region. Hair cells in this area respond preferentially to ipsilaterally directed acceleration due to the orientation of their polarization vectors[26] (FIG. 9). The resulting horizontal nystagmus with contraversive slow phases represents the appropriate compensatory eye movement for an ipsilateral acceleration. Stimulation of the entire utricular nerve in cats produces not only vertical and torsional but also horizontal eye movements that are always contraversive.[27,28] Conversely, the section of the utricular nerve in cats uniformly results in a transient horizontal nystagmus with slow phases that are directed toward the operated ear.[29,30] In summary, both our findings and data from animal experiments would support the hypothesis that input for the horizontal LVOR is provided exclusively by hair cells of the midlateral macular region that are activated by ipsilateral linear acceleration and induce contraversive eye movements.

Uchino and coworkers offered another model of the LVOR organization that is based on the identification of mono- and disynaptic connections from the utricle to the ipsilateral abducens nucleus.[31,32] This arrangement would be difficult to reconcile with the stimulation and lesion studies cited earlier, which uniformly suggest that the LVOR is transmitted to the contralateral abducens nerve. Moreover, these fast pathways should produce latencies comparable with the disynaptic AVOR, that is, between 5 and 15 ms,[33] whereas the horizontal LVOR needs 14–54 ms in humans.[1,34] The longer latencies reflect the polysynaptic modulation of the otolith signal to calibrate the LVOR

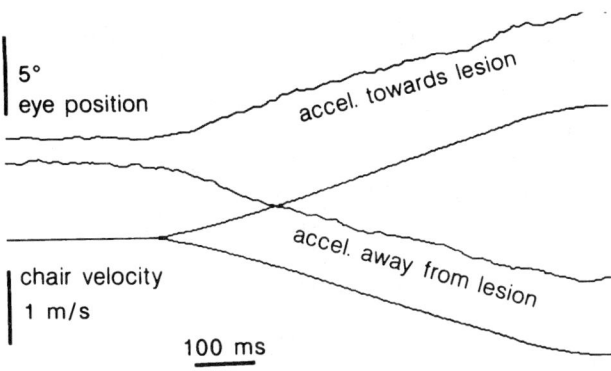

one week after surgery

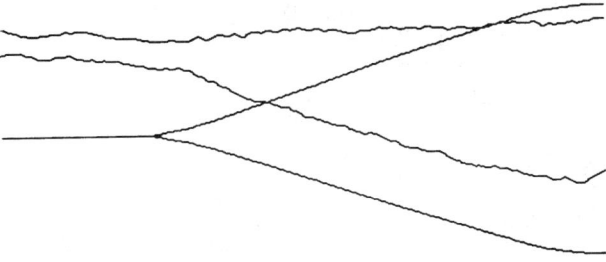

10 weeks after surgery

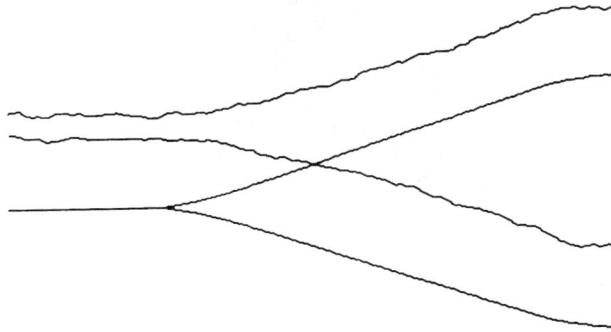

FIGURE 11. Horizontal LVOR in a patient before, one week after, and ten weeks after left-sided vestibular nerve section, averaged from five desaccaded recordings for each direction. Note postoperative asymmetry of the LVOR with weak responses for acceleration toward the operated ear. After ten weeks, responses are bilaterally restored. Chair-velocity feedbacks are presented for all test sessions.

according to viewing distance,[2,35] which involves processing within the vestibular nuclei and the cerebellum.[4,36] Uchino *et al.* suggested that these fast projections carry a baseline LVOR signal that is independent of the viewing distance.[32] This hypothesis, however, has not found support from findings in monkeys[35] and humans[37] where an early, nonvisually modulated otolith response has not been observed.

Symmetry of the LVOR was restored after 6 to 10 weeks in four out of five patients. Compensation appears to depend on input from the contralateral utricle. In unilaterally deafferented monkeys, LVOR compensation breaks down after subsequent contralateral utricular nerve section resulting in a permanent loss of the LVOR.[38] It is not known if compensation of the LVOR requires—similar to the angular VOR—restoration of spontaneous activity and dynamic responsiveness within the ipsilateral vestibular nucleus[39] mediated through the vestibular commissures. Alternatively, recovery of a bidirectional LVOR could be achieved by the contralateral vestibular nucleus alone if it utilizes inhibitory responses from lateral hair cells during acceleration in their off-direction. This principle, however, seems not to work in the acute stage of unilateral loss when resting activity tends also to be depressed in the contralateral vestibular nucleus due to cerebellar inhibitory output ("cerebellar shutdown"), which acts as an early defense mechanism to minimize vestibular asymmetry.[40] Due to the low resting discharge, otolith neurons on the intact side would be easily driven into inhibitory cutoff, explaining why in our patients responses were severely depressed in the off-direction at this stage. During compensation, however, contralateral activity may recover as observed in canal-related neurons.[39] If this also applies to otolith neurons, the push–pull arrangement between a single vestibular nucleus and the ocular motorneurons could be used to reestablish a bidirectional modulation of the LVOR.

LVOR IN PATIENTS WITH BENIGN POSITIONAL VERTIGO

According to current thinking, benign positional vertigo (BPV) is caused by dense particles that move within the endolymph of the posterior semicircular canal whenever head position is changed with respect to gravity.[41,42] These particles probably consist of utricular otoconia that have dislodged from the macula secondary to infection, ischemia, trauma, or ageing. It is still unclear, however, whether otoconial loss and utricular damage in BPV is sufficient to cause overt dysfunction of the otolith system.

Therefore, we measured LVORs in 14 patients (mean age = 50 years) with unilateral BPV. Ten patients were acutely suffering from BPV at the time of LVOR testing, whereas four were asymptomatic. LVORs were elicited both in the dark and in the light in this study. For LVOR testing in the light, patients fixated the center of a cardboard measuring 75×55 cm with black and white vertical stripes at 60-cm distance. Since LVORs interact with pursuit mechanisms in this condition, pursuit performance was studied in isolation by displaying a moving target of identical size, distance, and relative velocity on a screen while the patient remained stationary. The mean relative target velocity was 78.8 deg/s between 300 and 500 ms after onset of movement. Velocities and asymmetries for LVORs in the light and pursuit were measured in the same way as for LVORs in the dark (see earlier). Normative data for LVORs in the light and pursuit were recorded in 10 healthy subjects (mean age = 28 years).

Normal subjects had mean LVOR velocities of 10.3 deg/s in the dark and 72.7 deg/s in the light. Pursuit velocities averaged 53.4 deg/s, indicating a synergistic enhancement of pursuit and otolith mechanisms during head translation in the light. In patients, absolute LVOR velocities in the dark (mean = 10.0 deg/s) were similar to those of normals. Responses in the light, however, were mildly but significantly reduced (mean = 60.2 deg/s, $p < 0.05$). Pursuit velocities in patients were also slightly decreased (mean = 49.6 deg/s), but the difference was not statistically significant.

Asymmetries in normal subjects did not exceed 13% in the dark and 9% in the light, which were regarded as the upper limits of normal. Five patients had abnormal asymmetries in the dark ranging from 18% to 38%. There was no consistent relationship to the affected side: two patients had weaker and three had stronger responses when translated toward the side with BPV. Two patients had mild LVOR asymmetries in the light of 11 and 13%; one had a weaker and one had a stronger LVOR when moving toward the lesion side. These minor abnormalities were not related to age, preexisting labyrinthine disorders or current activity of BPV.

CONCLUSIONS

We could not detect any pronounced abnormalities of otolith-ocular responses in patients with BPV. Only LVORs in the light that depend largely on the pursuit system were mildly reduced. However, as there appeared to be also some age-related decrease of pursuit in patients, the contribution of otolith failure remains uncertain. Normal LVORs in BPV may be explained by central compensation of a unilateral utricular lesion, which is accomplished within 6–10 weeks after an acute lesion. Alternatively, when BPV is caused by head trauma, exposure to ototoxic drugs, or age-related degeneration, utricular damage is usually bilateral but otoconial loss is often incomplete,[43,44] which may leave the overall function of the macula unaffected.

ACKNOWLEDGMENT

This work was supported by a grant from Deutsche Forschungsgemeinschaft and by the CEC Human Capital and Mobility Program ERBFMGE CT 950068 (access to large-scale facilities).

REFERENCES

1. BRONSTEIN, A. M. & M. A. GRESTY. 1988. Short latency compensatory eye movement responses to transient linear head acceleration: a specific function of the otolith-ocular reflex. Exp. Brain Res. **71**: 406–410.
2. SCHWARZ, U. & F. A. MILES. 1991. Ocular responses to translation and their dependence on viewing distance. I. Motion of the observer. J. Neurophysiol. **66**: 851–863.
3. PAIGE, G. D., G. BARNES, L. TELFORD & S. H. SEIDMAN. 1996. Influence of sensorimotor context on the linear vestibulo-ocular reflex. Ann. N.Y. Acad. Sci. **781**: 322–331.
4. BALOH, R. W., Q. YUE & J. L. DEMER. 1995. The linear vestibuloocular reflex in normal subjects and patients with vestibular and cerebellar lesions. J. Vestib. Res. **5**: 349–361.
5. LEMPERT, T., C. GIANNA, G. BROOKES, M. A. GRESTY & A. M. BRONSTEIN. 1997. Effect of otolith dysfunction. Impairment of visual acuity during linear head motion in labyrinthine defective subjects. Brain **120**: 1005–1013.
6. LEMPERT, T., C. GIANNA, G. BROOKES, A. M. BRONSTEIN & M. A. GRESTY. 1998. Horizontal otolith-ocular responses in humans after unilateral vestibular deafferentation. Exp. Brain Res. **118**: 533–540.
7. ANASTASOPOULOS, D., T. LEMPERT, C. GIANNA, M. A. GRESTY & A. M. BRONSTEIN. 1997. Horizontal otolith-ocular responses to lateral translation in patients with benign positional vertigo. Acta Otolaryngol. **117**: 468–471.
8. BUSETTINI, C., F. A. MILES, U. SCHWARZ & J. R. CARL. 1994. Human ocular responses to translation of the observer and of the scene: dependence on viewing distance. Exp. Brain Res. **100**: 484–494.
9. BENSON, A. J. & G. R. BARNES. 1978. Vision during angular oscillation: the dynamic interaction of visual and vestibular mechanisms. Aviat. Space Environ. Med. **49**: 340–345.

10. FUKUSHIMA, K. & J. FUKUSHIMA. 1991. Otolith-visual interaction in the control of eye movement produced by sinusoidal vertical linear acceleration in alert cats. Exp. Brain Res. **85**: 36–44.
11. BOREL, L. & M. LACOUR. 1992. Functional coupling of the stabilizing eye and head reflexes during horizontal and vertical linear motion in the cat. Exp. Brain Res. **91**: 191–206.
12. BALOH, R., K. BEYKIRCH & V. HONRUBIA. 1988. Eye movements induced by linear acceleration on a parallel swing. J. Neurophysiol. **60**: 2000–2013.
13. SHELHAMER, M. & L. R. YOUNG. 1994. The interaction of otolith organ stimulation and smooth pursuit tracking. J. Vestib. Res. **4**: 1–15.
14. DAUNTON, N. & D. THOMSEN. 1979. Visual modulation of otolith-dependent units in cat vestibular nuclei. Exp. Brain Res. **37**: 173–176.
15. XERRI, C., J. BARTHELEMY, L. BOREL & M. LACOUR. 1988. Neuronal coding of linear motion in the vestibular nuclei of the cat. III. Dynamic characteristics of visual-otolith interactions. Exp. Brain Res. **70**: 299–309.
16. BRONSTEIN, A. M., M. A. GRESTY & G. B. BROOKES. 1991. Compensatory otolithic slow-phase eye movement responses to abrupt linear head motion in the lateral direction. Findings in patients with labyrinthine and neurological lesions. Acta Otolaryngol. (Suppl.) **481**: 42–46.
17. BALOH, R. W., J. OAS, V. HONRUBIA & D. M. MOORE. 1992. Preservation of the electrical-evoked vestibuloocular reflex and otolith-ocular reflex in two patients with markedly impaired canal-ocular reflexes. Ann. N.Y. Acad. Sci. **656**: 811–813.
18. TOKITA, T., M. MIYATA, M. MASAKI & S. IKEDA. 1981. Dynamic characteristics of the otolithic oculomotor system. Ann. N.Y. Acad. Sci. **374**: 56–68.
19. ISRAEL, I. & A. BERTHOZ. Contribution of the otoliths to the calculation of linear displacement. 1989. J. Neurophysiol. **62**: 247–263.
20. LATHAN, C. E., C. WALL III & L. R. HARRIS. 1995. Human eye movement response to z-axis linear acceleration: the effect of varying the phase relationships between visual and vestibular inputs. Exp. Brain Res. **103**: 256–266.
21. EWALD, E. J. R. 1892. Physiologische Untersuchungen über das Endorgan des Nervus Octavus. Bergmann. Wiesbaden.
22. FERNANDEZ, C., J. M. GOLDBERG & W. K. ABEND. 1972. Response to static tilts of peripheral neurons innervating otolith organs of the squirrel monkey. J. Neurophysiol. **35**: 996–1008.
23. ROSENHALL, U. 1972. Vestibular macular mapping in man. Ann. Otol. **81**: 339–351.
24. FERNANDEZ, C. & J. M. GOLDBERG. 1976. Physiology of peripheral neurons innervating otolith organs in the squirrel monkey. I. Response to static tilts and to long-duration centrifugal force. J. Neurophysiol. **39**: 970–984.
25. FLUUR, E. & A. MELLSTRÖM. 1970. Utricular stimulation and oculomotor reactions. Laryngoscope **80**: 1701–1712.
26. GOLDBERG, J. M., G. DESMADRYL, R. A. BAIRD & C. FERNANDEZ. 1990. The vestibular nerve of the chinchilla. V. Relation between afferent discharge properties and peripheral innervation patterns in the utricular macula. J. Neurophysiol. **63**: 791–804.
27. SUZUKI, J. I., K. TOKAMASU & K. GOTO. 1969. Eye movements from single utricular nerve stimulation in the cat. Acta Otolaryngol. **68**: 350–362.
28. TOKAMASU, K., J. I. SUZUKI & K. GOTO. 1971. A study of the current spread on electric stimulation of the individual utricular and ampullary nerves. Acta Otolaryngol. **71**: 313–318.
29. FERNANDEZ, C. R. ALZATE & J. R. LINDSAY. 1960. Experimental observations on postural nystagmus in the cat. Ann. Otol. **69**: 816–829.
30. FLUUR, E. & J. SIEGBORN. 1973. The otolith organs and the nystagmus problem. Acta Otolaryngol. **76**: 438–442.
31. SCHWINDT, P. C., A. RICHTER & W. PRECHT. 1973. Short latency utricular and canal input to ipsilateral abducens motoneurons. Brain Res. **60**: 259–262.
32. UCHINO, Y., H. IKEGAMI, M. SASAKI, K. ENDO, M. IMAGAWA & N. ISU.1994. Monosynaptic and disynaptic connections in the utriculo-ocular reflex arc of the cat. J. Neurophysiol. **71**: 950–958.
33. MAAS, E. F., W. P. HUEBNER, S. H. SEIDMAN & R. J. LEIGH. 1989. Behavior of human horizontal vestibulo-ocular reflex in response to high acceleration stimuli. Brain Res. **499**: 153–156.
34. GIANNA, C. C., M. A. GRESTY & A. M. BRONSTEIN. 1997. Eye movement response to linear acceleration steps in humans: visual enhancement and suppression [Abstr.] J. Physiol. **499**: 87p.

35. SNYDER, L. H. & W. M. KING. 1992. Effect of viewing distance and location of the axis of head rotation on the monkey's vestibuloocular reflex. I. Eye movement responses. J. Neurophysiol. **67**: 861–874.
36. SNYDER, L. H. & W. M. KING. 1989. Modulation of gaze velocity Purkinje (GVP) cells during vestibuloocular reflex (VOR) with near and far visual targets [Abstr.] Soc. Neurosci. Abstr. **15**: 807.
37. GIANNA, C. C., M. A. GRESTY & A. M. BRONSTEIN. 1997. Eye movements induced by lateral acceleration steps: effect of visual context and and acceleration levels. Exp. Brain Res. **114**: 124–129.
38. TAKEDA, N., M. IGARASHI, I. KOIZUKA, S. CHAE & T. MATSUNAGA. 1990. Recovery of the otolith-ocular reflex after unilateral deafferentation of the otolith organs in squirrel monkeys. Acta Otolaryngol. **110**: 25–30.
39. NEWLANDS, S. D. & A. A. PERACHIO. 1990. Compensation of horizontal canal related activity in the medial vestibular nucleus following unilateral ablation in the decerebrate gerbil. I. Type I neurons. Exp. Brain Res. **82**: 359–372.
40. MCCABE, B. F.; J. H. RYU & T. SEKITANI. 1972. Further experiments on vestibular compensation. Laryngoscope **82**: 381–396.
41. BRANDT, T. & S. STEDDIN. 1993. Current view of the mechanism of benign paroxysmal positioning vertigo: cupulolithiasis or canalolithiasis? J. Vest. Res. **3**: 373–382.
42. LEMPERT, T., C. WOLSLEY, R. DAVIES, M. A. GRESTY & A. M. BRONSTEIN. 1997. Three hundred sixty-degree rotation of the posterior semicircular canal for treatment of benign positional vertigo: a placebo controlled trial. Neurology **49**: 729–733.
43. JOHNSSON, L. G., C. G. WRIGHT, R. E. PRESTON & P. J. HENRY. 1980. Streptomycin-induced defects of the otoconial membrane. Acta Otolaryngol. **89**: 401–406.
44. IGARASHI, M., R. SAITO, K. MIZUKOSHI & B. R. ALFORD. 1993. Otoconia in young and elderly persons: a temporal bone study. Acta Otolaryngol. (Suppl.) **504**: 26–29.

Vestibular–Pursuit Interactions: Gaze-Velocity and Target-Velocity Signals in the Monkey Frontal Eye Fields

KIKURO FUKUSHIMA,[a,c] JUNKO FUKUSHIMA,[b] AND TOSHIKAZU SATO[a]

[a]*Department of Physiology, School of Medicine, Hokkaido University, West 7, North 15, Kitaku, Sapporo 060-8638, Japan*

[b]*College of Medical Technology, Hokkaido University, Sapporo 060-0812, Japan*

ABSTRACT: Visual information about a moving object is obtained by accurate tracking with the eyes using the smooth pursuit system, which must interact with the vestibular system during head movement. Such pursuit–vestibular interactions require calculation of gaze (i.e., eye in space) in order to match eye velocity in space to actual target velocity, using vestibular, retinal-image velocity, and eye-velocity information. To understand the role the frontal eye fields (FEFs) play in pursuit–vestibular interactions, we examined responses of pursuit-related neurons near the arcuate sulcus in head-stabilized monkeys during visual tracking tasks that dissociate eye movement in the orbit from that in space. The activity of the majority of pursuit-related neurons was related to gaze velocity. They also responded to passive body rotation in complete darkness. When the monkeys fixated the stationary target, similar modulation was observed, reflecting the velocity signal of a second test target. Muscimol infusion into the FEF pursuit areas severely impaired smooth gaze tracking. These results suggest that the region near the arcuate sulcus coordinates its various inputs to provide signals for target velocity in space and accurate gaze-velocity command during pursuit–vestibular interactions.

INTRODUCTION

Vestibular information, unlike other sensory systems, is not prominent in our consciousness. Nevertheless, it is important in virtually every aspect of our daily life because head acceleration information detected by semicircular canals and otolith organs is essential for our adequate behavior in three-dimensional space not only through vestibular reflexes that act constantly on somatic muscles and autonomic organs (see Wilson and Melvill Jones[1] for review) but also through various cognitive functions.[2,3]

Smooth pursuit eye movement has evolved with the development of the fovea in primates to track a small moving object with good visual acuity. This system does not work independently but interacts with the vestibular system during head movement to maintain the accuracy of eye movements in space (i.e., gaze). For example, ocular tracking

---

[c]To whom correspondence may be addressed. Phone: +81-11-706-5038; fax: +81-11-706-5041; e-mail: kikuro@med.hokudai.ac.jp

of a moving object accurately during whole-body rotation requires computation of actual target velocity in space using signals concerning eye and head (vestibular) velocity and retinal-image-slip velocity,[4,5] and command signals for eye velocity in space (i.e., gaze velocity) have to be generated in order to match it to actual target velocity.[5] An understanding of where and how these signals are generated may provide an important initial step for understanding neural mechanisms of pursuit–vestibular interactions.

The frontal eye fields (FEFs) have long been known as structures that control purposive saccades.[6] Recent studies have shown that this area, particularly the portion involving the posterior bank of the arcuate sulcus is involved in generation of smooth pursuit eye movement.[7-9] However, it is still unknown how FEF pursuit–related cells are involved in generation of smooth gaze tracking during head movement.

By assigning the monkeys the behavioral tasks that dissociate eye movement in the orbit from eye movement in space, we were able to show that the FEFs carry various velocity signals necessary to compute target velocity in space and accurate gaze-velocity command during pursuit–vestibular interactions. Consistent with this interpretation, we observed severe impairments of smooth gaze tracking following injection of a GABA agonist (muscimol) into the FEF pursuit areas.

METHODS

Two Japanese monkeys (Macaca fuscata) were used. All the procedures were evaluated and approved by the Animal Care and Use Committee of the Hokkaido University School of Medicine. Our methods for animal preparation, training, and recording are described elsewhere in detail,[9,10] and therefore are summarized here only briefly. Monkeys' heads were firmly restrained in the primate chair in the stereotaxic plane, and they were trained in darkness to track a laser spot (0.2 deg in diameter) back-projected onto a tangent screen positioned 75 cm in front of the animals' eyes. The monkey chair was fixed to the turntable that had two degrees of freedom of motion (horizontal and vertical rotation) under computer control. The interaural midpoint of the animals' head was brought close to the axis of vertical and horizontal rotation. The target moved either sinusoidally or stepwise. The chair was moved sinusoidally in the vertical, horizontal, or oblique directions by adding vertical and horizontal rotations. Vertical and horizontal components of eye movement were recorded using a scleral search-coil method. Reward circuits compared target-position signals with the monkeys' gaze-position signals. The latter signals were calculated electronically as the sum of eye position and turntable position signals. If the monkeys' gaze was within the error window of ± 1 deg for 0.5–1 s, a drop of apple juice was automatically delivered to the animal.

Extracellular recordings were made in and near the arcuate sulcus, particularly in the posterior part, while the monkeys tracked the target during sinusoidal whole-body rotation. Once responding single cells were encountered, pursuit responses were tested in four directions (vertical, horizontal, and two oblique directions at 45 deg angles) at 0.5 Hz (±10 deg) to determine an optimal direction for activation of each neuron without chair movement (FIG. 1).

To examine whether cell activity was related to eye movement in the orbit or in space during whole-body rotation, three tracking conditions were assigned to the monkeys using combinations of whole-body rotation and target movement in the same plane at 0.5 Hz (±10 deg). In the first, the monkeys tracked a target that moved with the same amplitude and phase as the chair in the same plane. This condition requires the monkeys to suppress the VOR so that gaze moved together with the chair without eye movement in the orbit (VOR suppression, FIG. 1). In the second condition, the target moved with the identical amplitude as, but in antiphase with the chair, requiring the monkeys to increase eye movement in the orbit to twice the chair movement but with the opposite phase (VOR

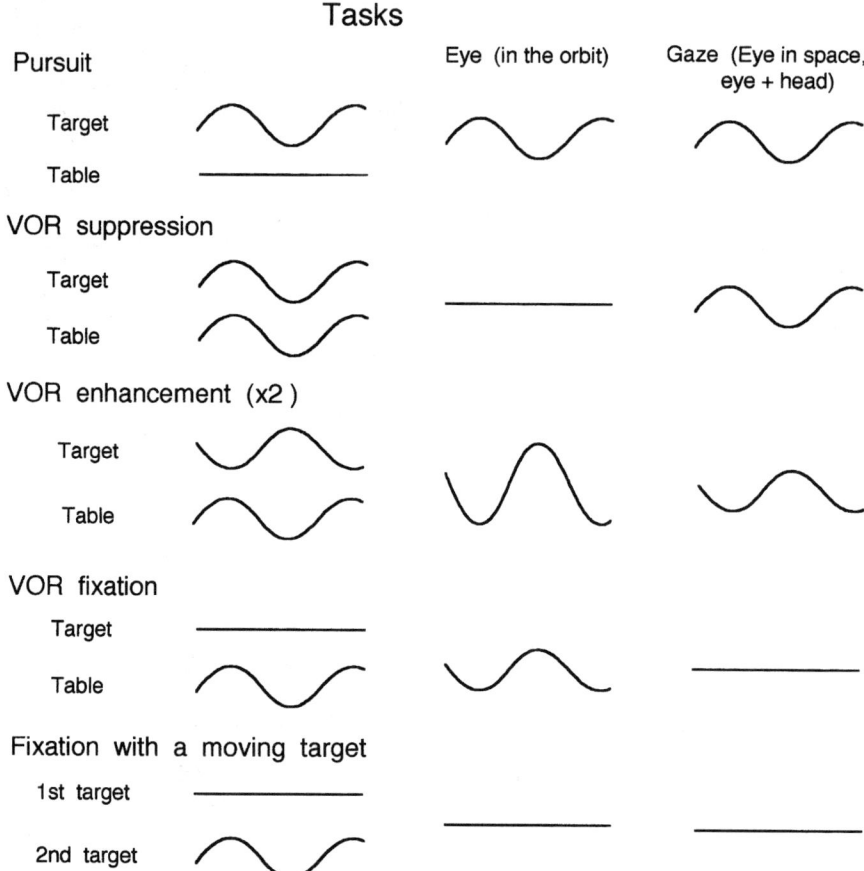

FIGURE 1. Behavioral paradigms used in this study. For further explanation, see text.

enhancement, ×2; FIG. 1), while the amplitude of gaze movement is the same as the amplitude during VOR suppression. These two conditions are useful to compare cell response, since they not only require qualitatively different eye movement but also require gaze movement with the opposite phase relative to that of the chair. In the third condition, the target stayed stationary in space during chair rotation, which requires perfect VOR without gaze movement (VOR fixation, ×1;257 FIG. 1). Chair rotations were tested in the four directions together with target movement in the same plane. Whole-body rotation was also tested in complete darkness.

To examine visual responses, the target was briefly extinguished (for 200 ms) during pursuit or VOR suppression ("blank") in the optimal direction. A longer-duration blank (800 ms) was also tested. Monkeys were also required to fixate the stationary laser spot (first target) while the second laser spot (0.6° in diameter) moved sinusoidally in the four directions to examine an optimal direction for visual responses for each neuron (FIG. 1).

Having mapped FEF pursuit areas, we examined the effects of chemical deactivation by injecting a GABA agonist muscimol (10–15 µg) dissolved in physiological saline

(10 µg/µl) unilaterally into a single area where we recorded many pursuit-related cells. Smooth pursuit and VOR suppression tasks were examined before and after muscimol infusion for both monkeys.

The data were analyzed off-line as previously described.[9-11] Cell discharge was discriminated with a dual time-amplitude-window discriminator and digitized together with eye-position, table-position, and target-position signals at 500 Hz using a 16-bit A/D board. Position signals were differentiated by analog circuits (DC-50 Hz, -12 dB/oct) to obtain velocity. Cycle rasters and histograms were constructed for the discharge of each neuron by averaging over 10 to 30 cycles after bursts or pauses, and associated fast eye movements were marked manually and deleted from the analysis. A least-squares method was used for fitting a sine function to cell- and eye-movement responses. Signal-to-noise (S/N) ratio of the response was defined as the ratio of the amplitude of the fitted fundamental frequency component to the root-mean-square (rms) amplitude of the third through eighth harmonics. Harmonic distortion (HD) was defined as the ratio of the amplitude of the second harmonic to that of the fundamental. The responses that had a HD of more than 50% or an S/N of less than 1.0 were discarded. Phase shift of cell or eye velocity response was calculated as the difference in phase between the peak upward (or rightward) stimulus velocity and the peak of the fundamental component of cell- or eye-velocity response. Gain was calculated as the peak amplitude of the fundamental component of cell- or eye-velocity response divided by that of the stimulus velocity. Recording locations were histologically confirmed by making iron deposits produced by positive current through the electrodes.

RESULTS

Gaze-Velocity Signals Carried by FEF Pursuit-responding Neurons

We recorded over 100 pursuit-responding neurons in two hemispheres near the right arcuate sulcus of two monkeys. A typical example of pursuit-related activity is shown in FIGURE 2. The optimal direction of activation was oblique for this neuron (FIG. 2A–2E). Optimal directions for individual neurons distributed almost equally for all directions (FIG. 2F).[9] For these neurons, saccadic responses were weak or nonexistent as reported elsewhere.[7-9]

The great majority (82%) of pursuit responding neurons also responded during VOR suppression (FIG. 1), and the optimal direction of each neuron during VOR suppression was similar to its optimal pursuit direction. For example, neurons shown in FIGURES 3 and 4 increased their activity during VOR suppression when the gaze moved downward (FIG. 3B) or rightward (FIG. 4A2), similar to their responses during smooth pursuit (FIGS. 3A, 4A1). During VOR enhancement ($\times 2$ condition, FIG. 1), they showed phase reversal with the response magnitude similar to that observed during VOR suppression (FIG. 3C). During VOR fixation ($\times 1$, FIG. 1), they showed little response (FIG. 3D).

Mean gains ($\pm$SD) of discharge modulation (relative to stimulus velocity at 0.5 Hz) during suppression, enhancement ($\times 2$), and fixation ($\times 1$) for 73 neurons tested were 0.39 ($\pm$ 0.24), 0.43 ($\pm$ 0.23), and 0.16 ($\pm$ 0.17) spikes/s/deg/s, respectively, with phases of the majority of responses near stimulus velocity, indicating that discharge modulation during suppression and enhancement ($\times 2$) was similar and much higher than that during VOR fixation ($\times 1$). Thus, the behavior of the great majority of pursuit-related neurons during these tasks is consistent with the interpretation that their activity is related to gaze velocity (FIG. 1). This conclusion is supported by the significant correlation between amplitude of modulation and peak gaze velocity for two representative neurons during VOR suppression, enhancement ($\times 2$), and fixation ($\times 1$) at different stimulus frequencies (FIG. 3H).

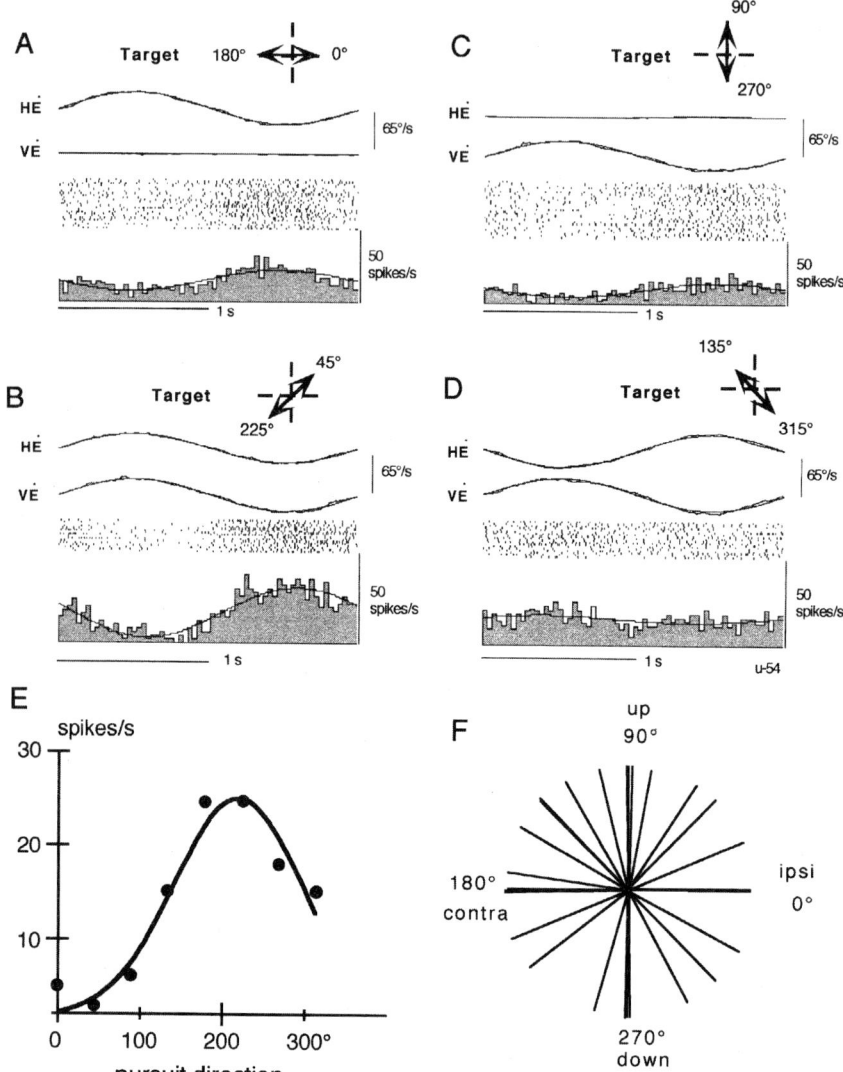

FIGURE 2. Optimal direction (**A**)–(**D**) and the directional tuning (**E**) of a FEF pursuit-responding cell. Records in A–D show rasters and histograms (below) of a single cell together with horizontal and vertical eye velocity (HĖ, VĖ) that were desaccaded and averaged. (**F**) summarizes optimal directions of individual pursuit-related neurons.

To examine whether responses during smooth ocular tracking were induced, in part, by visual input, we turned off the target briefly (for 200 ms) during pursuit or VOR suppression for 20 neurons. Half of them showed no clear change in their activity during the blanking periods (FIG. 4A1, 4A2). Although in the remaining half, their activity decreased

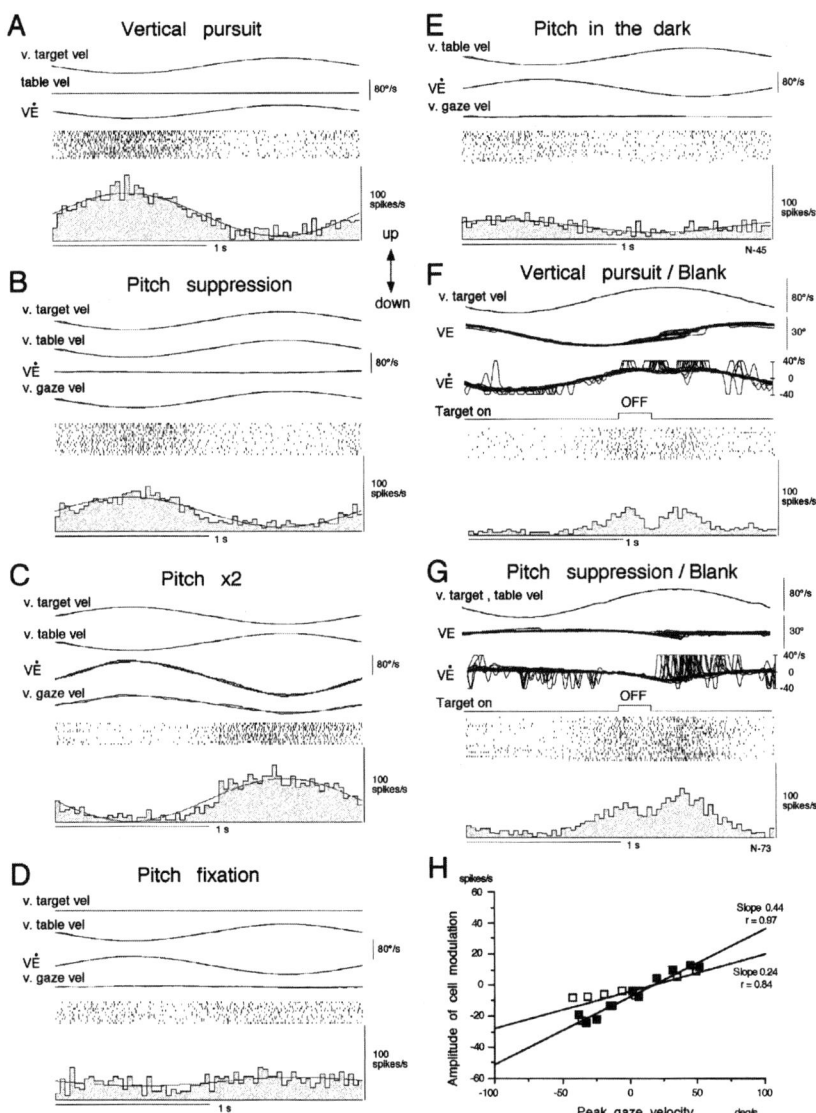

FIGURE 3. Responses of pursuit-responding neurons during vertical pursuit and whole-body rotation. Records in (A)–(E) show rasters and histograms (below) of a single cell together with stimulus, and eye velocity (VĖ) and gaze velocity that were desaccaded and averaged. (**A**) pursuit; (**B**)–(**D**) VOR suppression, enhancement (×2) and fixation (×1), respectively. (**E**) VOR in complete darkness. Records in (**F**)–(**G**) show rasters and histograms of another neuron together with stimulus velocity and superimposed eye position (VE) and velocity (VĖ) when the target was briefly (200 ms) extinguished during pursuit (F) and VOR suppression (G). (**H**) shows amplitudes of modulation of 2 representative neurons plotted against peak gaze velocity during VOR suppression, enhancement (×2), and fixation (×1) at different stimulus frequencies. Amplitudes of cell- and eye-velocity responses with the same phase during VOR suppression were plotted as positive; those with opposite phase were plotted as negative according to the convention used by Lisberger and Fuchs.[16] A linear regression is shown for each neuron with slope and correlation coefficient (*r*).

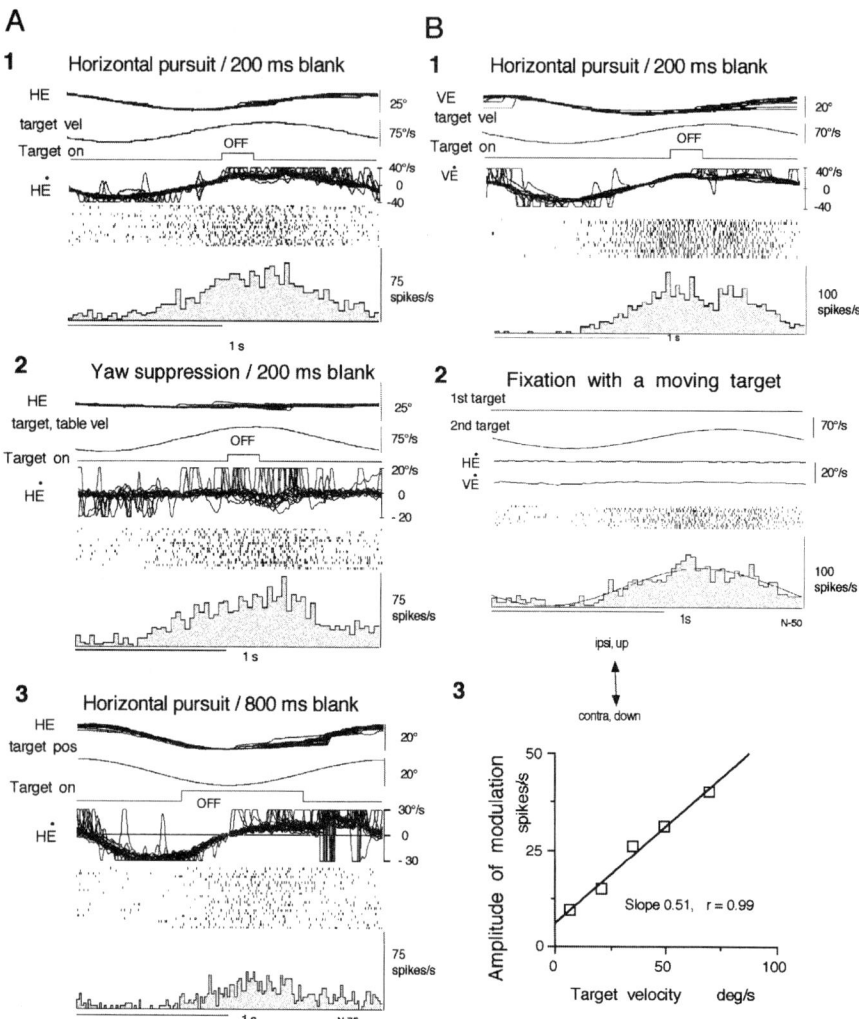

FIGURE 4. Similar arrangements for two frontal pursuit-related neurons (**A**)–(**B**) during pursuit (A1, A3, B1) and VOR suppression (A2) while the target was briefly (200 ms or 800 ms) extinguished. B2 shows responses of the same cell (B) to the second laser spot moved sinusoidally at 0.5 Hz (±10 deg) while the monkey fixated the first target. (*Note*: Different velocity scales are used for vertical and horizontal components of eye velocity (HĖ, VĖ) to reveal any possible response in B2.) B3 shows amplitudes of modulation of this cell against velocity of the second target at different frequencies while the monkey fixated the first target. A linear regression is shown with slope and correlation coefficient (*r*).

after 120–150 ms following target offset (FIG. 3F), the decreased activity was associated with a decrease in eye velocity, particularly during pursuit (FIG. 3F; also FIG. 4B1). These results indicate that the activity of these neurons during ocular tracking is maintained by extraretinal input during at least the early blanking period (<150 ms).[9]

To examine further whether these neurons are activated during voluntary smooth gaze tracking even without a target, we applied 800 ms long-duration blank before the target changed the direction during smooth pursuit and VOR suppression tasks. An example is shown in FIGURE 4A3 for the pursuit task. Our monkeys were well trained, and they performed smooth eye movement with change of direction even without the target in complete darkness. All five frontal pursuit cells tested increased activity associated with smooth eye movement without the target (FIG. 4A3).

In the neurons that showed decreased activity during late blanking periods (>150 ms, FIG. 3F), the effects of extinguishing the target were different during smooth pursuit and VOR suppression (FIG. 3F–3G). Compared to the clear decrease in discharge rate after about 150 ms during pursuit (FIG. 3F), the decrease in their activity was much less and often not as obvious during VOR suppression (FIG. 3G). Since discharge is maintained during VOR suppression, the activity of these neurons cannot simply reflect pursuit-related activity that might be used to cancel out the VOR during whole-body rotation. Instead, it may partially reflect vestibular-related activity. Indeed, the great majority of pursuit-related neurons tested (29/37=78.4%) responded weakly but clearly (FIG. 3E) to whole-body rotation in complete darkness without the target. The mean response gain (relative to stimulus velocity) for the 20 neurons examined were 0.20 (± 0.14 S.D.) spikes/s/deg/s. We also recorded a total of five neurons in this area that responded to vestibular stimulation but had no clear correlation to eye movement. These results suggest that neurons in this area receive vestibular inputs. Response phase of these neurons over wide frequencies (0.1–1.0 Hz) suggest mostly canal (rather than otolith) inputs.

Target Velocity Signals Carried by Gaze Velocity FEF Cells

To examine whether the present gaze velocity neurons near the arcuate sulcus receive target-velocity information, we tested cell activity with a second laser spot moved sinusoidally while the monkeys were rewarded by fixating the stationary laser spot (first target). Of the 39 pursuit-responding neurons tested, about half of them (19/39= 49%) showed no consistent response to the second target. However, a clear modulation was observed in the remaining half (n =20, 51%, FIG. 4B2) with the optimal direction similar to the direction for pursuit and vestibular sensitivities and with similar phases and magnitudes (FIGURE 4B1-4B2). Since in this situation eye velocity was minimal (FIG. 4B2), the modulation of cell activity cannot reflect eye velocity. FIGURE 4B3 plots amplitude of discharge modulation against target velocity for this neuron while the monkey fixated the first target. The activity of many of these neurons followed target velocity at higher than 100 deg/s, confirming that their activity reflected the retinal image velocity of the second target.

Effects of Muscimol Infusion into FEF Gaze-Velocity Areas

In order to obtain further information on how the FEFs could be involved in smooth gaze tracking with or without chair movement (i.e., VOR suppression and smooth pursuit tasks, respectively), we injected muscimol into the region where we recorded many of our gaze-velocity neurons. Results obtained in two monkeys were similar. An example is shown in FIGURE 5. After muscimol infusion, eye velocity during pursuit decreased and catch-up saccades frequently appeared (FIG. 5A vs. 5D). Mean eye gain decreased to nearly half after muscimol infusion.

Effects of muscimol infusion was also tested when a long-duration (800 ms) blank was applied before the target changed the direction. Before infusion, the monkeys performed smooth eye movement with change of direction fairly well even without the target in complete darkness (FIG. 5B). After muscimol infusion, however, the monkeys were virtually unable, in many trials, to generate smooth eye movement to the direction ipsilateral to the infusion side without the target. Instead, they performed the task with saccades. In some trials, they generated slow eye velocity with the mean of only 38% of the control value obtained before muscimol infusion (FIG. 5E).

The monkeys' performance of the VOR suppression task was also severely impaired after muscimol infusion (FIG. 5C vs. 5F). Before infusion, VOR was not induced during

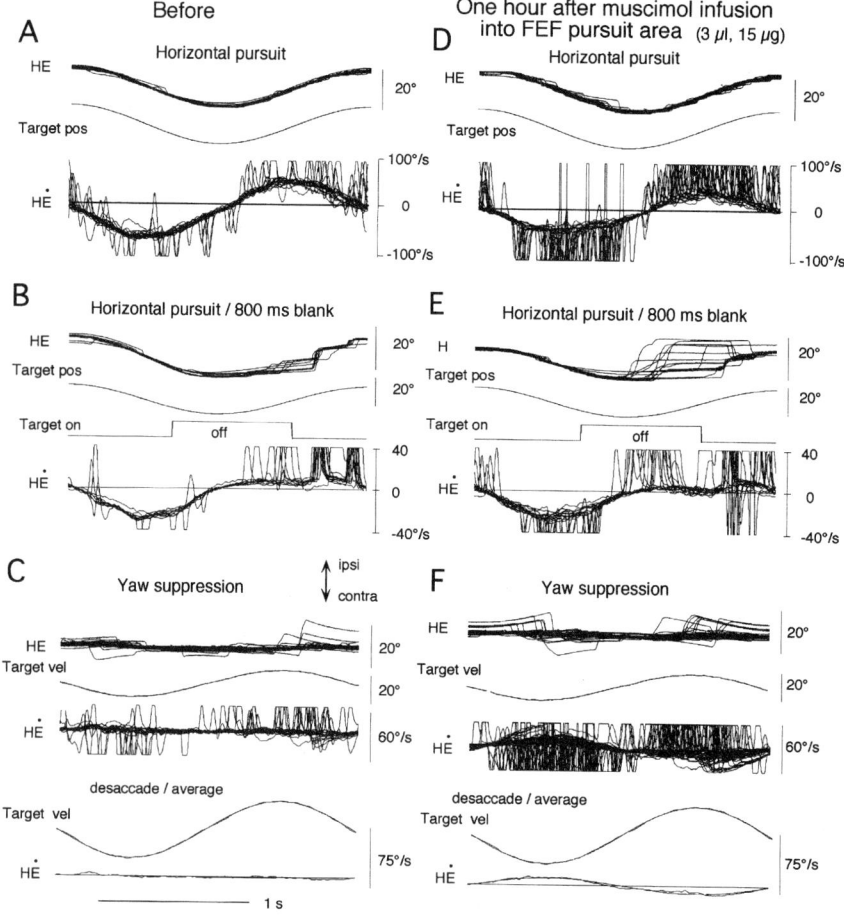

FIGURE 5. Effects of muscimol infusion into unilateral FEF where gaze-velocity cells were recorded. Traces in (**A**)–(**F**) are superimposed horizontal eye position (HE), horizontal target position, target velocity, horizontal eye velocity (HĖ), and gaze velocity. For further explanation, see text.

the suppression task (gain ~0.01). After infusion, however, they were unable to suppress the VOR (gains ~0.3, compare desaccaded/averaged traces), and corrective saccades appeared frequently to compensate for impaired VOR suppression (FIG. 5F). In contrast, the monkeys' performance of the VOR fixation (×1) task was not clearly affected by muscimol infusion.

DISCUSSION

Using the behavioral tasks that involve pursuit–vestibular interactions, the present study shows that FEF pursuit neurons receive signals related to vestibular and retinal image velocity in addition to eye velocity and that the great majority of them carry gaze-velocity signals with optimal directions of individual neurons distributed almost equally for all directions. Thus, actual target velocity in space[4,5] could be computed by signals carried by these FEFs neurons.

Gaze-movement-related signals were demonstrated in the posterior parietal cortex [medial superior temporal (MST) area[12,13]] and the central thalamus[14] in addition to the cerebellar flocculus, ventral paraflocculus, vermis, and vestibular nuclei.[15–19] Such signals are intermediate signals in sensory-motor transformations used in visual tracking performance during head movement, providing either target velocity information (e.g., MST,[20,21] vermis[22]) or motor-related gaze-velocity signals (e.g., the floccular lobe[16]). Retinal-image-velocity signals found in gaze-velocity neurons in this study may come from the posterior parietal regions,[23] and these FEF neurons could provide an estimate of actual target velocity information during head and/or whole-body movement[4,5] (FIG. 1A).

In a separate series of experiments, we recorded FEF pursuit-related neurons while the monkeys performed a step-ramp pursuit task.[9] The activity of the majority of pursuit-related neurons preceded the onset of eye movement with a median latency of −12 ms;[9] these results are consistent with recent reports by Gottlieb et al.[8] It has also been reported that electrical microstimulation of these areas induces smooth pursuit eye movement.[8] These results together with the present results showing that frontal pursuit-related cells increased activity associated with smooth pursuit even without the target (FIG. 4A3) suggest an involvement of these neurons in generating voluntary smooth ocular tracking. Consistent with this interpretation, chemical deactivation of these areas in this study severely impaired smooth pursuit generation and VOR suppression (FIG. 5).[3,24]

Optimal directions of gaze-velocity signals found in the present study are different from those reported in the cerebellar floccular lobe, since optimal directions of gaze velocity Purkinje cells are either vertical or horizontal,[16,25,26] similar to those of premotor and ocular motor neurons in the brainstem.[27,28] Such difference in spatial properties may suggest that the responses of FEF gaze-velocity neurons are organized in spatial coordinates (cf. reference 29), similar to MST neurons,[13] as opposed to those of floccular gaze-velocity Purkinje cells assumed to be organized in the cranial coordinates defined by the vestibular apparatus and eye muscles.[26,30,31]

In summary, our results suggest that the region near the arcuate sulcus coordinates its various inputs to provide signals for target velocity in space[4,5] and accurate gaze-velocity command during pursuit-vestibular interactions.

ACKNOWLEDGMENTS

Supported in part by CREST of JST, the Japanese Ministry of Education, Science and Culture (09268201, 09680806, 10164202, 09670971), and Marna Cosmetics. We gratefully thank Dr. Chris R.S. Kaneko of the Primate Center, University of Washington, Seattle, U.S.A., for his valuable comments on the early version of the manuscript.

REFERENCES

1. WILSON, V. J. & G. MELVILL JONES. 1979. Mammalian Vestibular Physiology. Plenum Press. New York.
2. BÜTTNER, U. & V. HENN. 1981. Circularvection: psychophysics and single-unit recordings in the monkey. Ann. N.Y. Acad. Sci. **374**: 274–283.
3. FUKUSHIMA, K. 1997. Corticovestibular interactions: anatomy, electrophysiology, and functional considerations. Exp. Brain Res. **117**: 1–16.
4. ROBINSON, D. A. 1982. A model of cancellation of the vestibulo-ocular reflex. In Functional Basis of Ocular Motility Disorders, G. Lennerstrand, D. S. Zee, and E. L. Keller, Eds.: 5–13. Pergamon press, Oxford.
5. YOUNG, L. R. 1971. Pursuit eye tracking movements. In The Control of Eye Movements, P. Bach-y-rita, C. C. Collins, and J. E. Hyde, Eds.: 429–443. Academic Press. New York.
6. GOLDBERG, M. E. & M. A. SEGRAVES. 1989. In The Neurobiology of Saccadic Eye Movements, R. H. Wurtz and M. E. Goldberg, Eds. : 283–313. Elsevier. Amsterdam.
7. MACAVOY, M., J. P. GOTTLIEB & C. BRUCE. 1991. Smooth-pursuit eye movement representation in the primate frontal eye field. Cerebral Cortex **1**: 95–102.
8. GOTTLIEB, P. J., M. G. MACAVOY & C. J. BRUCE. 1994. Neural responses related to smooth-pursuit eye movements and their correspondence with electrically elicited smooth eye movements in the primate frontal eye field. J. Neurophysiol. **72**: 1634–1653.
9. TANAKA, M. & K. FUKUSHIMA. 1998. Neuronal responses related to smooth pursuit eye movements in the periarcuate cortical area of monkeys. J. Neurophysiol. **80**: 28–47.
10. FUKUSHIMA, K., J. FUKUSHIMA, S. CHIN, H. TSUNEKAWA & C. R. S. KANEKO. 1996. Cross axis vestibulo-ocular reflex induced by pursuit training in alert monkeys. Neurosci. Res. **25**: 255–265.
11. FUKUSHIMA, K., T. OHASHI, J. FUKUSHIMA & C. R. C. KANEKO. 1995. Discharge characteristics of vertical vestibular and saccade neurons in the rostral midbrain of alert cats. J. Neurophysiol. **73**: 2129–2143.
12. KAWANO, K., M. SASAKI & M. YAMASHITA. 1984. Response properties of neurons in posterior parietal cortex of monkey during visual-vestibular stimulation. J. Neurophysiol. **51**: 340–351.
13. THIER, P. & R. G. ERICKSON. 1992. Responses of visual-tracking neurons from cortical area MST-l to visual, eye and head motion. Eur. J. Neurosci. **4**: 539–553.
14. SCHLAG, J. & M. SCHLAG-REY. 1986. Role of the central thalamus in gaze control. Prog. Brain Res. **64**: 191–201.
15. MILES, F. A. & J. H. FULLER. 1975. Visual tracking and the primate flocculus. Science **189**: 1000–1003.
16. LISBERGER, S. G. & A. F. FUCHS. 1978. Role of primate flocculus during rapid behavioral modification of vestibuloocular reflex. I. Purkinje neuron activity during visually guided horizontal smooth-pursuit eye movements and passive head rotation. J. Neurophysiol. **41**: 733–763.
17. SUZUKI, D. A. & E. L. KELLER. 1988. The role of the posterior vermis of monkey cerebellum in smooth-pursuit eye movement control. I. Eye and head movement-related activity. J. Neurophysiol. **59**: 1–18.
18. SATO, H. & H. NODA. 1992. Posterior vermal Purkinje cells in macacques responding during saccades, smooth pursuit, chair rotation and/or optokinetic stimulation. Neurosci. Res. **12**: 583–595.
19. SCUDDER, C. A. & A. F. FUCHS. 1992. Physiological and behavioral identification of vestibular nucleus neurons mediating the horizontal vestibuloocular reflex in trained rhesus monkeys. J. Neurophysiol. **69**: 244–264.
20. KAWANO, K., M. SHIDARA, Y. WATANABE & S. YAMANE. 1994. Neural activity in cortical area MST of alert monkey during ocular following responses. J. Neurophysiol. **71**: 2305–2324.
21. NEWSOME, W. T., R. H. WURTZ & H. KOMATSU. 1988. Relation of cortical areas MT and MST to pursuit eye movements. II. Differentiation of retinal from extraretinal inputs. J. Neurophysiol. **60**: 604–620.
22. KASE, M., H. NODA, D. A. SUZUKI & D. C. MILLER. 1979. Target velocity signals of visual tracking in vermal Purkinje neurons of the monkey. Science **205**: 717–720.
23. TUSA, R. J. & L. G. UNGERLEIDER. 1988. Fiber pathways of cortical areas mediating smooth pursuit eye movement in monkeys. Ann. Neurol. **23**: 174–183.

24. SHI, D., H. R. FRIEDMAN & C. J. BRUCE. 1997. Deficits in smooth pursuit eye movements from muscimol injection in the primate frontal eye field. Soc. Neurosci. Abstr. **23**: 473.
25. SHIDARA, M. & K. KAWANO. 1993. Role of Purkinje neurons in the ventral paraflocculus in short-latency ocular following responses. Exp. Brain Res. **93**: 185–195.
26. KRAUZLIS, R. J. & S. G. LISBERGER. 1996. Directional organization of eye movement and visual signals in the floccular lobe of the monkey cerebellum. Exp. Brain Res. **109**: 289–302.
27. FUKUSHIMA, K. 1991. The interstitial nucleus of Cajal in the midbrain reticular formation and vertical eye movement. Neurosci. Res. **10**: 159–187.
28. ROBINSON, D. A. 1982. The use of matrices in analyzing the three-dimensional behavior of the vestibulo-ocular reflex. Biol. Cybern. **46**: 53–66.
29. ANDERSEN, R. A. 1996. Coordinate transformation and motor planning in posterior parietal cortex. *In* The Cognitive Neurosciences, M.S. Gazzaniga, Ed.: 519–533. MIT Press, Cambridge.
30. FUKUSHIMA, K., E. V. BUHARIN & J. FUKUSHIMA. 1993. Responses of floccular Purkinje cells to sinusoidal vertical rotation and effects of muscimol infusion into the flocculus in alert cats. Neurosci. Res. **17**: 297–305.
31. FUKUSHIMA, K., S. CHIN, J. FUKUSHIMA & M. TANAKA. 1996. Simple-spike activity of floccular Purkinje neurons responding to sinusoidal vertical rotation and optokinetic stimuli in alert cats. Neurosci. Res. **24**: 275–289.

Short-Latency Visual Stabilization Mechanisms that Help to Compensate for Translational Disturbances of Gaze

F. A. MILES[a]

Laboratory of Sensorimotor Research, National Eye Institute, National Institutes of Health, Building 49, Room 2A50, 49 Convert Drive, Bethesda, Maryland 20892, USA

ABSTRACT: Recent studies in primates have revealed short-latency visual tracking mechanisms that help to stabilize the eyes during translational disturbances of the observer, and so operate as backups to otolith-mediated vestibulo-ocular reflexes. One such mechanism generates version eye movements to help stabilize gaze when the moving observer looks off to one side, utilizing binocular disparity to help single out the images in the plane of fixation (*ocular following*). Two others generate vergence eye movements to help maintain binocular alignment on objects that lie ahead: one responds to the radial patterns of optic flow (*radial-flow vergence*) and the other to the changes in binocular parallax (*disparity vergence*). Accumulating evidence suggests that, despite their short latency, all are mediated by the medial superior temporal area of cortex.

INTRODUCTION

Most studies concerned with the visual stabilization of gaze have used stimuli in the form of an optokinetic drum or a planetarium-like rotating array of spots, with the object of simulating the optic flow associated with shortcomings of the canal–ocular reflexes during head rotations. The presumption that *rotational* disturbances were the adequate stimulus for the visual reflexes as well as for the canal–ocular reflex was nicely reinforced by the finding in rabbits that the optokinetic system was organized in canal coordinates,[1] an arrangement that was assumed to foster the veridical summation of the visual and vestibular inputs. It is possible that the geometry is in fact dictated by the pulling directions of the eye muscles rather than the orientations of the canals[2] but, regardless, there seems to be no question that the optokinetic system of the lateral-eyed rabbit is organized in rotational coordinates. Curiously, such a shared organization for canal–ocular and optokinetic reflexes has yet to be demonstrated in primates, an oversight worthy of a doctoral thesis. In recent years, however, it has become apparent that primates like ourselves with frontal eyes have (in addition?) visual and vestibular reflexes for which *translational* disturbances are the adequate stimulus. While my major concern here is with these newly discovered visual mechanisms in primates, it is useful to first touch on some aspects of vestibulo-ocular function in primates during translational disturbances.

---

[a]Address for telecommunication: Phone: 301/496-2455; fax: 301/402-0511; e-mail: fam@lsr.nei.nih.gov

THE TRANSLATIONAL VESTIBULO-OCULAR REFLEXES

A Visual or Gaze-Centric View

The canal–ocular and otolith–ocular reflexes are often referred to as the angular and linear vestibulo-ocular reflexes (VORs) in accordance with the adequate stimulus for activation of the endorgans, angular and linear accelerations, respectively. There are two linear VORs, however, one compensating for translations and the other for changes in the orientation of the head with respect to gravity, and I shall refer to them as the translational vestibulo-ocular reflex (TVOR) and the orientational vestibulo-ocular reflex (OVOR), respectively. In keeping with these functional descriptors, I shall refer to the canal-ocular reflex as the rotational vestibulo-ocular reflex (RVOR).

The available evidence suggests that the TVOR operates in all directions,[3–5] stimuli and responses typically being defined in head-centric coordinates with naso–occipital, interaural, and dorsoventral components. Of course, this is simply a convenient descriptive convention, and the reference frame used by the nervous system is unknown. However, the visual backup mechanisms appear to be organized in a *gaze-centric* reference frame, and it simplifies matters to assume that this organization is shared by the otolith reflexes dealing with translational disturbances. One visual backup, *ocular following*, generates conjugate (version) eye movements to compensate for translational disturbances that are orthogonal to the direction of gaze (as when the moving observer looks off to one side) and, I suggest, works in concert with a labyrinthine mechanism that I shall term the gaze-orthogonal-TVOR (GO-TVOR). A second visual backup, *radial-flow vergence*, generates disjunctive (vergence) eye movements to compensate for translational disturbances that are isogonal to the direction of gaze (as when the moving observer looks in the direction of heading) and, I suggest, works in concert with a labyrinthine mechanism that I shall term the gaze-isogonal-TVOR (GI-TVOR). Note that version (Vs), the average position of the two eyes [(L + R)/2], and vergence (Vg), the difference in the positions of the two eyes [L − R], provide a complete description of binocular eye movements, the position of the left eye being given by [Vs + (Vg/2)] and of the right eye by [Vs - (Vg/2)]. While drawing a clear distinction between version and vergence mechanisms, I do not wish to imply that they are totally independent subsystems.[6]

Dependence on Viewing Distance

In the case of the RVOR, if we ignore the separation between the axes of eye rotation and the axis of head rotation, perfect compensation would require that any head rotation (the input) be exactly matched by an equal and opposite eye rotation (the output), in which event the gain (given by the ratio, output/input) would be simply unity for *all* head rotations. The desired eye rotations here are always conjugate, hence the expected output is always version. In contrast, for the TVOR to be optimally effective, its gain should accord with the proximity of the object of interest, nearby objects necessitating much greater compensatory eye movements than distant ones in order for their retinal images to be stabilized during translation. If the observer looks off to one side during translation so that his or her gaze is orthogonal to the direction of heading (pure activation of the GO-TVOR), then the required compensatory eye movements are conjugate (version) and should have a gain that is inversely proportional to the viewing distance. There is ample evidence that this is, indeed, the case.[3,5,7–12] On the other hand, when the observer looks ahead during translation so that his/her (cyclopean) gaze is in the direction of heading (pure activation of the GI-TVOR), then the required compensatory eye movements are disconjugate (vergence) and should have a gain that is inversely proportional to the square of the viewing distance. Although it is intuitively obvious that the instantaneous vergence angle required for binocular alignment is inversely proportional to the viewing distance, it is not so readily apparent that, as the observer moves toward or away from the object of regard, this vergence angle must change at a rate that is inversely proportional to the square of the viewing distance: if we assume an observer with interpupillary separation, S, fixating an object at a distance, D, then as the observer moves a distance, ΔD, toward that object, the required rate of increase in the vergence angle with respect to time is approximated by $[(s/(D-\Delta D))-s/D]/\Delta t$, which simplifies in the limit to $(s/D^2)* \Delta D/\Delta t$. Perusal of the available data suggests that this might be the case,[3] although no formal attempt has been made to fit such a function to the data. I shall return to this issue of the dependence on viewing distance later.

OPTIC FLOW PATTERNS ASSOCIATED WITH ROTATIONS AND TRANSLATIONS

The vestibular system's decomposition of head movements into rotational and translational components results directly from the contrasting physical properties of the endorgans, the semicircular canals and the otolith organs, respectively.[13] However, there is no such decomposition of the optic flow by the visual endorgans: the two retinas see the full consequences of both translational and rotational disturbances of the observer. Any visual decomposition must be done by signal processing within the CNS, presumably utilizing the different *patterns* of optic flow associated with rotational and translational disturbances of the observer. An observer who undergoes pure rotation (without compensating) experiences *en masse* motion of his or her entire visual world, the direction and the speed of the optic flow at all points being dictated solely by the observer's motion. The overall pattern of optic flow resembles the lines of latitude on a globe (FIG. 1A) but, of course, the

---

FIGURE 1. Patterns of optic flow experienced by a (passive) moving observer. (**A**) The retinal optic flow can be considered to be distributed over the surface of a sphere and created by projection through a vantage point at the center. Here, the observer rotates about this vantage point and the pattern of flow resembles the lines of latitude on a globe. In reality things are never as simple as this, voluntary head turns occurring about an axis some distance behind the eyes so that the latter always undergo some slight translation. Such second-order effects are ignored here. (From Miles *et al.*,[19] with permission.) (**B**) A cartoon showing the observer's limited field of view and the kind of motion experienced during rotation about a vertical axis as the observer looks straight out to the side. The speed of optic flow is greatest at the center ("equator") and decrements as the cosine of the angle of latitude. However, both the pattern and the speed of the optic flow at all points are determined entirely by the observer's motion—the 3-D structure of the scene is irrelevant. (From Miles,[25] with permission.) (**C**) The pattern of optic flow experienced by the translating observer resembles the lines of longitude on a globe. (From Miles *et al.*[19] with permission.) (**D**) A cartoon showing the centrifugal pattern of optic flow experienced by the observer who looks in the direction of heading—the black dot at the foot of the mountain. (From Busettini *et al.*,[34] with permission.) (**E**) The optic flow experienced by the moving observer who looks off to the right but makes no compensatory eye movements so that the visual scene appears to pivot about the distant mountains (effective infinity). The speed of image motion is inversely proportional to the viewing distance. (After Miles *et al.*,[21] with permission.) (**F**) Again, the observer looks off to one side but here attempts to stabilize the retinal image of a particular object in the middle ground (tree), necessitating that he or she track to compensate for his or her own motion, thereby reversing the apparent motion of the more distant objects and creating a swirling pattern of optic flow. The scene now appears to pivot about the tree. (After Miles *et al.*,[21] with permission.)

observer's restricted field of view means that only a portion will be visible at any given time (as in FIG. 1B). In contrast, the optic flow experienced by the observer who undergoes pure translation (without compensating) consists of streams of images emerging from a focus of expansion straight ahead and disappearing into a focus of contraction behind, the overall pattern resembling the lines of longitude on a globe (see FIG. 1C). As during rotational disturbances, the *direction* of flow at any given point during translational disturbances depends solely on the motion of the observer. However, the *speed* of the flow at

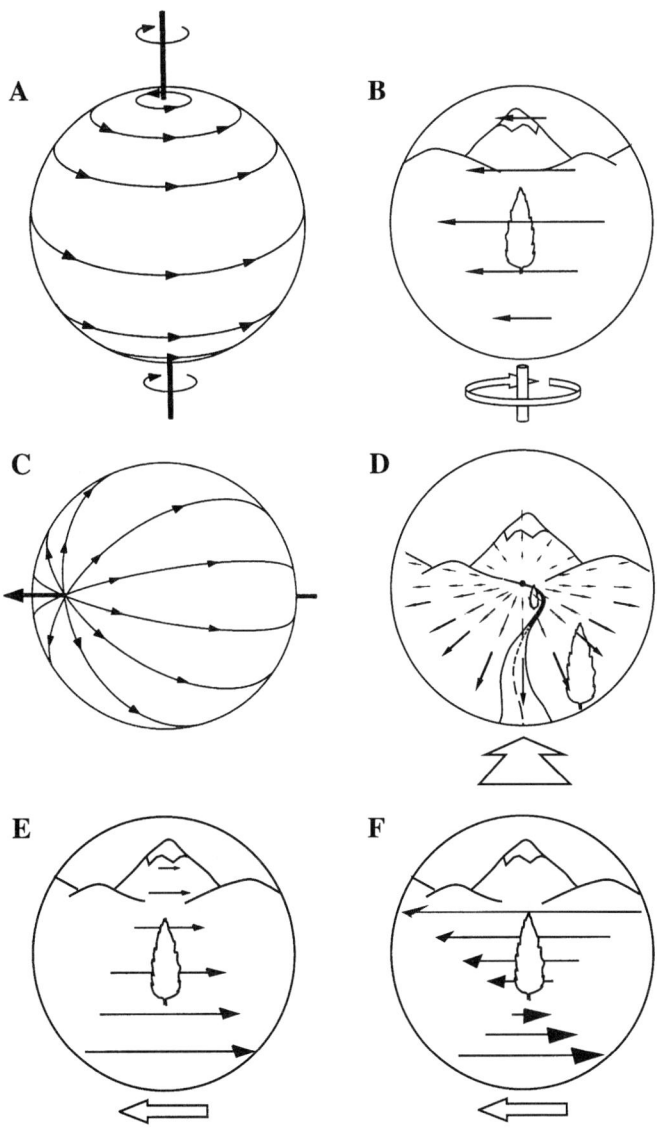

any given point during translations depends not only on the observer's speed of motion but also on the *viewing distance* at that location as nearby objects move across the field of view much more rapidly than more distant ones: *motion parallax*.[14,15] A consequence of this is that, during translation, eye movements can stabilize only the images in one particular depth plane, and if the observer makes no active attempt to compensate, then only the images of the most distant objects are stable. Again, given the observer's restricted field of view, the pattern of motion actually experienced depends on the direction in which the observer chooses to look: straight ahead the observer sees a radially expanding world (as in FIG. 1D), whereas off to one side the sensation is of the visual world pivoting around the far distance as images move across the field of view in inverse proportion to their viewing distance (as in FIG. 1E). If the observer undergoing translation wishes to fixate an object in the middleground off to one side, then he or she must now compensate for the motion and, in so doing, his or her visual world will now rotate about this new object (as in FIG. 1F).

VISUAL STABILIZATION

The traditional approach to the visual stabilization of gaze is represented by the optokinetic response, which has two distinct components: an early component (OKNe) with brisk dynamics and a delayed component (OKNd) with sluggish dynamics.[16]

Images that Lie Off to One Side: Ocular Following

Recent studies of OKNe have often employed large moving patterns backprojected onto a translucent tangent screen facing the observer because it offers much better control of the stimulus parameters and the responses evoked in this situation, which have very short latencies (<60 ms in monkeys and <85 ms in humans), have been termed "ocular following."[7,17] My colleagues and I have suggested that OKNd evolved as a visual backup to the RVOR, dealing with residual disturbances of gaze associated with rotations of the observer, whereas ocular following/OKNe evolved as a visual backup to the TVOR, dealing with residual disturbances of gaze associated with lateral translations of the observer (such as in FIG. 1E, 1F).[11,18–26] Initial support for this idea rested largely on two observations: first, changes in the gain of the RVOR (resulting from exposure to magnifying or minifying spectacles) were associated with proportional changes in the gain of OKNd but *not* of OKNe.[27] Second, changes in the gain of the TVOR (resulting from changes in the viewing distance) were associated with proportional changes in the gain of ocular following.[7,11,18] Such changes in the gains of the visually driven responses were attributed to changes in central pathways that are shared with the vestibular reflexes, presumably reflecting functional synergies between the RVOR and OKNd on the one hand, and the GO-TVOR and OKNe on the other. The block diagrams in A and B of FIGURE 2 illustrate the two hypothesized visuovestibular mechanisms dealing independently with rotational and translational disturbances, and indicate the shared gain elements: in the case of the RVOR this gain element mediates long-term adaptive gain control, whereas in the case of the GO-TVOR this gain element gives the reflex its dependence on (the inverse of) the current viewing distance. Of course, this latter assumes some internal representation of viewing distance, perhaps using efference copy of cues such as vergence and/or accommodation.

Recent experiments indicate that ocular following has special built-in features for dealing with the visual problems posed when the moving observer looks off to one side, as in parts E and F of FIGURE 1. The visual task confronting the visual stabilization mechanisms here is to single out the motion of particular elements in the scene, such as the mountain in FIGURE 1E and the tree in FIGURE 1F, and ignore all of the competing motion elsewhere. One way to achieve this would be to use attentional focusing mechanisms to spotlight the target of interest. Such mechanisms exist and are used by the so-called *pursuit system,* but have the limitation that they require high-level executive decisions to select the image to be tracked, and this of necessity is very time-consuming. A more rapid alternative might be to use low-level stereomechanisms that perform rapid parallel processing of binocular images, effectively sorting them on the basis of the depth plane that they occupy. This idea uses the fact that the object on which the two eyes are aligned (such as the mountain in FIG. 1E or the tree in FIG. 1F), which is said to reside in the plane of fixation, is imaged at corresponding positions on the two retinas. In contrast, objects that are nearer or farther than the plane of fixation have images that occupy noncorresponding positions on the two retinas and are said to have "binocular disparity." Clearly, a highly reliable algorithm for stabilizing gaze on objects in the plane of fixation would be to track only those images that lack disparity and to ignore all others. This would require neurons sensitive to both motion and binocular disparity, a combination known to be commonplace in the dorsal stream of cortex.[28] Early support for a stereo algorithm was the finding that optokinetic responses deteriorate when the driving images have binocular disparity.[29,30] However, high-level processing, perhaps involving selective attention, may have been a factor in these studies, which examined the closed-loop, steady-state responses. More recent experiments indicate that the early ocular following responses of both monkeys and humans can be disrupted by disparity before there can have been time for attentional mechanisms to operate.[31] This is consistent with the notion that the motion detectors driving ocular following are also disparity selective. Thus, the ocular-following system helps to stabilize gaze on objects of interest not by selecting a particular one but by stabilizing the image of any object that happens to lie close to the plane of fixation, an implicit assumption therefore being that this plane contains the objects likely to be of most interest. This means that the time-consuming process of selecting the object of interest rests with the oculomotor subsystems that bring images into the plane of fixation, that is, the saccadic system working in concert with the vergence system. These latter systems redirect gaze to objects using higher-level criteria, whereas ocular following relies on low-level rapid parallel filters. Thus, the general concept is of low-level reflex systems stabilizing whatever images the high-level systems happen to bring into the plane of fixation.

There is strong evidence that ocular following derives at least some of its input from the medial superior temporal (MST) region of the cortex. Thus, chemical lesions in MST result in impairments of even the earliest components of ocular following[32] and single-unit recordings in this region indicate the presence of many directionally selective neurons that discharge in close relation to the large-field, high-speed motion stimuli that are optimal for eliciting these motor responses.[33]

Images that Lie Ahead: Radial-Flow Vergence

The visual challenge considered in the previous section on ocular following was that confronting the moving observer who looks off to one side. I now consider the gaze stability problems of the moving observer who looks in the direction of heading and so experiences radial patterns of optic flow such as that featured in FIGURE 1D. Insofar as the radial pattern of flow is associated with a change in viewing distance, it should be accompanied by increases in the vergence angle of the two eyes in order for the object of interest in the scene ahead to stay imaged on both foveas. Recent experiments on humans[34] have indicated that radial optic flow elicits vergence eye movements at latencies that are closely comparable with the ultrashort values mentioned earlier for human

A: rotation

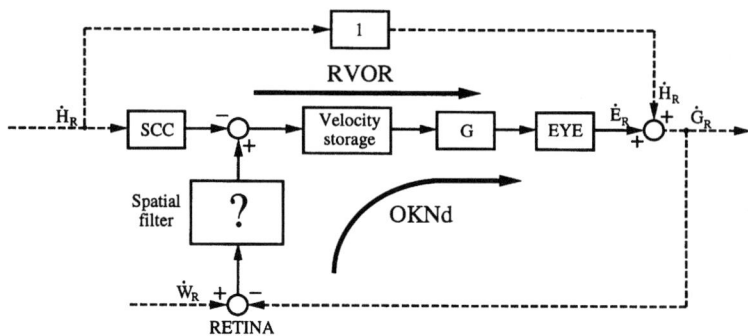

B: translation orthogonal to direction of gaze

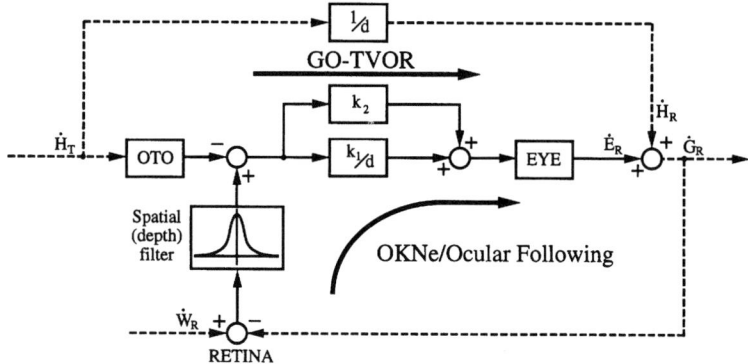

C: translation isogonal to direction of gaze

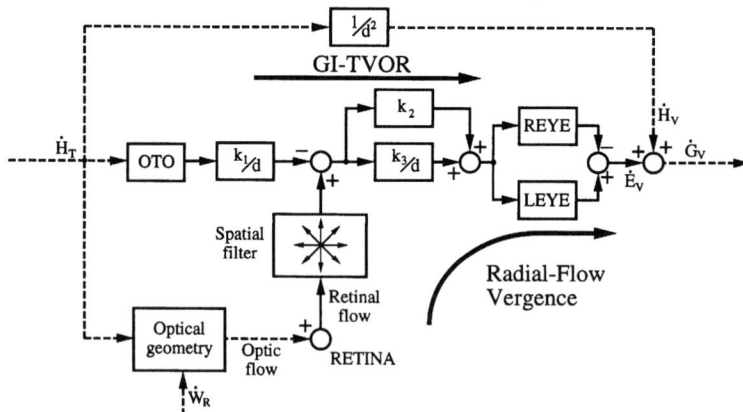

FIGURE 2. Block diagrams showing the proposed linkages between the visual and vestibular reflexes operating to stabilize gaze. (**A**) The open-loop RVOR and the closed-loop OKNd generate version eye movements, $\dot{E}_R$, that compensate for rotational disturbances of the head, $\dot{H}_R$. These reflexes share (a) a velocity storage element, which is responsible for the slow build-up in OKN and the gradual decay in RVOR with sustained rotational stimuli, and (b) a variable gain element, G, that mediates long-term regulation of RVOR gain. The characteristics of the spatial filter remain to be determined in primates. (After Miles et al.,[20] with permission.) (**B**) The open-loop GO-TVOR and the closed-loop OKNe/ocular following generate version eye movements, $\dot{E}_R$, that compensate for translational disturbances of the head that are orthogonal to the direction of gaze, $\dot{H}_T$, and affect gaze in inverse proportion to the viewing distance, d. These reflexes share (a) a variable gain element, k_1/d, that gives them their dependence on proximity, and (b) a fixed gain element, k_2, that generates a small response irrespective of proximity. The spatial filter is centered on the plane of fixation, effectively rejecting image motion in other depth planes. (After Schwarz et al.,[11] with permission.) (**C**) The open-loop GI-TVOR and the open-loop radial-flow–vergence mechanism generate vergence eye movements to help maintain binocular alignment on the object(s) of regard during translational disturbances of the head in the direction of gaze, $\dot{H}_T$. The GI-TVOR is assumed to have an overall gain that is inversely proportional to the square of the viewing distance, as required by the optical geometry, and this is achieved by two gain elements in series, each having a gain inversely proportional to viewing distance, d. Radial-flow vergence affects gaze in inverse proportion to the viewing distance, d, and this is achieved by having the radial-flow input share the later part of the GI-TVOR pathway. A fixed gain element, k_2, once more generates a small response (this time, radial-flow vergence) irrespective of proximity. The spatial filter is tuned to radial patterns of optic flow. (After Yang et al.,[38] with permission.) *Dashed lines* represent physical links: $\dot{H}_T$, head velocity in linear coordinates; $\dot{H}_R$, $\dot{E}_R$, $\dot{G}_R$ and $\dot{W}_R$, velocity of head, eyes (in head), gaze and visual surroundings, respectively, in angular coordinates; $\dot{H}_V$, $\dot{E}_V$, and $\dot{G}_V$, velocity of head, eyes (in head) and gaze in depth (equivalent vergence) coordinates. SCC, semicircular canals; OTO, otolith organs.

ocular following (<85 ms). Centrifugal (expanding) flow, which signals a forward approach and hence a decrease in the viewing distance, resulted in increased convergence, and centripetal (contracting) flow, which signals the converse, resulted in decreased convergence.[34] Interestingly, the apparent changes in the sizes of objects as the observer moves closer or farther away also elicit convergence, but only at much longer, pursuitlike latencies generally estimated to be in excess of 200 ms.[34–36]

The clear suggestion here is that the brain is able to sense the radial pattern of optic flow and to infer from this that there has been a change in viewing distance. However, a characteristic of the ocular responses to these radial-flow patterns is that each eye always moves in the direction of the net motion vector in the nasal hemifield, and this allows an alternative and less interesting explanation for the responses: the vergence might result from monocular tracking, in which each eye tracks only the motion that it sees and with a preference for the nasal hemifields. For example, with centrifugal flow the net motion vector in the nasal hemifields is toward the nose and each eye moves in that direction: hence, the increased convergence. That this was *not* the explanation was apparent from the observation that binocular vergence responses persisted, albeit weaker, when the nasal hemifields were masked off, consistent with the idea that the vergence responses result from a true parsing of the radial pattern of flow. Because latencies are so short, it is reasonable to assume that the system must depend on parallel processing to sense the pattern of flow. This is consistent with the idea that there are neurons or networks that act like templates or tuned filters so as to respond to radial patterns of optic flow.[37] Once again the general concept is of a relatively low-level reflex system responding appropriately to whatever region of the optic flow field is brought into view by the high-level saccadic system.

Recent experiments have shown that radial-flow vergence responses are a linear function of the preexisting vergence angle, and hence would be expected to share ocular following's dependence on the reciprocal of the viewing distance.[38] The etiology of this dependence on viewing distance is assumed to be similar to that proposed for ocular following—central pathways that are shared with the synergistic translational vestibular reflexes, here the GI-TVOR. The block diagram in C of FIGURE 2 illustrates this hypothesized linkage. The GI-TVOR is assumed to have an overall gain that is inversely proportional to the square of the viewing distance as discussed earlier, and this is achieved by two gain elements in series, each having a value that varies with the inverse of the viewing distance, d. However, because the radial-flow vergence responses modulate only in inverse proportion to the viewing distance, it is assumed that only the later part of the GI-TVOR pathway is shared.

There is extensive evidence that area MST in the monkey's cortex contains neurons that are selectively sensitive to radial optic flow patterns such as those now known to evoke vergence eye movements at ultrashort latencies.[39–47] In fact, MST is the *first* stage in the so-called dorsal (motion) pathway at which *global* flow is encoded at the level of single cells: at earlier stages, such as MT, individual cells have much smaller receptive fields and encode only local motion.[44,48–52] It is tempting to assume that radial-flow vergence, like its sister reflex, ocular following, is also mediated by MST.

Other Short-Latency Mechanisms?

There is at least one more short-latency visual tracking mechanism that can be revealed with large-field stimuli. This third mechanism generates vergence eye movements in response to binocular disparity and so would be expected to help maintain binocular alignment during motion in depth.[53] However, this is a secondary function rather than a primary one because this disparity vergence mechanism operates to maintain the vertical—as well as the horizontal—alignment of the two eyes, a function clearly unrelated to motion of the observer *per se*. Nonetheless, this reflex has much in common with the other two that have been discussed, in addition to an ultrashort latency. Thus, like ocular following[7,54] and radial-flow vergence,[34] this disparity–vergence mechanism shows dependence on a prior saccade, whereby stimuli presented in the immediate wake of a saccade are much more effective than the same stimuli presented some time later; this post-saccadic enhancement is largely visual, resulting from the shift in the image of the world on the retina induced by the saccade and can be simulated by saccadelike shifts of the visual scene. There is also evidence from single-unit recordings that MST contains neurons that discharge closely in relation to these disparity vergence responses.[55] Last, as with ocular following and radial-flow vergence, the ultrashort latency suggests that subjects respond before they can even be aware that there has been a stimulus, that is, the responses are independent of perception. In this regard, it has recently been shown that the short-latency disparity–vergence responses can be produced by applying the disparity steps to patterns that do not give rise to the perception of depth because they have opposite contrast at the two eyes,[56] so-called anticorrelated images.[57] This is consistent with the idea that these short-latency vergence responses derive their visual input from an early stage of cortical processing prior to the level at which depth percepts are elaborated.

CLOSING REMARKS

The three visual stabilization mechanisms share a number of features in addition to their ultrashort latencies, and it has been argued that all are involved in generating eye movements to compensate for translational disturbances, two being highly specialized for this purpose. This has led to the suggestion that these mechanisms constitute a family of reflexes, and Table 1 summarizes our current knowledge of their fundamental similarities and differences.

TABLE 1. **Major features of the three visual stabilization mechanisms**

	X/Y-Translation	Z-Translation (Depth)	
	Ocular following	Radial flow vergence	Disparity vergence
Function	Stabilizes gaze against motion in fixation plane (tolerates position errors)	Stabilizes gaze against motion in depth (tolerates position errors)	Eliminates residual vergence errors
Input	Binocular motion in plane of fixation	Radial optic flow (monocular/binocular)	Binocular disparity (local matches)
Output (binocular)	Horizontal/vertical version	Transient horizontal vergence	Horizontal/vertical vergence
Control type	Velocity feedback	Velocity feedforward	Position (?) feedback
Latency	<60 ms monkeys <85 ms humans	? ms monkeys <85 ms humans	<60 ms monkeys <85 ms humans
Independent of perception	?	?	Yes
Postsaccadic enhancement	Yes (part visual reafference)	Yes (part visual reafference)	Yes (part visual reafference)
Adaptive gain control	Yes	?	?
Dependence on viewing distance	$\propto \dfrac{1}{\text{Distance}}$	$\propto$ Vergence angle	?
Neural mediation	MST	MST?	MST?

Source: After Miles,[26] with permission.

REFERENCES

1. SIMPSON, J. I. 1984. The accessory optic system. Ann. Rev. Neurosci. **7**: 13–41.
2. SIMPSON, J. I. & W. GRAF. 1985. The selection of reference frames by nature and its investigators. In Adaptive Mechanisms in Gaze Control: Facts and Theories, A. Berthoz and G. Melvill Jones, Eds.: 3–16. Elsevier/North-Holland. Amsterdam.
3. PAIGE, G. D. & D. L. TOMKO. 1991. Eye movement responses to linear head motion in the squirrel monkey. II. visual-vestibular interactions and kinematic considerations. J. Neurophysiol. **65**: 1183–1196.
4. TOMKO, D. L. & G. D. PAIGE. 1992. Linear vestibuloocular reflex during motion along axes between nasooccipital and interaural. Ann. N.Y. Acad. Sci. **656**: 233–241.
5. TELFORD, L., S. H. SEIDMAN & G. D. PAIGE. 1997. Dynamics of squirrel monkey linear vestibuloocular reflex and interactions with fixation distance. J. Neurophysiol. **78**: 1775–1790.
6. COVA, A. & H. L. GALIANA. 1995. Providing distinct vergence and version dynamics in a bilateral oculomotor network. Vision Res. **35**: 3359–3371.
7. BUSETTINI, C., F. A. MILES, U. SCHWARZ & J. R. CARL. 1994. Human ocular responses to translation of the observer and of the scene: dependence on viewing distance. Exp. Brain Res. **100**: 484–494.
8. BUSH, G. A. & F. A. MILES. 1996. Short-latency compensatory eye movements associated with a brief period of free fall. Exp. Brain Res. **108**: 337–340.
9. GIANNA, C. C., M. A. GRESTY & A. M. BRONSTEIN. 1997. Eye movements induced by lateral acceleration steps. Effect of visual context and acceleration levels. Exp. Brain Res. **114**: 124–129.

10. PAIGE, G. D. 1989. The influence of target distance on eye movement responses during vertical linear motion. Exp. Brain Res. **77**: 585–593.
11. SCHWARZ, U., C. BUSETTINI & F. A. MILES. 1989. Ocular responses to linear motion are inversely proportional to viewing distance. Science **245**: 1394–1396.
12. SCHWARZ, U. & F. A. MILES. 1991. Ocular responses to translation and their dependence on viewing distance. I. Motion of the observer. J. Neurophysiol. **66**: 851–864.
13. GOLDBERG, J. M. & C. FERNANDEZ. 1975. Responses of peripheral vestibular neurons to angular and linear accelerations in the squirrel monkey. Acta Otolaryngol. **80**: 101–110.
14. GIBSON, J. J. 1950. The Perception of the Visual World. Houghton Mifflin. Boston.
15. GIBSON, J. J. 1966. The Senses Considered as Perceptual Systems. Houghton Mifflin. Boston.
16. COHEN, B., V. MATSUO & T. RAPHAN. 1977. Quantitative analysis of the velocity characteristics of optokinetic nystagmus and optokinetic after-nystagmus. J. Physiol. (Lond.) **270**: 321–344.
17. MILES, F. A., K. KAWANO & L. M. OPTICAN. 1986. Short-latency ocular following responses of monkey. I. Dependence on temporospatial properties of the visual input. J. Neurophysiol. **56**: 1321–1354.
18. BUSETTINI, C., F. A. MILES & U. SCHWARZ. 1991. Ocular responses to translation and their dependence on viewing distance. II. Motion of the scene. J. Neurophysiol. **66**: 865–878.
19. MILES, F. A., U. SCHWARZ & C. BUSETTINI. 1991. The parsing of optic flow by the primate oculomotor system. *In* Representations of Vision: Trends and Tacit Assumptions in Vision Research, A. Gorea, Ed.: 185–199. Cambridge Univ. Press. Cambridge.
20. MILES, F. A., C. BUSETTINI & U. SCHWARZ. 1992. Ocular responses to linear motion. *In* Vestibular and Brain Stem Control of Eye, Head and Body Movements, H. Shimazu and Y. Shinoda, Eds.: 379–395. Springer-Verlag/Japan Scientific Societies Press. Tokyo.
21. MILES, F. A., U. SCHWARZ & C. BUSETTINI. 1992. The decoding of optic flow by the primate optokinetic system. *In* The Head-Neck Sensory-Motor System, A. Berthoz, W. Graf and P. P. Vidal, Eds.: 471–478. Oxford Univ. Press. New York.
22. MILES, F. A. & C. BUSETTINI. 1992. Ocular compensation for self motion: visual mechanisms. *In* Sensing and Controlling Motion: Vestibular and Sensorimotor Function. B. Cohen, D. L. Tomko and F. Guedry, Eds.: 220–232. Ann. N.Y. Acad. Sci. New York.
23. MILES, F. A. 1993. The sensing of rotational and translational optic flow by the primate optokinetic system. *In* Visual Motion and Its Role in the Stabilization of Gaze. F. A. Miles, J. Wallman, Eds.: 393–403. Elsevier. Amsterdam.
24. Miles, F. A. 1995. The sensing of optic flow by the primate optokinetic system. *In* Eye Movement Research: Mechanisms, Processes and Applications. J. M. Findlay, R. W. Kentridge and R. Walker, Eds.: 47–62. Elsevier. Amsterdam.
25. MILES, F. A. 1997. Visual stabilization of the eyes in primates. Curr. Opin. Neurobiol. **7**: 867–871.
26. MILES, F. A. 1998. The neural processing of 3-D visual information: Evidence from eye movements. Eur. J. Neurosci. **10**: 811–822.
27. LISBERGER, S. G., F. A. MILES, L. M. OPTICAN & B. B. EIGHMY. 1981. Optokinetic response in monkey: underlying mechanisms and their sensitivity to long-term adaptive changes in vestibuloocular reflex. J. Neurophysiol. **45**: 869–890.
28. MAUNSELL, J. H. R. & D. C. VAN ESSEN. 1983. Functional properties of neurons in middle temporal visual area of the macaque monkey. II. Binocular interactions and sensitivity to binocular disparity. J. Neurophysiol. **49**: 1148–1167.
29. HOWARD, I. P. & E. G. GONZALEZ. 1987. Human optokinetic nystagmus in response to moving binocularly disparate stimuli. Vision Res. **27**: 1807–1816.
30. HOWARD, I. P. & W. A. SIMPSON. 1989. Human optokinetic nystagmus is linked to the stereoscopic system. Exp. Brain Res. **78**: 309–314.
31. BUSETTINI, C., G. S. MASSON & F. A. MILES. 1996. A role for stereoscopic depth cues in the rapid visual stabilization of the eyes. Nature **380**: 342–345.
32. KAWANO, K., Y. INOUE, A. TAKEMURA, T. KITAMA & F. A. MILES. 1997. A cortically mediated visual stabilization mechanism with ultra-short latency in primates. *In* Parietal Lobe Contributions to Orientation in 3D Space, P. Thier and H.-O. Karnath, Eds.: 185–199. Springer-Verlag. Heidelberg.
33. KAWANO, K., M. SHIDARA, Y. WATANABE & S. YAMANE. 1994. Neural activity in cortical area MST of alert monkey during ocular following responses. J. Neurophysiol. **71**: 2305–2324.

34. BUSETTINI, C., G. S. MASSON & F. A. MILES. 1997. Radial optic flow induces vergence eye movements with ultra-short latencies. Nature **390**: 512–515.
35. ERKELENS, C. J. & D. REGAN. 1986. Human ocular vergence movements induced by changing size and disparity. J. Physiol. **379**: 145–169.
36. COHEN, G. A. & S. G. LISBERGER. 1996. Motion disparity and looming cues form hierarchical inputs for smooth pursuit of targets moving in 3 dimensions. Soc. Neurosci. Abstr. **22**: 964.
37. PERRONE, J. A. 1992. Model for the computation of self-motion in biological systems. J. Opt. Soc. Am. A **9**: 177–194.
38. YANG, D.-S., E. J. FITZGIBBON & F. A. MILES. 1999. Short-latency vergence eye movements induced by radial optic flow in humans: dependence on ambient vergence level. J. Neurophysiol. **81**: 945–949.
39. SAITO, H., M. YUKIE, K. TANAKA, K. HIKOSAKA, Y. FUKADA & E. IWAI. 1986. Integration of direction signals of image motion in the superior temporal sulcus of the macaque monkey. J. Neurosci. **6**: 145–157.
40. TANAKA, K. & H. SAITO. 1989. Analysis of motion of the visual field by direction, expansion/contraction, and rotation cells clustered in the dorsal part of the medial superior temporal area of the macaque monkey. J. Neurophysiol. **62**: 626-641.
41. TANAKA, K., Y. FUKADA & H. SAITO. 1989. Underlying mechanisms of the response specificity of expansion/contraction and rotation cells in the dorsal part of the medial superior temporal area of the macaque monkey. J. Neurophysiol. **62**: 642–656.
42. DUFFY, C. J. & R. H. WURTZ. 1991. Sensitivity of MST neurons to optic flow stimuli. I. A continuum of response selectivity to large-field stimuli. J. Neurophysiol. **65**: 1329–1345.
43. DUFFY, C. J. & R. H. WURTZ. 1991. Sensitivity of MST neurons to optic flow stimuli. II. Mechanisms of response selectivity revealed by small-field stimuli. J. Neurophysiol. **65**: 1346–1359.
44. LAGAE, L., H. MAES, S. RAIGUEL, D.-K. XIAO & G. A. ORBAN. 1994. Responses of macaque STS neurons to optic flow components: a comparison of areas MT and MST. J Neurophysiol. **71**: 1597–1626.
45. DUFFY, C. J. & R. H. WURTZ. 1995. Response of monkey MST neurons to optic flow stimuli with shifted centers of motion. J. Neurosci. **15**: 5192–5208.
46. LAPPE, M., F. BREMMER, M. PEKEL, A. THIELE & K.-P. HOFFMANN. 1996. Optic flow processing in monkey STS: a theoretical and experimental approach. J. Neurosci. **16**: 6265–6285.
47. PEKEL, M., M. LAPPE, F. BREMMER, A. THIELE & K.-P. HOFFMANN. 1996. Neuronal responses in the motion pathway of the macaque monkey to natural optic flow stimuli. NeuroReport **7**: 884-888.
48. VAN ESSEN, D. C., J. H. R. MAUNSELL & J. L. BIXBY. 1981. The middle temporal visual area in the macaque: myeloarchitecture, connections, functional properties and topographic organization. J. Comp. Neurol. **199**: 293–326.
49. MAUNSELL, J. H. R. & D. C. VAN ESSEN. 1983. Functional properties of neurons in middle temporal visual area of the macaque monkey. I. Selectivity for stimulus direction, speed, and orientation. J. Neurophysiol. **49**: 1127–1147.
50. ALBRIGHT, T. D. & R. DESIMONE. 1987. Local precision of visuotopic organization in the middle temporal area (MT) of the macaque. Exp. Brain Res. **65**: 582–592.
51. KOMATSU, H. & R. H. WURTZ. 1988. Relation of cortical areas MT and MST to pursuit eye movements. I. Localization and visual properties of neurons. J. Neurophysiol. **60**: 580–603.
52. ALBRIGHT, T. D. 1989. Centrifugal directional bias in the middle temporal visual area (MT) of the macaque. Visual Neurosci. **2**: 177–188.
53. BUSETTINI, C., F. A. MILES & R. J. KRAUZLIS. 1996. Short-latency disparity vergence responses and their dependence on a prior saccadic eye movement. J. Neurophysiol. **75**: 1392–1410.
54. KAWANO, K. & F. A. MILES. 1986. Short-latency ocular following responses of monkey. II. Dependence on a prior saccadic eye movement. J. Neurophysiol. **56**: 1355–1380.
55. TAKEMURA, A., Y. INOUE, K. KAWANO & F. A. MILES. 1997. Short-latency discharges in medial superior temporal area of alert monkeys to sudden changes in the horizontal disparity. Soc. Neurosci. Abstr. **23**: 1557.
56. MASSON, G. S., C. BUSETTINI & F. A. MILES. 1997. Vergence eye movements in response to binocular disparity without depth perception. Nature **389**: 283–286.
57. CUMMING, B. G. & A. J. PARKER. 1997. Responses of primary visual cortical neurons to binocular disparity without depth perception. Nature **389**: 280–283.

Linear Vestibular Self-Motion Signals in Monkey Medial Superior Temporal Area

F. BREMMER,[a] M. KUBISCHIK, M. PEKEL, M. LAPPE,
AND K.-P. HOFFMANN

*Department of Zoology & Neurobiology, Ruhr University Bochum,
D-44780 BOCHUM, Germany*

ABSTRACT: The present study was aimed at investigating the sensitivity to linear vestibular stimulation of neurons in the medial superior temporal area (MST) of the macaque monkey. Two monkeys were moved on a parallel swing while single-unit activity was recorded. About one-half of the cells (28/51) responded in the dark either to forward motion ($n = 10$), or to backward motion ($n = 11$), or to both ($n = 7$). Twenty cells responding to vestibular stimulation in darkness were also tested for their responses to optic flow stimulation simulating forward and backward self-motion. Forty-five percent (9/20) of them preferred the same self-motion directions, that is, combined visual and vestibular signals in a synergistic manner. Thirty percent (6/20) of the cells were not responsive to visual stimulation alone. The remaining 25% (5/20) preferred directions that were antialigned. Our results provide strong evidence that neurons in the MST area are at least in part involved in the processing of self-motion.

INTRODUCTION

When one is moving through a natural environment a number of sensory signals are generated that allow the nervous system to infer self-motion parameters. Many authors suggest a dominant role of cortical–visual motion areas MT and MST in the processing of visual self-motion signals.[1–3] Area MST integrates the MT output in a way that might be useful for the purpose of estimating the direction one is heading.[4] For a system involved in estimating self-motion, however, the ongoing rotational and linear head accelerations should also be taken into account. If this is true, vestibular information arising from the labyrinth organs (signaling head rotation) and from the otolith organs (signaling linear head acceleration) should also be present in area MST. Integration of head rotation signals in area MST has already been described.[5] The work reported here is the first to investigate the modulation of the responses of MST neurons during linear self-motion, that is, during linear vestibular stimulation.

---

[a]To whom correspondence may be addressed. Phone: 49 234 700 4369; fax: 49 234 709 4278; e-mail: bremmer@neurobiologie.ruhr-uni-bochum.de

METHODS

Animal Preparation and Data Acquisition

The procedures for monkey training, electrophysiological recordings, and data analysis were described in detail in a previous paper.[6] Briefly, monkeys were surgically prepared for recordings: under general anesthesia and sterile surgical conditions, each animal was implanted with a head-holding device. Two scleral search coils were implanted in order to monitor eye position according to the method published by Judge et al.[7] and were connected to a plug on top of the skull. A recording chamber for microelectrode penetrations was placed in a para-sagittal plane with the recording chamber tilted 60 deg with respect to the vertical. Recording chamber, eye coil plug, and head holder were all embedded in dental acrylic, which itself was connected to the skull by self-tapping screws. Analgesics were applied postoperatively and recording started no sooner than one week after surgery. All procedures were in accordance with published guidelines on the use of animals in research (European Communities Council Directive 86/609/EEC).

During experiments, animals were seated in a primate chair with the head fixed. Behavioral paradigms and data acquisition were controlled by a PC running the NABEDA software package (Dr. Martin Pekel). Location of area MST was based on electrode and chamber position, guided by prior NMR scans and by physiological criteria.[8] MST-specific response properties that were used included large receptive fields, often extending into the ipsilateral hemifield, direction selectivity to large-field stimuli, and responses to optic-flow stimuli. While one animal is still used in experiments, histological analysis from the first monkey verified that recording sites had been located in area MST.

Experimental Setup

A parallel swing was used to move the monkey (see FIG. 1A). Monkey chair and magnetic field were placed on the swing platform. A guide rail stabilized the whole system on the ground for precise backward and forward swinging. The sinusoidal forward–backward swing movement had a frequency of 0.25 Hz. Peak-to-peak amplitude was 0.5 m, with a maximum acceleration of 1.22 m/s2. The up–down movement caused by the parallel swing mechanism had a peak-to-peak amplitude of 0.031 m and a maximum acceleration of 0.076 m/s2.

A tangent screen for presenting visual stimuli was placed in front of the animal. It covered the central 90 deg by 90 deg of the visual field when the swing was closest (0.48 m) to the screen.

Vestibular and Visual Stimuli

Pure vestibular stimulation was tested in total darkness. All room lights were shut off and the swing was separated from the remaining part of the lab with a light-tight curtain. In some sessions, the monkey had to fixate a central target that was moved together with the swing (chair-mounted LED) in order to suppress the translational vestibulo-ocular reflex (tVOR). For the optic-flow stimulation, computer-generated visual stimuli (Performer 2.1 software running on a Silicon Graphics Workstation) were back-projected by a video projector (Electrohome 4100) while the monkey had to fixate a central target. One set of stimuli was an exact virtual replica of the swing movement (see below). Another set of stimuli simulated self-motion in either the horizontal or sagittal plane in a way that allowed the testing of all possible movement directions in a single trial (see FIG. 1B). These stimuli were a

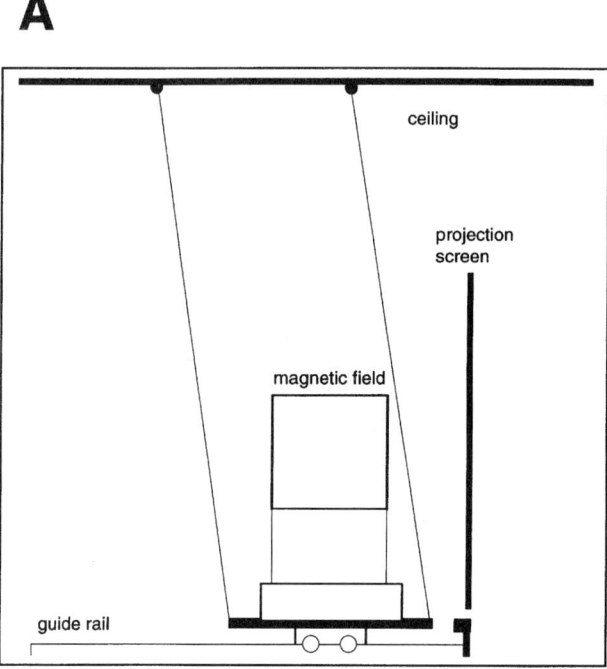

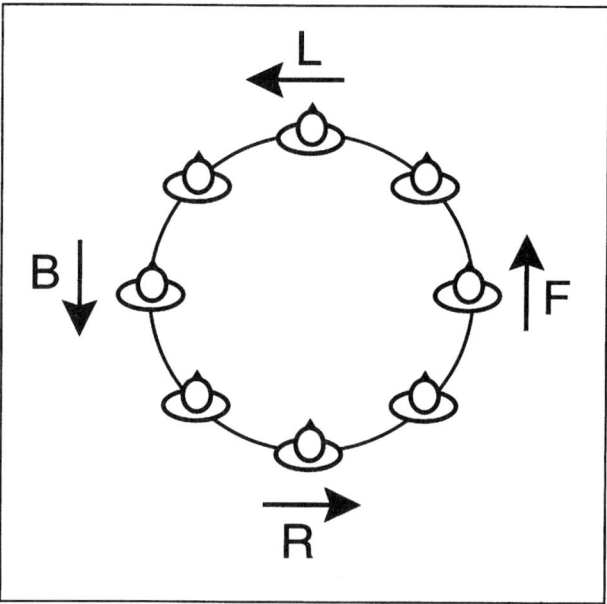

FIGURE 1. Vestibular and visual stimulation. **(A)** This panel shows schematically the experimental setup for vestibular stimulation. During the experiments, the head-fixed monkey, sitting in a primate chair, was placed on the swing with the head centered in the magnetic field. **(B)** This scheme illustrates the circular pathway paradigm for movement in the horizontal plane. The optic flow pattern presented in this case mimicked movement of an observer on the indicated trajectory in front of a large-field random-dot pattern.

three-dimensional (3-D) adaptation of a method used previously to map frontoparallel motion selectivity.[6] A detailed description can be found in that reference. Briefly, the stimulus mimicked movement through a 3-D environment on a circular path but with a constant orientation of the viewing direction. Thus, stimulus direction changed smoothly and constantly throughout the trial. These paradigms allowed determining the full 3-D directional tuning of the cells by looking at the response rate at different times during the trial.

In a subset of cells we also studied visual vestibular interactions. This was done by testing cells for their responsiveness to vestibular stimulation in darkness and in light, and comparing this to responses to pure visual stimulation that simulated the monkey's movement on the swing. In more detail, the procedure was as follows: after testing vestibular stimulation in darkness, swing movement was performed while a stationary dot pattern was presented on the tangent screen. This gave a combination of visual and vestibular stimulation. The visual stimulation consisted of expansion and contraction of the overall pattern with respect to the monkey's field of view, since the animal was moved toward and away from the tangent screen during the swing. A slight vertical motion caused by the movement characteristics of the parallel swing (see earlier) accompanied this expansion and contraction of the visual stimulus pattern. In both conditions (vestibular stimulation in darkness and visual–vestibular stimulation in light) the animal had to fixate a central target in order to suppress the tVOR. Finally, the pure visual stimulation simulated the monkeys self-motion, reproducing exactly the same pattern movement on the screen (sinusoidal expansion or contraction, accompanied by a vertical displacement) while the monkey was stationary in space.

Data Analysis

A distribution-free ANOVA was used to compare at least three different temporal intervals during the responses to vestibular stimulation. The Mann-Whitney U-test was used to test for differences between stimulus-driven activity and spontaneous activity in the visual paradigm.

RESULTS

Fifty-one neurons from two hemispheres of two macaque monkeys were tested for their responsiveness to linear vestibular stimulation. Twenty-eight neurons (55%) showed a statistically significant response ($p<0.05$). An example is shown in FIGURE 2. The histogram in the upper panel reveals clearly that this cell preferred forward movement over backward movement in total darkness ($p<0.0001$). The lower panel indicates the position of the swing during the trial.

The very same cell was also tested for its responsiveness to visual stimulation using optic flow fields that simulated self-motion (FIG. 3). The left column of this figure shows the result for stimulation in the horizontal plane, as histogram (upper panel) and as polar plot (lower panel). As already described in the Methods section, this stimulation method simulated a self-motion in which the movement direction changed continuously (circular

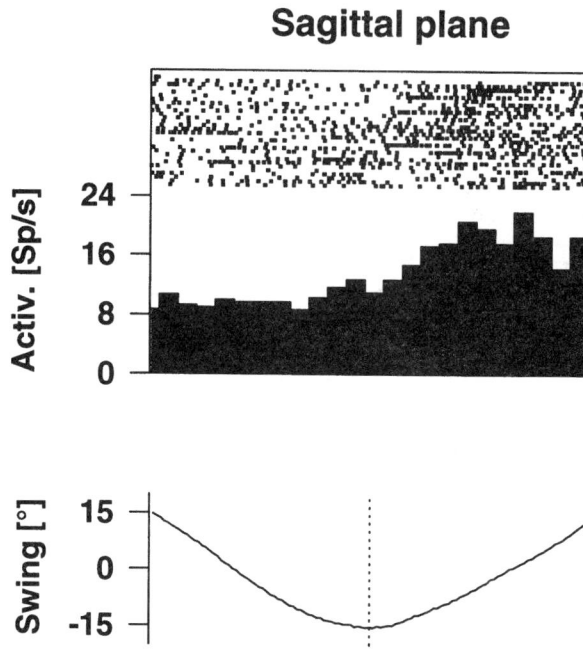

FIGURE 2. Response of a MST neuron to vestibular stimulation in darkness. The histogram indicates the firing rate of the neuron during sinusoidal backward and forward motion. The lower panel indicates the position of the parallel swing during the trial.

pathway stimulation). The right column shows the result of testing the cell with circular pathway stimulation in the sagittal plane, again as histogram (upper panel) and as polar plot (lower panel). It is very obvious from both stimulations that this cell preferred exclusively forward motion, that is, an expansion stimulus. Thus, for this neuron the preferred self-motion directions for visual and vestibular stimulation were equally directed in space.

Fifty-five percent of the cells (28/51) revealed a significant response to linear vestibular stimulation. This result was obtained regardless of whether or not the animal had to suppress the tVOR by fixating a chair-mounted LED. As shown in FIG. 4, 10 (36%) cells preferred forward motion, while 11 (39%) cells preferred backward motion. Seven (25%) cells showed a bidirectional response characteristic. Twenty cells that were responsive to real physical movement in darkness were also tested for their response to visually simulated self-motion. As shown in FIGURE 5, almost half of the cells (9/20) preferred the same direction of self-motion in space in both conditions (pure visual and pure vestibular stimulation) and therefore were synergistically organized. Five (25%) cells had different preferred directions for visual and vestibular stimulation and thus were nonsynergistically organized, while 6 (30%) cells were not responsive to visual stimulation.

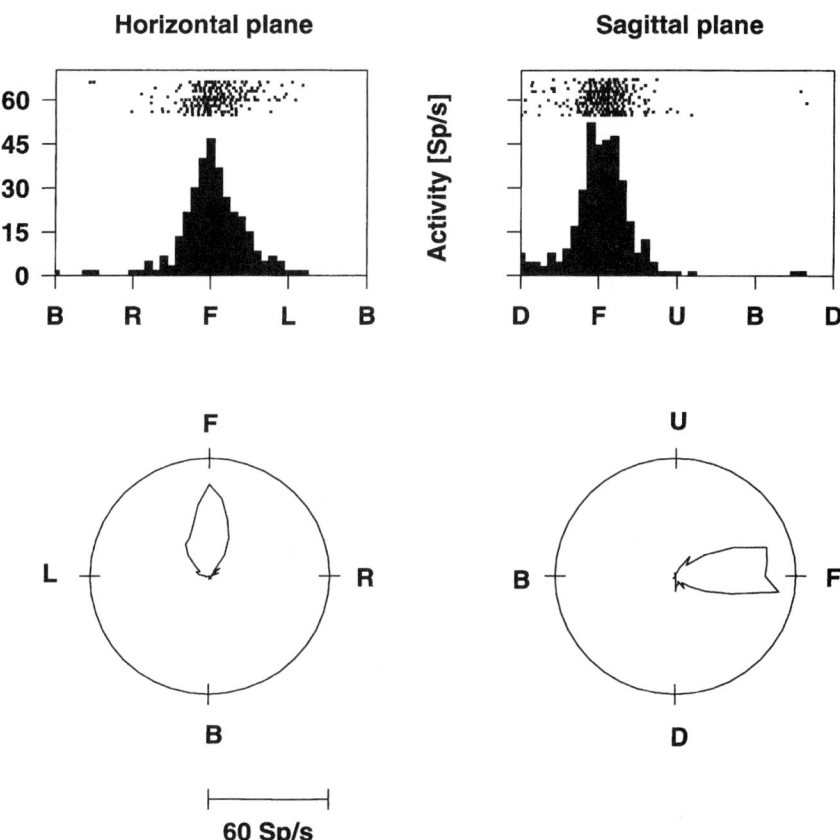

FIGURE 3. Response of the very same neuron to visual stimulation. The left and right columns show the response of the neuron from FIGURE 2 to optic flow fields simulating self-motion. The **left column** indicates the responses for simulated movement in the horizontal plane. The **right column** shows responses for simulated movement in the sagittal plane. Movement directions changed continuously throughout the trial in both stimulus conditions. Movement directions indicated specifically: F = forward; L = left; B = backward; R = rightward; U = upward; D = downward.

Finally, a small subset of cells ($n = 11$) was tested completely for visual-vestibular interactions, that is, neuronal activity was recorded for vestibular stimulation in darkness and in light as well as for pure visual stimulation. The visual stimulation used in this case exactly mimicked the swing movement, while the monkey was stationary in space. An even more complex response pattern emerged from these experiments. Only a few of the cells preferred the same self-motion directions in space in all three experimental conditions. Others had the same response characteristic for vestibular stimulation in darkness and light, while visual stimulation could not drive the cell or caused a different response profile. Finally, we found cells that did not respond to pure visual or vestibular stimulation alone, but only for a combined visual vestibular stimulation. Yet, the small number of neurons recorded does not allow any population analysis.

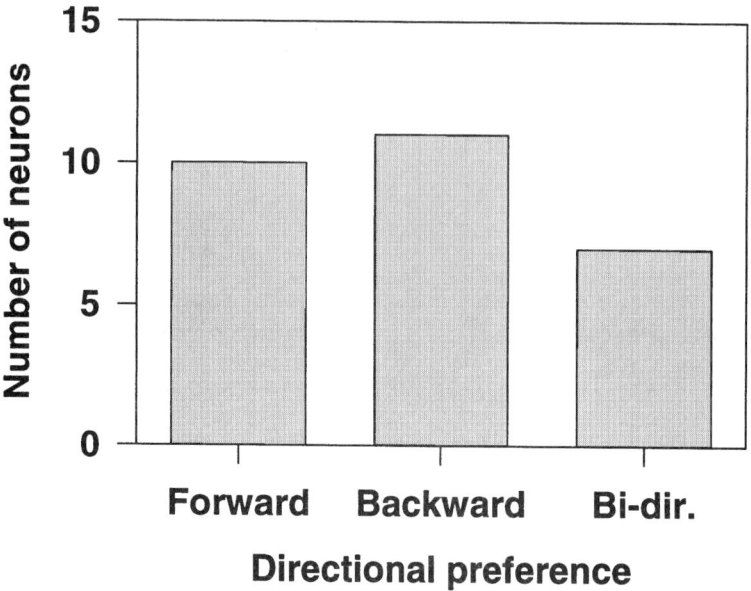

FIGURE 4. Distribution of preferred directions for vestibular stimulation. The bars in this histogram indicate the numbers of neurons being responsive for forward movement, backward movement, or movement into both directions (bi-dir.).

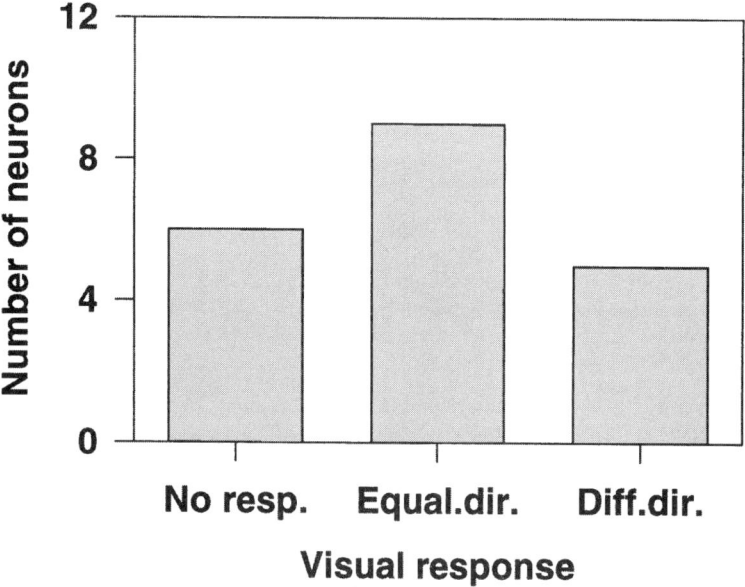

FIGURE 5. Distribution of visual vestibular interactions. The bars in this histogram indicate the number of vestibular responsive neurons in which the preferred direction for visual stimulation were either in the same (*middle bar*) or in the opposite (*right bar*) direction of the vestibular preferred direction. Some of the vestibular responsive neurons had no significant visual response (*left bar*).

DISCUSSION

Response to Linear Vestibular Stimulation

Our results show that MST neurons respond to linear vestibular stimulation. Seventy-five percent of the vestibular responsive cells responded selectively to only one movement direction (forward or backward), while the remaining cells had a bidirectional response profile. Almost half of the vestibular responsive cells (45%) preferred the same self-motion direction in space when simulated by optic flow fields. However, the remaining 55% of the cells responded either for the opposite direction or were not responsive to visual stimulation. This indication of cross-modal interactions was confirmed in a study on a subset of cells that was tested for vestibular stimulation in darkness and light as well as for pure visual stimulation. Thus, from our experiments it can be concluded that visual vestibular interactions take place in the MST area and that both sensory signals are not always combined in a synergistic manner. Instead, for about half of the cells visual response characteristics seemed to be independent of vestibular responsiveness.

Vestibular Signals in the Cortex

Responses to vestibular stimulation have already been described for several cortical areas in the macaque. It was first found in area 2v adjacent to the somatosensory face field. Subsequently, vestibular responsiveness was found in area 3a within the somatosensory arm field, in the PIVC region, area 7, MST area, VIP area, area T3, and also the FEF.[9-15] Most neurons in these areas respond to head velocity and receive converging visual, vestibular, and somatosensory input. Interestingly, all neurons in the VIP area as well as many neurons in the other cortical areas reveal a nonsynergistic behavior in that their preferred directions for visual and vestibular stimulation are coaligned.[12,13] The functional role of this preference for these nonsynergistic stimuli is not clear.

It is not only the dynamic context of head *rotation* that is represented in these cortical areas. It was also shown before that the angular head *position* in space influences the activity of parietal cortex neurons.[16] For many of these neurons head and eye position influenced neuronal activity in a synergistic manner, that is, neuronal activity was, for example, increased by having the eyes or the head (or both) in an eccentric position relative to the straight-ahead direction of the trunk.

To our knowledge, however, all studies on vestibular responsiveness so far involved exclusively rotational components thereby stimulating the semicircular canals. The representation of linear vestibular signals from the otoliths has not been shown yet. Our results therefore show for the first time responsiveness of the macaque cortical system to linear vestibular stimulation. Moreover, this responsiveness is located in an area that has been associated with the processing of self-motion based on its optic flow response properties.[1-4]

The Functional Role of Vestibular Signals in the Cortex

The functional role of cortical vestibular responsiveness is best understood by results coming from lesion studies. Several studies have shown that lesions of the parietovestibular regions influence the VOR.[17] Often an asymmetry in the VOR is observed, with a lower gain of the slow phases directed to the ipsiversive side, and sometimes a lesion in this cortical area leads to a spontaneous nystagmus away from the lesioned side. Lesions of the identified vestibular areas can modify normal cognitive behavior like locomotor navigation or the percept of visual spatial constancy. Straube and Brandt[18] could show that a patient with lesions of the vestibular cortex did not have the impression of vection when visual stimulation was restricted to the ipsilateral part of the visual field with respect to the lesion side. In addition, vestibular signals can also be used for spatial memory processes in humans[19,20] such as path integration.

SUMMARY

Our results show that many neurons in the MST area are activated during real self-motion, which generates linear vestibular signals arising from the otoliths. About the same number of cells responded to either forward or backward motion. About half of the vestibular responsive cells responded to visual stimuli simulating self-motion in a synergistic manner. The hypothesis that the MST area is involved in the processing of self-motion information is supported by our present findings. These results point toward a sensory-integrating rather than a purely visual function of the MST area.

ACKNOWLEDGMENTS

This work was supported from grants of the DFG (SFB 509) and the Human Frontier Science Program (RG 71/96B). We thank W. Zinke for data analysis and Dr. C. Distler for help with the anatomy.

REFERENCES

1. TANAKA, K. & H. A. SAITO. 1989. Analysis of motion of the visual field by direction, expansion/contraction, and rotation cells clustered in the dorsal part of the medial superior temporal area of the macaque monkey. J. Neurophysiol. **62**: 626–641.
2. DUFFY, C. J. & R. H. WURTZ. 1991. Sensitivity of MST neurons to optic flow stimuli. I: A continuum of response selectivity to large-field stimuli. J. Neurophysiol. **65**: 1329–1345.
3. PEKEL, M., et al. 1996. Neuronal responses in the motion pathway of the macaque monkey to natural optic flow stimuli. NeuroReport **7**: 884–888.
4. LAPPE, M., et al. 1996. Optic flow processing in monkey STS: a theoretical and experimental approach. J. Neurosci. **16**: 6265–6285.
5. THIER, P. & R. G. ERICKSON. 1992. Responses of visual-tracking neurons from cortical area MST-I to visual, eye and head motion. Eur. J. Neurosci. **4**: 539–553.
6. BREMMER, F., et al. 1997. Eye position effects in monkey cortex. I: Visual and pursuit related activity in extrastriate areas MT and MST. J. Neurophysiol. **77**(2): 944–961.
7. JUDGE, S. J., et al. 1980. Implantation of magnetic search coils for measurement of eye position: An improved method. Vision Res. **20**: 535–538.
8. CELEBRINI, S. & W. T. NEWSOME. 1994. Neuronal and psychophysical sensitivity to motion signals in extrastriate area MST of the macaque monkey. J. Neurosci. **14**: 4109–4124.
9. GRÜSSER, O.-J., et al. 1990. Localization and responses of neurons in the parieto-insular vestibular cortex of awake monkeys (*Macaca fascicularis*). J. Physiol. **430**: 537–557.
10. KAWANO, K., et al. 1980. Vestibular input to visual tracking neurons in the posterior parietal association cortex of the monkey. Neurosci. Lett. **17**: 55–60.
11. KAWANO, K., et al. 1984. Response properties of neurons in posterior parietal cortex of monkey during visual-vestibular stimulation. J. Neurophysiol. **51**: 340–351.
12. BREMMER, F., et al. 1995. Supramodal encoding of movement space in the ventral intraparietal area of macaque monkeys. Soc. Neurosci. Abstr. **21**: 282.
13. GRAF, W., et al. 1996. Visual-vestibular interaction in the ventral intraparietal area of macaque monkeys. Soc. Neurosci. Abstr. **22**: 1692.
14. JONES, E. G. & H. BURTON. 1976. Areal differences in the laminar distribution of thalamic afferents in cortical fields of the insular, parietal and temporal opercular regions of primates. J. Comp. Neurol. **168**: 197–247.

15. FUKUSHIMA, K., et al. 1997. Properties of neurons encoding vestibular related signals in and near the frontal eye fields of alert monkeys. Soc. Neurosci. Abstr. **23**: 473.
16. BROTCHIE, P. R., et al. 1995. Head position signals used by parietal neurons to encode locations of visual stimuli. Nature **375**: 232–235.
17. ESTANOL, B., et al. 1980. Oculomotor and oculovestibular functions in a hemispherectomy patient. Arch. Neurol. **37**: 365–368.
18. STRAUBE, A. & T. BRANDT. 1987. Importance of the visual and vestibular cortex for self-motion perception in man (circularvection). Hum. Neurobiol. **6**: 211–219.
19. BERTHOZ, A., et al. 1995. Spatial memory of body linear displacement: What is being stored? Science **269**: 95–98.
20. ISRAEL, I., et al. 1997. Spatial memory and path integration studied by self-driven linear displacement. I. Basic properties. J. Neurophysiol. **77**: 3180–3192.

The Contributions of Vestibular Signals to the Representations of Space in the Posterior Parietal Cortex

RICHARD A. ANDERSEN,[a] KRISHNA V. SHENOY, LAWRENCE H. SNYDER, DAVID C. BRADLEY, AND JAMES A. CROWELL

Division of Biology, California Institute of Technology, Pasadena, California 91125, USA

ABSTRACT: Vestibular signals play an important role in spatial orientation, perception of object location, and control of self-motion. Prior physiological research on vestibular information processing has focused on brainstem mechanisms; relatively little is known about the processing of vestibular information at the level of the cerebral cortex. Recent electrophysiological experiments examining the use of vestibular canal signals in two different perceptual tasks are described: computation of self motion and localization of visual stimuli in a world-centered reference frame. These two perceptual functions are mediated by different parts of the posterior parietal cortex, the former in the dorsal aspect of the medial superior temporal area (MSTd) and the latter in area 7a.

INTRODUCTION

The vestibular system consists of two sets of organs, the otoliths and the semicircular canals. The otoliths respond to linear accelerations of the head, the canals to rotations of the head. Psychophysical studies have established that these organs play an important role in human spatial orientation.[1,2]

Our research has focused on the canal signals related to head rotation, and examines the cerebral cortical structures and representations that utilize this vestibular cue. More specifically, we are attempting to answer the following two questions: (1) Which structures in the primate cerebral cortex use vestibular canal signals to encode spatial locations of objects and the direction of an animal's self-motion? and (2) How is this information represented and combined with visual signals about self-motion and object location? It is possible that the rules for processing otolith signals at the cortical level will be similar to those we have found for processing canal signals, which are reviewed below.

It has been thought for some time that the posterior parietal cortex (PPC) processes vestibular inputs for spatial awareness. Some support for this notion comes from clinical studies of patients with parietal-lobe damage. These patients exhibit a phenomenon known as "neglect": although their early visual pathways are undamaged, they are unaware of objects and events within the contralesional field. Caloric stimulation of the contralateral

---

[a]To whom correspondence may be addressed. Phone: 626/395-8336; fax: 626/795-2397; e-mail: andersen@vis.caltech.edu

ear with cold water (or the ipsilateral ear with warm water) can transiently reduce the patents' neglect, suggesting that vestibular signals are important in constructing spatial representations that are compromised with parietal damage.[3]

Prior electrophysiological work has identified specific areas within the PPC that receive vestibular inputs. Grusser and colleagues[4] identified what appears to be the primary vestibular sensory area in the parietal lobe of monkeys, which they named the parieto-insular vestibular cortex (PIVC). About two-thirds of the neurons in this area responded to angular acceleration of the head. They also found a small number of cells in area 7 that had vestibular-related activity.[5] In the medial superior temporal area (MST) neurons have also been found that are responsive to vestibular-canal stimulation.[6–8]

Until recently there has been relatively little work on the nature of the representation of vestibular information in the cerebral cortex. In the studies reviewed here we examine two different types of vestibular/visual interactions in two cortical areas. First we describe research on the use of vestibular-canal angular velocity signals—related to the rate at which the head turns—in conjunction with visual motion signals to estimate the direction of an animal's motion through space. These studies were performed in area MSTd. Then we will examine the use of canal signals related to the position of the head in localizing seen objects in a world-centric reference frame; these studies were performed in area 7a.

THE PROBLEM OF RECOVERING SELF-MOTION DURING GAZE ROTATIONS

Figure 1 shows an example of optical flow patterns generated while driving an automobile. The focus of the expansion (FOE) corresponds to the direction of heading. As diagrammed in FIGURE 2, however, if the eye is rotating, either due to a smooth pursuit eye movements or smooth head movements, then a laminar motion is introduced on the eyes opposite to the direction of gaze rotation. This laminar motion disrupts the focus of expansion. In the special case of approaching a wall (FIG. 2) the focus is displaced in the direction of the eye movement. However, we know from a number of psychophysical studies in humans that the true direction of self-motion can still be perceived during eye movements.[9,10] More recently we have demonstrated that correct heading perception is also recovered during gaze rotations that result from head movements in human observers.[11] In simple terms, the laminar motion field due to gaze rotation appears to be subtracted from the motions due to translation of the observer through the environment. This subtraction is afforded by an extraretinal signal since the same retinal motions that occur in the active gaze-rotation case, but without gaze rotation, lead to no compensation. In the case of pursuit, the source of this extraretinal signal is not known. However, during head movements, we have recently found that the gaze-rotation signal is derived from a combination of vestibular, proprioceptive, and efference copy sources.[11]

NEURAL CORRELATES OF HEADING PERCEPTION DURING PURSUIT EYE MOVEMENTS

The dorsal aspect of area MSTd (FIG. 3) has been found to contain neurons that are selective for motion patterns that are generated during self-motion.[12–27] Different cells are sensitive to expansions, contractions, rotations, spirals, and laminar motions. The largest fraction of cells are selective for expansions and laminar motions.[26] These different pattern selectivities appear to be anatomically organized within cortical columns.[28,29] Cells in MSTd are also active during smooth pursuit eye movements and demonstrate selectivity for pursuit direction.[19,30] Thus area MSTd has the appropriate signals for encoding self motion perception.

FIGURE 1. Driving down a street generates an optical flow of motion signals. Surrounding objects seem to radiate out from the focus point, expanding toward the edges of the field of vision. Processing of these complex signals occurs in area MSTd. (Photograph by David Bradley.)

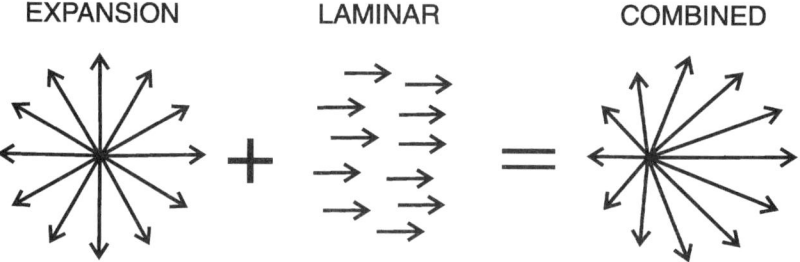

FIGURE 2. When we move forward, the visual world appears to expand. If the eyes are still, the focus position tells us our direction of heading. However, leftward eye movement adds rightward laminar flow motion to the retinal image, which shifts the focus. To recover the heading direction, we must correct for this focus shift. (From Bradley et al.[31] Reproduced by permission.)

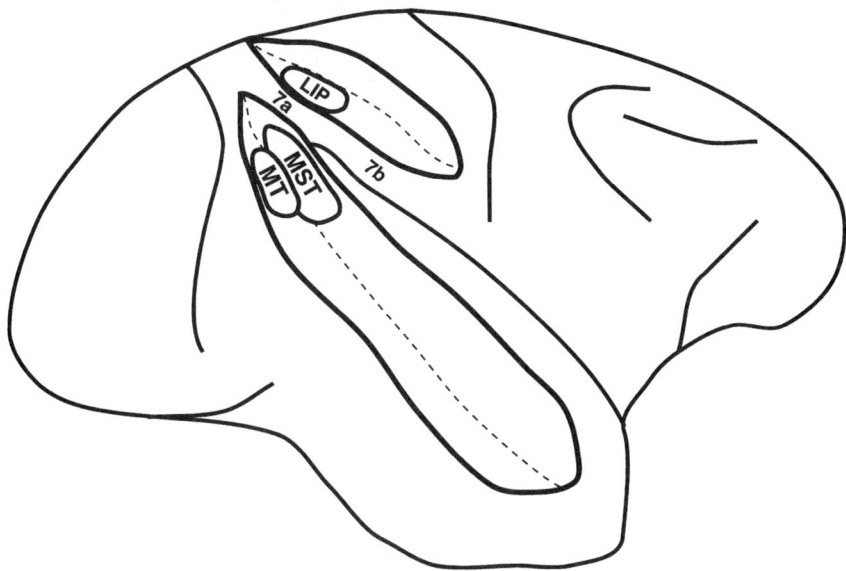

FIGURE 3. Locations of areas discussed in the text drawn on a lateral view of a macaque monkey brain. (From Andersen et al.[34] Reproduced by permission.)

Recently we examined the neural mechanism that allows heading direction to be recovered during pursuit eye movements.[31] Duffy and Wurtz[44] have shown that many MSTd neurons are tuned for the location of the focus of expansion. We mapped MSTd focus tuning curves when the monkeys' eyes were stationary, much as Duffy and Wurtz had done, and then again during smooth pursuit eye movements. We found that a large fraction of area MSTd neurons shift their focus tuning curves in order to compensate for the pursuit direction (see FIG. 4, upper panels). This compensation was shown to be due to an extraretinal signal related to the eye pursuit. When the same retinal image was presented on the screen that the animal experienced during pursuit eye movements, but with the eyes stationary (the so-called simulated pursuit condition), then no compensation occurred (FIG. 4, bottom panels). The source of this extraretinal signal is not known, but it could be derived from an efference copy of the eye pursuit command, or proprioceptive signals from the eye muscles and periorbital tissue. These results are very consistent with the human psychophysical results mentioned earlier, and may form the neural basis for correct self-motion perception during pursuit eye movements.

NEURAL CORRELATES OF HEADING PERCEPTION USING VESTIBULAR SIGNALS DURING GAZE ROTATION GENERATED BY HEAD MOVEMENT

Of course we often move our heads as well as our eyes when pursuing objects. This is certainly true during locomotion, where the gaze movements are often quite large in amplitude, requiring head movements. As mentioned before, recent findings from our lab indicate that gaze rotations during head movements are also taken into account to accurately perceive self-motion.[11] However, compensation only occurs if the vestibular signals are present with neck proprioceptive or efference copy head movement signals.

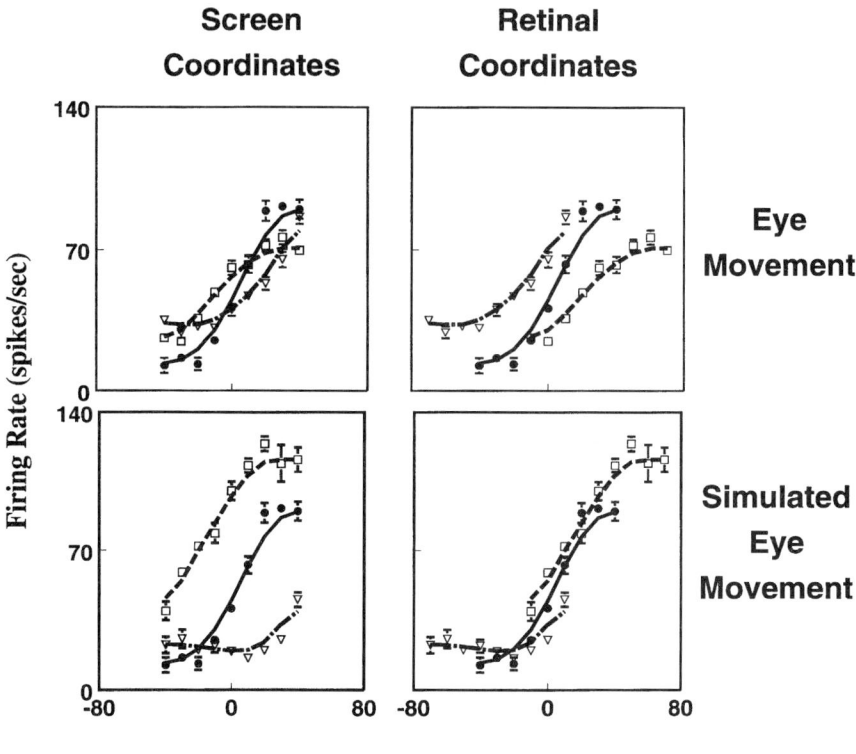

FIGURE 4. An MSTd heading cell. In all panels, the *solid lines* and *solid circles* represent fixed-eye focus tuning (identical in all four graphs), the *dashed lines* and *open squares* are preferred-direction eye movements (real or simulated), and the *dot-and-dashed lines* and *open triangles* are antipreferred-direction eye movements (real or simulated). Data in the left and right columns are identical, except that pursuit curves in the right column were shifted by 30 degrees relative to screen coordinates (thus giving retinal coordinates). The moving-eye focus tuning curves align in screen coordinates (**top left panel**), and thus encode the direction of heading. However, for simulated eye movements, the curves align in retinal coordinates (**bottom right panel**). Smooth curves are five-point moving averages of the data. Data points are shown as the mean ± SEM for four replicates, where each replicate is the mean firing during the middle 500 ms of the stimulus-presentation interval. (From Bradley *et al.*[31] Reproduced by permission.)

We have recently examined whether the focus-tuning curves of MSTd neurons show compensation during smooth head rotations where the vestibular ocular reflex is canceled (VORc). The monkeys sat in a vestibular chair, and the animal's head was fixed to the chair. In the VORc task, the chair was rotated with the fixation point moving in the world as if it was attached to the chair. To successfully pursue the target, the monkeys had to cancel the VOR, either by suppressing it, or by adding a pursuit signal in the opposite direction to cancel the reflex. Using this task, we found that a large number of area MSTd neurons were active for the direction of smooth head rotation during VORc in the yaw–pitch plane.[32,33] This result corroborates reports of similar activity in MST.[6-8] The novel aspect of our experiment was that we then had the animals produce the gaze rotations during VORc across optic flow fields. A similar paradigm to the eye pursuit experiments was used.[31] We first mapped the focus-tuning curves of individual MSTd neurons with the eyes stationary. Next we had the animal pursue the fixation point across the optic flow displays, either with pursuit eye movements when the chair was stationary, or by canceling the VOR when the chair was moving. We found that there was similar compensation in both the eye movement and VORc conditions.[32] Moreover, cells that compensated for one condition generally compensated for the other condition, and by approximately the same amount.[33] The simulated gaze rotation condition, in which the retinal stimulation was the same as the VORc and pursuit conditions, but the eyes were stationary, produced considerably less compensation.

These results indicate that area MSTd compensates for gaze rotations regardless of whether they are produced by eye or head movements. The compensation during VORc must at some stage utilize vestibular cues. From the current experiments we do not know if compensation in MSTd is due to the direct action of vestibular signals on flow-sensitive neurons. Kawano et al.[6,7] and Thier and Ericksen[8] both found that rotating the animal in the dark often produced modulation of activity, but it was usually less than that seen during VORc. This result indicates that there are vestibular-canal-derived signals present in area MST, but we do not yet know if these same signals are the ones responsible for the compensation. An alternative explanation is that a pursuit signal is generated to cancel the VOR, and it is the efference copy of the pursuit command that is used for compensation. If this is the case, then it is possible that compensation during eye movements and VORc result from the same efference copy signal. Finally, it is possible that a combination of both efference copy and vestibular cues account for the VORc result.

REPRESENTING STATIC LOCATIONS IN NONRETINAL COORDINATES

The preceding experiments examined the integration of eye and head movements with visual motion for accurate spatial perception. The next set of experiments to be described examine the interaction of static eye and head position signals with static retinal signals to again produce accurate spatial perceptions of where static stimuli are located in space. An example of such a computation is the accurate reaching to an object, which can be accomplished independent of the exact retinal location of the stimulus, or the direction in which the eyes are looking.

In a series of experiments from our lab over the last decade we have examined how eye and retinal position signals are integrated in areas LIP and 7a of the posterior parietal cortex[34] (see FIG. 3). We routinely found that these cells had receptive fields with fixed retinal locations, but their activity was gain modulated by eye position[35-37] (see FIG. 5). In modeling experiments we designed neural networks that receive eye and retinal position signals and yield locations in head-centered coordinates.[38] The middle layer units that produced the transformation from retinal (at the input) to head-centered (at the output) coordinates were very similar to the PPC neurons; that is, they had retinal-receptive fields gain modulated by eye position. These modeling studies demonstrated that the cells in PPC can represent head-centered locations in a distributed manner. Thus PPC can simultaneously represent both retinal and head centered coordinates in the same population of neurons.

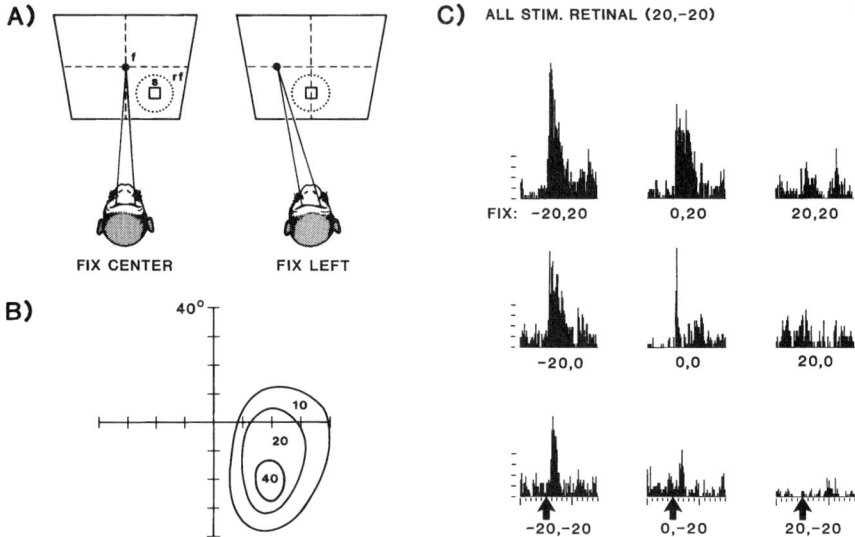

FIGURE 5. (**A**) Method of determining spatial gain fields of area 7a neurons. The animal fixates point *f* at different locations on the screen with his head fixed. The stimulus, *s*, is always presented in the center of the receptive field, r.f. (**B**) Receptive field of a neuron plotted in coordinates of visual angle determined with the animal always fixating straight ahead (screen coordinates 0,0). The *contours* represent the mean increased response rates in spikes per second. (**C**) Spatial gain field of the cell in (A). The poststimulus histograms are positioned to correspond to the locations of the fixations on the screen at which the responses were recorded from retinotopically identical stimuli presented in the center of the receptive field (histogram ordinate, 25 spikes per division, and abscissa, 100 ms per division; *arrows* indicate onset of stimulus flash) modified from Andersen *et al.*[36]

HEAD-POSITION GAIN FIELDS

To go back to our example of accurate reaching, eventually the brain must know where the object is with respect to the hand to make an accurate reaching movement, a computation that requires information about head and limb position as well as eye and retinal position. Thus we next addressed how head position might influence the activity of PPC neurons. To examine this issue we trained monkeys to orient their direction of gaze to different locations with respect to their bodies, by either making eye or head movements. For instance, we would have them look 20 degrees to the left by having them move their heads to the left, or by moving their eyes to the left with the head straight forward. In the former case, the eyes are straight ahead in the orbits, and in the latter the eyes are deviated to the left in the orbits. For many cells in both areas LIP and 7a we found gain fields for head position.[39] The cells demonstrating head-position gain fields also had eye-position gain fields, and the gain modulation functions were the same for head and eye. An example is shown in FIGURE 6. For this neuron, when the monkey looked more to the right, the activity increased for the same retinal location of stimulation (and same saccade) regardless of whether the head (FIG. 6A) or the eye (FIG. 6B) was used to orient the gaze position. The graph of the mean firing rates in the two conditions in FIGURE 6C illustrates that the gain functions for head and eye position for this cell were virtually identical. This similarity between eye and head gain fields was found for the population of cells examined in this study, showing that this category of neurons is modulated by gaze direction, independent of whether that gaze direction is produced by orienting the head or the eyes.

VESTIBULAR CONTRIBUTIONS TO HEAD-POSITION GAIN FIELDS

As with the flow experiments, the head gain fields could be derived from vestibular, neck proprioceptive, or efference copy signals. An interesting distinction can be made between the vestibular source and the other two sources in terms of frame of reference. A vestibularly derived gain would require the integration of a vestibular velocity signal to code head position. This integrated signal would indicate the orientation of the head in the world. On the other hand, the neck proprioceptive signals would likely indicate the orientation of the head on the body. Thus vestibular gain fields would be useful for representing locations of stimuli in world coordinates, whereas neck proprioceptive gain fields would be useful for representing locations in body-centered coordinates.

We have recently begun to address whether vestibular or neck proprioceptive signals contribute to the head gain fields we see in PPC. The vestibular contribution was tested by rotating the entire animal on a vestibular turntable, with the head always straight ahead with respect to the body. Neck proprioceptive contributions were tested by changing the orientation of the body with respect to the head, with the head always in the same orientation with respect to the recording room. The results were extremely interesting. Although one potential outcome of the experiments would have been that both vestibular and neck proprioceptive signals could contribute to the gain fields of single neurons, this was rarely found. Cells generally had gain fields only for one of the two manipulations.[40,41] Even more interesting was the finding that the type of gain field was segregated by cortical area. Area 7a was found to have predominantly vestibularly derived gain fields, whereas area LIP had primarily neck gain fields. This segregation makes sense in terms of the outputs of these two areas. Area 7a projects to the presubiculum and parahippocampal gyrus, and both these structures are closely associated with the hippocampal formation. The finding of "place cells" in this area in rats suggest that at least one important reference frame used by the hippocampus is world-centered.[42] The outputs of LIP are to eye movement centers. Given that shifts in gaze direction are often combined with head and even trunk movements, it is perhaps not surprising to see gain fields consistent with a body-centered reference frame.

CONCLUSIONS

The studies discussed demonstrate that vestibular information is used by the PPC for the perception of self-motion and the spatial location of objects in body and world reference frames. No doubt vestibular cues also contribute to a number of other spatial functions in PPC. Until recently[43] work on vestibular functions in PPC has centered on canal, and not otolith, signals. Otolith signals may be used by parietal areas to perceive orientation with respect to gravity and the perception of linear translation.

Vestibular gain fields may be a common method for integrating the vestibular modality with other senses such as vision. For instance, even in MSTd, many of the cells that did not show compensation did have gain modulation by vestibular inputs.[32] It has been proposed that similar eye pursuit gain-modulations in MSTd may be an intermediate step to the shifting focus-tuning curves that are seen for many cells.[31] Similarly, the vestibular gain-modulated non-shifting fields may be an intermediate step to the cells which show shift compensation under the VORc condition.[32]

Of course, the findings reviewed herein are probably just the tip of the iceberg in terms of the influence that vestibular signals have on cerebral cortical processing. No doubt the questions being asked at the cortical level will be very different from those asked at the brainstem level. Whereas research at lower levels have been concerned with motor control, adaptation, learning, and vestibular motor reflexes, future experiments at the cortical level will probably be most successful in examining perceptual and cognitive processes related to vestibular function. This relatively new field of cortical vestibular research should provide a rich and interesting area of inquiry in the years to come.

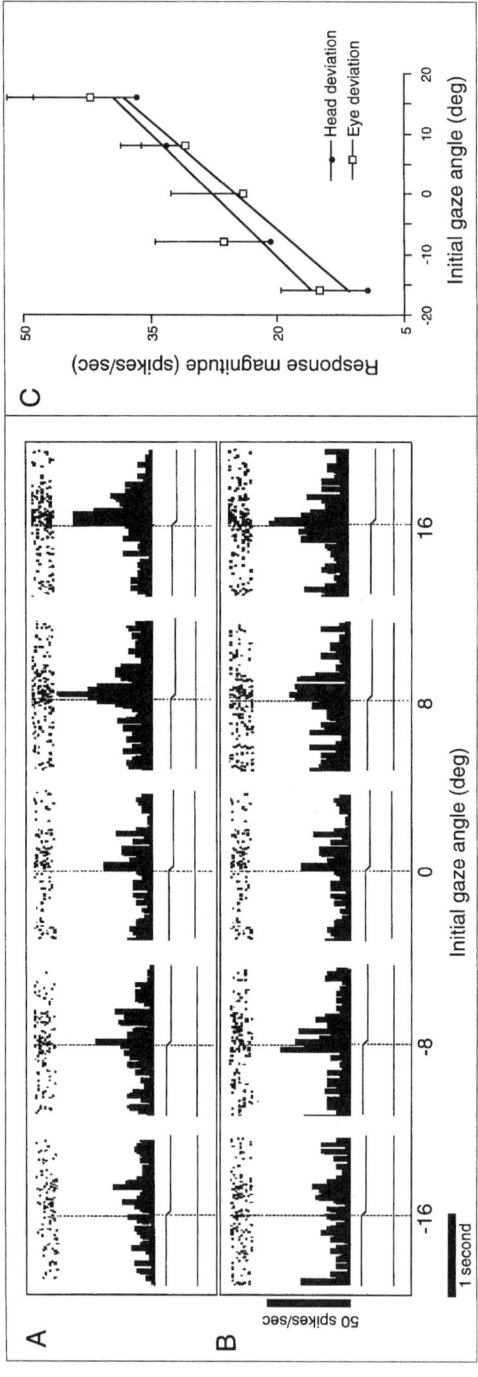

FIGURE 6. Activity of a cell while the monkey is making identical saccades to the left from a fixation point placed at five different gaze positions on the horizontal plane. All trials are aligned with the onset of the saccade, indicated by the *vertical broken lines*. (**A**) The animal has its head oriented toward each of the fixation points, with the eyes centered in their orbits before each saccade. (**B**) The head of the animal is directed toward the center of the screen, with the eyes deviated toward each of the fixation points before each saccade. (**C**) Magnitude of the cell's activity around the time of the saccade (25 ms before to 75 ms after onset of the saccade) as it varies with initial gaze position. *Bars* indicate one standard error. A linear relationship of activity with gaze direction was confirmed by a significant linear regression and by a nonsignificant analysis of variance of the regression residuals. (From Brotchie *et al.*[39] Reproduced by permission.)

ACKNOWLEDGMENTS

The work from our lab, reviewed here, was supported by the National Eye Institute, the Office of Naval Research, the Human Frontiers Scientific Program, and the James G. Boswell Foundation. We thank Betty Gillikin and Viktor Shcherbatyuk for technical assistance and Cierina Reyes for editorial assistance.

REFERENCES

1. ISRAEL, I. & A. BERTHOZ. 1989. Contribution of the otoliths to the calculation of linear displacement. J. Neurophysiol. **62**: 247–263.
2. BLOOMBERG, J., G. M. JONES & B. SEGAL. 1991. Adaptive modification of vestibularly perceived rotation. Exp. Brain Res. **84**: 47–56.
3. KARNATH, H.-O. 1997. Neural encoding of space in egocentric coordinates?—Evidence for and limits of a hypothesis derived from patients with parietal lesions and neglect. *In* Parietal Lobe Contributions to Orientation in 3D Space, P. Thier and H.-O. Karnath, Eds. Springer-Verlag. Heidelberg.
4. GRUSSER, O.-J., M. PAUSE & U. SCHRIETER. 1990. Localtiztion and responses of neurones in the parieto-insular vestibular cortex of awake monkeys (macaca fascicularis). J. Physiol. **430**: 537–557.
5. GRUSSER, O.-J., M. PAUSE & U. SCHREITER. 1982. Neural responses in the parieto-insular vestibular cortex of alert java monkeys (maccaca fascicularis). In Physiological and Pathological Aspects of Eye Movements. A. Roucoux and M. Crommelinck Eds., Dr. W. Junk. The Hague.
6. KAWANO, K., M. SASAKI & M. YAMASHITA. 1980. Vestibular input to visual tracking neurons in the posterior parietal association cortex of the monkey. Neurosci. Lett. **17**: 55.
7. KAWANO, K., M. SASAKI & M. YAMASHITA. 1984. Response properties of neurons in posterior parietal cortex of monkey during visual-vestibular stimulation. I. Visual tracking neurons. J. Neurophysiol. **51**: 340.
8. THIER, P. & R. C. ERICKSON. 1992. Responses of visual tracking neurons from cortical area MST-l to visual, eye and head motion. Eur. J. Neurosci. **4**: 539.
9. ROYDEN, C. S., M. S. BANKS & J. A. CROWELL. 1992. The perception of heading during eye movements. Nature **360**: 583–585.
10. ROYDEN, C. S., J. A. CROWELL & M. S. BANKS. 1994. Estimating heading during eye movements. Vision Res. **34**: 3197–3214.
11. CROWELL, J. A., M. S. BANKS, K. V. SHENOY & R. A. ANDERSEN. 1998. Visual self-motion perception during head turns. Nat. Neurosci. **1**: 732–737.
12. SAKATA, H., H. SHIBUTANI, K. KAWANO & T. HARRINGTON. 1985. Neural mechanisms of space vision in the parietal association cortex of the monkey. Vision Res. **25**: 453.
13. SAKATA, H., H. SHIBUTANI, Y. ITO, K. TSURUGAI, S. MINE & M. KUSUNOKI. 1994. Functional properties of rotation-sensitive neurons in the posterior parietal association cortex of the monkey. Exp. Brain Res. **101**: 183–202.
14. SAITO, H., M. YUKIE, K. TANAKA, K. HIKOSAKA, Y. FUKADA & E. IWAI. 1986. Integration of direction signals of image motion in the superior temporal sulcus of the macaque monkey. J. Neurosci. **6**: 145.
15. TANAKA, K., K. HIKOSAKA, H. SAITO, M. YUKIE, Y. FUKADA & E. IWAI. 1986. Analysis of local and wide-field movements in the superior temporal visual areas of the macaque monkey. J. Neurosci. **6**: 134.
16. TANAKA, K. & H. A. SAITO. 1989. Analysis of motion of the visual field by direction, expansion/contraction, and rotation cells clustered in the dorsal part of the medial superior temporal area of the macaque monkey. J. Neurophysiol. **62**: 626–641.
17. TANAKA, K., Y. FUKADA & H. A. SAITO. 1989. Underlying mechanisms of the response specificity of expansion/contraction and rotation cells in the dorsal part of the medial superior temporal area of the macaque monkey. J. Neurophysiol. **62**: 642–656.
18. KOMATSU, H. & R. H. WURTZ. 1988. Relation of cortical areas MT and MST to pursuit eye-movements. I. Localization and visual properties of neurons. J. Neurophysiol. **60**: 580.

19. KOMATSU, H & R. H. WURTZ. 1988. Relation of cortical areas MT and MST to pursuit eye-movements. III. Interaction with full-field visual stimulation. J. Neurophysiol. **60**: 621.
20. LAGAE, L., H. MAES, S. RAIGUEL, D. K. XIAO & G. A. ORBAN. 1994. Responses of macaque STS neurons to optic flow components—a comparison of MT and MST. J. Neurophysiol. **71**: 1597.
21. LAPPE, M., F. BREMMER, M. PEKEL, A. THIELE & K.-P. HOFFMANN. 1996. Optic flow processing in monkey STS: A theoretical and experimental approach. J. Neurosci. **16**: 6265–6585.
22. DUFFY, C. J. & R. H. WURTZ. 1991. Sensitivity of MST neurons to optic flow stimuli. I. A continuum of response selectivity to large-field stimuli. J. Neurophysiol. **65**: 1329.
23. DUFFY, C. J. & R. H. WURTZ. 1991. Sensitivity of MST neurons to optic flow stimuli. II. Mechanisms of response selectivity revealed by small-field stimuli. J. Neurophysiol. **65**: 1346.
24. DUFFY, C. J. & R. H. WURTZ. 1997. Planar directional contributions to optic flow responses in MST neurons. J. Neurophysiol. **77**: 782–796.
25. DUFFY, C. J. & R. H. WURTZ. 1997. Medial superior temporal area neurons respond to speed patterns in optic flow. J. Neurosci. **17**: 2839–2851.
26. GRAZIANO, M. S. A., R. A. ANDERSEN & R. J. SNOWDEN. 1994. Tuning of MST neurons to spiral motions. J. Neurosci. **14**: 54–67.
27. RAIGUEL, S., M. M. VAN HULLE, D.-K. XIAO, V. L. MARCAR, L. LAGAE & G. A. ORBAN. 1997. Size and shape of receptive fields in the medial superior temporal area (MST) of the macaque. NeuroReport **8**: 2803–2807.
28. GEESAMAN, B. J., R. T. BORN, R. A. ANDERSEN & R. B. TOOTELL. Maps of complex motion selectivity in the superior temporal cortex of alert macaque monkey: a double-label 2-deoxyglucose study. Cereb. Cortex **7**: 749–757.
29. BRITTEN, K. H. 1998. Clustering of response selectivity in the medial superior temporal area of extrastriate cortex in the macaque monkey. Visual Neurosci. **15**: 553–558.
30. NEWSOME, W. T., R. H. WURTZ & H. KOMATSU. 1988. Relation of cortical areas MT and MST to pursuit eye-movements. II. Differentiation of retinal from extraretinal inputs. J. Neurophysiol. **60**: 604.
31. BRADLEY, D. C., M. A. MAXWELL, R. A. ANDERSEN, M. S. BANKS & K. V. SHENOY. 1996. Mechanisms of heading perception in primate visual cortex. Science **273**: 1544–1547.
32. SHENOY, K. V., D. C. BRADLEY & R. A. ANDERSEN. 1999. Influence of gaze rotation on the visual response of primate MSTd neurons. J. Neurophysiol. In press.
33. SHENOY, K. V., J. A. CROWELL, D. C. BRADLEY & R. A. ANDERSEN. 1997. Perception and neural representation of heading during gaze-rotation. Soc. Neurosci. Abstr. **23**.
34. ANDERSEN, R. A. 1997. Multimodal representation of space in the posterior parietal cortex and its use in planning movements. Ann. Rev. Neurosci. **20**: 303–330.
35. ANDERSEN, R. A. & R. M. BRACEWELL, S. BARASH, J. W. GNADT & L. FOGASSI. 1990. Eye position effects on visual, memory, and saccade-related activity in areas LIP and 7a of macaque. J. Neurosci. **10**: 1176–1196.
36. ANDERSEN, R. A., G. K. ESSICK & R. M. SIEGEL. 1985. Encoding of spatial location by posterior parietal neurons. Science **230**: 456–485.
37. ANDERSEN, R. A. & V. B. MOUNTCASTLE. 1983. The influence of the angle of gaze upon the excitability of the light sensitive neurons of the posterior parietal cortex. J. Neurosci. **3**: 532–548.
38. ZIPSER, D. & R. A. ANDERSEN. 1988. A back-propagation programmed network that simulates response properties of a subset of posterior parietal neurons. Nature **331**: 679–684.
39. BROTCHIE, P. R., R. A. ANDERSEN, L. H. SNYDER & S. J. GOODMAN. 1995. Head position signals used by parietal neurons to encode locations of visual stimuli. Nature **375**: 232–235.
40. GRIEVE, K. L., L. H. SNYDER, J. XING, & R. A. ANDERSEN. 1997. Frames of reference in primate parietal cortex: separation between 7a and LIP; physiology and simulation. Soc. Neurosci. Abstr. **23**.
41. SNYDER, L. H., K. L. GRIEVE, P. R. BROTCHIE & R. A. ANDERSEN. 1998. Separate body- and world-referenced representations of visual space in parietal cortex. Nature. **394**: 887–891.
42. MCNAUGHTON, B. L., C. A. BARNES & J. O'KEEFE. 1983. The contributions of position, direction and velocity to single unit activity in the hippocampus of freely moving rats. Exp. Brain Res. **52**: 41–49.
43. DUFFY, C. J. 1996. Real movement responses of optic flow neurons in MST. Soc. Neurosci. Abstr. **22**.
44. DUFFY, C. J. & R. H. WURTZ. 1995. Response of monkey MST neurons to optic flow stimuli with shifted centers of motion. J. Neurosci. **15**: 5192–5208.

The Vestibular Cortex
Its Locations, Functions, and Disorders

THOMAS BRANDT[a] AND MARIANNE DIETERICH

Department of Neurology, Klinikum Grosshadern, Ludwig Maximillians University, Marchioninistr.15, 81366 Munich, Germany

ABSTRACT: Evidence is presented that the multisensory parieto-insular cortex is the human homologue of the parieto-insular vestibular cortex (PIVC) in the monkey and is involved in the perception of verticality and self-motion. Acute lesions (patients with middle cerebral artery infarctions) of this area caused contraversive tilts of perceived vertical, body lateropulsion, and, rarely, rotational vertigo. Brain activation studies using positron emission tomography or functional magnetic resonance tomography showed that PIVC was activated by caloric irrigation of the ears or by galvanic stimulation of the mastoid. This indicates that PIVC receives input from both the semicircular canals and otoliths. PIVC was also activated during small-field optokinetic stimulation, but not when the nystagmus was suppressed by fixation. Activation of vestibular cortex areas, visual motion-sensitive areas, and ocular motor areas exhibited a significant right-hemispheric dominance. The vestibular cortex intimately interacts with the visual cortex to match the two 3-D orientation maps (perception of verticality, room-tilt illusion) and mediates self-motion perception by means of a reciprocal inhibitory visual–vestibular interaction. This mechanism of an inhibitory interaction allows a shift of the dominant sensorial weight during self-motion perception from one sensory modality (visual or vestibular) to the other, depending on which mode of stimulation prevails: body acceleration (vestibular input) or constant velocity motion (visual input).

INTRODUCTION

Vestibular pathways run from the VIIIth nerve and the vestibular nuclei through ascending fibers such as the medial longitudinal fasciculus to the ocular motor nuclei and the supranuclear integration center in the rostral midbrain. From there, they reach several vestibular cortex areas through thalamic projections. Many neurologists will admit to having only a vague concept of the vestibular cortex: its locations, functions, and disorders. The multiplicity of representations of vestibular cortical areas as identified in electrophysiological animal experiments raises doubts about the existence of a primary vestibular cortex comparable to the visual or auditory cortex, and most vestibular functions are not specifically vestibular. They also rely on visual and somatosensory input. All three systems—vestibular, visual, and somatosensory—provide us with redundant information about the position and motion of our body relative to the external space. The two major cortical functions of the vestibular system are perception of verticality and perception of self-motion. Perception of verticality relies mainly on otolith input; perception of self-motion involves otolith and semicircular canal input.

[a]To whom correspondence may be addressed. Phone: 49 89 7095 2570; Fax: 49 89 7095 8883; e-mail: tbrandt@brain.nefo.med.uni-muenchen.de

We will present evidence that the multisensory parieto-insular cortex in humans is involved in both of these functions. This evidence is based on lesional studies in patients with acute unilateral infarctions and on cerebral activation using optokinetic, caloric, and galvanic stimulation in PET and fMRI studies.

MULTIPLE VESTIBULAR CORTEX AREAS

Animal studies have identified several distinct and separate areas of the parietal and temporal cortices that receive vestibular afferents, such as area 2v at the tip of the intraparietal sulcus,[1–3] area 3aV (neck, trunk, and vestibular region of area 3a) in the central sulcus,[4] the parieto-insular vestibular cortex (PIVC) at the posterior end of the insula,[5,6] and area 7 in the inferior parietal lobule[7] (FIG. 1). Not only do these areas receive bilateral vestibular input from the vestibular nuclei, but they in turn directly project down to the vestibular nuclei.[8–10] Thus, corticofugal feedback may modulate vestibular brainstem function.

Our knowledge about vestibular cortex function in humans is less precise. It is derived mainly from stimulation experiments reported anecdotally in the older literature and from recent brain activation studies with PET[11–13] and fMRI.[14,15] It is not always possible to extrapolate from monkey species to human cortex, as Andersen and Gnadt[16] demonstrated for Brodmann's area 7 in rhesus monkey and humans. Area 2v corresponds best to the vestibular cortex, as described by Foerster.[17] The PIVC corresponds best to a region from which Penfield and Jasper[18] were able to induce vestibular sensations by electrical stimulation with a depth electrode within the sylvian fissure, medial to the primary acoustic cortex.

NO PRIMARY VESTIBULAR CORTEX

Microelectrode recordings from the so-called vestibular areas all demonstrate that the neurons are multisensory: they respond not only to vestibular but also to somatosensory and optokinetic stimuli (PIVC,[5,6] area 2v,[3] area 3a[4,19–20]). The same is true for area 7 of the inferior parietal cortex,[7,21–22] a major multisensory integration center for spatial orientation and visuomotor function. Area 7 also receives vestibular input, and vestibular neurons accumulated in part of area 7 (*7ant*) may represent the homologue to area 2v of the macaque brain.[23] The cytoarchitectonic structure of these vestibular cortical areas is also more characteristic of a multisensory or sensorimotor rather than a primary (unimodal) sensory cortex. Pandya and Sanides[24] emphasized the cytoarchitectonic homology of the retroinsular parietal cortex (*reipt*) in the monkey and the vestibular area anterior to the suprasylvian sulcus (ASSS) in the cat.[25,26]

If separate primary cortices for visual and auditory signals were established by evolution only to be abandoned later to provide the vestibular system with a primary area, what effect would this have on the conscious perception of different sensory modalities? Why is there no purely unimodal vestibular sensation? Shape and color of a presented object are detected by vision alone, and analysis within the visual cortex does not require proprioception or other senses. The same is true for the differentiation of tones and melodies. In contrast, natural stimulation of the vestibular system during head motion and locomotion is always multisensory (visual, vestibular, somatosensory). Redundant information for spatial orientation and postural control is provided by different sensory cues, among which vestibular signals may play a dominant role. Clinical vertigo syndromes with disorientation and falls serve as an example. Due to multisensory interaction, vestibular stimulation causes sensations that have vestibular, somatosensory, and visual qualities. Unlike the complexity of visual and auditory stimuli, the physical characteristics of vestibular stimuli are sufficiently defined by direction and amount of acceleration applied to the head. Their concurrent consequences for the visual and somatosensory systems make a separate evaluation of the unimodal vestibular quality in the cortex unnecessary.

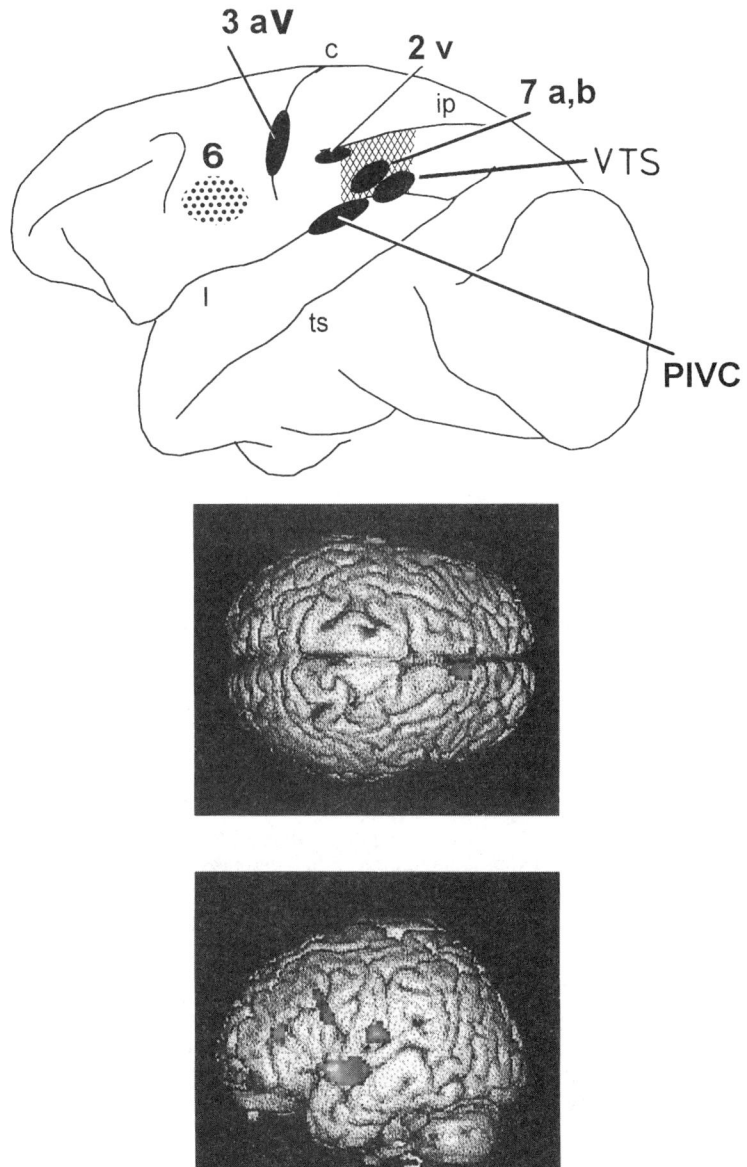

FIGURE 1. Schematic representation of a monkey brain (**top**) with the experimentally established areas that receive vestibular input: area 2v at the anterior part of the intraparietal sulcus, area 3a in the central sulcus, area 6, multisensory area 7 at the inferior parietal cortex, the PIVC deep in the posterior end of the insula, and the visual temporal sylvian VTS area, a temporal region adjacent to the PIVC. PET activation study in humans (**bottom**): Comparison of rCBF following caloric irrigation ($n = 9$) with 44° warm water (right ear) versus irrigation with water of body temperature. All voxels displayed were above a significance threshold of $p < 0.01$. It is possible to separate several distinct temporoparietal activations, the posterior part of the insula, the superior temporal gyrus, the praecentral gyrus, which may correspond to PIVC, areas 6 and 7, possibly 2v, and the praefrontal cortex.

THE PARIETO-INSULAR VESTIBULAR CORTEX

In view of the strong interconnections between PIVC and other vestibular cortex areas (mainly 3aV and 2v) as well as the vestibular brainstem nuclei, Guldin and Grüsser[9] postulate that it is the core region within the vestibular cortical system. About 50% of the neurons in this region respond to vestibular stimulation in addition to somatosensory, optokinetic, or visual stimulation. This area is involved not only in the processing of vestibular, somatosensory, and visual information, which is generated whenever the position of the body changes in relation to the extrapersonal space,[9] but also when stationary human subjects perform optokinetic nystagmus.[14,15]

The terminology and parceling of the insula and its surrounding opercula according to cytoarchitectonic structure vary for different species. The overlapping areas of infarctions in patients with pathological tilts of the subjective vertical center on the posterior insula (FIG. 2),[27] a region that Pandya and Sanides[24] termed "reipt" (retroinsular parietal cortex) in the rhesus monkey and that Grüsser and coworkers[5,6] delineated as PIVC, a part of the retroinsular cortex (Ri). The PIVC probably extends into the cervical representation of the secondary somatosensory area (SII) of the parietal operculum.[28] The neighboring area caudal to the PIVC is a secondary auditory area. This corresponds with findings of stimulation experiments in humans[18] and clinical experience[29] that shows that tinnitus and contralateral paresthesia may precede or accompany vertigo in vestibular epilepsy. Penfield and Kristiansen[30] localized the "epileptogenic lesions" in patients with vestibular auras to be roughly in the center of the posterior part of the temporal lobe. A focal activation of regional cerebral blood flow (rCBF) in the superior temporal region posterior to the auditory area was found during caloric vestibular stimulation in humans.[31] The same areas in the posterior insula, and the temporoparietal cortex, the putamen, and the anterior cingulate gyrus were activated in PET studies during caloric vestibular stimulation in humans.[11,12] Surprisingly the posterior insula (PIVC) was deactivated during large-field visual motion stimulation, which induces apparent self-motion ("reciprocal inhibitory visual–vestibular interaction").[13] Furthermore, PIVC is activated in fMRI with a significantly right-hemispheric dominance, when optokinetic nystagmus is performed (FIG. 3). This is not seen if optokinetic nystagmus is suppressed by fixation.[14] Activation of PIVC during optokinetic stimulation is thus related to oculomotor functions rather than to self-motion perception. When cerebral activation is investigated with fMRI during galvanic stimulation of the mastoid, three different sensory systems (vestibular, auditory, and nociceptive) can be visualized at the insular-thalamic level, including PIVC (FIG. 4).[32]

Speculations about the existence of a vestibular cortex function are based on the multiple projections from the thalamus to a kind of "inner circle" of the vestibular cortical representations, mainly PIVC areas (Ri, reipt), 7ant (2v), and 3aV, that have been analyzed by intracortical retrograde tracer injections.[33] The "proprioceptive vestibular area 3aV" receives its major thalamic projection from the oral and the superior ventroposterior nucleus (VPo). As the dominant cortical vestibular area, the PIVC receives its main input from the vestibular parts of the ventroposterior complex and the medial pulvinar. Area 7ant receives its major input from the posteromedial pulvinar. Cortico–cortical connections involve frontal and several parietal cortical areas as well as the parietotemporal association area T3, which receives optokinetic information via the medial, lateral, and inferior pulvinar.[33]

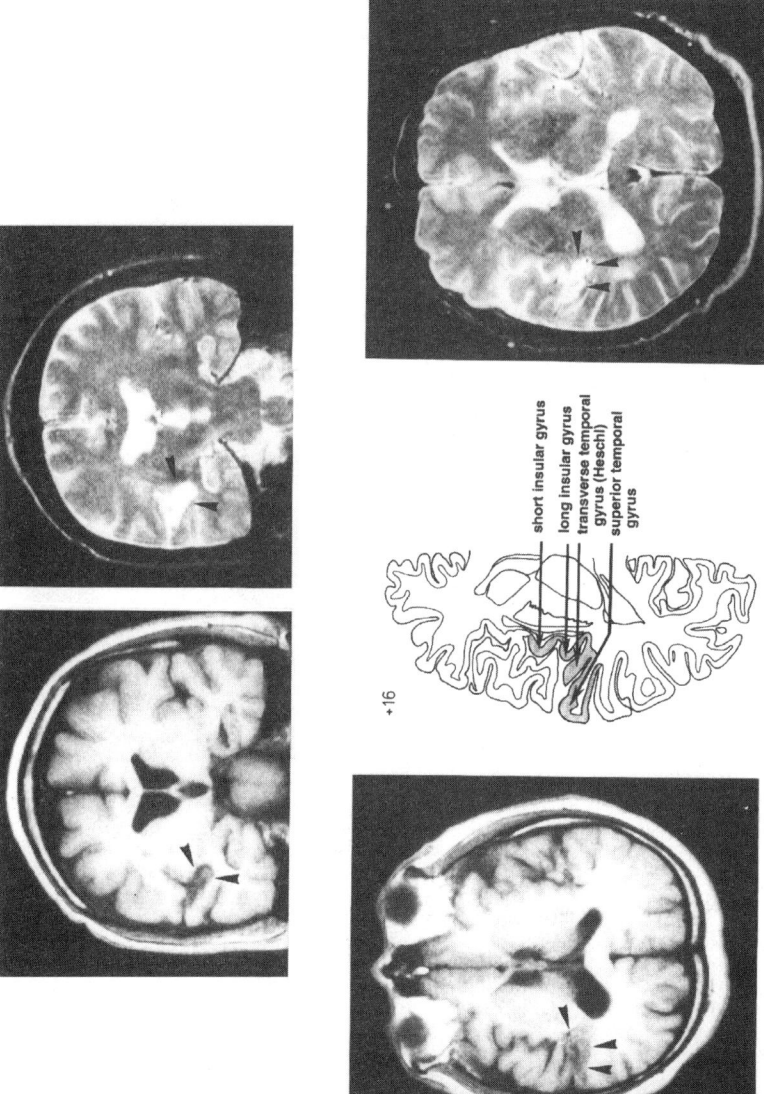

FIGURE 2. Magnetic resonance scan of a patient with regional infarction of the temporal branches of the right middle cerebral artery causing contraversive tilt of perceived vertical. (**Top**) The frontal sections show the vertical extent of the lesion (*arrowheads*), which does not involve the superficial temporal or parietal cortex (T1-weighted sequence on left, T2-weighted sequence on right). (**Bottom**) Transversal sections at a level about 16 mm above the anterior–posterior commissure (AC–PC) plane show an ischemic lesion (*arrowheads*) of the posterior part of the insula involving long insular gyral and retroinsular regions (T1-weighted sequence on left, T2-weighted sequence on right). For identification of the structures forming the posterior insula, see the appropriate section of the atlas of Duvernoy (**middle**).[27]

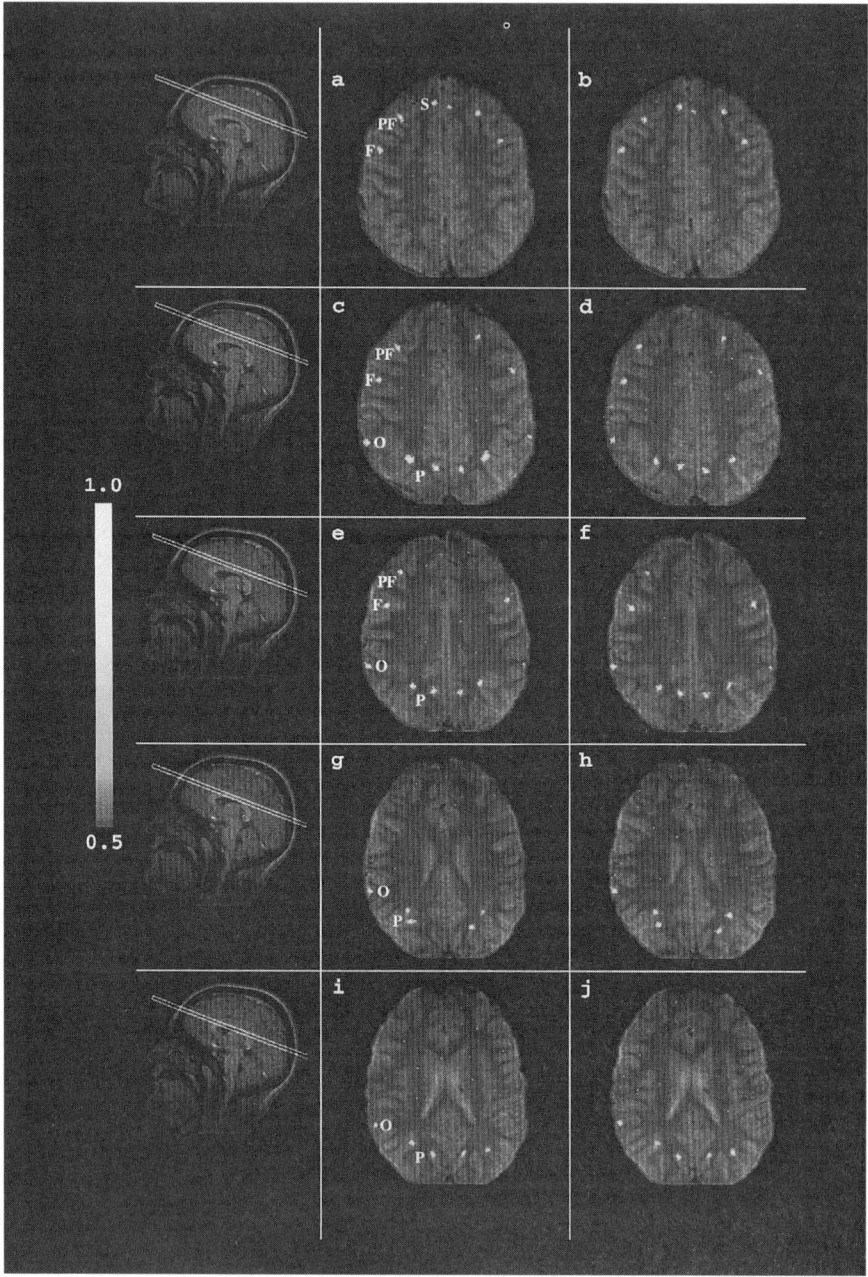

FIGURE 3. fMRI during optokinetic stimulation. T_2*-weighted coronal MR images of five cortical sections with superimposed activation map associated with optokinetic nystagmus induced by right (a, c, e, g, i) and left rotation of a drum (b, d, f, h, j) of a 27-year-old control subject (TR/TE=63/30 ms, (α=10 deg). The color-coded correlation coefficient scale ranges from 0.5 to a maximum of 1.0. During OKN bilaterally activated areas are the medial part of the superior frontal

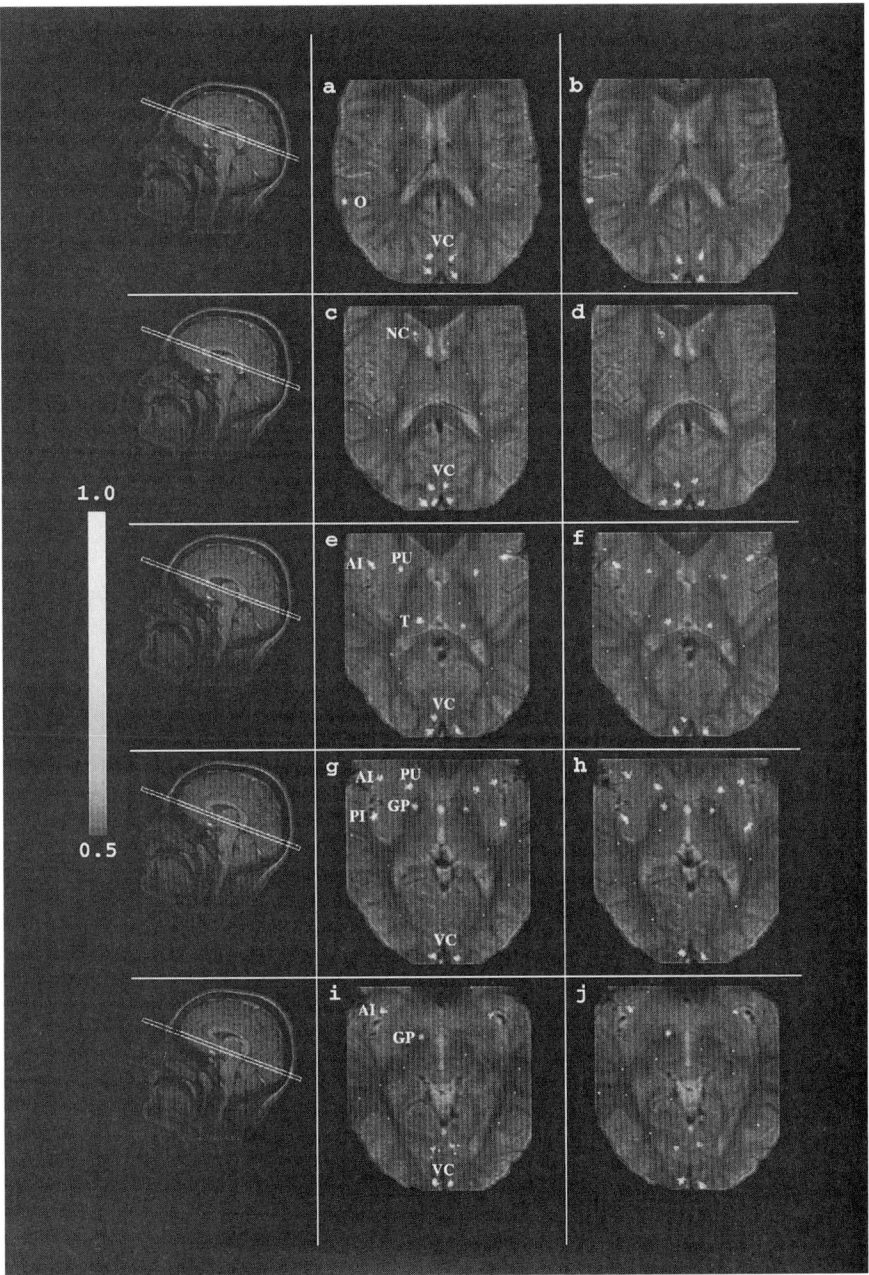

gyrus (supplementary eye field; S; a, b), the prefrontal cortex (PF; a–f), the precentral and posterior median frontal gyrus (frontal eye fields; F; a–f), and parts of the parietal cortex including the parietal eye field (P; c–j). The lateral occipitotemporal cortex (O; c–j) shows a strong asymmetric activity, mainly in the right hemisphere. Note that there is no difference in the anatomical location and extent of activation for both directions of object motion.[15]

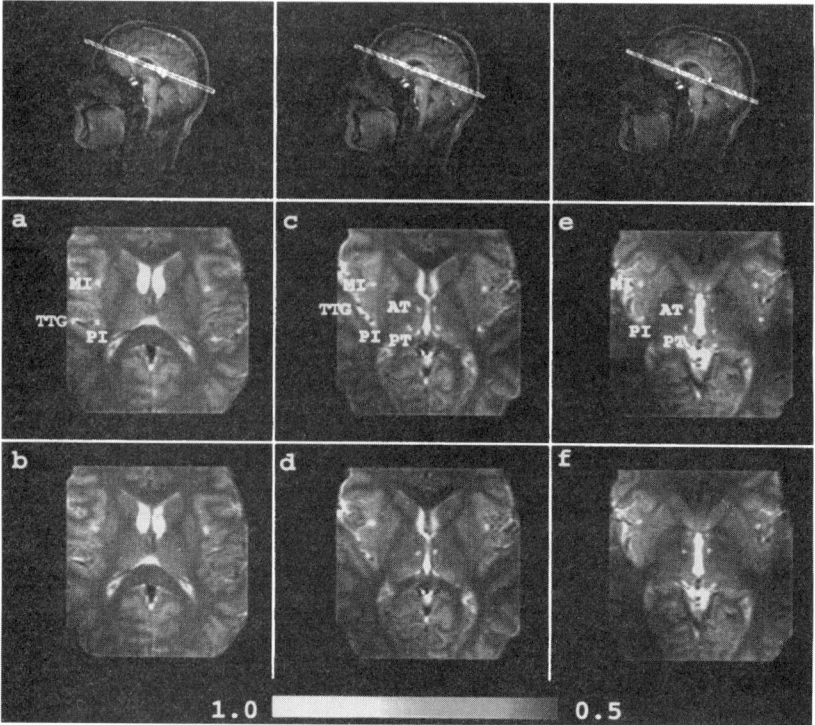

FIGURE 4. Galvanic vestibular stimulation: magnified cortical and subcortical activation maps superimposed on the corresponding coronal T_2*-weighted anatomical images of three sections of a 31-year-old subject (repetition time/echo time = 63/30 ms, α = 10 deg). The superimposed activation maps are associated with galvanic stimulation at the mastoid level (a, c, and e) and galvanic stimulation at the C4–5 level (b, d, and f) for control. The color-coded correlation coefficient scale ranges from 0.5 to a maximum of 1.0. Stimulation at the mastoid level (vestibular stimulation and cutaneous pain stimulation) caused activation in the medial part of the insula (MI; a, c, and e), the posterior part of the insula (PI; a, c, and e), the transverse temporal gyrus (TTG; a and c), the anterior median thalamus (AT; c and e), and the posterior median thalamus (PT; c and e). In contrast, stimulation at the C4–5 level (cutaneous pain stimulation) was associated with activity in the medial part of the insula (b, d, and f) and in the anterior medial thalamus (d and f) only.[32]

MULTIMODAL SENSORIMOTOR VESTIBULAR CORTEX FUNCTION AND DYSFUNCTION

The vestibular, the visual, and the somatosensory systems cooperate to determine our internal representation of space and subjective body orientation in unique 3-D coordinates, which are either egocentric (body-centered) or exocentric (world-centered). This is not a trivial process, since two of the sensory systems are anchored in the head, which moves relative to the trunk. Retinal coordinates—dependent as they are on gaze and head position—and fixed-head labyrinthine coordinates would require continuous updating of the particular eye and head positions in order to deliver reliable input for adequate ocular motor and motor exploration of space. Nature seems to have solved this impossible sensorimotor control of a multilink and multiaxis system by multisensory coding of space in either common egocentric or exocentric rather than retinotopic or head-centered coordinates. This has been demonstrated for posterior parietal neurons.[34,35] Spatial information in nonretinal coordinates allows us to determine body position relative to visual space, which is a necessary prerequisite for accurate motor response. To obtain such a frame of reference, information coded in coordinates of the peripheral sensory organs (retina, otoliths, semicircular canals, and proprioceptors such as muscle spindles) must be transformed and integrated.[36] This function is most probably subserved by the posterior parietal cortex, a lesion of which produces a visuospatial hemineglect. Karnath et al.[37,38] argued that neglect in brain-damaged patients is caused by a disturbance of the central transformation process that converts the sensory input coordinates from the periphery into an egocentric, body-centered coordinate system. The importance of the vestibular input for spatial orientation and the continuous updating of our internal representation of space becomes evident by the deficient spatial memory in microgravity during spacecraft missions. Large errors are made during prolonged microgravity when pointing at memorized targets, and it is the lack of knowledge of target position, not limb position, that is causative.[39]

In patients an inappropriate vestibular input due to peripheral or central dysfunction can cause paroxysmal "room-tilt illusions," the result of a mismatch of the two 3-D visual and vestibular coordinate maps. Furthermore, a plane- and direction-specific tilt of static spatial orientation occurs in disorders of the vestibulo-ocular reflex, such as downbeat and upbeat nystagmus. Adjustments of subjective straight-ahead exhibit an upward shift in downbeat nystagmus and a downward shift in upbeat nystagmus.[40] Here the tilt of perceived straight-ahead is elicited by the asymmetric vestibular tone in the pitch plane in the brainstem, which reaches the cortex by ascending projections. Vestibular syndromes caused only by cortical lesions have not yet been well defined.

Static cortical spatial disorientation may occur as

- Paroxysmal room-tilt illusion in parietal or frontal-lobe lesions,
- Contralateral spatial hemineglect in inferior parietal or frontal-lobe lesions,
- Vertical neglect below the horizontal meridian in bilateral parieto-occipital lesions, and
- Tilts of perceived vertical (mostly contraversive) and body lateropulsion in unilateral PIVC lesions.

Dynamic cortical spatial disorientation with apparent motion or rotational vertigo may occur

- In vestibular or rotatory epilepsy with temporoparietal foci, and
- Rarely as a transient vertigo in acute lesions of the vestibular cortex.

PAROXYSMAL ROOM-TILT ILLUSION

Room-tilt illusions are transient upside-down vision or apparent 90-degree tilts of the visual scene. In central vestibular disorders, they can be induced by either acute vestibulocerebellar brainstem lesions or cortical dysfunction. The two causative cortical regions are the parieto-occipital area[41–43] and rarely the frontal lobe.[44] A lesion in these regions can also cause spatial hemineglect. Solms et al.[44] have carefully reviewed 21 previously reported cases of transient upside-down vision. Tiliket et al.[45] reported that a sudden 90-deg room-tilt illusion could be elicited in three patients with unilateral brainstem lesions following vestibular stimulation by off-vertical rotary chair rotation. Furthermore, patients with peripheral bilateral vestibular failure may report transient room-tilt illusions on awakening in the morning. The latter observation may be comparable to the reports of astronauts about occasional upside-down vision in microgravity.[46]

We believe that room-tilt illusions are transient mismatches of the cortical visual and the vestibular 3-D coordinate maps that occur in 90- or 180-deg steps as the erroneous result of the attempted cortical match (FIG. 5).[47] Perceived tilt in one plane causes the afflicted person to attribute upright to horizontal or even down. The visual scene, which itself contains empirical spatial cues for upright, will then in turn "dominate and correct" spatial orientation. Vision will tell the vestibular systems where upright is. Therefore, room tilt illusions are transient: you cannot perceive two verticals at once. They should not be confused with the frequent tilts of subjective visual vertical, as they occur with unilateral peripheral vestibular, brainstem, thalamus, or cortex lesions. The matching of two separate sensory 3-D coordinate maps must be plastic in order to compensate for visual–vestibular tone imbalances and/or to adapt to unusual environments (microgravity). The plasticity of this visual–vestibular interaction has been best demonstrated when wearing reversing prisms.[48]

SELF-MOTION PERCEPTION: THE MECHANISM OF RECIPROCAL INHIBITORY VISUAL–VESTIBULAR INTERACTION

The vestibular system—a sensor of head accelerations—cannot detect motion at constant velocity, and thus requires supplementary visual information. Visual perception of self-motion induced by large-field optokinetic stimulation [circularvection, (CV); linearvection, (LV)] is essential.[49] Vestibular stimuli invariably lead to the sensation of body motion. Stimuli of visual motion, however, can always have two perceptual interpretations: either self-motion or object-motion.[50] The subject who observes moving stimuli may perceive either himself as being stationary in space (egocentric motion perception) or the actually moving surroundings as being stable while he is being moved (exocentric motion perception). Visual self-motion can be perceived while gazing at moving clouds or a train moving on the adjacent track in a train station. Vestibular information about motion is elicited only through acceleration or deceleration; it ceases when the cupulae within the semicircular canals or the otoliths have returned to their resting position during constant velocity. Our perception of self-motion during constant-velocity car motion is completely dependent on optokinetically induced vection.

To determine the unknown cortical visual–vestibular interaction during CV, we conducted a PET activation study on CV in human volunteers.[13] The PET images of activated cortical areas during visual motion stimulation without CV were subtracted from those with CV. It was shown that CV not only bilaterally activates a medial parieto-occipital visual area separate from motion-sensitive areas MT/MST, but simultaneously deactivates the parieto-insular vestibular cortex (FIG. 6). This finding supports a new functional interpretation: reciprocal inhibitory visual–vestibular interaction as a basic sensorimotor mechanism for adequate self-motion perception. Such a mechanism protects visual perception of self-motion from potential vestibular mismatches caused by involuntary head accelerations during locomotion.[13] This mechanism allows a shift of the dominant sensorial weight during self-motion perception from one sensory modality (visual or vestibular) to the other.

Depending on the mode of stimulation, perception of self-motion is dominated by either vestibular input (head acceleration) or visual input (constant velocity), or both. Quantitative visual–vestibular interaction is not a simple but rather a complex process; it reflects not only the pattern of motion stimulation but also particular active postural and locomotor tasks. The mechanism of reciprocal inhibitory visual–vestibular interaction (FIG. 7) makes it possible to relate the dominant perception of self-motion to the actual input of one of the two sensory modalities so as to avoid perceptual ambiguity. As a functional consequence, the concurrent deactivation of the vestibular cortex during CV should decrease the vestibular

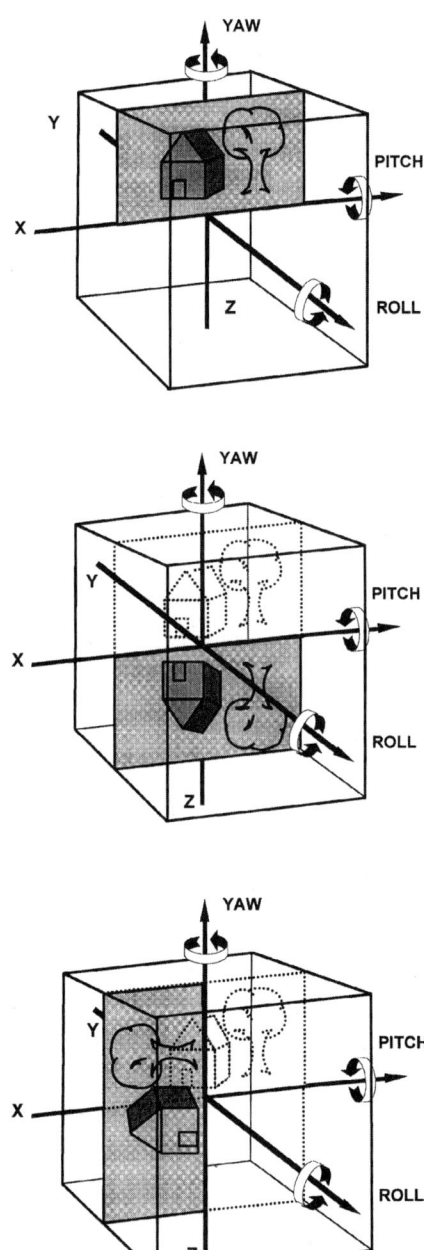

FIGURE 5. Schematic representation of the head as a cube with the cortical matching of the vestibular and the visual 3-D coordinate maps. The three major planes of action of the vestibular system are the frontal roll, the horizontal yaw, and the sagittal pitch about the x, y, and z axes, respectively. (**Top**) Visual scene matched with the vestibular coordinates. (**Middle**) Room tilt illusion with 180-deg-tilted visual scene in the pitch plane (*upside-down vision*). (**Bottom**) Room tilt illusion with 90-deg tilt in the roll plane.[47]

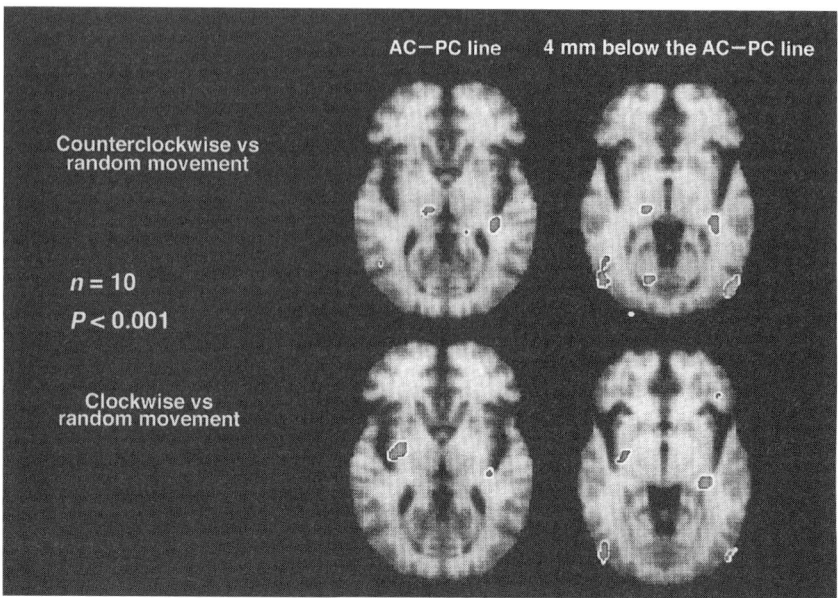

FIGURE 6. PET activation study during large-field visual stimulation inducing circularvection (CV). Comparison of the relative rCBF decreases under both conditions that induce CV (clockwise and counterclockwise) compared to the control condition without CV (random movement). All voxels shown are significantly above the statistical threshold ($p<0.001$ corrected for multiple comparisons). The transversal images illustrate deactivation of the posterior insula (PIVC) and of V5 (this deactivation is only relative to the control condition; compared to baseline, there is an rCBF increase). Thus, the vestibular cortex is deactivated during visually induced apparent self-motion. An inhibitory visual–vestibular interaction must be assumed during visual self-motion perception.[13]

system's sensitivity to head accelerations. This would make the perception of visually induced CV more robust and largely insensitive to visual–vestibular mismatches occurring during involuntary head accelerations in spatial planes different from the main direction of locomotion or transportation. The actual horizontal direction and speed perceived during constant velocity of car motion are transduced only by the relative optic flow of the surroundings. Concurrent vertical vestibular stimulations caused by car motion and secondary involuntary head accelerations provide vestibular information that is inadequate or even misleading with respect to self-motion perception in the horizontal direction. It is desirable that they be suppressed by deactivation of the vestibular system.[13] This hypothesis is supported by earlier findings that thresholds for detecting vestibular body accelerations (vestibular system) are significantly increased during optokinetically induced CV (visual system) in a combined rotatory chair-drum system.[51]

In light of these considerations, an earlier observation can be interpreted as the vestibulo–visual pendant for self-motion perception. In a PET study using caloric vestibular irrigation, activation of the vestibular cortex caused a significant, bilateral decrease of rCBF in the occipital visual cortex that covers Brodmann areas 17, 18, and 19 (FIG. 8).[52] Elementary visual hallucinations have been evoked by caloric stimulation in humans,[53] which might be related to hemodynamic disturbance of the visual system. Using transcranial Doppler sonography, we detected a reduction of cerebral blood flow velocity during caloric vestibular stimulation.[54] These findings suggested to us that deactivation of the visual cortex is beneficial to the organism during vestibular stimulation, since it suppresses visual motion input (e.g., distressing oscillopsia, owing to the retinal slip of the visual scene during vestibular nystagmus).

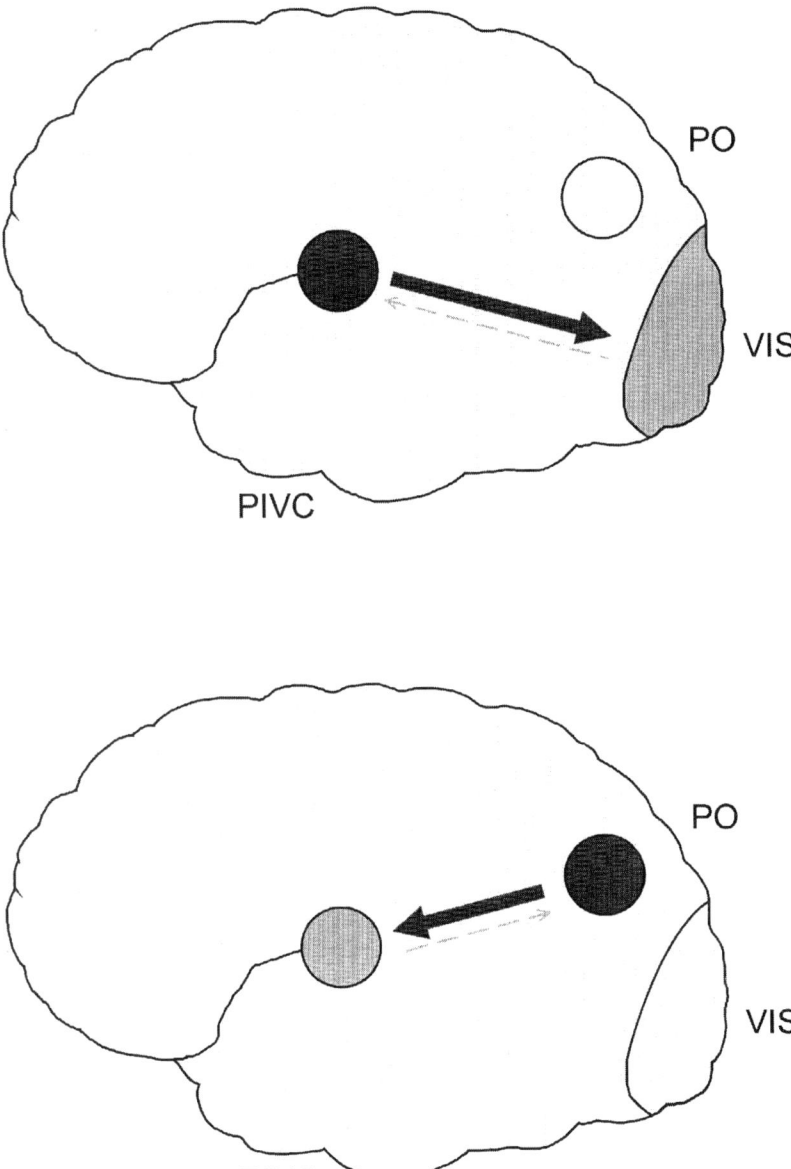

FIGURE 7. Schematic representation of the human brain, illustrating reciprocal inhibitory visual–vestibular interaction for self-motion perception. If vestibular stimulation dominates during body accelerations, the PIVC is activated and the primary visual cortex (VIS) is deactivated (**top**). This suppresses distressing oscillopsia due to vestibular nystagmus and reduces the visual sensorial weight for self-motion perception. If self-motion is elicited by large-field visual motion stimulation (vection), this activates the medial parieto-occipital cortex (PO) (**bottom**), but deactivates the vestibular cortex (PIVC). In functional terms this means that if self-motion perception is dominated by vision, perception of concurrent vestibular stimuli is partially suppressed.

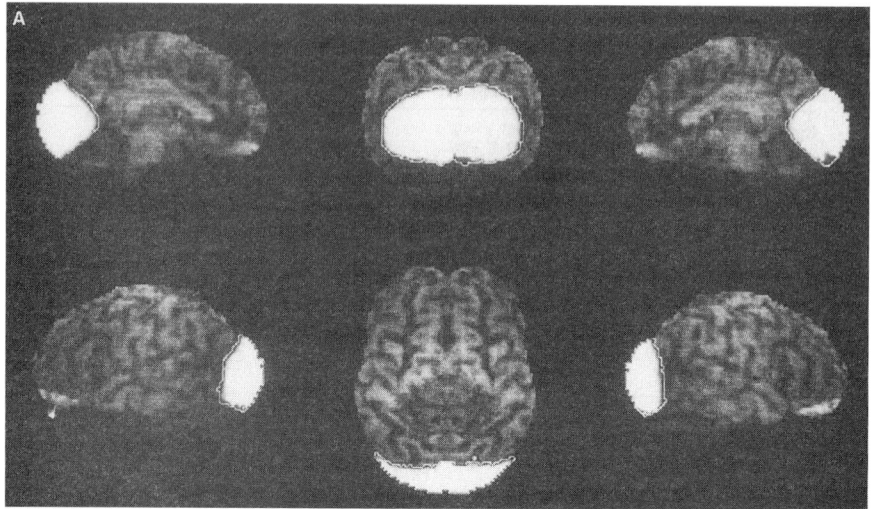

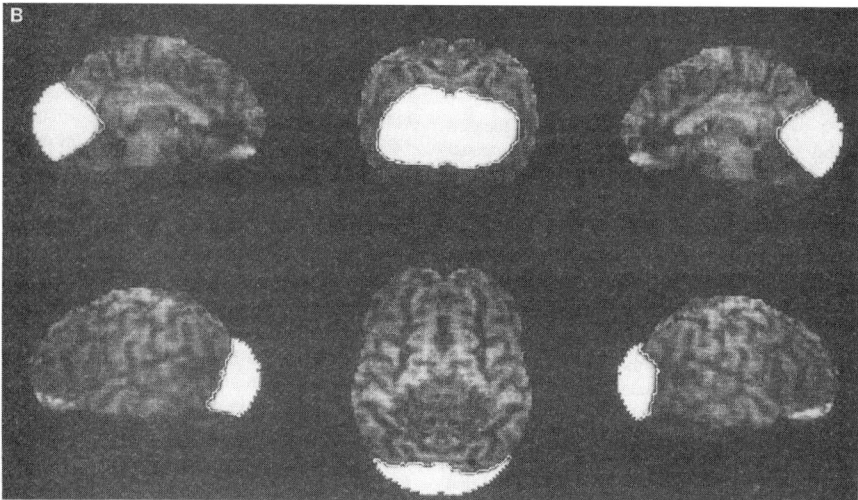

FIGURE 8. PET activation/deactivation study following caloric vestibular stimulation. Comparison of the activation conditions versus rest. Significant decreases of rCBF following vestibular irrigation ($n=6$). All significant pixels above the adjusted statistical threshold ($Z>4.24$ in A) and ($Z>4.27$ in B) are displayed. All pixels displayed white indicate a Z score >5. Cold water irrigation of the right ear (**A**) and of the left ear (**B**): top row, from left to right, medial view of the left hemisphere, posterior view, medial view of the right hemisphere; bottom row, lateral view of the corresponding hemispheres and view from above. Shaded areas of the normalized MRI indicate areas outside the field of view. There is a marked decrease of rCBF restricted to the occipital cortex under both stimulation conditions.[52]

In the same way that deactivation of the visual cortex largely protects the vestibular system from conflicting visual motion input, deactivation of the vestibular cortex prevents visually induced CV from conflicting with vestibular input.

Besides this reciprocal inhibitory interaction, it is very likely that visual and vestibular cortices have other forms of interaction depending on the actual stimulation, the required dynamic spatial orientation, and the intended motor tasks. Furthermore, somatosensory information about motion must also be integrated (visual–vestibular–somatosensory interaction). Activation of both cortices is required for adequate perception of self-motion and for postural control in stimulus situations involving unexpected, multidirectional transitions between body accelerations and motion at constant velocity. Both hemispheres must be activated in stimulus situations in which, for example, both visual fields have contradictory information about motion. When sitting in a train reading, information about the constant velocity of self-motion is provided by optokinetic stimulation in one visual hemifield, whereas the other is filled with contrasts of the stationary train. Since you cannot perceive two different states of body motion at the same time, the two hemispheres have to correspond to each other, and by an internal shift of attention, determine an actual and unique perception of motion or absence of motion.

All prestriatal visual areas have transcallosal interhemispheric connections for this kind of bilateral interaction. In fact, we recently demonstrated with fMRI that during horizontal optokinetic stimulation the occipito-temporal motion-sensitive areas MT/MST were activated in both hemispheres of patients with acute complete homonymous hemianopia (FIG. 9). The most likely explanation for this phenomenon is that the unaffected hemisphere is either activated transcallosally (FIG. 10) or via ipsilateral extrastriatal pathways from the geniculate bodies or the superior thalamic pulvinar colliculus route.[55] MT/MST provide us with intermediate information about visual motion in extraretinal coordinates, which can be used for motor control of not only the eyes but also the body. Interhemispheric interaction requires neural and interhemispheric synchronization that is mediated by corticocortical connections.[56,57]

CONCLUSIONS

We have tried to present evidence that the parieto-insular cortex

- Is the human homologue of the major multisensory vestibular cortex (PIVC) in the monkey;
- Is involved in the perception of verticality and self-motion;
- Can be activated by otolith (galvanic) or semicircular canal input (caloric or galvanic) and optokinetic stimulation;
- If lesioned, may cause perceived tilt, lateropulsion, or rotational vertigo;[58]
- Intimately interacts with the visual cortex for matching the two 3-D orientation maps (room tilt illusion) and mediating self-motion perception (inhibitory reciprocal visual-vestibular interaction).

Electrophysiological single-unit recordings and tracer studies in monkeys, lesional studies in neurological patients, and brain-activation studies in human subjects using PET and fMRI have taught us a lot about the vestibular cortex, its locations, functions, and disorders. Compared to the visual system, however, research on the vestibular cortex is still in its infancy.

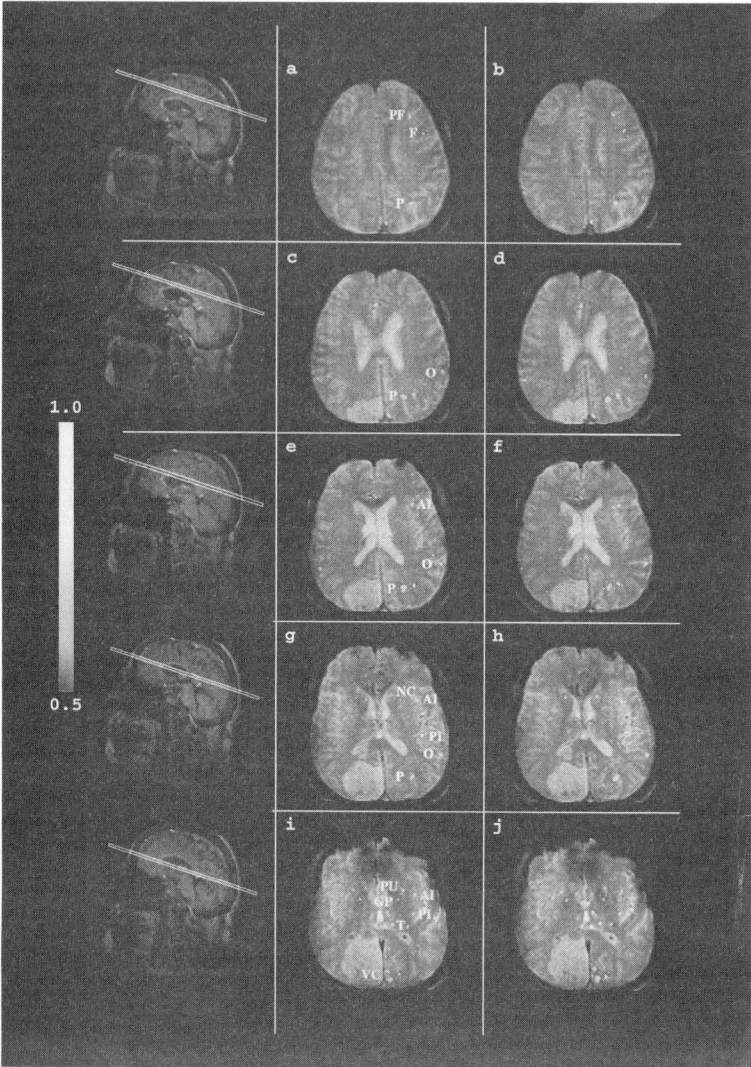

FIGURE 9. T_2*-weighted coronal MR images of five cortical sections of a 63-year-old patient with infarction of the right posterior cerebral artery that caused complete hemianopia (TR/TE=63/30 ms, α=100 deg). The superimposed activation maps are associated with optokinetic nystagmus induced by right (a, c, e, g, i) and left motion stimulation (b, d, f, h, j). The color-coded correlation coefficient scale ranges from 0.5 to a maximum of 1.0. Although motion stimulation was restricted to the left hemisphere, bilateral activation of the motion-sensitive areas MT/MST was found (O, e–h). All other cortical areas including the prefrontal cortex (PF; a,b), the precentral and posterior median frontal gyrus (frontal eye fields; F; a,b), parts of the parietal cortex including the parietal eye field (P; c–h), the anterior part (AI; g–j) and posterior part (PI; g–i) of the insula, and the primary visual cortex (VC, e–j) were not activated on the infarcted hemisphere. Thalamic activity was not seen in the infarcted hemisphere, while other subcortical areas such as the putamen, globus pallidus, caudate nucleus showed bilateral activation. Note that there is no difference in the anatomical location and extent of activation for both directions of object motion.[55]

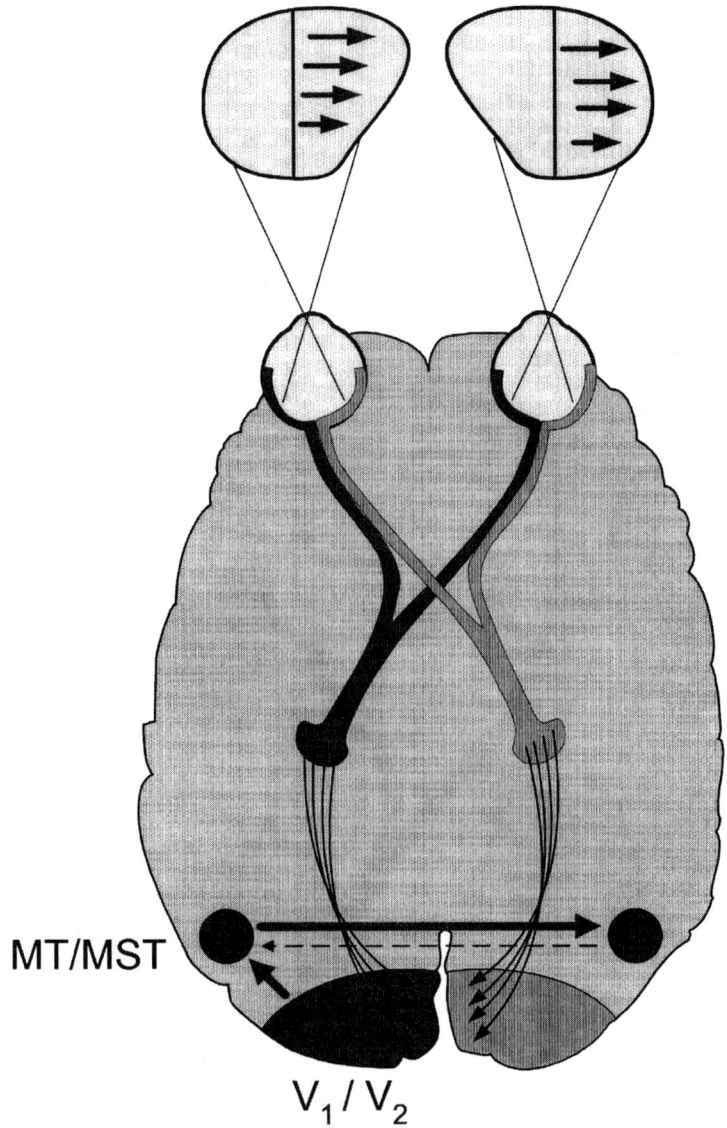

FIGURE 10. Schematic representation of the visual pathways and stimulation of the cortex in complete homonymous hemianopia due to infarction of the right occipital cortex. Full-field visual motion stimulation will then lead to stimulation of only the unaffected left hemisphere (V1/V2). Motion-sensitive visual areas (MT/MST) are activated by ipsilateral pathways on the unaffected hemisphere. Simultaneous activation of the contralateral MT/MST is due either to ipsilateral extrastriatal pathways from the geniculate body or the superior thalamic pulvinar colliculus route or by interhemispheric transcallosal connections (*long horizontal arrow*). Interhemispheric interaction is required for adequate visual self-motion perception, which is best mediated by transcallosal cortico–cortical connections.

REFERENCES

1. SCHWARZ, D. W. F. & J. M. FREDRICKSON. 1971. Rhesus monkey vestibular cortex: a bimodal primary projection field. Science **172**: 280–281.
2. FREDRICKSON, J. M., *et al.* 1966. Vestibular nerve projection to the cerebral cortex of the rhesus monkey. Exp. Brain Res. **2**: 318–327.
3. BÜTTNER, U. & U. W. BUETTNER. 1978. Parietal cortex area 2v neuronal activity in the alert monkey during natural vestibular and optokinetic stimulation. Brain Res. **153**: 392–397.
4. ÖDKVIST, L. M., *et al.* 1974. Projection of the vestibular nerve to the area 3a arm field in the squirrel monkey (*Saimiri sciureus*). Exp. Brain Res. **21**: 97–105.
5. GRÜSSER, O. J., M. PAUSE & U. SCHREITER. 1990. Localization and responses of neurons in the parieto-insular vestibular cortex of the awake monkeys (*Macaca fascicularis*). J. Physiol. **430**: 537–557.
6. GRÜSSER, O. J., M. PAUSE & U. SCHREITER. 1990. Vestibular neurons in the parieto-insular cortex of monkeys (*Macaca fascicularis*): visual and neck receptor responses. J. Physiol. **430**: 559–583.
7. FAUGIER-GRIMAUD, S. & J. VENTRE. 1989. Anatomic connections of inferior parietal cortex (area 7) with subcortical structures related to vestibulo-ocular function in a monkey (*Macaca fascicularis*). J. Comp. Neurol. **280**: 1–14.
8. AKBARIAN, S., O.-J. GRÜSSER & W. O. GULDIN. 1994. Corticofugal connections between the cerebral cortex and brainstem vestibular nuclei in the macaque monkey. J. Comp. Neurol. **339**: 421–437.
9. GULDIN, W. & O.-J. GRÜSSER. 1996. The anatomy of the vestibular cortices of primates. *In* Le Cortex Vestibulaire, M. Collard, M. Jeannerod, and Y. Christen, Eds.: 17–26. Ipsen. Boulogne, France.
10. JEANNEROD, M. 1996. Vestibular cortex. A network from directional coding of behavior. *In* Le Cortex Vestibulaire, M. Collard , M. Jeannerod and Y. Christen, Eds.: 5–15. Ipsen. Boulogne, France.
11. BOTTINI, G., *et al.* 1994. Identification of the central vestibular projections in man: a positron emission tomography activation study. Exp. Brain Res. **99**: 164-169.
12. DIETERICH, M., *et al.* 1999. Differential effects of warm and cold water caloric irrigation on cerebral activation (a PET study). (submitted).
13. BRANDT, T., *et al.* 1998. Reciprocal inhibitory visual–vestibular interaction: visual motion stimulation deactivates the parieto-insular vestibular cortex. Brain **121**: 1749–1758.
14. DIETERICH, M., *et al.* 1998. Horizontal or vertical optokinetic stimulation activates visual motion sensitive ocularmotor and vestibular cortex areas with right hemispheric dominance: an fMRI study. Brain **121**: 1479–1495.
15. BUCHER, S. F., *et al.* 1997. Sensorimotor cerebral activation during optokinetic nystagmus: an fMRI study. Neurology **49**: 1370–1377.
16. ANDERSEN, R. A. & J. W. GNADT. 1989. Posterior parietal cortex. *In* Reviews in Oculomotor Research, Vol. 3, The Neurobiology of Saccadic Eye Movements, R. H. Wurtz and M. E. Goldberg, Eds.: 315–335. Elsevier. Amsterdam.
17. FOERSTER, O. 1936. Sensible Kortikale Felder. *In* Handbuch der Neurologie, Vol Vl, O. Bumke and O. Foerster, Eds.: 358–449. Springer-Verlag. Berlin.
18. PENFIELD, W. & H. JASPER. 1954. Epilepsy and the Functional Anatomy of the Human Brain. Little, Brown. Boston.
19. PHILLIPS, C. G., T. P. S. POWELL & M. WIESENDANGER. 1971. Projection from low threshold muscle afferents of hand and forearm to area 3a of Baboon's cortex. J. Physiol. (Lond.) **217**: 419–446.
20. SCHWARZ, D. W. F., L. DEECKE & J. M. FREDRICKSON. 1973. Cortical projection of group I muscle afferents to areas 2, 3a and the vestibular field in the rhesus monkey. Exp. Brain Res. **17**: 516–526.
21. ANDERSEN, R. A. 1987. Inferior parietal lobule function in spatial perception and visuomotor integration. *In* Handbook of Physiology. Section I: The Nervous System, Vol. V, V. B. Mountcastle, F. Plum and S. R. Geiger, Eds.: 483–518. American Physiological Society. Bethesda, Md.
22. KAWANO, K., M. SASAKI & M. YAMASHITA. 1980. Vestibular input to visual tracking neurons in the posterior parietal association cortex of the monkey. Neurosci. Lett. **17**: 55–60.

23. GULDIN, W. O., S. AKBARIAN & O.-J. GRÜSSER. 1992. Cortico-cortical connections and cytoarchitectonics of the primate vestibular cortex: a study in squirrel monkeys (*Saimiri sciureus*). J. Comp. Neurol **324**: 1–27.
24. PANDYA, D. N. & F. SANIDES. 1973. Architectonic parcellation of the temporal operculum in rhesus monkey and its projection pattern. Z. Anat. Entwicklungsg. **139**: 127–161.
25. WALZL, E. M. & V. B. MOUNTCASTLE. 1949. Projection of vestibular nerve to cerebral cortex of cat. Am. J. Physiol. **159**: 595.
26. MICKLE, W. A. & H. W. ADES. 1952. A composite sensory projection area in the cerebral cortex of the cat. Am. J. Physiol. **170**: 682–689.
27. BRANDT, T., M. DIETERICH & A. DANEK. 1994. Vestibular cortex lesions affect the perception of verticality. Am. Neurol. **35**: 403–412.
28. JONES, E. G. & H. BURTON. 1976. Areal differences in the laminar distribution of thalamic afferents in cortical fields of insular, parietal and temporal opercular regions of primates. J. Comp. Neurol. **168**: 197–247.
29. SMITH, B. H. 1960. Vestibular disturbance in epilepsy. Neurology **10**: 465–469.
30. PENFIELD, W. & K. KRISTIANSEN. 1951. Epileptic seizure patterns. Charles L. Thomas. Springfield, Ill.
31. FRIBERG, L., et al. 1985. Focal increase of blood flow in the cerebral cortex of man during vestibular stimulation. Brain **108**: 609–623.
32. BUCHER, S. F., et al. 1998. Cerebral functional magnetic resonance imaging of vestibular, auditory, and nociceptive areas during galvanic stimulation. Ann. Neurol. **44**: 120–125.
33. AKBARIAN, S., O.-J. GRÜSSER & W. O. GULDIN. 1992. Thalamic connections of the vestibular cortical fields in the squirrel monkey (*Saimiri sciureus*). J. Comp. Neurol. **325**: 1–19.
34. ANDERSEN, R. A., G. K. ESSICK & R. M. SIEGEL. 1985. Encoding of spatial location by posterior parietal neurons. Science **230**: 456–458.
35. GALLETTI, C., P. P. BATTAGLINI & P. FATTORI. 1993. Parietal neurons encoding spatial orientations in craniotopic coordinates. Exp. Brain Res. **96**: 221–229.
36. KARNATH, H.-O. 1994. Subjective body orientation in neglect and the interactive contribution of neck muscle proprioception and vestibular stimulation. Brain **117**: 1001–1012.
37. KARNATH, H.-O., P. SCHENKEL & B. FISCHER. 1991. Trunk orientation as the determining factor of the "contralateral" deficit in the neglect syndrome and as the physical anchor of the internal representation of body orientation in space. Brain **114**: 1997–2014.
38. KARNATH, H.-O., K. CHRIST & W. HARTJE. 1993. Decrease of contralateral neglect by neck muscle vibration and spatial orientation of trunk midline. Brain **116**: 383–396.
39. WATT, D. G. D. 1997. Pointing at memorized targets during prolonged microgravity. Aviat. Space Environ. Med. **68**: 99–103.
40. DIETERICH, M., W. M. GRÜNBAUER & T. BRANDT. 1998. Direction-specific impairment of motion perception and spatial orientation in downbeat and upbeat nystagmus.Neurosci. Lett. **245**: 29–32.
41. GERSTMANN, J. 1926. Über eine eigenartige Orientierungsstörung im Raum bei zerebraler Erkrankung. Wien. Med. Wochenschr. **76**: 817–818.
42. HALPERN, F. 1930. Kasuistischer Beitrag zur Frage des Verkehrtsehens. Z. Gesamte Neurol. Psychiatr. **126**: 246–252.
43. KLOPP, H. 1951. Über Umgekehrt- und Verkehrtsehen. Deutsch. Z. Nervenheilk. **165**: 231–260.
44. SOLMS, M., et al. 1988. Inverted vision after frontal lobe disease. Cortex **24**: 499–509.
45. TILIKET, C., et al. 1996. Room tilt illusion. A central otolith dysfunction. Arch. Neurol. **53**: 1259–1264.
46. GLASAUER, S. & H. MITTELSTAEDT. 1992. Determinants of orientation in microgravity. Acta Astronaut. **27**: 1–9.
47. BRANDT, T. 1997. Cortical matching of visual and vestibular 3-D coordinate maps. Ann. Neurol. **42**: 983–984.
48. KOHLER, I. 1956. Die Methode des Brillenversuches in der Wahrnehmungspsychologie mit Bemerkungen zur Lehre der Adaptation. Z. Exp. Angew. Psychol. **3**: 381–417.
49. DICHGANS, J. & T. BRANDT. 1978. Visual-vestibular interaction: effects on self-motion perception and postural control. *In* Handbook of Sensory Physiology, Vol VIII, Perception, R. Held, H. W. Leibowitz, and H.-L. Teuber, Eds.: 755–804. Springer-Verlag. Berlin.

50. BRANDT, T., J. DICHGANS & E. KOENIG. 1973. Differential effects of central versus peripheral vision on egocentric and exocentric motion perception. Exp. Brain Res. **16**: 476–491.
51. PROBST, T., A. STRAUBE & W. BLES. 1985. Differential effects of ambivalent visual-vestibular-somatosensory stimulation on the perception of self motion. Behav. Brain Res. **16**: 71–79.
52. WENZEL, R., et al. 1996. Deactivation of human visual cortex during involuntary ocular oscillations: a PET activation study. Brain **119**: 101–110.
53. KOLEV, O. I. 1995. Visual hallucinations evoked by caloric vestibular stimulation in normal humans. J. Vestib. Res. **5**: 19–23.
54. TIECKS, F. P., et al. 1996. Reduction in posterior cerebral artery blood flow velocity during caloric vestibular stimulation. J. Cerebr. Blood Flow Metab. **16**: 1379–1382.
55. BRANDT, T., et al. 1998. Bilateral functional MRI activation of the basal ganglia and middle temporal/medial superior temporal motion sensitive areas. Arch. Neurol. **55**: 1126–1131.
56. NOWAK, L. G., et al. 1995. Structural basis of cortical synchronization. I. Three types of interhemispheric coupling. J. Neurophysiol. **74**: 2379–2400.
57. MUNK, M. H. J., et al. 1995. Structural basis of cortical synchronization. II. Effects of cortical lesions. J. Neurophysiol. **74**: 2401–2414.
58. BRANDT, T., et al. 1995. Rotational vertigo in embolic stroke of the vestibular and auditory cortices. Neurology **45**: 42–44.

Cortical Areas Activated by Bilateral Galvanic Vestibular Stimulation

ELIE LOBEL,[a,c] JUSTUS F. KLEINE,[b] ANNE LEROY-WILLIG,[c]
PIERRE-FRANÇOIS VAN DE MOORTELE,[c] DENIS LE BIHAN,[c]
OTTO-JOACHIM GRÜSSER,[b,e] AND ALAIN BERTHOZ[a,d]

[a]*Laboratoire de Physiologie de la Perception et de l'Action, Collège de France, Paris, France*

[b]*Department of Physiology, Freie Universität, Berlin, Germany*

[c]*Service Hospitalier Frédéric Joliot, C.E.A., Orsay, France*

ABSTRACT: The brain areas activated by bilateral galvanic vestibular stimulation (GVS) were studied using functional magnetic resonance imaging. In six human volunteers, GVS led to activation in the region of the temporoparietal junction, the central sulcus, and the anterior interior intraparietal sulcus, which may correspond to macaque areas PIVC, 3aV, and 2v, respectively. In addition, activation was found in premotor regions of the frontal lobe, presumably analogous to areas 6pa and 8a in the monkey. Since these areas were not detected in previous studies using caloric vestibular stimulation, they could be related to the modulation of otolith afferent activity by GVS. However, the simple paradigm used did not allow separation of the otolithic and semicircular canal effects of GVS. Further studies must be performed to clarify the question of cortical representation of the otolithic information in the human and monkey brain.

INTRODUCTION

In the present work we have studied the areas of the cerebral cortex involved in the perceptual and motor effects of stimulation of the vestibular organs using galvanic vestibular stimulation (GVS) with functional magnetic resonance imaging (fMRI). This is the first time that GVS has been used with fMRI, because we have been able to implement GVS in this electromagnetic environment without any danger to the subjects.

GVS has been used for a long time for the assessment of vestibular function. Early in this century, clinicians were interested in GVS for the functional exploration of vestibular organs[1,2] and they studied some specific pathologies to show that the galvanically induced postural effects had a vestibular origin.[3] More recently the clinical relevance of GVS was evaluated[4-7] and quantitative methods for measuring the postural effects, attributed to vestibulospinal or vestibulo–reticulo–spinal descending influences, were proposed.[8-12] Some limits to the interest of the method were established[13] and in contrast to caloric vestibular stimulation (CVS), nowadays it is not extensively used in the clinic.

However, GVS has several advantages over CVS. First, it is generally recognized that GVS stimulates both canals and otolith afferents, as compared to CVS, which mainly stimulates the semicircular canals.[14] GVS is also appropriate for use during a neuroimaging

[d]To whom correspondence may be addressed: e-mail: aber@ccr.jussieu.fr
[e]Deceased on October 17, 1995.

experiment because it has a fast reaction time (VS builds up after the onset and ceases after the offset of current with a 1- to 2-s lag) and the intensity of the delivered current can be adjusted individually to reach a precise level of VS. Neurophysiological studies have provided some information concerning the mechanisms of action of GVS.[15] It now seems certain that the main effect of GVS is on irregular afferent nerve fibers[14] and it has been shown that this effect induces modification both in the canal-induced vestibulo-ocular reflex and in the otolith-induced nystagmic eye movements during OVAR.[16,17]

However, no information is yet available concerning the cortical areas involved in the processing of galvanic stimulation and that may be implied in the various perceptual or motor effects. The purpose of the present study was therefore to use GVS in order to obtain a more complete description of the areas involved in vestibular processing. We used a specially designed stimulator using sinusoidal galvanic stimulation at a frequency of 1 Hz.[18,19] In this first study we chose binaural stimulation in order to minimize the cutaneous or nociceptive stimulation. The perceptive, postural, and oculomotor effects of the stimulation were tested for each subject before the fMRI experiment.

It is clear that in the present set of data both somatosensory and vestibular stimulation were present as in normal clinically applied galvanic stimulation. In any case, most of the vestibular areas recognized so far have been found to be activated by multisensory inputs. Further studies are in progress to try to separate these two components of the stimulation.

MATERIAL AND METHODS

This study was approved by the Institutional Ethics Committee for Biomedical Research. Six right-handed male volunteers, aged 20 to 30, healthy and devoid of ear problems, participated in this experiment after giving their informed consent.

Galvanic Stimulation

Stimulus waveforms were generated by a function generator (Hewlett-Packard 33120 A) and fed via optocouplers into two channels of a specially constructed battery-driven current-controlled amplifier (Dipl.-Ing. Nitert, FU Berlin). The current signals reached the subjects via high-impedance carbon-fiber electrodes (Bruker, Germany). To increase the contact area at the electrode sites and thereby reduce somatosensory side effects, three triplets of electrodes were used; these were placed on either mastoid and on the back between the scapulae, making up two independently controlled circuits. The stimuli in all experiments consisted of 1.0-Hz sinusoids, which were 180 deg out of phase in the two circuits, resulting in synergistic stimulation of the left and right labyrinth. Current flow was measured as the voltage drop across resistors placed in series with the test subjects and continuously monitored on oscilloscopes during the experiments. This device was made compatible with the MRI equipment using LC filters and carbon fiber electrodes.[20] Several preliminary experiments were performed to ensure the safety and the quality of the installation.

Psychophysical Experiments

In order to determine individual susceptibility to galvanically induced vestibular sensations and the sensitivities for undesired side effects, such as pain at the electrode sites or nausea, all participants were subjected to pretests one day prior to the fMRI experiment. To accustom them to the stimulation, subjects were first tested while sitting upright in an armchair equipped with a head support in a dark room with their eyes closed. Sinusoidal GVS was applied (see earlier) and the stimulus amplitude slowly increased from 0 mA until the subject first had a clear perception of body movement, and was then further increased to the level where heat sensation at the electrode sites started to become unpleasant or painful. Most subjects perceived a slight metallic taste at high stimulus amplitudes, but major nausea did not occur. Finally, stimulus intensity was lowered again until the vestibular sensation disappeared. By repeating the entire procedure several times, approximate threshold values for vestibular sensations and painful side effects could reliably be established in all subjects. Also, threshold values were determined with the subjects lying in the supine position, their head free on a firm mattress. Vestibular thresholds in the seated and the supine position differed by no more than 0.2 mA in all subjects and ranged from 1.2 to 1.6 mA (mean 1.5 mA). Pain thresholds were somewhat more variable, and ranged from 3.5 to 4.7 mA (mean 4.1 mA). The stimulus intensity for the fMRI-experiment was set at 0.5 mA less than the individual pain threshold. Thus, for each subject, stimulus amplitude in the fMRI-experiment was at about 2.5 times, at least 2 times, the vestibular threshold value as determined during the pretest.

fMRI Experiments

Experiments were performed on a 3-tesla whole-body system (Bruker, Germany), equipped with a quadrature birdcage RF coil and a head-gradient coil insert designed for echoplanar imaging (EPI). Involuntary head movements inside the magnet were prevented by taping the subject's forehead and firmly restraining it on either side with foam pads. For noise protection, the subjects wore earplugs and ear protectors. During the functional experiment, the volunteers were placed in complete darkness, with their eyes closed, and submitted for 5 min to alternations of periods with and without GVS (32.7 s per period). To keep attention constantly focused, the subjects continuously performed rhythmic fingertapping with their right index finger throughout both stimulation and rest periods. After image acquisition, they were asked to give detailed descriptions of what they had perceived during the experiment.

Data Acquisition

Sets of high-resolution images (voxel: $1 \times 1 \times 2.5$ mm^3) were acquired for anatomical identification. Functional images were acquired with a T2*-weighted single-shot gradient-echo EPI sequence (TE = 40 ms, TR = 4670 ms, voxel = $4 \times 4 \times 5$ mm,^{3} 14 to 18 slices, 63 repetitions). The functional slab was centered on the lateral sulcus: the field of view of the analysis we performed (see below) includes the inferior half of the frontal and parietal gyri, as well as the superior temporal gyrus and the superior part of the occipital lobe.

Data Analysis

Data were analyzed as a group using the SPM96 software (Wellcome Department of Neurology, London). The first four functional volumes were discarded to make sure that the steady-state signal was reached. The functional images were first corrected for movement, normalized into the standard space defined by the Montreal National Institute template, and spatially smoothed with a 8-mm FWHM Gaussian filter.[22] Statistical parametric maps were then calculated using a multilinear regression analysis based on a hemodynamic modeling of the two states of the experiment, and including global signal change and low frequencies (< 1/120 Hz) as counfounding covariates.[21] Significantly activated clusters were obtained for the contrast (GVS–control) using the following thresholds: cluster size > 5 voxels, corrected *p*-value (based on *z*-value and cluster size) < 0.05.[22]

RESULTS

Psychophysical Results

The vestibular sensations induced by GVS during the pretests were similar in all subjects. The participants consistently reported experiencing small-amplitude (5 to 15 deg) oscillations of the entire body at stimulus frequency about an approximately midsagittal axis, which was variously perceived as passing through the body somewhere between the head and the upper abdomen. The orientation of this axis relative to the body did not essentially change when the body was brought from the seated to the supine position. These illusory vestibular phenomena are equivalent to those reported in previous studies making use of sinusoidal GVS.[23,24]

All participants had qualitatively similar sensations inside the magnet during fMRI as in the pretests. However, the perception of illusory body oscillation was generally less intense. The subjects spontaneously attributed this effect to the rigid head fixation and the noise generated during image acquisition, which was still prominent despite our sound-protection measures. Nevertheless, all subjects reported having had distinctly felt sensations of body movement during the periods of GVS and not during the resting periods. The buildup and the cessation of the vestibular sensation were perceived as quick, and took less than 2 s from the onset and offset, respectively, of stimulation in all subjects. Somatosensory side effects at the electrode sites were present in all subjects and were mostly described as a moderate warming sensation. All subjects except one perceived a slight metallic taste during GVS. As in the pre-tests, nausea, did not occur.

fMRI Results

Comparison of the GVS condition with the control condition showed significant activation during GVS in seven cortical areas. Areas with significant signal change are shown in FIGURE 1, and are presented in TABLE 1 with their location and their statistical significance.

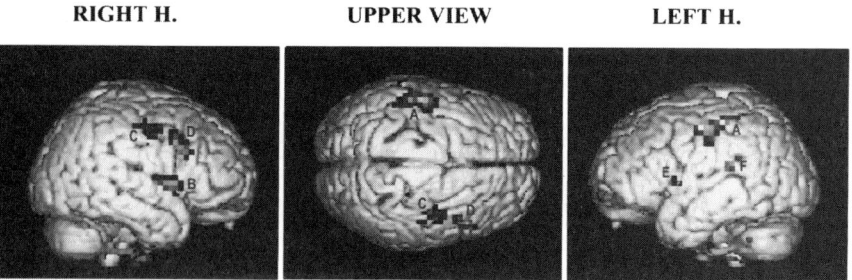

FIGURE 1. Areas of activation (group analysis). Foci of activation are reported on a 3-D reconstruction of a standardized brain. **Left to right**: Right hemisphere view, upper view, left hemisphere view. *Letters* correspond to codes in TABLE 1. Talairach coordinates for each area given in TABLE 1.

TABLE 1. Areas of significant signal change in the comparison of the GVS condition with the control condition (group analysis)

Region of activation	Code	Talairach x,y,z	Corrected cluster p-value
Left postcentral gyrus and intraparietal sulcus	A	−48, −12, 48 −52, −28, 52	0.001
Right inferior precentral and frontal gyri	B	56, 12, 0	0.002
Right central sulcus	C	40, −12, 44	0.002
Right middle precentral and frontal gyri	D	48, 16, 36	0.004
Left inferior frontal gyrus	E	−64, 8, 4	0.02
Left posterior lateral sulcus	F	−64, −36, 20	0.06

Note: Talairach x, y, z correspond to the coordinates of the most significant voxels in the cluster.

Temporoparietal Junction Activation

A significant activation (corrected p-value = 0.06) was found in the left temporoparietal junction (area F), but was not seen in the right hemisphere. This activation was located in the posterior part of the lateral sulcus, on its external banks, and overlapped both the posterior part of the superior temporal gyrus and the inferior part of the supramarginal gyrus in the parietal lobe.

Central Sulcus Activation

Activation was found in the right central sulcus (area C). The most significantly activated voxel was located in the depth of the central sulcus (approximately 2 cm inside the sulcus), at about the level of the middle frontal gyrus. The area of activation also extended into the adjacent postcentral and precentral gyri. In the left hemisphere, area A showed a highly significant activation extremum at (−48,−12,48), which is located adjacent to the central sulcus, in the anterior part of the left postcentral gyrus.

Parietal Lobe Activation

Activation was found in only one area of the parietal lobe, unilaterally in the left hemisphere (area A). It was a very large area of activation, presenting two significant extrema (see TABLE 1). This large area was mainly located on the lateral surface of the left postcentral gyrus, but also extended posteriorly, and the second extremum was found in the most anterior part of the intraparietal sulcus. As previously stated, area C in the right hemisphere extended into the postcentral gyrus, and we cannot exclude bilateral activation of the postcentral gyrus. Area A extended much more posteriorly, however, and the activation in the intraparietal sulcus was observed in the left hemisphere only.

Frontal Lobe Activations

Two main activated areas were found in the frontal lobe.
1. *Inferior frontal gyrus.* Bilateral activation was found on the lateral surface of the frontal lobe (area B in the right hemisphere, area E in the left). In both hemispheres, the

activation was located in the most inferior and the most posterior part of the inferior frontal gyrus, that is, in the pars opercularis. In the right hemisphere, the activation was wider and also appeared to extend into the frontal operculum, as well as into the inferior part of the precentral gyrus. In both hemispheres, however, the most significantly activated voxels were located inside the inferior frontal gyrus (see coordinates in TABLE 1).

2. *Intersection of precentral sulcus and inferior frontal sulcus.* In the right hemisphere activation was also found dorsal to the previous area, in the vicinity of the intersection of the precentral sulcus and of the inferior frontal sulcus (area D). This area was located approximately 1 cm inside the sulcus and overlapped the cortex of the precental gyrus and of the inferior and middle frontal gyri.

DISCUSSION

The results of this study indicate that a large set of areas is activated by galvanic vestibular stimulation (GVS) in humans, including expected areas such as the temporoparietal junction and the intraparietal sulcus, and also new areas in premotor regions of the frontal lobes. Since GVS is a complex stimulation that induces both somatosensory and vestibular effects, and although we used a small current intensity and large surface electrodes in order to limit the somatosensory sensations of GVS, both types of effects have to be taken into account in the interpretation of these activated areas.

Temporoparietal Junction Activation

In the group analysis we performed (area F), activation in the TPJ appeared in the left hemisphere. This area corresponds closely to a polysensory vestibular field, which was recently discovered in several primate species and designated parieto-insular vestibular cortex (PIVC) by Grüsser and coworkers.[25–28] The PIVC is located deep in the lateral sulcus, posterior to the insular cortex. It is one of three areas where neurons responding to vestibular stimulation have been found in monkeys. In particular, neurons in the PIVC respond to natural semicircular canal stimulation, as well as to optokinetic and/or somatosensory stimulation.

In a PET study in humans, Bottini *et al.*[29] showed that there were two distinct areas in the TPJ region contralaterally activated by cold caloric vestibular stimulation: a posterior insula region, and a more superficial posterior lateral sulcus activation. They proposed that the posterior insula region could be the human's equivalent of PIVC. It was, to our knowledge, the first time that a functional imaging study pointed to a potential role for the human's posterior insula in vestibular signal processing.

Until then, human experiments concerning the vestibular cortex had pointed to the involvement of the superficial part of the superior temporal or inferior parietal gyri. Penfield[30] and Smith,[31] using electrical stimulation of epileptic patients during their presurgical checkup, reported that vestibular sensations could be evoked when the electrode was placed on the external banks of the lateral sulcus, at a depth that excluded insular stimulation. Friberg *et al.*,[32] using 133Xe SPECT and warm vestibular caloric stimulation, showed contralateral activation coming from the superficial part of the TPJ as well.

In our data, no significant insular activation was found. However, area F is close to the area found by Bottini *et al.*[29] in the lateral sulcus. It cannot be confused with the auditory cortex, which is located almost 2 cm deeper in the lateral sulcus, in Heschl's gyrus according to several studies, nor with temporoparietal areas involved in speech processing, which are located 1 to 2 cm lower or posterior.[33–36] Area F, that is, the superficial part of the posterior lateral sulcus, could therefore be the equivalent in humans of the monkey's PIVC.

Central Sulcus Activation

In the right hemisphere, the group analysis showed an activation centred in the depth of the central sulcus, and extending into the adjacent postcentral and precentral gyri (area C). In the left hemisphere, activation was also found in the postcentral gyrus (anterior part of area A). Such an activation could be expected. In several species, including monkeys there is a bilateral vestibular projection to the central sulcus, inside area 3a.[37–39] This vestibular area is small and is located within the arm field of the somatosensory cortex.

Activation in areas A and C cannot be due to the finger-tapping task (FT) in which subjects were involved during the whole experiment, because (1) FT was performed continuously, and should be subtracted out by the comparison of the two conditions, and (2) even if differences in rhythm or in attention during the performance of FT could have led to some difference in the comparison of the two conditions, such FT-activations would be located unilaterally in the left motor cortex, since FT was performed with the right index finger only.

Areas A and C are distinct from the frontal eye fields (FEF), which are located 1.4 cm more anterior in the right hemisphere and 1.9 cm more anterior in the left (review in Paus[40]). They are also distinct from the neck somatosensory area, which is situated above the hand area, according to Penfield and Rasmussen.[41]

Thus it appears that the region we found is the homologue in humans of the vestibular projection to area 3a described in animals. The specific role played by this area is not well understood, but its high degree of convergence of vestibular and proprioceptive input could make it an important relay to ensure correct performance of directional movements.

Parietal Lobe Activation

An interesting result of this study is the unilateral activation we found in the left posterior parietal lobe. Activation in this area was expected since it has already been shown by Foerster[42] that electrical stimulation of the anterior intraparietal sulcus in epileptic patients typically results in the vestibular sensation of rolling with the apparent motion of the visual surroundings. The existence of a short-latency vestibular projections to the parietal lobe was then demonstrated in the cat by Walzl and Mountcastle.[43] In the monkey, it was also found within the lower bank of the anterior tip of the intraparietal sulcus in a specific architectonic area called 2v, which lies posterior to SI neurons responding to hand or mouth stimulation.[44] Many neurons in area 2v receive deep somatic afferents[45] or respond to optokinetic stimulation.[46] In cats and in monkeys, vestibular projections to area 2v are mainly contralateral, although a weak ipsilateral projection also exists.[44,47]

The strongly asymmetrical pattern of activation in the parietal lobe is an unexpected result of our study. The functional significance of this asymmetry is unclear, but it could indicate a specific human left-hemispheric specialization in the cortical processing of self-motion information in the parietal lobe. A similar asymmetry has been found in a PET study that showed left IPS activation when comparing blood flow during coherent wide-field visual motion with blood flow during incoherent wide-field visual motion.[48] Such an interpretation is speculative, however, and apparently contradicted by a wealth of evidence pointing to the right hemisphere specializing in spatial information processing.

The region of the intraparietal sulcus (IPS) is generally known to be involved in coordinate transformations, from retinotopic to craniotopic or spatiotopic coordinates, and also from sensory to motor coordinates (review in Andersen[49]). Our results indicate that the most anterior part of the IPS is a specific region where vestibular information is processed, and where it can probably be integrated with other available information such as visual motion. This integration would enable the computation of spatiotopic coordinates, and consequently spatial behavior under visual control.

Frontal Activations

One of the most important results of this study is the discovery of bilateral activation in premotor regions of the frontal lobe during galvanic vestibular stimulation, since no such activations have yet been reported in previous PET experiments on vestibular stimulation.

In monkeys, it was shown that two frontal areas (area 6pa and area 8a) are involved in the cortical processing or modulation of vestibular information.[50–52] In humans, Israël et al.[53] reported that lesions of the middle and inferior posterior frontal gyri, as well as lesions of the medial anterior frontal gyrus (SEF), result in specific impairments in a task of vestibular memory-guided saccades in the dark, while they do not result in any impairment in a task of visual memory-guided saccades. Furthermore, in a recent study of evoked potentials induced by selective stimulation of the vestibular nerve in humans during surgery, it was shown that 3 dipoles could be modeled in the frontal lobe 10 ms after the stimulation.[54]

Based on old[55] and recent[56] cytoarchitectonic studies, the activations we found correspond to Brodmann's area 44 (BA44) or to the most ventral part of BA6, with a possible anterior extension of area D onto the intersection with BA9 / BA44. In the left hemisphere, these activations are close to what is classically thought of as Broca's area, and whose lesion provokes Broca's aphasia. However, functional imaging studies show that activation foci related to language tasks are located 1 cm to 2 cm superior (during speech generation) or anterior (during semantic processing) to our activation focus,[35,36,57,58] and are lateralized to the left hemisphere in males.[59] These activations are also different from the FEF, localized in the depth of the precentral sulcus at its intersection with the superior frontal sulcus,[40,60] more than 2.5 cm away from our activation foci.

Since inferior frontal activation has not been detected in previous studies using caloric vestibular stimulation, this area is a strong candidate for being part of the representation of otholitic signal in the human cortex. It is also important to note the asymmetrical pattern of activation our bilateral stimulation evoked in this area: it may be related to the general specialization of the right hemisphere in spatially oriented behavior, and can be compared to the right-sided hemispheric dominance of lesions provoking spatial hemineglect (review in Bisiach and Vallar[61]). In the right hemisphere, lesions to this particular region can indeed result in visual hemineglect.[62] Furthermore, recent fMRI experiments have shown that in normals the computation of the subjective midsagittal plane, a basic egocentric spatial reference that can be strongly disturbed in spatial hemineglect, involves a bilateral activation of the same region of the lateral premotor frontal cortex with a much stronger activation in the right hemisphere.[63] Altogether, these results suggest that the inferior lateral premotor area we identified plays an important role in the transformations for directing attention and for planning upper-limb movements to explore contralateral space in spatiotopic coordinates.

ACKNOWLEDGMENTS

This work was supported by a grant "ACC Sciences de la Vie" from the French Ministry of Research. One of the authors (E. L.) was supported in part by a grant, from the Institut Lilly. This work is supported by grants 'Action Concertée Sciences de la Vie' and 'Galilée' from the French Ministry of Research. The authors thank Dr. W. Guldin for helpful comments, Dipl. Ing. H. Nitert for the construction of the stimulator, E. Giacomini for help in adapting the stimulator to the MRI environment.

REFERENCES

1. BABINSKI, J. 1903. Sur le mécanisme du vertige voltaïque. C. R. Soc. Biol. (Paris) **55**: 350–353.
2. DOHLMANN, G. 1929. Experimentelle Untersuchungen über die galvanische vestibularische Reaktion. Acta Otolaryngol. (Stockh.) **8**(Suppl.): 48.
3. BLONDER, E. J. & L. DAVIS. 1936. The galvanic falling reaction in patients with verified intracranial neoplasms. J. Am. Med. Ass. **107**: 411–412.
4. WATANABE, Y., M. ASAI, A. KINOSHITA & K. MIZUKOSHI. 1989. Retro-labyrinthine disorders detected by galvanic body sway responses in routine equilibrium examinations. Acta Otolaryngol. (Stockh.) **468**: 343–348.
5. SEKITANI, T. & M. TANAKA. 1975. Test for galvanic vestibular responses—Survey of experimental and clinical investigation for the last 20 years. Bull. Yamagushi Med. Sch. **22**: 439–452.
6. PFALTZ, C. R. & Y. KOIKE. 1968. Galvanic test in central vestibular lesions. Acta Otolaryngol. (Stockh.) **65**: 161–168.
7. HAHN, R. & P. MENZIO. 1966. La valeur diagnostique de la stimulation vestibulaire galvanique dans certains syndrômes otoneurologiques. Confin. Neurol. **28**: 327–332.
8. ILES, J. F. & J. V. PISINI. 1992. Vestibular-evoked postural reactions in man and modulation of transmission in spinal reflex pathways. J. Physiol. (Lond.) **455**: 407–424.
9. BALDISSERA, F., P. CAVALLARI & G. TASSONE. 1990. Effects of transmastoid electrical stimulation on the triceps brachii EMG in man. NeuroReport **1**: 191-193.
10. NJOKIKTJIEN, C. & J. F. FOLKERTS. 1971. Displacement of the body's centre of gravity at galvanic stimulation of the labyrinth. Confin. Neurol. **33**: 46–54.
11. SPIEGEL, E. A. & N. P. SCALA. 1943. Response of the labyrinthine apparatus to electrical stimulation. Arch. Otolaryngol. **38**: 131–138.
12. JOHANSSON, R., M. MAGNUSSON & P. A. FRANSSON. 1995. Galvanic vestibular stimulation for analysis of postural adaptation and stability. IEEE Trans. Biomed. Eng. **BME-42**: 282–292.
13. COATS, A. C. 1972. Limit of normal of the galvanic body-sway test. Acta Otolaryngol. (Stockh.) **81**: 410–416.
14. GOLDBERG, J. M., C. E. SMITH & C. FERNANDEZ. 1984. Relation between discharge regularity and responses to externally applied galvanic currents in vestibular nerve afferents of the squirrel monkey. J. Neurophysiol. **51**: 1236–1256.
15. COURJON, J. H., W. PRECHT & D. W. SIRKIN. 1987. Vestibular nerve and nuclei unit responses and eye movement responses to repetitive galvanic stimulation of the labyrinth in the rat. Exp. Brain Res. **66**: 41–48.
16. ANGELAKI, D. E. & A. A. PERACHIO. 1993. Contribution of irregular semicircular canal afferents to the horizontal vestibuloocular response during constant velocity rotation. J. Neurophysiol. **69**: 996–999.
17. ANGELAKI, D. E., A. A. PERACHIO, M. MUSTARI & C. STRUNK. 1992. Role of irregular otoliths afferents in the steady-state nystagmus during off-vertical axis rotation. J. Neurophysiol. **68**: 1895–1900.
18. NASHNER, L. M. & P. WOLFSON. 1974. Influence of head position and proprioceptive cues on short latency postural reflexes evoked by galvanic stimulation of the human labyrinth. Brain Res. **67**: 255–268.
19. DZENDOLET, E. 1963. Sinusoidal electrical stimulation of the human vestibular apparatus. Percept. Mot. Skills **17**: 171–185.
20. LOBEL, E., J. F. KLEINE, D. LE BIHAN, E. GIACOMINI, A. BERTHOZ & A. LEROY-WILLIG. 1997. FMRI of the vestibular cortex with galvanic stimulation: a feasibility study (Abstract). Proc. Soc. Magn. Reson. **1**: 450.
21. WORSLEY, K. J. & K. J. FRISTON. 1995. Analysis of fMRI time-series revisted—Again. Neuroimage **2**: 173–181.
22. FRISTON, K. J., K. J. WORSLEY, R. S. J. FRACKOWIAK, J. C. MAZZIOTTA & A. C. EVANS. 1994. Assessing the significance of focal activations using their spatial extent. Hum. Brain Mapp. **1**: 214–220.
23. VON ROMBERG, G., E. VON HOLST & W. DODEN. 1951. Über Zusammenhänge zwischen Wahrnehmung und objektivem Geschehen bei Wechselstromreizung des Vestibularapparates und optokinetischer Pendelreizung der Retina. Pflügers Archiv **254**: 98–106.
24. GRÜSSER, O.-J. & J. F. KLEINE. 1995. The effect of binaural galvanic stimulation in man. Soc. Neurosci. Abstr. **21**: 135.

25. GRÜSSER, O.-J., M. PAUSE & U. SCHREITER. 1990. Localization and responses of neurones in the parieto-insular vestibular cortex of awake monkeys (*Macaca fascicularis*). J. Physiol. **430**: 537–557.
26. GRÜSSER, O.-J., M. PAUSE & U. SCHREITER. 1982. Neuronal response in the parieto-insular vestibular cortex of alert Java monkeys (*Maccaca fascicularis*). *In* Physiological and Pathological Aspects of Eye Movements, A. Roucoux and M. Crommelinck, Eds.: 251–270. W. Junk. The Hague.
27. GRÜSSER, O.-J., M. PAUSE & U. SCHREITER. 1983. A new vestibular area in the primate cortex. Soc. Neurosci. Abstr. **9**: 749.
28. GULDIN, W. O. & O.-J. GRÜSSER. 1987. Single unit responses in the vestibular cortex of squirrel monkeys. Soc. Neurosci. Abstr. **13**: 1224.
29. BOTTINI, G., R. STERZI, E. PAULESU, G. VALLAR, S. F. CAPPA, F. ERMINIO, R. E. PASSINGHAM, C. D. FRITH & R. S. J. FRACKOWIAK. 1994. Identification of the central vestibular projections in man: a positron emission tomography activation study. Exp. Brain Res. **99**: 164–169.
30. PENFIELD, W. 1957. Vestibular sensation and the cerebral cortex. Ann. Otol. Rhinol. Laryng. **66**: 691–698.
31. SMITH, B. H. 1960. Vestibular disturbances in epilepsy. Neurology **10**: 465–469.
32. FRIBERG, L., T. S. OLSEN, P. E. ROLAND, O. B. PAULSON & N. A. LASSEN. 1985. Focal increase of blood flow in the cerebral cortex of man during vestibular stimulation. Brain **108**: 609–623.
33. FIEZ, A. J., M. E. RAICHLE, D. A. BALOTA, P. TALLAL & S. E. PETERSEN. 1996. PET activation of posterior temporal regions during auditory word presentation and verb generation. Cereb. Cortex **6**: 1–10.
34. JOHNSRUDE, I. S., T. PAUS, R. J. ZATORRE, D. W. PERRY, G. P. WARD & A. C. EVANS. 1997. The location of auditory activation foci relative to Heschl's gyri. (Abstr.) Neuroimage **5**: S177.
35. ZATORRE, R. J., A. C. EVANS, E. MEYER & A. GJEDDE. 1992. Lateralization of phonetic and pitch discrimination in speech processing. Science **256**: 846–849.
36. WISE, R., F. CHOLLET, U. HADAR, K. J. FRISTON, E. HOFFNER & R. S. J. FRACKOWIAK. 1991. Distribution of cortical neural networks involved in word comprehension and word retrieval. Brain **114**: 1803–1817.
37. ÖDKVIST, L. M., D. W. F. SCHWARZ, J. M. FREDRICKSON & R. HASSLER. 1974. Projection of the vestibular nerve to the area 3a arm field in the squirrel monkey. Exp. Brain Res. **21**: 97–105.
38. ÖDKVIST, L. M., A. M. RUBIN, D. W. F. SCHWARZ & J. M. FREDRICKSON. 1973. Vestibular cortical projections in the rabbit. J. Comp. Neurol. **149**: 117–120.
39. ÖDKVIST, L. M., A. M. RUBIN, D. W. F. SCHWARZ & J. M. FREDRICKSON. 1973. Vestibular and auditory cortical projection in the guinea pig (*Cavia porcellus*). Exp. Brain Res. **18**: 279–286.
40. PAUS, T. 1996. Location and function of the human frontal eye-field: a selective review. Neuropsychologia **34**: 475–483.
41. PENFIELD, W. & T. RASMUSSEN. 1950. The Cerebral Cortex of Man: A Clinical Study of Localization of Function. Macmillan. New York.
42. FOERSTER, O. 1936. Sensible corticale Felder. *In* Handbuch der Neurologie, O. Bumke and O. Foerster, Eds.: 358-449. Springer-Verlag. Berlin.
43. WALZL, E. M. & V. B. MOUNTCASTLE. 1949. Projection of vestibular nerve to cerebral cortex of cat. Am. J. Physiol. **159**: 595
44. FREDRICKSON, J. M., U. FIGGE, P. SCHEID & H. H. KORNHUBER. 1966. Vestibular nerve projection to the cerebral cortex of the rhesus monkey. Exp. Brain Res. **2**: 318–327.
45. SCHWARZ, D. W. F. & J. M. FREDRICKSON. 1971. Rhesus monkey vestibular cortex: a bimodal primary projection field. Science **172**: 280–281.
46. BUETTNER, U. & U. W. BÜTTNER. 1978. Parietal cortex activity in the alert monkey during natural vestibular and optokinetic stimulation. Brain Res. **153**: 392–397.
47. KORNHUBER, H. H. & J. S. DAFONESCA. 1964. Optovestibular integration in the cat's cortex: a study of sensory convergence on cortical neurons. *In* The Oculomotor System, M. B. Bender, Ed.: 239–279. Harper & Row (Hoeber). New York.
48. CHENG, K., H. FUJITA, I. KANNO, S. MIURA & K. TANAKA. 1995. Human cortical regions activated by wide-field visual motion: an $H_2^{15}O$ PET study. J. Neurophysiol. **74**: 413–427.
49. ANDERSEN, R. A. 1994. Coordinate transformations and motor planning in posterior parietal cortex. *In* The Cognitive Neurosciences, M. S. Gazzaniga, Ed.: 519–532. MIT Press. Cambridge.

50. AKBARIAN, S., O.-J. GRÜSSER & W. O. GULDIN. 1993. Corticofugal projections to the vestibular nuclei in squirrel monkeys: further evidence of multiple cortical vestibular fields. J. Comp. Neurol. **332**: 89–104.
51. AKBARIAN, S., O.-J. GRÜSSER & W. O. GULDIN. 1994. Corticofugal connections between the cerebral cortex and brainstem vestibular nuclei in the macaque monkey. J. Comp. Neurol. **339**: 421–437.
52. GULDIN, W. O., S. AKBARIAN & O.-J. GRÜSSER. 1992. Cortico-cortical connections and cytoarchitectonics of the primate vestibular cortex: a study in squirrel monkeys (*Saimiri sciureus*). J. Comp. Neurol. **326**: 375–401.
53. ISRAEL, I., S. RIVAUD, A. BERTHOZ & C. PIERROT-DESEILLIGNY. 1992. Cortical control of vestibular memory-guided saccades. Ann. N.Y. Acad. Sci. **656**: 472–484.
54. BAUDONNIERE, P. M., C. DE WAELE, P. TRAN BA HUY & P.-P. VIDAL. 1996. Réponses évoquées vestibulaires avant et après neurectomie vestibulaire unilatérale chez l'homme. *In* Le Cortex Vestibulaire, M. Collard, M. Jeannerod, and Y. Christen, Eds.: 95–108. Irvinn. Boulogne.
55. BRODMANN, K. 1909. Vergleichende Lokalisationslehre der Grosshinrinde in ihren Prinzipien dargestellt auf Grund des Zellenbaues. Barth, Leipzig.
56. ZILLES, K., G. SCHLAUG, S. GEYER, G. LUPPINO, M. MATELLI, M. QU, A. SCHLEICHER & T. SCHORMANN. 1996. Anatomy and transmitter receptors of the supplementary motor areas in the human and nonhuman primate brain. *In* Supplementary Sensorimotor Area, H. O. Luders, Ed.: 29–43. Lippincott-Raven. Philadelphia.
57. PETERSEN, S. E., P. T. FOX, M. I. POSNER, M. MINTUN & M. E. RAICHLE. 1988. Positron emission tomographic studies of the cortical anatomy of single-word processing. Nature **331**: 585–589.
58. PAULESU, E., C. D. FRITH & R. S. J. FRACKOWIAK. 1993. The neural correlates of the verbal component of working memory. Nature **362**: 342–345.
59. SHAYWITZ, B. A., S. E. SHAYWITZ, K. R. PUGH, R. TODD CONSTABLE, P. SKUDLARSKI, R. K. FULBRIGHT, R. A. BRONEN, J. M. FLETCHER, D. P. SHANKWEILER, L. KATZ & J. C. GORE. 1995. Sex differences in the functional organization of the brain for language. Nature **373**: 607–609.
60. LOBEL, E., A. BERTHOZ, A. LEROY-WILLIG & D. LE BIHAN. 1996. fMRI study of voluntary saccadic eye movements in humans. (Abstr.) Neuroimage **3**: S396.
61. BISIACH, E. & G. VALLAR. 1988. Hemineglect in humans. *In* Handbook of Neuropsychology, F. Boller and J. Grafman, Eds.: 195–222. Elsevier. Amsterdam.
62. HUSAIN, M. & C. KENNARD. 1996. Visual neglect associated with frontal lobe infarction. J. Neurol. **243**: 652–657.
63. VALLAR, G., E. LOBEL, G. GALATI, A. BERTHOZ, L. PIZZAMIGLIO & D. LE BIHAN. 1999. A fronto-parietal system for computing the egocentric spatial frame of reference in humans. Exp. Brain Res. **124**: 281–286.

The Interaction of Otolith and Proprioceptive Information in the Perception of Verticality

The Effects of Labyrinthine and CNS Disease

ADOLFO M. BRONSTEIN[a]

Medical Research Council, Human Movement and Balance Unit, Institute of Neurology, National Hospital for Neurology and Neurosurgery, Queen Square, London WC1N 3BG, UK

ABSTRACT: A review of recent experiments in patients with labyrinthine and neurological disorders assessing the subjective postural vertical (SPV) and the subjective visual vertical (SVV) is presented. The SPV was measured with subjects (Ss) seated in a motorized flight simulator tilting at 1.5 deg/s in roll and pitch; the Ss' task was to indicate when they entered and left self-verticality. The SVV was measured by Ss adjusting a straight line to what they perceived as gravitational upright. Clear dissociations between the SVV and SPV were found, for example, patients with acute unilateral vestibular disorders had marked tilts of the SVV toward the side of the lesion but a "lean" (bias, tilt) of the SPV was never found. Dissociations of the SPV and SVV could also be induced in normal subjects by roll-plane visual motion stimuli: the SVV was tilted in the direction of motion, but the SPV was not. Prolonged lateral body tilt did, however, bias the SVV (the "A" effect) and the SPV, but these effects are likely to be mediated by somatosensory rather than otolithic input. Evidence for the latter came from (i) findings in patients with absent vestibular function, who showed an enhanced "A" effect, and (ii) from a patient with a thalamic infarction, who showed absence of the "A" effect when leaning on the hemihypesthetic side. In separate experiments where normal Ss indicated space-vertical and space-horizontal with saccadic eye movements, we found differences between these percepts, that is, subjective external space lost orthogonality. The findings in these various experiments can be interpreted if we abandon the idea of a single, "internal representation" of verticality. Different sensory modalities convey different and sometimes conflicting messages about verticality. Otolithic and somatosensory signals can have opposite sign effects during verticality estimates while tilted. In man, somatosensory cues have a prominent role in verticality perception.

INTRODUCTION

The otolith organs of the labyrinth are suited to detect linear acceleration. These organs must therefore play a major part in sensing gravitational input, in particular the perception of uprightness. In order to examine this function in man, patients with labyrinthine and neurological lesions were investigated as to their ability to sense uprightness. The concept of uprightness, or verticality, involves the detection of one's self-verticality (subjective postural vertical, or SPV), the verticality of objects as inspected by vision (subjective visual vertical,

---

[a]Address for telecommunication: Phone: (+44) (171) 837 3611, ext. 4111; fax: (+44) (171) 837 7281; e-mail: A.Bronstein@ion.ucl.ac.uk

or SVV) or touch (haptic vertical), and the preservation of orthogonality (the relationship between verticality and horizontality). Considering the many sensory, motor, and cognitive activities likely to be involved in these tasks, we intuitively realize that the otolithic contribution can only be just one of many. In this review I will summarize the results of our recent studies with normal human subjects and patients with CNS and labyrinthine lesions that illustrate dissociation between different aspects of verticality perception and the relative contribution of the vestibular and somatosensory system in these tasks.

THE SUBJECTIVE POSTURAL VERTICAL IN CENTRAL AND PERIPHERAL VESTIBULAR DISEASE

Clinical tests of vestibular dysfunction focus largely on motor phenomena, despite reports that their specificity and sensitivity to disease is less than for perceptual tests of vestibular disorder.[1] This prompted us to study spatial orientation in patients with vestibular disease, in particular the subjective postural vertical (SPV), and how this function compared with motor phenomena.

A priori one would consider accurate perception of postural vertical to be important to maintain upright stance and gait and that, when disordered, it may lead to clinical unsteadiness. Patients with unilateral peripheral lesions or CNS lesions with vertical (up- or downbeating) nystagmus would be particularly important, since it has been suggested that the abnormal sway present in these patients may be related to their perception of body orientation in space.[2] Schematically, one would anticipate two types of abnormality in the perception of the postural vertical: a loss of sensitivity in this perception—or enlargement of the cone of subjective verticality—and a tilt or bias—a 'lean' of the cone of verticality (FIG. 1).

For this experiment the subjects were seated in a padded chair in a motorized gimbal with the head and torso restrained and their eyes closed (FIG. 1; reference 3). The gimbal was externally controlled and executed 7–10 cycles of 15-deg tilt to either side around the vertical at 1.5-deg/s constant velocity in pitch and after a few minutes rest in roll. This slow velocity does not provide significant semicircular canal input.[4] The (machine) vertical was defined when the plane of the seat was orthogonal to the gravitational vertical. Subjects were simply asked to indicate when they entered and left upright, rather than a single position of uprightness. This protocol would therefore identify "uprightness" as perceived within a zone of tilt ("cone of verticality," FIG. 1) and minimize the possibility that subjects could derive verticality perception through rational pondering. Subjective reports from normal subjects and patients indicated that estimates were based on compelling bodily sensations rather than on rational estimates.

After rejection of the first cycle of tilt, the angles of entering and exiting vertical for each subsequent cycle were averaged to give a mean position of SPV (negative values would indicate backwards or leftwards bias or 'lean' of the SPV). The width of the sectors is a parameter for the sensitivity of perception of uprightness (larger values indicate wider "cone of verticality," FIG. 1). This is obtained from subtracting angle of entering upright-angle of exit, and was similarly averaged over cycles of tilt. In patients groups with unilateral lesions the roll data were normalized for left lesions. Statistical comparisons were carried out with both the two-tailed t-test and nonparametric statistics (Mann-Whitney and Wilcoxon) in all cases. There were no differences between the outcome of these two analyses, so significance values given are those of the t-test.

Fifty-two normal subjects, mean age 40.4 years (range 21–80) were tested, as well as (1) eight bilateral labyrinthine defective subjects (LDS), (2) eight acute unilateral peripheral vestibular lesions, (3) six asymptomatic patients with chronic stable unilateral absence of peripheral vestibular function, (4) nine patients with typical benign

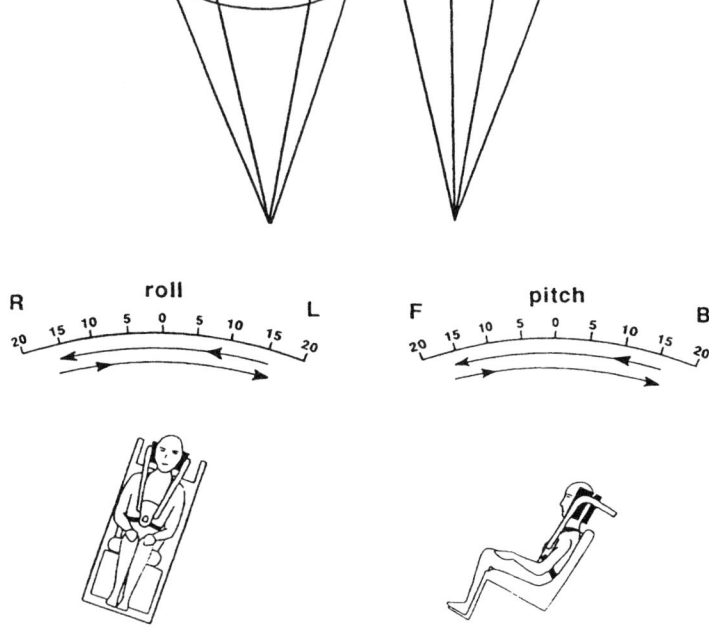

FIGURE 1. Diagram illustrating the theoretically expected findings in different patient groups in the SPV experiment. Patients could conceivably suffer from a loss of sensitivity (widening of the sector or "cone" of verticality) or a bias of the mean SPV (a "lean") either in roll or pitch. (From Bisdorff et al.[3] Reproduced with permission from the Editor of Brain.)

paroxysmal positional vertigo (BPPV), (5) 12 patients with Ménière's disease suffering frequent or severe attacks of vertigo, (6) four patients tested before and after (< 1 week) unilateral vestibular nerve section for refractory vertigo, and (7) 15 patients with CNS disease (usually cerebellar) with vertical (up- or downbeat) nystagmus. Five of the latter patients had no position-dependent modulation of their nystagmus, but 10 showed persistent (tonic) or transient (dynamic) changes in intensity of their nystagmus when adopting different head positions.

FIGURE 2A shows results of the mean position of the SPV ("bias" or "lean" of the SPV), and FIGURE 2B shows sectors size ("sensitivity"). In normal subjects the width of the sector of SPV in pitch was 5.8 SD 2.1 deg, and in roll 5.9 SD 2.1 deg with mean positions of the SPV close to zero. The patients with bilateral absence of vestibular function had significantly ($p \leq 0.01$) larger sectors in roll (8.2 deg) and pitch (9.2 deg), that is a non-plane-specific loss of sensitivity, but the mean position of SPV was normal. The patients with acute unilateral vestibular lesions had enlarged sectors in both planes (pitch 9.7 deg, roll

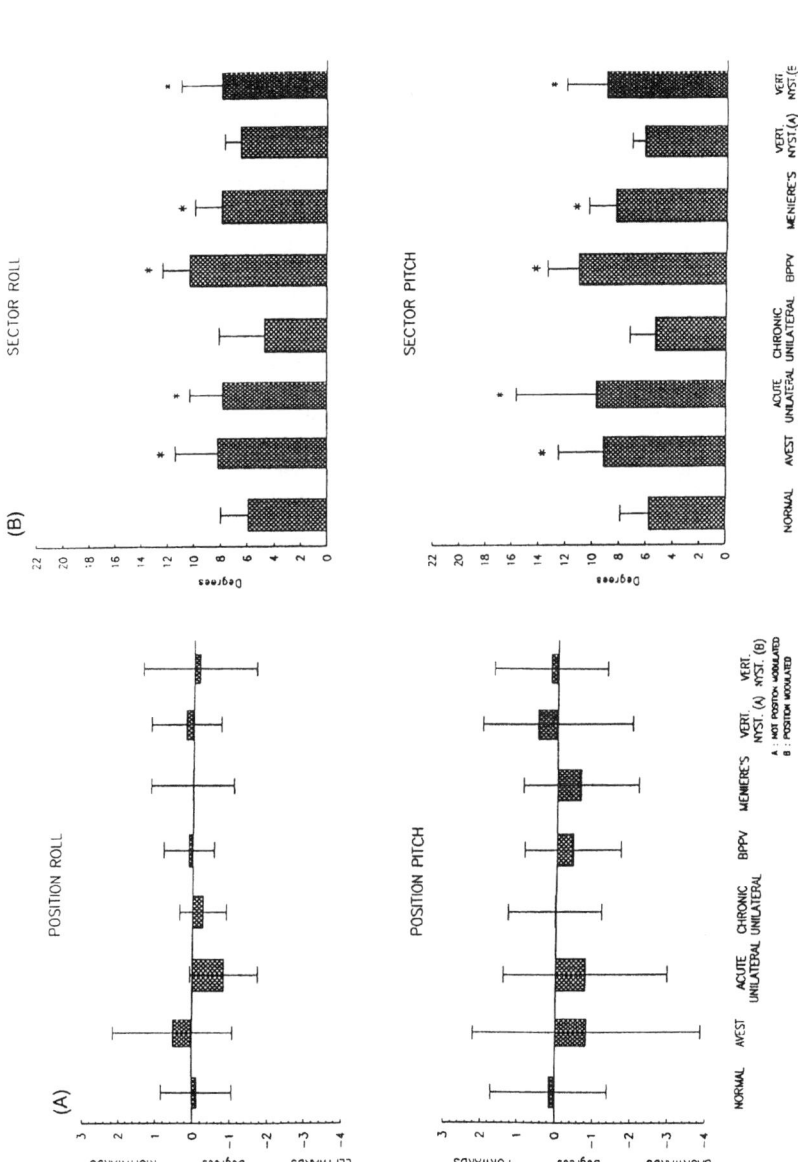

FIGURE 2. Summary of SPV results in the patient groups tested. (**A**) The mean position of the SPV in roll and pitch; note lack of significant bias or lean in any patient group. (**B**), the sensitivity of the SPV. Note that patients with absent or unstable vestibular input have abnormally large sectors or "cones" of verticality. *Asteriscs* indicate different from normal subjects at $p < 0.05$. Avest = absent vestibular function; Acute/chronic unilateral = acute/chronic unilateral absence of vestibular function; Vert nyst = vertical (up or downbeat) nystagmus due to CNS lesion, A = vertical nystagmus modulated by head position with respect to gravity, B = not modulated by head position. (From Bisdorff et al.[3] Reproduced with permission from the Editor of Brain.)

7.8 deg; $p \leq 0.02$). There was a slight trend to directional bias in the direction of the lesion for the mean position angle of the SPV (p = 0. 051 in the two-tailed t-test), but the biases were under 1 deg in size (FIG. 2A). Patients with chronic unilateral peripheral vestibular lesions had normal sectors in pitch and in roll and no directional bias. In the group of neurological patients with vertical nystagmus, the presence of positional modulation of the nystagmus had an effect on the SPV estimates; the patients without modulation were normal, but the group of patients with positioning or positional modulation of the nystagmus had enlarged sectors in both planes (pitch 9 deg, roll 7.9 deg; $p \leqq 0.01$). Contrary to the expectation, separate evaluation of up- and downbeat patients did not reveal a directional bias of the mean SPV in particular not in the pitch plane. The patients with benign paroxysmal positional vertigo (BPPV) and Ménière's disease had a loss of sensitivity, that is, enlarged sectors in both planes but no directional bias in either plane. Of note, in the four patients studied before and after vestibular neurectomy, the tilt estimates of the SPV did not change. In these patients the subjective visual vertical was also measured pre- and postoperatively, and this changed significantly with an average tilt of 8.5 deg in the direction of the lesion (FIG. 3).

In summary, the abnormality found in some patient groups was an increase in sector width of the subjective postural vertical. This was observed in patients with recent, active vertigo (acute unilateral lesions, benign positional vertigo, Ménière's disease), patients with positionally modulated up-/downbeat nystagmus, and patients with absent vestibular function. The type and degree of abnormality, essentially a diminished sensitivity in estimating body uprightness, was similar across these groups. Patients with chronic unilateral lesions and patients with nonpositionally modulated vertical nystagmus showed entirely normal results. The findings imply that, while seated, the somatosensory system can provide an essentially accurate *mean* estimate of body uprightness (mean SPV). However, the sensitivity of such estimate, that is, the width of the SPV sector, can be improved if reliable otolithic input is available.

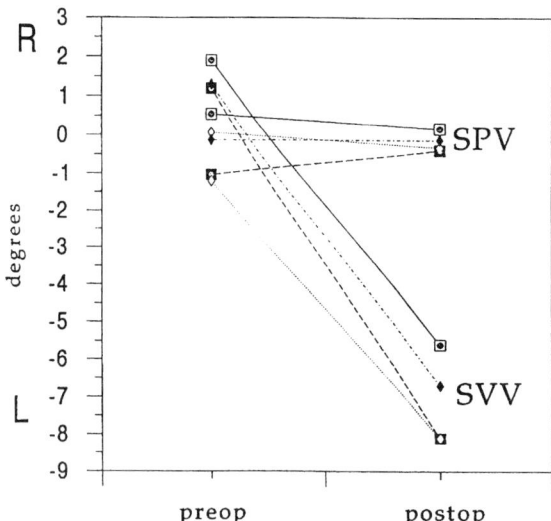

FIGURE 3. Mean SPV and SVV in four patients, before and < 1 week postvestibular nerve section. The operations have been normalized to the left side. Note the large postop dissociation between SPV and SVV. (From Bisdorff *et al.*[3] Reproduced with permission from the Editor of Brain.)

THE EFFECTS OF VESTIBULAR AND SOMATOSENSORY LESIONS ON TILT-INDUCED BIASES IN VERTICALITY PERCEPTION.

The task most frequently used for the assessment of the perception of verticality is the SVV. In this task subjects set a straight line to the perceived verticality (or horizontality) in the absence of relevant visual cues. Normal subjects are extremely accurate in this task, usually setting the line within less than two degrees of "true" gravitational vertical.[5] It has been known for some time that acute unilateral peripheral vestibular lesions produce a tilt of the line settings in the direction of the lesion.[6] However, this ipsilateral tilt of the SVV is largely due to the presence of ipsilesional ocular tilt[7,8] rather than secondary to a bias of a global verticality percept[3] (see FIG. 3). Tilts of the SVV can also be observed in CNS lesions presumably involving vestibular pathways,[6,9] including cases without ocular tilt.[10] Visual motion stimuli rotating about the line of sight (roll-motion) also induce a powerful deviation of line settings in the direction of motion, and it has been postulated that this bias in the SVV is due to central visuovestibular interaction.[11,12] However, consistent and large tilts of the SVV in normal subjects are induced by simply asking subjects to set the SVV while lying 80–90 deg to one side. In this position normal subjects invariably set the line tilted in the direction of body tilt by some 10–30 deg. This is called the "Aubert" or "A" effect,[13–15] which can be easily experienced when lying sideways in a dark bedroom if light from outside comes in around doors or window frames. (In this case true vertical or horizontal lines appear to be rotated in the opposite direction of body tilt so that when subjects are asked to set it vertically, they introduce a bias in the same direction of body tilt.) The origin of the A effect has been debated.[5] It has been postulated that this effect is somatosensory rather than vestibularly mediated,[16] and that it represents a compromise between a tendency to set the line to true gravitational vertical and a trend to set the line parallel to the longitudinal axis of the body (the idiotropic vector.)[15,17]

Recently we studied the effects of body tilt and visual roll-motion on the perception of visual verticality in bilaterally labyrinthine defective subjects (LDS).[18] Eight LDS and 24 normal controls (and 24 patients with Parkinson's disease) were examined as to the ability to set a straight line to perceived vertical against a dotted background that was either stationary or rotating about the line of sight. The task was conducted with the subjects sitting upright and repeated while lying approximately 90 deg on the right-hand side.

FIGURE 4 summarizes the results of this experiment for when seated upright (FIG. 4A) and lying on the right-hand side (FIG. 4B). It can be seen that in the upright position with a static visual background, LDS (and Parkinsonian patients) do not differ from normal subjects. LDS have larger deviations of the visual vertical under visual background rotation, presumably because the effect of the rotating visual stimulus is unopposed by the lack of otolith input (FIG. 4A). More relevant to this review, however, are the results obtained when lying sideways. FIGURE 4B shows that all group of subjects have a deviation of the vertical line to the right (A-effect), but it can be seen that this effect is nearly twice as large in the LDS, amounting to a mean of 40-deg deviation from true vertical.

The finding of an increased A-effect in LDS is a clear indication that tilt-mediated effects on the visual vertical are primarily nonvestibular. They must be mediated by the somatosensory system, but the relative contribution of contact, proprioceptive, or visceral cues is unknown at this stage. As in the preceding section, the results point out that the importance of the somatosensory system in the perception of verticality and visual orientation have previously been overlooked.

More direct evidence for a significant role of the somatosensory system in verticality perception came from the examination of a patient with a unilateral hemisensory loss. The patient was a 21-year-old woman who had suffered an ischaemic stroke involving the right thalamus two years earlier. She had a severe hemianesthesia on the left, including absence of position sense on the ankle and wrist. The MRI showed an area of infarction in the right thalamus, the etiology of which was never clarified. The SVV when the patient was seated

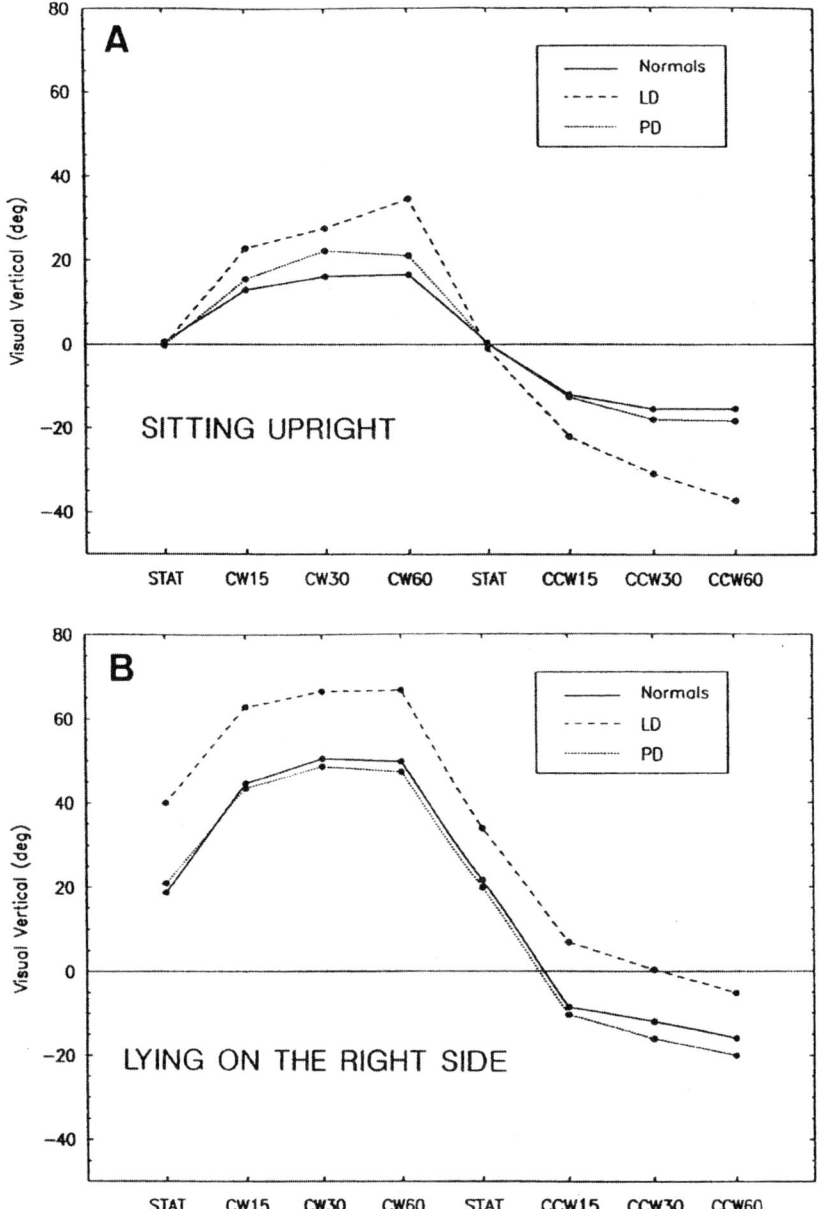

FIGURE 4. Mean position of the SVV in normal subjects, bilateral labyrinthine-defective subjects (LD), and patients with Parkinson's disease. Data shown are for sitting upright and lying on the right-hand-side conditions, both statically (STAT) and during visual roll-motion at 15, 30, and 60 deg/s clockwise (CW) and counterclockwise (CCW). The SVV in the LD subjects tilts more than normal in response to the visual motion stimuli and while lying sideways (enhanced A-effect). From Bronstein et al.[18] Reproduced with permission from the Editor of Neurology.)

upright was within normal limits. However, strong right–left asymmetries appeared when investigated under body tilt. When the visual vertical was investigated with the patient lying on the right side, a normal A-effect of approximately 18 deg was observed. When lying on the left, hemianesthetic side no consistent deviation of the line with respect to the upright settings was detected and her settings became erratic (the coefficient of variation lying on her right side was 40, whereas on the left, was 1939.[19] Two other patients with similar findings are described in this volume by Anastasopoulos *et al.*[20]

SENSORY STIMULI PARTICIPATING IN THE PERCEPTION OF VERTICALITY

The perception of verticality is subserved by multisensory inputs. Incoming otolith signals convey the orientation of the head with respect to gravitational vertical.[21] When verticality is assessed by means of a visual task, the orientation of the retina with respect to the head is clearly important, as shown by studies in patients with ocular tilts[9,7]. A contribution by visceral receptors has been indicated by Mittelstaedt.[17] The studies reported in this review provide support for a role of the somatosensory system in the perception of both the visual and postural vertical.

The results we obtained with SPV in patients were somewhat surprising. We thought we would find a bias of the SPV related to the clinical abnormality. Patients with up- or downbeat nystagmus or those with acute unilateral vestibular lesions were expected to show a tilt of the SPV in corresponding directions, that is, a bias or "lean" of the SPV. Presumably SPV settings would have been biased in the same direction to the slow phase of the nystagmus, since such slow phase normally occurs with tilt or turns in the opposite direction. This was never observed, and neither galvanic nor visual stimuli, delivered while subjects were tested in the tilting device, induced a bias of the SPV, and nor indeed did unilateral vestibular neurectomy.[3]

In marked contrast, clear biases in the SPV were created in normal subjects by asymmetric tilt during the procedure. Offsetting the cycles by as little as 5 deg induced an ipsilateral bias of the SPV of 2.1 deg in normal subjects, and the bias increased by larger magnitude off-sets.[3] The latter effect is, however, more likely to be of somatosensory rather than vestibular origin. Thus, in the upright position, when somatosensory receptors are not subjected to strong asymmetric input, the perception of the mean SPV can be mediated by somatosensory cues. This is also supported by the findings reported by Anastasopoulos *et al.*[20] that some patients with significant hemisensory loss show a small bias in the mean SPV. The mean SPV is quite robust and capable of resisting strong vestibular asymmetries, but even in the upright position, the sensitivity (accuracy) of the SPV can be "fine-tuned" by vestibular input, provided that this is stable. This was shown by the fact that patients with recent/repetitive vertigo, acute lesions, positionally sensitive nystagmus, and absent vestibular function had a larger "cone" sector of the SPV, whereas patients with chronic unilateral absence of vestibular function were entirely normal. Thus, one labyrinth appears to be sufficient to judge SPV correctly in every respect, at least in stable, predictable conditions. If measurements of the SPV in patients with unilateral vestibular lesions are taken during whole-body vibration, a bias can appear (Aoki, Burchil, and Gresty, personal communication), but it is not clear if vibration interferes with somatosensory, vestibular, or visceral mechanisms.

As already mentioned, asymmetric somatosensory input—for example, lying sideways or offsetting the cycles of tilt—may cause a bias of the SPV and SVV. Otolith signals are stable and veridical with respect to gravity, and the tilt-induced biases can be explained by asymmetric somatosensory input. The finding of an abnormally large A-effect in subjects with absent vestibular function indicates that this particular tilt-mediated effect is of somatosensory origin.[18] The unilateral lack of A-effect in our hemianesthetic patient, and the bilateral absence in a patient with severe polyneuropathy[16] lead to the same conclusion. These somatosensory-mediated effects are likely to be mediated by adaptation, and this is specifically supported by the finding that the magnitude of the A-effect increases with duration of tilt.[22] Aftereffects observed in the visual vertical (e.g., the tilt of the SVV observed after a preceding head tilt) decay with time, and this has been interpreted as proprioceptive adaptation.[23] Tilt-induced errors in the SPV of LDS are also time-dependent and therefore also explained on similar grounds.[24]

The fact that somatosensory adaptation may be responsible for introducing biases in SPV and SVV makes it actually difficult to think of a possible useful role for this sensory system in verticality perception. Indeed, one could argue that the visual vertical settings of the hemianaesthetic patients are more veridical (closer to true gravitational vertical) when lying on the anesthetic side than when on the normal side. It should be noted that it is only the *average* value that looks veridical because the responses while lying on the anesthetic side varied greatly. This variability must be a reflection of the lack of reliable input from the down (contact) side. Therefore, it can be argued that reliable somatosensory input, conveying information on the body–support-surface interface, is required during body tilt to estimate the SVV reliably. Information on the inclination and physical characteristics (e.g. compliance) of the supporting surface may not seem necessary to estimate the orientation of a straight line, but it could be vital to an individual who has to orient with respect to the environment and take immediate action on the basis of this orientation. The tilt-induced biases may be an inevitable trade-off, due to intrinsic adaptation in the somatosensory system, in order to improve accuracy both in SPV and SVV.

This review summarized some examples of dissociation between perceptual and motor phenomena and between visual and postural vertical. An example of the former is the dissociation between directional sway or nystagmus (e.g., in acute peripheral lesions or up-/downbeat nystagmus patients), which was not reflected in a directionally biased SPV in the various patient groups. An example of the latter was the profound dissociation observed between the subjective visual and postural vertical after vestibular neurectomy. Similarly, stimuli that induced directional sway, vection illusions, and tilts of the visual vertical (e.g., roll optokinetic stimulation) did not induce consistent changes in the SPV.[3] More recently, we investigated the ability of human subjects to indicate the earth gravitational horizontal and vertical by means of self-paced saccadic eye movements executed in total darkness.[25] We found that, at large angles of body tilt (45 and 90 deg), subjects were significantly less accurate in signalling horizontality than verticality. These experiments thus identified yet another dissociation in the perception of space, a dissociation between verticality and horizontality, that is, a loss of orthogonality during body tilt. Craniocentric saccades remained normal during tilt, indicating that internal space remained unchanged and that the saccadic changes could not be explained by the effect of ocular counterrolling.

Taken together, the findings summarized in this review of our work indicate that the perception of verticality or uprightness is not a single, unified concept. Different sensory channels can convey different and occasionally conflicting messages as to the perception of verticality. A prominent role of the somatosensory system for the perception of body uprightness and, at least when lying sideways, for visual orientation with respect to gravity, has also emerged. It is likely that the coexistence of contradictory messages about space perception in individuals with labyrinthine or CNS lesions underlies the disorientation reported by such patients.

REFERENCES

1. KANAYAMA, R., A. M. BRONSTEIN, M. A. GRESTY, G. B. BROOKES, M. E. FALDON & T. NAKAMURA. 1995. Perceptual studies in patients with vestibular neurectomy. Acta Otolaryngol (Stockh.) Suppl. **520**:408–411.
2. BUCHELLE, O. W., T. BRANDT & D. DEGNER. 1983. Ataxia and oscillopsia in downbeat-nystagmus vertigo syndrome. Adv. Oto-Rhino-Laryngol. **30**: 291–297.
3. BISDORFF, A. R., C. J. WOLSLEY, D. ANASTASOPOULOS, A. M. BRONSTEIN & M. A. GRESTY. 1996. The perception of body verticality (subjective postural vertical) in peripheral and central vestibular disorders. Brain **119**: 1523–1534.
4. BENSON, A. J. & S. F. BROWN. 1989. Visual display lowers detection threshold of angular, but not linear, whole-body motion stimuli. Aviat. Space Environ Med. **60**: 629-633.
5. HOWARD, I. P. 1982. Human Visual Orientation: 412. Wiley. Chichester.
6. FRIEDMAN, G. 1970. The judgement of the visual vertical with peripheral and central vestibular lesions. Brain **93**: 313–328.
7. CURTHOYS, I. S., M. J. DAI & G. M. HALMAGYI. 1991. Human ocular torsional position before and after vestibular neurectomy. Exp. Brain Res. **85**: 218–225.
8. DAI, M. J., I. S. CURTHOYS & G. M. HALMAGYI. 1989. Linear acceleration perception in the roll plane before and after unilateral vestibular neurectomyu. Exp. Brain Res. **77**: 315–328.
9. DIETERICH, M. & T. BRANDT. 1992. Wallenberg's syndrome: lateropulsion, cyclorotation and subjective visual vertical in 36 patients. Ann. Neurol. **31**: 399–408.
10. BRANDT, T., M. DIETERICH & A. DANEK. 1994. Vestibular cortex lesions affect the perception of verticality. Ann. Neurol **35**: 403–412.
11. DICHGANS, J., R. HELD, L. R. YOUNG & T. BRANDT. 1972. Moving visual scenes influences the apparent direction of gravity. Science **178**: 1217–1219.
12. DICHGANS, J., T. BRANDT & R. HELD. 1975. The role of vision in gravitational orientation. Fortschr. Zool. **23**: 255–263.
13. AUBERT, H. 1861. Scheinbare bedeutende Drehung von Objecten bei Neigung des Kopfes nach rechts oder links. Virchows Arch. **20**: 381–393.
14. MÜLLER, G. E. 1916. Ueber das Aubertsche Phaenomenon. Z. Psychol. Physiol. Sinnesorg. **49**: 109–246.
15. MITTELSTAEDT, H. 1988. The information processing stricture of the subjective vertical. A cybernetic bridge between its psychophysics and its neuro-biology. *In* Processing Structures for Perception and Action, H. Marko, G. Hauske & A. Struppler. Eds.: 217–263. VCH-Verlagsgesellschaft. Weinheim.
16. YARDLEY, L. 1990. Contribution of somatosensory information to perception of the visual vertical with body tilt and rotating visual field. Percept. Psychophys. **48**(2): 131–134.
17. MITTLESTAEDT, H. 1992. Somatic versus vestibular gravity reception in man. Ann. N.Y. Acad. Sci. **656**: 124–139.
18. BRONSTEIN, A. M., L. YARDLEY, A. P. MOORE & L. CLEEVES. 1996. Visually and posturally mediated tilt illusion in Parkinson's disease and in labyrinthine defective subjects. Neurology **47**: 651–656.
19. ANASTASOPOULOS, D. & A. M. BRONSTEIN. Submitted.
20. ANASTASOPOULOS, D., A. BRONSTEIN, T. HASLWANTER, M. FETTER & J. DICHGANS. 1999. The Role of the Somatosensory Input for the Perception of Verticality. This volume.
21. FERNANDEZ, C. & J. M. GOLDBERG. 1976. Physiology of peripheral neurons innervating otolith organs of the squirrel monkey. I: Response to static tilts and long-duration centrifugal force; II: Directional sensitivity and force-response relations. J. Neurophysiol. **39**: 970–995.
22. WADE, N. J. 1970. Effect of prolonged tilt on visual orientation. Q. J. Exp. Psychol. **22**: 423–439.
23. DAY, R. H. & N. J. WADE. 1966. Visual spatial aftereffect from prolonged head-tilt. Science **154**: 1201–1202.
24. CLARK, B. & A. GRABIEL. 1962. Visual perception of the horizontal during prolonged exposure to radial acceleration on a centrifuge. J. Exp. Psychol. **63**: 294–302.
25. PETTOROSSI, V. E., D. BAMBAGIONI, A. M. BRONSTEIN & M. A. GRESTY. 1998. Assessment of the perception of verticality and horizontality with self-paced saccades. Exp. Brain Res. **121**: 46–50.

The Role of the Otoliths in Perception of the Vertical and in Path Integration

HORST MITTELSTAEDT[a]

Max-Planck-Institut für Verhaltensphysiologie, D-82319 Seewiesen, Germany

ABSTRACT: The role of the otoliths in essential performances of human orientation is analyzed. The following interactions of the otoliths are considered:

1. The otoliths cooperate with graviceptors in the trunk in the perception of body posture. The truncal graviceptors turn out to yield on average 60% of the total gain.

2. The otoliths cooperate with proprioceptors in the head-to-trunk coordinate transformation. However, under static conditions, proprioceptors in the legs, although effective in the control of posture, neither affect the perception of posture nor of the visual vertical.

3. In contrast to the perception of posture, the perception of the visual vertical (SVV) receives the necessary gravity information exclusively from the otoliths. However, their output appears to be affected by a central nervous component that tends to rotate the SVV into the z-axis of head and trunk. A theory of vectorial summation of this component, the "idiotropic vector," with the otolithic vector is able to explain the cause of the A- and E- effects, the increase of the variance of the SVV with the tilt angle, and the asymmetrical effect of rotatory visual flow.

4. Finally, it is shown that the otoliths, by the separation of the effects of tilt from those of translation, play an essential role in navigation by path integration.

INTRODUCTION

As we know more about a man when we come to know his friends, his foes, and his teammates, we can learn more about the role and the survival value of a sense organ when we study its interaction with sensory input that harmonizes with, or antagonizes, or complements, its own input. It is rather unexpected, even perplexing, what we have found in the course of such an approach in the case of an organ like the otolith, whose functional significance does not appear difficult to understand. In the absence of all other spatial information, and under purely static conditions, the otolith system permanently produces a set of afferents from which a virtually veridical representation of the vertical with respect to the head may be derived. Hence it appears ideally suited to serve as the common reference for

---

[a] Address for telecommunication: Phone: 49(0)8157-932-371; fax: 49(0)8157-932-209; e-mail: h.mittelstaedt@mpi-seewiesen.mpg.de

all afferents that contain spatial information, and to provide an identical basis for perception as well as control of posture.

It turned out, however, that the subjective postural vertical deviates systematically from the subjective visual vertical, and that postural perception is governed by different afferents and kinds of processing than postural control.

Perception of Posture

We begin with perception of static posture, with the exclusion of vision. This perception was traditionally thought to be based on vestibular information and supplemented by somatosensory information originating from proprioceptors. We tried to separate the effect of the latter from that of the former by placing the subject on the earth-horizontal platform of a variable-radius centrifuge.[1] When the subject is oriented horizontally, and (as in FIG. 1) so that the centrifugal force acts along her z-axis, extravestibular graviceptors must reveal themselves if they have any effect on the perception of posture. Because, if the centrifuge axis is collinear with the binaural axis, the otoliths are not affected by the centrifugal force, and hence indicate a horizontal posture. But graviceptors in the body are then, in addition to gravity, subjected to a force that is directed toward the feet, and hence indicate a head-up deviation from the horizontal.

Our 23 normal subjects felt horizontal when the centrifuge axis was between the binaural axis and the area of the last ribs, with a mean distance (d) of the centrifuge axis from the binaural axis of $d \approx -28$ cm. Subjects deprived of otolith function due to neuromectomy felt horizontal when the centrifuge axis was in the area of the last ribs.[2,3] Hence the mass centroid of the unknown graviceptors must lie there, that is, at a distance from the meatus of $d \approx -45$ cm. Taken together, and compared with control experiments on a tilt table, the results indicate an averaging interaction of otoliths and truncal graviceptors, under an idiosyncratically varying proportion of their gains, with an average of about 60% in favor of the latter.

Experiments with paraplegic patients under this paradigm[2,3] excluded receptors in the skin or between the vertebrae as sources of this graviceptive information. Rather, they led to the conclusion that it is provided by two distinctly localized inputs, the first entering the spinal cord at the eleventh thoracic segment, and the second reaching the brain cranial of the sixth cervical segment, presumably via the n. phrenicus or the n. vagus.

The effect of the first-named input is abolished after bilateral nephrectomy. This shows that the kidneys do affect gravity perception. But whether they function like statoliths or in another way cannot yet be decided.

For the second input, however, the results show unequivocally that it yields gravity information through the inertia of a mass in the body. It is hypothesized that this mass may be that of the blood in the large vessels. This is corroborated by the effect of shifting blood craniad, or caudad, respectively, by means of positive or negative pressure to the entire lower body, or only to the legs.[3,4] It is inferred that the inertial forces are measured by mechanoreceptors in the ligaments that mechanically support the large vessels, and, possibly, also by baroreceptors. Furthermore, an effect of the mass of the entire viscera, as suggested by Henning v. Gierke and Donald Parker,[5] is not yet excluded.

The Role of Proprioceptors

The results of these experiments also demand a qualification of the role of the proprioceptors. They are clearly essential for the coordinate transformations of head-fixed information to the trunk and the legs and vice versa, as well as for the control of posture. However, unexpectedly in our experiments, which separate perception from control, the

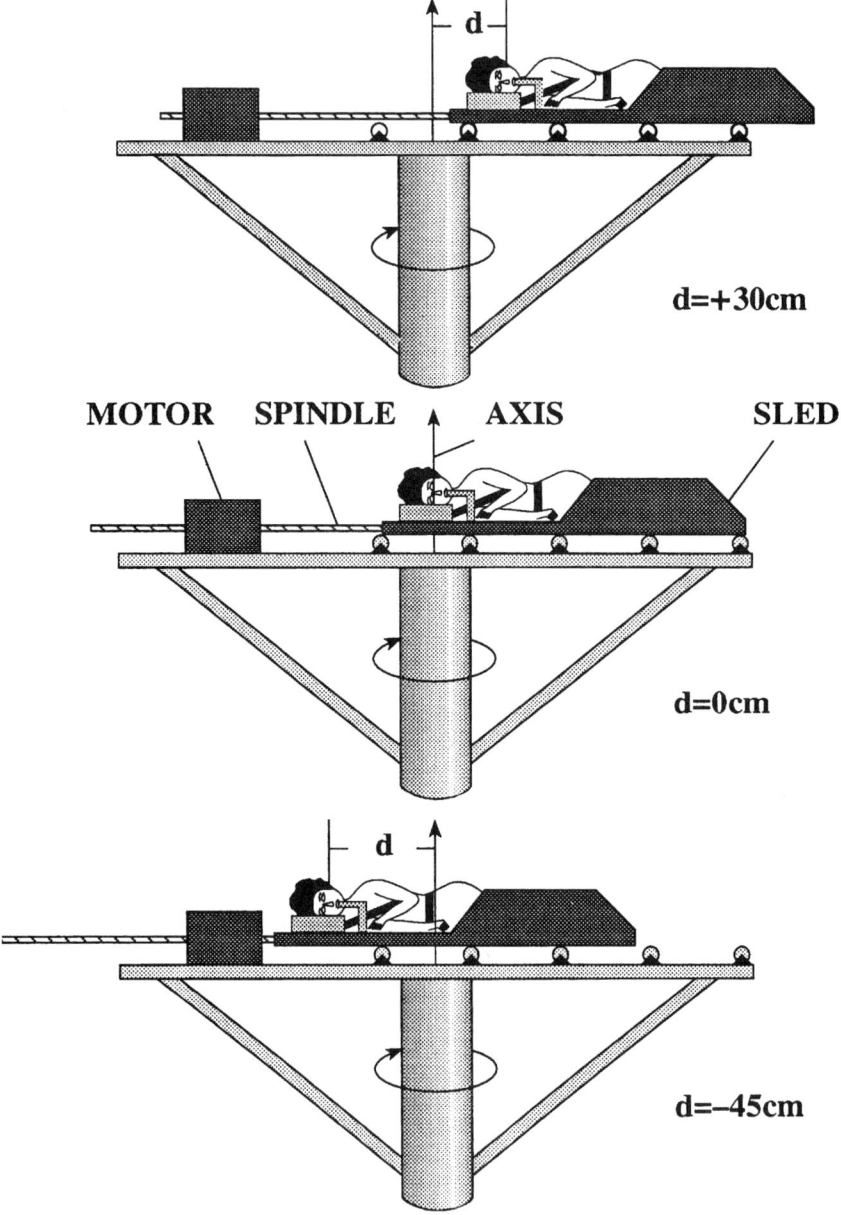

FIGURE 1. Test of the subjective horizontal posture on a rotating variable-radius centrifuge. The subject is placed right ear down on a sled that can be moved radially along the subject's spinal (*z*-) axis via remote control. In total darkness the centrifuge rotates with a constant velocity of 36 rev/min. If the otoliths were the only graviceptors, then the subject, due to the additional acceleration, would feel tilted upward when the rotation axis (RA) is cranial of the binaural axis (BA), feel tilted head-down when the RA is caudal of the BA, and horizontal when the axes coincide. In fact, subjects feel horizontal when the RA is caudal of the BA, on average by 28 cm.

mechanoreceptors in the legs had no effect on the perception of posture. This is demonstrated negatively by the missing effect of applying mechanical forces to the legs of normals,[1] and, positively, by the residual capability of those paraplegics who are merely deprived of all spinal afference from the legs.[2-4] Even an effect of leg flexion was still fully present in experiments on the tilt table with those paraplegics, and hence is supposedly caused by unknown abdominal afferents.

Leg proprioceptors, or perhaps even proprioceptors in general, may have no <u>direct</u> access to the perception of posture. They clearly mediate perception and control of joint position, but appear to affect the perception of posture only by implementing or modulating the output of a graviceptor, as in the coordinate transformations, or as in the previously mentioned effect of leg flexion. This seems to be at variance with the documented effect of leg muscle vibration[6] on posture and perception in freely standing subjects. Yet closer analysis of its mechanism may lead, as in the present case, to a qualification and revision.

A way to reconcile the ostensibly diverging interpretations may be obtained from the results of Alexander Bisdorff and colleagues,[7] who asked normal subjects to indicate their subjectively vertical posture under galvanic stimulation of the vestibulum, and under optokinetic stimulation by means of a rotating dome. When standing freely the subjects evinced body sway, and reported sensations of tilt or vection, respectively. Strapped onto a padded tiltable chair, however, the subjects were able to indicate a perfectly upright posture also under galvanic and optokinetic stimulation, but, paradoxically, sensed sway and vection simultaneously with feeling upright! The authors report that patients without vestibular function show merely a mild increase of variance in this test. They infer that proprioception alone may be sufficient to provide a veridical postural vertical, yet overlook the existence of a second candidate, namely, the truncal graviceptors. Similar hints at their influence on postural perception and control is accumulating in the recent literature (reviewed in reference 3; see also references 8 and 9).

Perception of the Visual Vertical

Perception of the visual vertical was tested in the standard way by means of a luminous pendulum in otherwise complete darkness. As in the perception of posture, neither on the tilt-table nor on the centrifuge did we find an effect of proprioceptors in the skin or the legs.[1] In contrast to postural perception, however, the subjective visual vertical (SVV) was unaffected by the truncal graviceptors as well.[10] This was not only tested under centripetal force along the z-axis, but also by placing the subject, with the centrifuge axis passing through the centroid of the labyrinths, at a 45-degree roll tilt with respect to the axis. Now half of the centrifugal force acted along the y-axis, but again without producing any change in the SVV.

Under these conditions, then, the SVV obtains the necessary gravity information about the roll tilt exclusively from the otoliths. However, the large deviations of the subjective from the objective horizontal seen at 90-degree roll tilts (A-effect) stand in sharp contrast to the almost veridical settings of the postural horizontal of the same subjects under the same conditions. Our explanation of this discrepancy assumes that the SVV is not only determined by the otoliths but also by an internal agent, the "idiotropic vector," that tends to rotate the SVV into the subject's z-axis.[11-13] According to the theory, the SVV is determined by the resultant of the tilt-independent idiotropic, with the tilt-dependent graviceptive vector given by orthogonal components of the utricle and the saccule (FIG. 2).

The mathematical formulation of the theory (cf. the Appendix) shows that the A- and E-effects are compelling consequences of this processing structure, and quantitatively determined by the magnitude of the idiotropic and the relative gains of the utricular and saccular components.[14] The theory also explains, as an inevitable corollary, why the variance of the SVV increases with the tilt angle—a well-known fact that

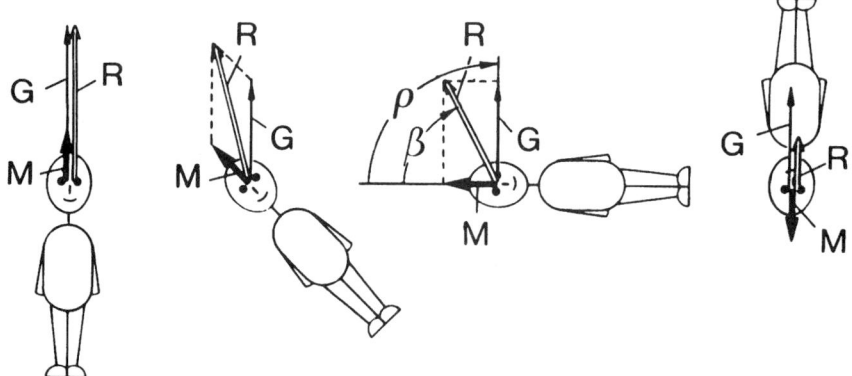

FIGURE 2. Theory of the SVV as a result of summation of two vectors: (1) a force-dependent vector G (*thin arrow*), yielded by the utricles and the saccules, and directed vertically, here exactly, but actually only more or less (see Appendix); (2) a force-independent vector M (*fat arrow*) that is permanently directed along the subject's z-axis (therefore named *idiotropic vector*). The resultant R (*double arrow*) determines the direction β of the subjective visual vertical, and, by its magnitude, the confidence of the subject in her or his estimate (as documented by the inverse relation of the variance of the SVV to the magnitude of R). Note the dramatic decrease of R with increasing roll tilt ρ. (As defined in the Appendix, ρ and β are positive in the right-ear-down attitude.)

seemed irreconcilable with the tilt-independence of the variance of the otolithic afferents. As shown in FIGURE 2, the magnitude of the resultant vector decreases with increasing tilt. Hence the variance of the SVV must then increase, because, necessarily, it is inversely related to the magnitude of the resultant. The theory may be valid beyond the SVV: As shown by Theodore Raphan and colleagues (reference 15, Fig. 6, p.150), this vectorial summation appears to also exist in the determination of the rotation axis of the slow phase of the optokinetic nystagmus at roll tilt.

Furthermore, the theory opens a way to explain a peculiar effect of rotatory visual flow on the SVV discovered in 1974 by Johannes Dichgans and colleagues.[16] Visual flow about the x-axis of constant velocity causes an angular deviation of the SVV in the same sense as the flow. In the upright posture it is independent of rotation sense. At roll tilt of the head, however, the deviation is larger if the flow has the opposite sense as the tilt than if it has the same sense as the tilt, and increasingly so with increasing tilt angle. This asymmetry seems to be irreconcilable with the anatomical and functional symmetry of the utricles. But it turns out to be an inevitable consequence of the effect of the idiotropic if the visual flow is fitted to the vectorial structure of the theory.[17] This can be done in a straightforward way by multiplying a representation of the signed flow velocity with the utricular component as well as with the saccular component, and adding the product with the utricule to the saccular component, while subtracting the product with the saccule from the utricular component (see Eq. A5 in the Appendix). As a consequence, the resultant vector, and hence the SVV, will rotate by an angle dependent on the gains of the flow, the otoliths, and the idiotropic. This algorithm can be fitted, by maximally two free parameters, to the data of Dichgans *et al.*,[16] as well to our own (as yet unpublished) data. Because, according to the theory, the asymmetry is caused by, and depends monotonically on, the idiotropic vector, it should be correlated with a subject's A-effect; and this is indeed found. Hence it should be missing if the A-effect is absent. In fact, in experiments of Lucy Yardley,[18] a patient who suffers from total loss of the myelinated proprioceptive afference below the neck, evinced no A-effect and no asymmetry at a 90 degree roll tilt. Thus proprioception, if indeed its loss has caused this result, alters the magnitude of the idiotropic vector, but does not provide a measurement of the force vector's direction; corroborating that proprioception affects the apparent vertical, if at all, in a nondirected (parametric) way, or only indirectly, as in the head-to-trunk coordinate transformation.[11] Another example is discussed in the Appendix.

The Role of the Otoliths in Navigation by Path Integration

It has been shown that mammals are able to home under conditions that are designed to exclude all external cues (for a review, see reference 19). This capability must then be based on the integration and storage of the animal's own movements. In 1973 we defined spatial information that can *only* be gained by the agent's active or passive movement as idiothetic information.[20] It could thus be gained in two basically different ways:

(1) From sense organs like the semicicular canals and the otolithic or truncal graviceptors, hence termed inertial idiothesis;

(2) From those proprioceptors or efference copies whose signals are normally correlated with, and hence allow inferences about, the agent's course with respect to ground, air, or water, hence termed substratal idiothesis.[19]

In either case, the rotatory components of the movement must be integrated over the translatory displacement along the path in order to obtain the coordinates of the agent's location with respect to his or her starting point. Hence we have named this performance *Wegintegration*[20] or path integration.[21,22]

The question of whether the otoliths are in fact involved in path integration has been investigated in experiments on the Mongolian gerbil. Gerbils are able to home by idiothesis from any place in a circular arena of 1.30 m diameter.[21,22] By rotating the arena with varying acceleration profiles it was shown that the rotatory component of the homing performance correlates well with the known properties of the semicircular canals. By contrast, in two experimental paradigms that were designed to test the effect of the otoliths, the animal appeared to be unable to reckon with translatory acceleration. In one of the paradigms, however, the gerbil was accelerated while it was sitting. Now, unlike the process that summates the rotations, the path integrator need not be kept running when the location stays constant. It should better be shut off then to guard against drift. In the other paradigm the entire arena was shifted while the gerbil was moving. But due to the large mass of the apparatus, the acceleration profile may well have been outside the range the system can, or does, use.

Therefore a third paradigm was designed to obviate these objections.[19] Under purely idiothetic conditions, the gerbil was trained to pick up her young at the border opposite home when the arena was immobile, and tested when the arena was rotating with constant velocity. The rationale is that in the case of inertial navigation the animal would orient to Newtonian space rather than to the surface of the arena. The result is shown in FIGURE 3, where each path is represented twice, namely with reference to the arena by a solid line, whereas with respect to external space by a dashed line. In the immobile arena the gerbil takes a straight course from the nest at 180 deg to the young at 0 deg. If it did so in the clockwise rotating arena under exclusive substratal navigation, then the path would be straight with respect to the surface, yet curved to the right when recorded from a camera fixed to the ceiling of the room. FIGURE 3a shows how a straight path under exclusively substratal idiothesis would look at the three rotation velocities (v) tested, $v_1 = 14.4$, $v_2 = 28.8$, and $v_3 = 36.0$ degrees per second. FIGURE 3b shows how a straight path would look if it were guided exclusively by inertial idiothesis. FIGURE 3c shows the means of 121 runs

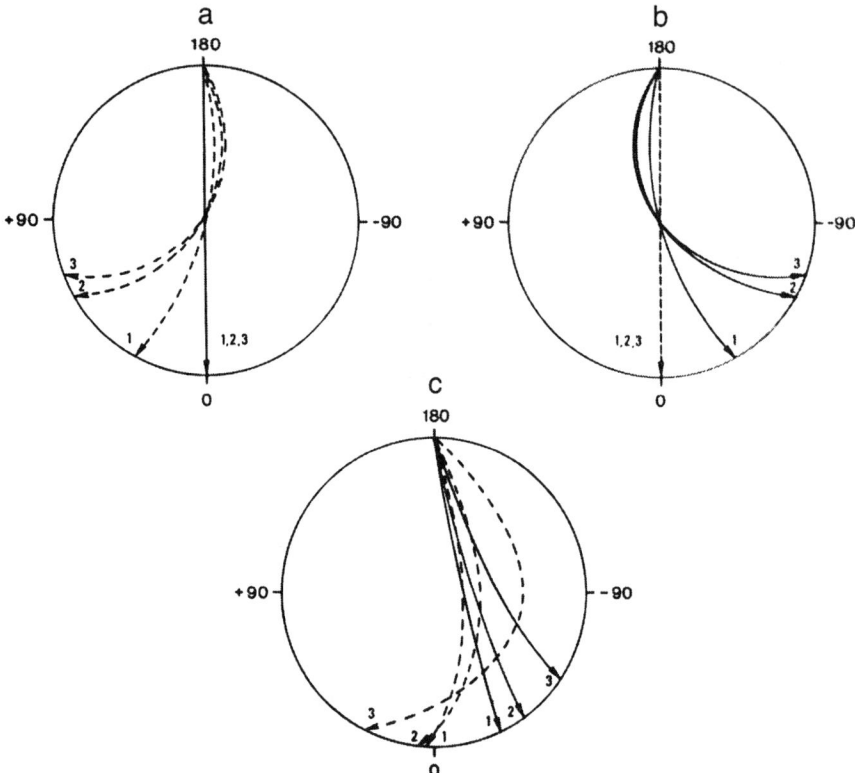

FIGURE 3. Path integration of a gerbil in total darkness on a disk rotating with constant velocity. While the disk (diameter: 1.30 m) was immobile, the gerbil has been trained to pick up her young from a point at the border (at 0 deg) that is opposite her nest (at 180 deg). Shown are theoretical (**a**), (**b**) and actual (**c**) paths on the disk rotating with $v_1 = 14.4$, $v_2 = 28.8$, and $v_3 = 36.0$ cm/s. All paths are represented twice: (1) as recorded from a camera rotating with the disk (*solid lines*), and (2) as recorded from a camera fixed in space (*dashed lines*). *Upper disks*: Paths under velocities 1,2,3, when the gerbil would run with a constant velocity on a straight path: (a) exclusively guided by substratal information, and (b) exclusively guided by inertial information. (c) Actual mean paths of 335 runs. Note that the gerbil takes intermediate courses, but that a target of the size of a young gerbil would more likely be hit if it were fixed in space at 0 deg than when it rotated with the disk.

of 11 animals at v_1, 127 runs of 14 animals at v_2, and 87 runs of 10 animals at v_3. The actual paths appear to be intermediates between those expected when guided by one kind of idiothesis alone.

The result points at an averaging interaction of the two idiothetic inputs—a rather plausible type of processing, since in normal life, in contrast to our experimental situation, the effect of the two inputs is synergetic and complementary. That rats are also able to use both inputs in navigation has recently been shown by Jan Bureš in place avoidance tasks on a slowly rotating disk.[23]

The essential finding in our present context is that the path integration system not only uses rotatory but also translatory inertial information. Inertial sensors like the semicircular canals cannot, however, discriminate immobility from constant rotation; and inertial sensors like the otoliths and the truncal graviceptors cannot, by principle, discriminate the effect of linear acceleration from that of gravity. Hence, the result indicates that under purely idiothetic conditions the central nervous processing is able to disambiguate their input, and to separate rotation, tilt, and translation in their messages.

We shall not discuss the present models of the separation (see reference 19, p.430; also reference 24 and 25), but turn to the results of Marie-Luise Mittelstaedt with human subjects who estimated their velocity after brief passive acceleration to a constant rotation under purely idiothetic conditions.[26] When they stood at the center of a rotating disk their subjective velocity declined exponentially, until they eventually felt being at rest. When they stood excentrically, gripping a radial bar with both hands, the duration of feeling rotated increased, as did the time constant. The latter turned out to be linearly dependent on the product of the radius and the objective rotatory velocity, that is, on the acting translatory acceleration. Eventually they felt at rest and slightly tilted. When the subjects were stopped after such a rotation, however, duration and time constant were independent of excentricity.

In principle, according to Newtonian mechanics, constant rotation can, in contrast to constant translation, be discriminated from rest. Therefore, the perceptual system should be able to gain an estimate of the velocity of a constant rotation purely by means of inertial information. Accelerated to constant rotation, the subject in fact continues to feel rotated. But the cupulae of the semicircular canals eventually return into their resting position, and the central nervous residues of their messages fade to zero also. The only organs that respond to perpetual rotation are those that are able to measure the centrifugal force, that is, the otoliths and the somatic graviceptors. However, since they cannot discriminate translation from tilt, the perceptual system must come to a reasonable decision on what is actually the case. Since a lasting change of the force vector's direction is, in everyday life, generally due to tilt, the subject should eventually feel tilted, and this indeed happens. However, at the end of the acceleration phase the cupulae still indicate a rotation about the subject's z-axis. Hence it makes sense for the perceptual system to resolve the output of the graviceptors into their orthogonal components, and conclude from the result that the subject is rotating excentrically in an upright attitude, rather than being tilted and at rest.

Taken together, all these findings speak in favor of the idea that the brain uses its "implicit knowledge" about the properties of the sense organs and the relevant physical laws in order to gain, as if through inductive reasoning (via *"unbewußte Schlüsse,"* to invoke Helmholtz' metaphor),[27] a best estimate of the percipient's real whereabouts—optimal only under the system's given constraints, yet trustworthy enough in everyday life.

REFERENCES

1. Mittelstaedt, H. & E. Fricke. 1988. The relative effect of saccular and somatosensory information on spatial perception and control. Adv. Oto-Rhino-Laryngol. **42**: 24–30.
2. Mittelstaedt, H. 1992. Somatic versus vestibular gravity reception in man. Ann. N.Y. Acad. Sci. **656**: 124–139.
3. Mittelstaedt, H. 1996. Somatic graviception. Biol. Psychol. **42**: 53–74.
4. Mittelstaedt, H. 1997. Interaction of eye-, head-, and trunk-bound information in spatial perception and control. J. Vestibular Res. **7**: 283–302.
5. Gierke, H. E. v. & D. E. Parker. 1994. Differences in otolith and abdominal viscera graviceptor dynamics: implications for motion sickness and perceived body position. Aviat. Space, Environ. Med. **65**: 747–751.
6. Hlavatska, F., M. Krizkova & F. B. Horak. 1995. Modification of human postural response to leg muscle vibration by electrical vestibular stimulation. Neurosci. Letters **189**: 9–12.
7. Bisdorff, A., A. Bronstein, M. Gresty & C. Wolsley. 1996. Subjective postural vertical inferred from vestibular-optokinetic vs. proprioceptive cues. Brain Res. Bull. **40**: 413–415.

8. MITTELSTAEDT, H. 1995. Evidence for somatic graviception from new and classical investigations. Acta Otolaryngol. (Stockh.) **520** (Suppl.): 186–187.
9. FITS, I. B. M. VAN DER, A. W. J. KLIP, L. A. VAN EYKERN & M. HADDERS-ALGRA. 1998. Postural adjustments accompanying fast pointing movements in standing, sitting and lying adults. Exp. Brain Res. **120**: 202–216.
10. MITTELSTAEDT, H. 1985. Subjective vertical in weightlessness. *In* Vestibular and Visual Control on Posture and Locomotor Equilibrium, M. Igarashi and O. Black, Eds.: 139–150. Karger. Basel.
11. MITTELSTAEDT, H. 1983 A new solution to the problem of the subjective vertical. Naturwissenschaften **70**: 272–281.
12. MITTELSTAEDT, H. 1986. The subjective vertical as a function of visual and extraretinal cues. Acta Psychol. **63**: 63–85.
13. MITTELSTAEDT, H. 1988. The information processing structure of the subjective vertical. A cybernetic bridge between its psychophysics and its neurobiology. *In* Processing Structures for Perception and Action, H. Marko, G. Hauske, and A. Struppler, Eds.: 217–263. VCH-Verlagsgesellschaft. Weinheim.
14. MITTELSTAEDT, H. 1995. The formation of the visual and the postural vertical. *In* Multisensory Control of Posture, T. Mergner and F. Hlavacka, Eds.: Plenum Press, New York and London **1995**: 147–155.
15. RAPHAN, T., M. DAI & B. COHEN. 1992. Spatial orientation of the vestibular system. Ann. N.Y. Acad. Sci. **656**: 140–157.
16. DICHGANS, J. M., H. C. DIENER, H. C. & T. H. BRANDT. 1974. Optokinetic-graviceptive interaction in different head positions. Acta Otolaryng. **78**: 391–398.
17. MITTELSTAEDT, H. 1991. The role of the otoliths in the perception of the orientation of self and world to the vertical. Zool. Jahrb. Abt. Physiol. **95**: 419–425.
18. YARDLEY, L. 1990. Contribution of somatosensory information to perception of the visual vertical with body tilt and rotating visual field. Percept. Psychophys. **48**: 131–134.
19. MITTELSTAEDT, M.-L. & S. GLASAUER. 1991. Idiothetic navigation in gerbils and humans. Zool. Jahrb. Abt. Physiol. **95**: 427–435.
20. MITTELSTAEDT, H. & M.-L. MITTELSTAEDT. 1973. Mechanismen der Orientierung ohne richtende Außenreize. Fortschr. Zool. **21**: 46–58.
21. MITTELSTAEDT, M.-L. & H. MITTELSTAEDT. 1980. Homing by path integration in a mammal. Naturwissenschaften **76**: 566.
22. MITTELSTAEDT, H. & M.-L. MITTELSTAEDT. 1982. Homing by path integration. *In* Avian Navigation, F. PAPI & H.-G. WALLRAFF, Eds.: 290–297. Springer-Verlag. Berlin.
23. BURES, J., A. A. FENTON, YU KAMINSKY & L. ZINYUK. 1997. Place cells and place navigation. Proc. Natl. Acad. Sci. USA **94**: 343–350.
24. GLASAUER, S. 1992 Interaction of semicircular canals and otoliths in the processing structure of the subjective zenith. Ann. N.Y. Acad. Sci. **656**: 874–849.
25. GLASAUER, S. & D. M. MERFELD. 1997. Modelling three dimensional vestibular responses during complex motion stimulation. *In* Three-Dimensional Kinematic Principles of Eye-, Head-, and Limb Movements in Health and Disease. M. Fetter, D. Tweed, and H. Misslisch, Eds.: 387-398. Harwood. Amsterdam.
26. MITTELSTAEDT, M.-L. & H. MITTELSTAEDT. 1996. The influence of otoliths and somatic graviceptors on angular velocity estimation. J. Vestib. Res. **6**: 355–366.
27. HELMHOLTZ, H. v. 1896. Handbuch der physiologischen Optik, III, Vol. 1. Kapitel. Hamburg and Leipzig.
28. PARKER D. E & R. L. POSTON. 1984. Tilt from a head-inverted position produces displacement of visual subjective vertical in the opposite direction. Percept. Psychophys. **36**: 461–465.
29. MITTELSTAEDT, H. 1995. New diagnostic tests for the function of utricles, saccules and somatic graviceptors. Acta Otolaryngol. (Stockh.) **520** (Suppl.): 188–193.
30. ANASTASOPOULOS, D., A BRONSTEIN, T. HASLWANTER, M. FETTER & J. DICHGANS. 1999. The role of the somato sensory input for the perception of verticality. This issue.
31. EGGERT T. 1998. Der Einfluß orientierter Texturen auf die subjektive Vertikale und seine systemtheoretische Analyse. Dissertation, Fakultät für Elektrotechnik und Informationstechnik der Technischen Universität München: 1–272.

APPENDIX

In order to demonstrate the essentials of the SVV-theory, only the case of pure roll shall be considered, where the noncompensated fraction of the ocular counterroll may be neglected, as well as the otolith-to-head coordinate transformation. Also, the effect of non-linear deviations of the otolithic primary afferents, which the postulated normalization cannot remove, and hence shows up mainly at large G-loads, will not be treated here (but see reference 13, esp. p. 235ff), as is the effect of neck receptors (but see reference 11 and 28). Furthermore, the settings of a luminous pendulum to subjective vertical are taken under exclusion of visual cues, and read off after cessation of all transients.

In pure roll, then, the apparent tilt (AT) of the luminous pendulum, under these idealized conditions, is proportional to the following function:

$$AT = \cos \beta * y/N - \sin \beta * (z/N + M), \tag{A1}$$

with $y = U_1 \sin \rho$, $z = S_1 \cos \rho$, and $N = \sqrt{y^2 + z^2}$,

where ρ is the deviation of the subject's z-axis from the objective vertical, and β is the deviation of the subject's z-axis from the luminous pendulum. Both are projections onto the y-z plane of the head in a right-handed coordinate system. U_1 is the gain of the utricle, S_1 the gain of the saccule, and M the magnitude of the z-component of the idiotropic vector.

The subject moves the pendulum until the AT is zero, that is, the function of Eq. A1 serves as a turning tendency ("attractor"). At equilibrium, then,

$$\tan \beta = \frac{y/N}{z/N + M}. \tag{A2}$$

This formula can be fairly well adapted to the mean SVV settings of normal subjects, with U_1 set to unity and M determined as the mean of the cotangents of β at $\rho = \pm 90$ deg. Thus only one free parameter remains, namely S_1. This latter turns out to be around 0.6, that is, S_1/U_1 is close to the relation of the number of saccular sense cells to those of the utricle.

The theory yields a straightforward explanation for the E- and A- effects. A crossover from over- to undercompensation (from E- to A-effect), or vice versa, should occur at a tilt angle (ρ_{cross}) if and when

$$\tan^2 \rho_{cross} = \frac{(S_1/U_1 - 1)^2}{M^2} - \left(\frac{S_1}{U_1}\right)^2. \tag{A3}$$

Hence, a crossover can only occur if

$$M^2 < \frac{(S_1/U_1 - 1)^2}{(S_1/U_1)^2}. \tag{A4}$$

The SVV crosses over from E- to A-effect at acute tilt angles if, as normally, $S_1/U_1 < 1$, but from A- to E- effect at obtuse tilt angles if $S_1/U_1 > 1$.

A "Ganzfeld" of random dots rotating with constant clockwise (positive) velocity v about the visual (x-) axis causes a clockwise deviation of the SVV (that is, a negative deviation of β), which increases with increasing tilt. When the field rotates in the opposite sense of the head tilt, this deviation is larger than when it rotates in the same sense of the head tilt; and this asymmetry also increases with increasing head tilt.[16]

This relation is yielded by the theory if representations $K_s v$ and $K_c v$ of the field's velocity v are introduced into Eqs. A1 and A2 in the following way (reference 17, esp. p. 423ff):

$$\tan \beta = \frac{y/N - K_s vz/N}{z/N + K_c vy/N + M}. \tag{A5}$$

with $N = \sqrt{y^2 + z^2}$

and with Ks/Kc varying idiosyncratically between 0.2 and 0.9 (own unpublished results with six probands).

The generalization of Eq. A5 is also able to account for the bias of the SVV in the acute phase after unilateral neurectomy (for details see reference 29, p.191). In fact, it may be generally applicable to centrally caused roll asymmetries that affect the SVV.

An example is found in the contribution of Dimitrios Anastasopoulos and colleagues to this volume,[30] who tested the SVV of two patients with unilateral hypesthesia due to cortical and thalamic ischemic infarction, respectively. In both patients the A-effect was missing, that is, the subjective coincided virtually with the objective vertical, when they were lying on their hypesthetic side. The authors conclude that "the origin of tilt-mediated effects on the visual vertical (A-effect) is primarily somatosensory, not otolithic." However, if the pressure receptors on the lower side of the body would provide roll-tilt information to the SVV, then their loss would increase the deviation of the subjective from the objective vertical, rather than reduce it to zero.

By contrast, Eq. A5 affords a rather plausible explanation of what happened to the SVV of these two patients. Because the magnitude of the idiotropic is determinable from the settings at 90-deg tilts, we obtain, for patients A and B, respectively,

$M = (\cotan \beta_{(+90)} - \cotan \beta_{(-90)})/2 = 0.108$ and 0.078.

Hence, the values of the two constants K_s and K_c (with v set to unity) can be determined independently of the otolithic gains, namely, $K_s = 0.072$ and -0.017, as well as $K_c = 0.165$ and -0.078, respectively.

Thus, according to the theory, the result is due to a bias in roll, combined with a reduction in the magnitude of the idiotropic vector to 40 and 30%, respectively, of that of the normal control group ($M=0.26$). This corroborates the conclusion drawn at the end of the third subsection of this paper. The infarction, apart from causing an ipsilateral bias, may have merely attenuated the idiotropic gain, either directly or by way of the hypesthesia. A plausible reason for the latter follows from a recent interpretation of the idiotropic tendency. Briefly, as proposed by Thomas Eggert,[31] the magnitude of the idiotropic vector is related to the *a priori* confidence of the "internal estimator" (cf. end of the fourth subsection) that the recipient is upright, independently of what the otoliths tell him or her. The loss of proprioceptive inflow is likely to reduce this confidence, and thus also the idiotropic gain.

As developed in References 12, 13, and 31, the theory is also able to account for the effect of static visual patterns, particularly of orthogonal stripes (as in the rod-and-frame test) on the subjective visual vertical.

Replication of Passive Whole-Body Linear Displacements from Inertial Cues

Facts and Mechanisms

R. GRASSO,[a,d] S. GLASAUER,[b] P. GEORGES-FRANÇOIS,[c] AND I. ISRAËL[c]

[a]*Human Physiology Section, Scientific Institute Santa Lucia, via Ardeatina 306, I-00179 Rome, Italy*

[b]*Center for Sensorimotor Research, Neurology Department, Ludwig-Maximilian-University of Munich, D-81377 Munich, Germany*

[c]*Laboratoire de Physiologie de la Perception & de l'Action, CNRS—Collège de France, F-75005 Paris, France*

ABSTRACT: **Using path integration, normal subjects should be able to compute the distance of a traveled path even from the sole inertial sensory input. Blindfolded subjects were submitted to a passive linear forward displacement along 2 to 10 m. Their task was to replicate the traveled distance, still blindfolded, by driving the vehicle they were seated upon using a joystick that controlled linear speed.**

Subjects replicated both the length and the velocity profile of the passive travel, suggesting that a dynamic record of experienced motion is stored in memory. Even when the replication of passive motion dynamics was made impossible, the subjects could still replicate the displacement.

The results are explained by a dynamic feedback model that performs a running comparison between the perceived instantaneous displacement of the ongoing motion and the displacement derived from a spatiotemporal record of perceived passive motion. A multimodal acceleration-related sensory input is transformed into a displacement-related perception through double time-integration.

INTRODUCTION

When an animal moves, various types of receptors are excited as a consequence of its displacement. The deriving signals are called *idiothetic*[1] and include optic flow, proprioception, inertial signals, and efference copies. All these signals may inform the animal about its current position through a process called *path integration*.[2,3] Path integration lies at the core of short-distance navigation in animals and is conceived as a neural mechanism able to produce a running estimate of the current subject position relative to the starting point, such that the homing trajectory can be computed at any time. Path integration may also participate in structuring long-term spatial knowledge by integrating external and internally generated signals.[4–6]

---

[d]To whom correspondence may be addressed. Phone: ++39.06.5150.1473; fax: ++39.06.5150.1477; e-mail: rgrasso@giannutri.caspur.it

In 1964, Barlow[7] first suggested that inertial signals, as sensed by the vestibular and somatosensory systems, may represent a relevant input for animal navigation. However, translational and rotational acceleration need to be integrated twice over time to yield an estimate of current position and orientation in space. Experimental results obtained from mammals returning home after moving along complex trajectories suggest that this may indeed be the case.[3,4,8,9]

Recent studies[10–16] have shown that humans also can estimate the traveled path solely from self-generated information, that is, without external signals (visual or acoustic landmarks). However, the contribution of inertial information is still to be clarified.

Passive linear displacement estimation in humans has been studied with a number of different paradigms: through verbal estimates,[17] saccadic eye movements,[18] or button-pushing responses,[14,19] and all of these studies suggest that the length of passive linear motion can be accurately estimated, a possible neural mechanism being that of a double time-integration of linear acceleration.[3,19]

We have previously reported,[20,21] however, that blindfolded subjects who were transported on a mobile robot along a linear path and then asked to drive the robot along the perceived distance, replicated the velocity profile of passive transport. This suggested that whole-body motion dynamics (along linear paths) are coded and stored in spatial memory. The replication of distance could therefore result as a side effect of the replication of velocity, and may not derive from an independent estimate of the length of the experienced travel.

In the present study, we prevented subjects from replicating the time course of the passive travel: rather than controlling the speed of the vehicle, subjects were either given an on–off switch device such that the linear velocity of the self-driven motion was constant or they were prevented from reaching the maximum speed attained during the passive displacement.

We will show that subjects can replicate the length of the passive travel in these experimental conditions as well, and we will develop and test a mathematical model of the sensory–motor integration occurring in the task. The relative contribution of vestibular and somatosensory cues will be assessed by analyzing the behavior of three patients affected by bilateral vestibular deficit and one patient with a complete spinal lesion (lacking extero- and proprioceptive sensory input from the lower trunk and limbs). An account of the results obtained in the previously published studies[20–21] is also included.

METHODS

Experimental Setup

A mobile robot, the Robuter™ (from Robosoft SA, France) with a race car seat fixed on it was used for this experiment (FIG. 1A). Two motor wheels ensure propulsion at a maximal linear velocity of 1.2 m/s, with a maximal acceleration of 1 m/s². Steering is obtained by controlling the relative speed of the two driving wheels. The robot can be controlled by a remote PC microcomputer through wireless modems, or by a joystick connected to the robot itself. For purely linear displacements, the joystick controls the robot velocity in steps of 0.05 m/s (velocity directly proportional to joystick angle) with a 0.2 s

FIGURE 1. Methods: (**A**) Experimental setup: the subject sits on the robot with seat belt fastened, wearing black goggles and earphones. To replicate passive transport he or she uses a joystick. Connection with the microcomputer is provided by wireless modems. (**B**) Procedure: a sample trial measured by odometry. The *top diagram* shows the stimulus (10 m) and response displacements from the starting point as functions of time. The *bottom diagram* displays the instantaneous linear velocity. For this trial, the distance, duration, and velocity profile of the stimulus were replicated during the response phase.

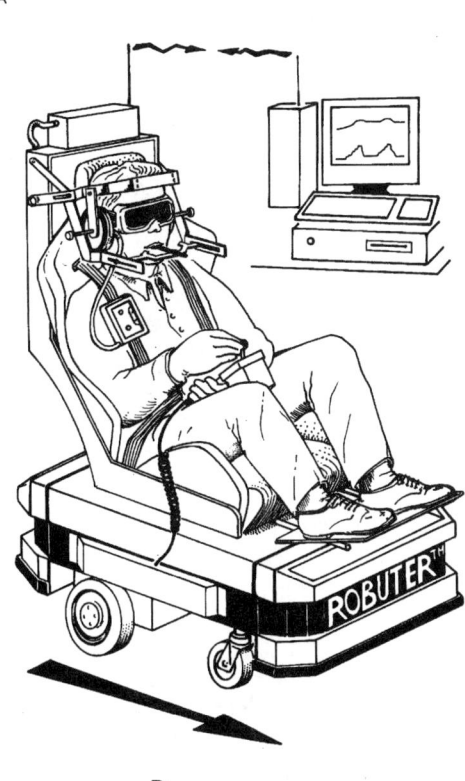

delay, due to hardware and software implementation of the joystick-mode control. Odometry (position on the X-axis and timing) was recorded by the robot during motion, at a 50-Hz sampling rate (FIG. 1B).

The subject was secured with three safety belts onto the seat of the robot (Fig. 1A). The head was restrained by two rigid cushions mounted onto the back of the seat to impede head translations and yaw rotations, and a bite bar also prevented pitch movements. The subjects wore headphones relaying a wide-band noise to prevent perception of external acoustic cues, and a pair of goggles with blacked out lenses to suppress visual information.

The joystick was set (by software) for the whole experiment so as to allow only linear movements of the robot along the X-axis, and all stimuli delivered by the PC were linear forward displacements in the range of speed of natural human locomotion. The experiment was performed within a 1.9-m-wide and 50-m-long corridor.

Experimental Procedure

Fifteen healthy volunteers, ranging in age from 20 to 50 and with no history of vestibular disorder, gave their informed consent to take part in the experiment, which was approved by the local ethical committee.

The subject first learned to manipulate the joystick by driving the robot freely in the corridor, with visual and auditory cues available. After about 5-min training, enough for the subject to feel confident with the apparatus, the headphones and black goggles were put on.

Condition 1: Basic Regular Condition (RE)

The subject was passively displaced along 2, 4, 6, 8, or 10 m, randomly. Velocity profiles of 13 out of 16 stimuli were triangular, that is, with equal values of acceleration and deceleration in the 0.06–0.5-m/s^2 range. The other 3 stimuli had square-shape velocity profiles. Peak velocity ranged from 0.6 m/s to 1 m/s^2. The triangular profile produces a continuous stimulation of the otoliths by linear acceleration. About 10 s after the end of the imposed displacement, the subject had to replicate as accurately as possible the distance traveled, controlling the robot with the joystick. The order of trials was randomly changed subject by subject.

Condition 2: Constant Duration Stimuli (CD)

In the RE condition, the total distance and duration of the passive transport were not independent of each other. Therefore, in order to prevent the subjects from using the duration of transport as a cue to replicate distance, we used a second condition where all the imposed distances were traveled for the same duration (16 s). For each given distance we varied the type of velocity profile: a rectangular velocity profile, (constant velocity), a trapezoid profile, and a triangular profile.

The same instruction as in the first condition was used, and the subjects (seven who also participated in the triangular velocity condition, and two additional ones) replicated the five distances, with the three velocity profiles (i.e., 15 trials) presented in completely random order.

Condition 3: Switch–Constant Speed, Constant Duration (SW)

In this condition, the joystick operated as an on-off switch, a fixed-velocity step (from 0 to 0.7 m/s) with acceleration 0.9 m/s^2 was generated when the stick was moved forward. Thus, the velocity profile of the replication phase resulted in a regular pulselike wave throughout the experimental session for all imposed distances (2,4,6,8, and 10 m). If the replication of a dynamic profile was necessary to replicate distance, then we would expect a net impairment in the stimulus–response distance relationship, if the replication of the profile were prevented. Six healthy volunteers participated in this experiment. Five of them had been previously tested in condition CD from which we drew the stimuli.

Condition 4: Limited Robot Speed (LS)

In the previous experiment subjects could learn the invariable transfer function of the joystick–robot control and thereby compute the inverse transformation to achieve a correct distance replication by a dynamic matching strategy (see model section). Therefore, we designed a final condition, in which the parameters of joystick A/D conversion (maximum attainable speed and acceleration) were changed from trial to trial so that subjects were prevented from building a model of the joystick–robot control. The changes were programmed in such a way that a simple replication of the passive dynamic profile or of the duration would lead to an inaccurate replication of distance. All stimuli utilized in this experiment (distance = 2,4,6,8, and 10 m) had a triangular velocity profile of variable duration (as in condition RE).

It should be noted that a partial matching of the stimulus velocity profile was allowed in this case, but in order to fulfill the task, the matching strategy had to be abandoned during the execution, as soon as the dynamic matching became impossible.

Six subjects were recruited from the first study (condition RE) to participate under this condition.

Patients

Three bilaterally vestibular-defective patients and one paraplegic (TABLE 1) participated under the basic RE condition; one vestibular patient also performed under the CD condition. Patients and normals gave their informed written consent before being included in the study.

TABLE 1

pat.	Age (yrs)	Deficit	Cause of Illness	Duration of Illness (yrs)	Clinical Tests Indicating Areflexia	Exp.
V1	41	Bilateral vestibular	Gentamicin assumption	5yrs	Caloric, OVAR	RE, CD
V2	28	Bilateral vestibular + Hypoacusia	Unknown (viral?)	3 mon	Caloric, OVAR	RE
V2	50	Bilateral vestibular + Hypoacusia	Postsurgery: bilateral VIII pair neurinoma	10 yrs	Caloric, OVAR	RE
P1	35	Paraplegia	Lumbar trauma (D11–D12)	10 yrs	Caloric, OVAR	RE

OVAR = off vertical axis rotation; RE = regular; CD = constant experiment.

Data Analysis

Normalized Root Mean Square Error (nrms) of Replicated Velocity Profiles

To compare the shape of the stimulus and response velocity profiles, regardless of errors in distance, duration, or peak-velocity replication, we first normalized the time scales of both stimulus and response from 0 to 100% through linear interpolation (corresponding to a low-pass filter with cutoff frequency of 5 to 10 Hz). Because the profiles were previously filtered at 5 Hz, this additional filtering did not markedly alter the relevant characteristics of the signals. Then, all of the normalized trials from the same subject or from the same stimulus were averaged, and the root mean square of the differences between the averaged stimulus and response velocity profiles (nrms) was taken as a quantitative index of mutual resemblance.

Statistical Analysis

Appropriate ANOVA designs were used to compare the data from individual trials between different conditions, with either response total magnitude (distance or duration) or algebraic error (response–stimulus) as dependent variables. The relative error: (response–stimulus)/Stimulus was used to quantify overall accuracy. Linear regressions of individual trials were performed to quantify the stimulus–response relationship from the subjects. A probability level of 0.05 was considered significant. Multiple regression analysis on response distance using stimulus distance, duration and peak velocity as predictors, was performed to estimate the statistical contribution of each stimulus parameter in determining the response.

RESULTS

Basic Regular Experiment

Subjects were able to replicate the distances with an overall accuracy of 25% (standard deviation of the pooled relative errors, $n = 232$). Means ± standard errors (SEs) of the mean are shown in FIGURE 2A. The linear regression between stimulus and response distance was calculated for each subject with all the 16 trials, and the correlation coefficient r was highly significant for all subjects ($p < 0.0001$). The replication of the shortest distance (2 m) led to a slight overshoot: 2.31 ± 0.12 m, while that of the longer distances exhibited an undershoot (9.21 ± 0.33 m for the 10-m stimulus).

Subjects also replicated the duration of the stimulus, although the instruction was to replicate the distance. The correlation coefficient between stimulus and response duration was highly significant for each subject ($p < 0.0001$, except for one: subject EC obtained $p < 0.0013$). This subject displayed a "step-strategy," that is, with joystick manipulations of short duration and displacement at high velocity, being nevertheless as accurate as the others in reproducing distance. For all subjects, the duration of the shortest stimulus (4 s) was replicated with an overshoot: 4.60 ± 0.32 s, and the longest (25 s) with an undershoot: 21.27 ± 1.0 s.

Most subjects also replicated stimulus peak velocity. The correlation coefficient between stimulus and response peak velocity was significant for all subjects ($p < 0.01$) but four. The average determination coefficient was therefore lower than that of distance and duration.

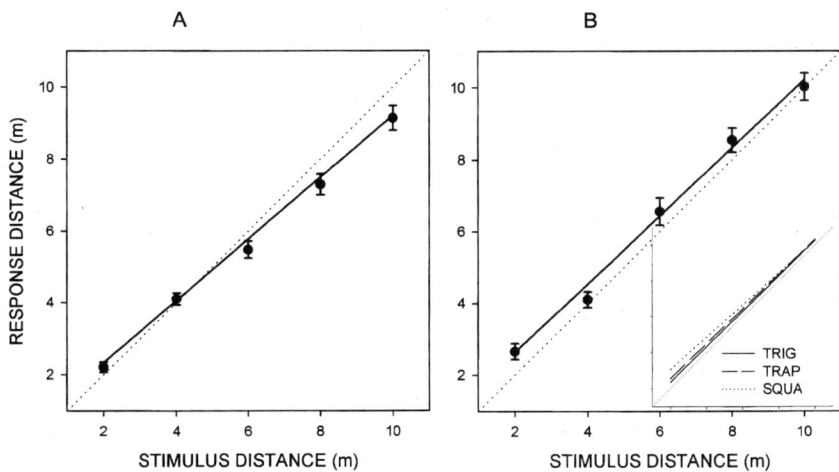

FIGURE 2. Condition RE and CD. (**A**) Means±SEs of responses for the condition RE ($N=15$). The total length of passive transport is reported on the X-axis, whereas the total length of active transport is reported on the Y-axis. *Thin dotted line*: line of identity; *thick line*: mean of individual regression lines. (**B**) Same as (**A**) for CD condition. Mean responses for the triangular (TRIG); trapezoidal (TRAP), and square-wave (SQUA) velocity profiles.

Multiple regression (and partial correlation) analysis showed that the only independent predictor of response distance was the stimulus distance, not the stimulus duration nor the peak velocity.

Constant Duration Condition

Subjects replicated the imposed distance as well as in the first condition (FIG. 2B): overall accuracy was 35% ($n = 132$), with 25% ($n = 44$) for the triangular profile, 28% for the trapezoid profile, and 47% for the rectangular profile. For the rectangular profile, the correlation coefficient r between stimulus and response distance was significant for all subjects but one (EC). For the seven subjects who also had participated in the RE condition, the mean stimulus–response regression in the CD condition, was the same as that obtained in RE.

In CD, the stimulus duration did not vary and the average response duration was very close to that of the stimulus: 14.14 ± 1.21 s ($n = 132$). There was no significant difference between the duration errors for the three profiles. The EC subject again exhibited a shorter response than the other subjects: his mean duration error was -7.40 s, while the error of the six remaining subjects was -0.52 ± 0.36 s. When deprived of temporal information correlated to distance, he apparently still applied his step-strategy without losing accuracy in fulfilling the task.

In this condition, where stimulus duration and distance were not correlated, peak stimulus velocity and distance were strongly interdependent for all velocity profiles. This could be one reason why there was a significant correlation between stimulus and response peak velocity in all subjects but two, for all velocity profiles ($p < 0.03$). Multiple regression analysis showed that distance was again the only independent predictor of distance replication.

In both the RE and CD conditions, subjects replicated the velocity profile of the stimuli (FIG 3)., The accuracy was greater (lower nrms) for longer distances. A global overshoot at the onset of the replication phase was seen principally in the triangular profiles. This could be due to the relatively long delay of joystick control (0.2 s). ANOVA revealed a significant difference among the nrms for the three types of profile ($F[2,8] = 5.61$, $p = 0.014$), and a *post hoc* comparison confirmed that the nrms error for the triangular profile was greater than that for the trapezoid and rectangular profiles (the errors of the latter two were not significantly different from one another). Therefore, in general the subjects replicated the rectangular and trapezoid profiles more accurately than the triangular one.

Switch Condition

Here, the velocity profile of the active phase was always rectangular, independent of the velocity profile of passive transport. FIGURE 4 shows the sample traces from one subject for all types of displacement imposed along a 6-m distance (experiment CD from which stimuli were drawn is here referred to as CONTROL). Note that the stimuli always had the same duration. The subjects succeeded in giving response distances proportional to the stimuli regardless of the velocity profile of passive transport. The mean regression coefficient r^2 from individual regression lines was just slightly lower than that of experiment CD (0.77 vs. 0.84), indicating that under SW condition response distance was also perfectly proportional to stimulus distance. Means±SEs are shown in FIGURE 5. The mean slope from the

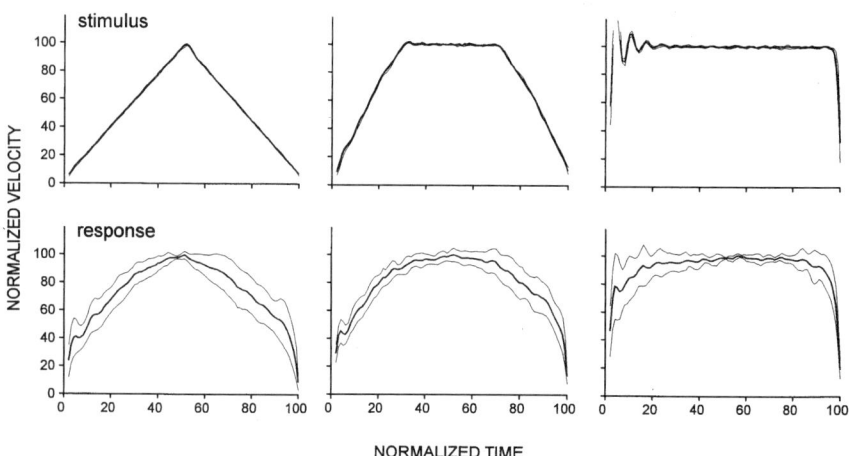

FIGURE 3. Condition CD. **Top row**: mean ± SD of normalized velocity profiles of passive transport (stimulus). **Bottom row**: mean ± SD of normalized velocity profiles of active transport (response).

FIGURE 5. Condition SW. Mean ± SE of replicated distance. The solid line is the mean of individual regression lines from 6 subjects. The *thick dashed-dotted line* corresponds to the mean stimulus–response curve from the CD condition that was used as control. The *thin dotted* line is the identity line.

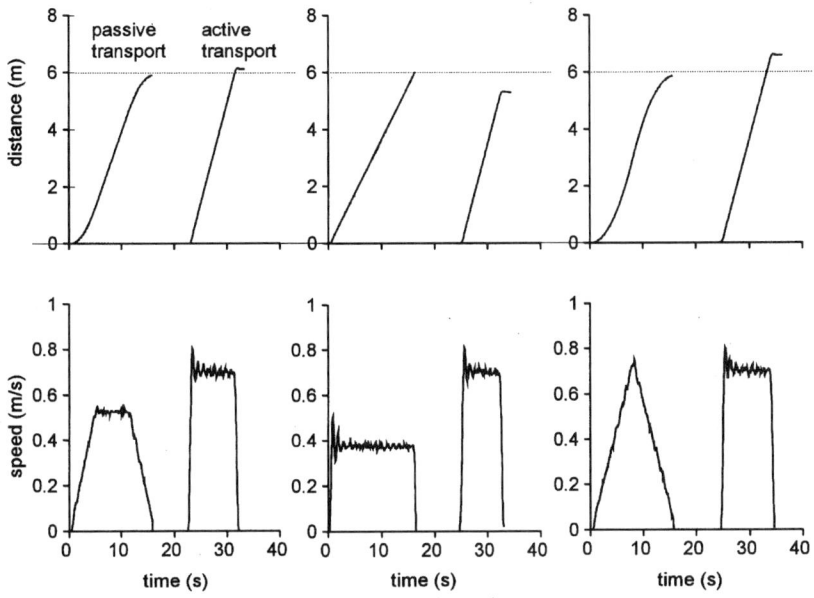

FIGURE 4. Condition SW. Passive transport along a 6-m linear path in three different trials with different displacement (**upper panels**) and velocity (**lower panels**) profiles (trapezoidal, pulse-like, and triangular; lower panels) for one subject. Active transport was constrained to a pulselike velocity profile. In this subject the imposed distance was replicated rather accurately in all conditions.

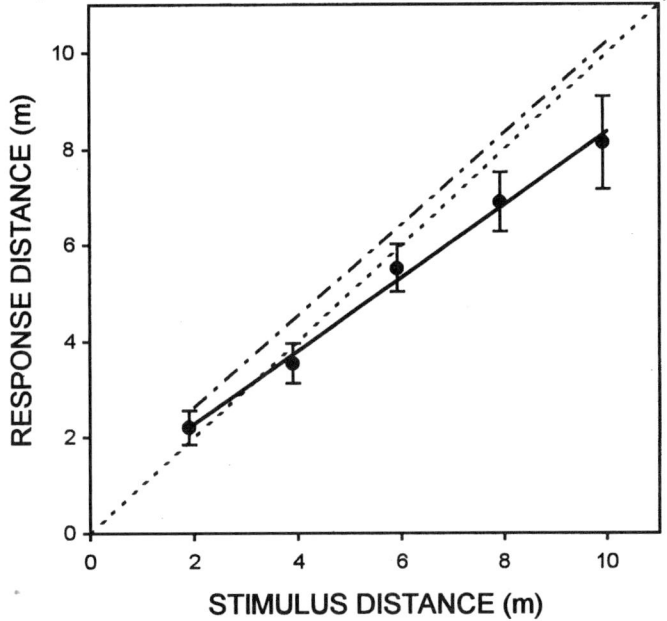

individual regression lines was also lower than in CD (0.77 vs. 0.95), as was the intercept (0.25 vs. 0.73). Three subjects consistently undershot the performance of the CONTROL experiment, whereas two others did not show a systematic bias toward under- or overshoot. An ANOVA (condition × stimulus distance × velocity profile) on response distance between CD and SW conditions showed a significant effect of condition ($F_{1,4} = 11.0, p < 0.03$). On average, the replicated distance was indeed significantly lower than in CD, and the average distance stimulus–response regression line showed a more gradual slope. The effect of stimulus distance was obviously highly significant ($F_{4,16} = 138.9, p < 0.0001$), but the type of velocity profile had no influence ($F_{2,8} = 1.6, p = 0.26$) on the replication of distance and the interaction was not significant.

Limited Speed Condition

All subjects roughly replicated the velocity profile of the stimulus in RE, but the limitation of the maximal robot speed during self-driven transport had different consequences depending on individuals. FIGURE 6 shows the sample recordings from three subjects (experiment RE is here referred to as CONTROL). Subject S1, for example, prolonged the duration of displacement until the stimulus distance was overshot. Interestingly, in this trial, the subject appeared to suddenly change strategy (from reproducing the velocity profile to another, as suggested by the dip in the plateau of the response velocity profile). The response profiles of subjects 2 and 3 had the same duration as the stimulus; thus the imposed distance was largely undershot. Subject 1 overshot all trials: the distance stimulus–response (S–R) regression was $R = 0.71*S + 2.45$, ($r^2 = 0.70$), whereas in the control condition it was $R = 0.79*S + 0.79$, ($r^2 = 0.82$); the intercept was therefore higher in the LS but not the slope of the regression line. Subject 3, on the other hand, largely undershot all trials ($R = 0.39*S + 0.71$, $r^2 = 0.66$ vs. $R = 0.75*S + 0.9$, $r^2 = 0.81$) due to a decrease in the regression slope and not of the intercept; subject 2 was intermediate and his performance did not differ systematically from the control condition, although the intertrial variability was higher in LS ($R = 0.68*S + 1.47$, $r^2 = 0.64$ vs. $R = 0.86*S + 0.82$, $r^2 = 0.88$).

In general, the effect of this task was nonsystematic: the joystick properties varied from trial to trial, and subjects admittedly seldom realized that matching the velocity profile of passive transport was not permitted. The slope of the mean interindividual regression line in LS (FIG. 7) was lower than in experiment RE (0.72 vs. 0.86), and the SE bars were very high. Also individual r^2 values were lower than in experiment RE (0.69 vs. 0.85), while the intercept was higher (1.05 vs. 0.62). When examining intraindividual comparisons between paired trials under RE and LS conditions, we found that only S1 and S3 showed a systematic over- and undershoot, respectively, relative to RE. A within-subjects 3-factors ANOVA (by condition, stimulus distance, and maximum stimulus speed) indicated no systematic effect due to the experimental condition ($F_{1,5}=0.04, p=0.85$). The response was, as usual, dependent on the stimulus distance ($F_{4,20}=81.7, p<0.0001$), whereas the effect of speed was not significant ($F_{2,10}=1.88, p=0.2$). The larger intra- and interindividual scatter displayed in this condition suggests that matching a scalar representation of distance was definitely not the subjects' preferred strategy to perform in our experimental paradigm.

Patients

All patients replicated the stimulus distance with an accuracy comparable to that of the control subjects included in the same paradigm (condition RE, FIG. 8A). Stimulus duration was also replicated. The correlation coefficient between the response and stimulus distance was always higher than that between stimulus and response duration. Multiple

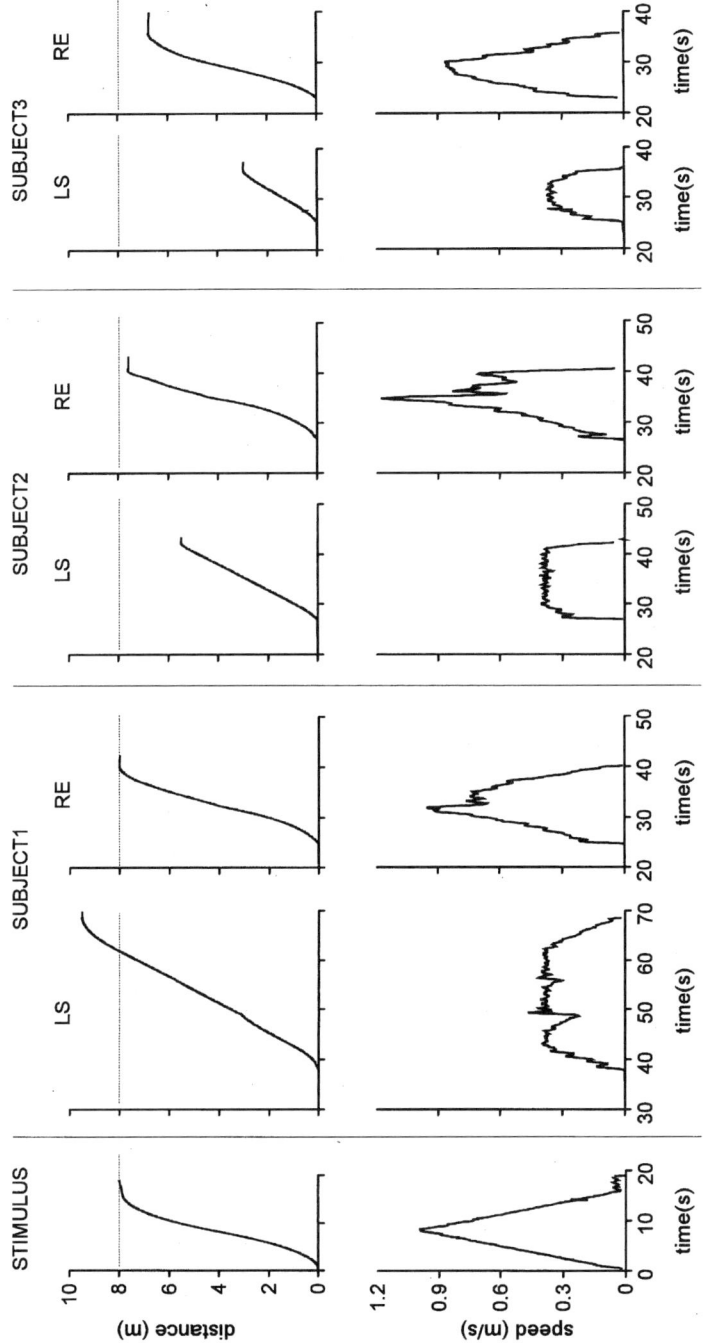

FIGURE 6. Condition LS. Distance and velocity response given in LS and RE condition to a 8-m stimulus with triangular velocity profile. Subject 1 in LS prolonged the response duration when the active maximum speed was lower than that of passive transport. Subjects 2 and 3 gave a response of the same duration as the stimulus.

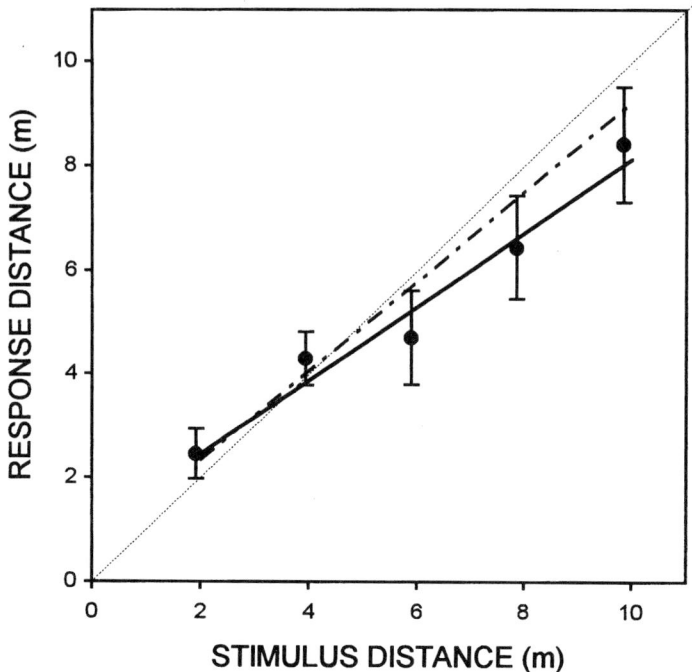

FIGURE 7. Condition LS. Mean ± SE of replicated distance. The *solid line* is the mean of individual regression lines from 6 subjects. The *thick dotted line* corresponds to the mean stimulus-response curve from the RE condition that was used as control. The *thin dotted line* is the identity line.

regression and partial correlation analyses confirmed that the only predictor independent of the response distance was the stimulus distance ($p < 0.01$). The relationship between stimulus and response peak velocities was only marginally significant, and stimulus peak velocity did not provide a significant contribution to multiple regression. Patient V1 also performed in condition CD (inset diagram of leftmost panel of FIG. 8A), in which all stimuli had the same duration: no significant difference was found between CD and RE or between his performance and that of healthy subjects.

The question arose as to whether patients and normals differed in the replication of the stimulus velocity profiles. As mentioned before, the stimuli always had a triangular velocity profile, but only the first vestibular defective patient (who was nonetheless areflexic on the off-vertical-axis-rotation test) exhibited a waveform similar to that of the stimulus (FIG. 8B). All the other patients showed more or less trapezoidal responses.

Modeling the Brain Processing Mechanism

The results indicate that the distance of the passive travel was replicated *per se*, independent of the other passive motion parameters such as duration, peak velocity, or velocity, profile. The replication of the spatiotemporal profile of passive motion, however, appears as the preferred spontaneously selected strategy to solve the task. When the

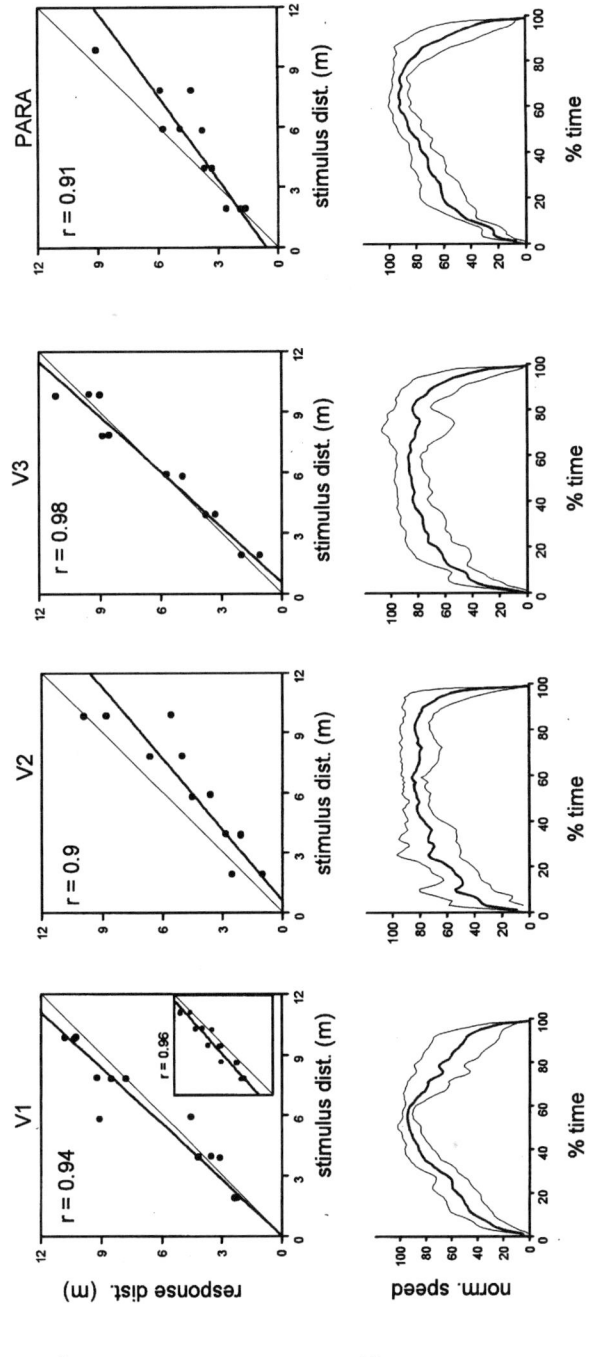

FIGURE 8. Patients. (A) Individual distance stimulus–response diagrams for all patients included in the study. V1–V3: bilaterally vestibular-defective patients. P1: paraplegic patient. V1 underwent both experiments RE and CD (shown in the right-bottom inset). (B) Patients. Mean normalized velocity profiles of patients in experiment RE (all stimuli had a triangular velocity profile). Only patient V1 shows a symmetric triangular profile.

spatiotemporal matching of the active transport to the passive transport is made impossible, some subjects could nevertheless replicate the passively traveled distance, although some others failed and replicated basically the temporal features. These findings suggest contrasting interpretations about the mechanisms of distance perception, because they indicate that, while the dynamics of passive motion are stored and available for further use, total distance may be independently estimated.

It is possible that subjects replicated the dynamic characteristics of the passive transport because the self-controlled transport was not goal-directed but amplitude-coded, and the task itself was ambiguous because "distance" is a static parameter, whereas "replication" may implicitly denote a dynamic task. Also, the joystick did not control the displacement magnitude but the robot speed, which might have induced the subjects to exploit motion dynamics. Moreover, the subjects might have felt that this strategy could help fulfill the requested task.

It might also be argued that the observed independence in distance replication does not necessarily imply that an accurate, independent, static internal estimate of total path-length was produced and memorized: we developed a theoretical model of the internal processing of inertial signals that is able to achieve the task in all experimental conditions and may explain the role of dynamic cues.

General Assumptions

To build the model we have made the following assumptions: (1) since the subjects replicated the spatiotemporal features of passive motion, the model is thought to be a dynamic one (see also Israel *et al.*[21]); (2) since it was impossible to disentangle vestibular from somatosensory contributions, the sensory motion-related input was summarized in one single transfer function, mainly accounting for the otolith component (we will show that this is not a restrictive assumption, as the system dynamics will depend mostly on its closed loop structure); (3) since the current experiment dealt with one-dimensional translatory motion, we restricted the model to one dimension; (4) short-term memory was assumed, as a first approximation, to be ideal without loss.

Model Description

(1) Passive transport phase (Fig 9): the input to the CNS during passive transport is the linear acceleration mediated via acceleration sensors, which will be modeled as the otolith transfer function (T) used in previous studies.[19,22,23] Passive displacement is then stored in memory as an acceleration, velocity, or position profile.

(2) Active phase (replication): the stored profile is played back in real time from memory and compared to the actual profile by subtraction. The resulting error is used to correct the velocity of the robot. The robot itself is modeled as having a low-pass transfer function (R) to account for the robot inertia. The robot velocity is differentiated to yield linear acceleration, which feeds the otoliths (T).

The model can now be formulated in terms of systems theory by the following equation:

$$O = R \cdot (T \cdot I - T \cdot O), \tag{1}$$

where I is the input acceleration during the passive phase, O the output acceleration during the active phase, T the transfer function of the acceleration sensors together with the conversion to the stored profile (i.e., single integration if a velocity profile is stored;

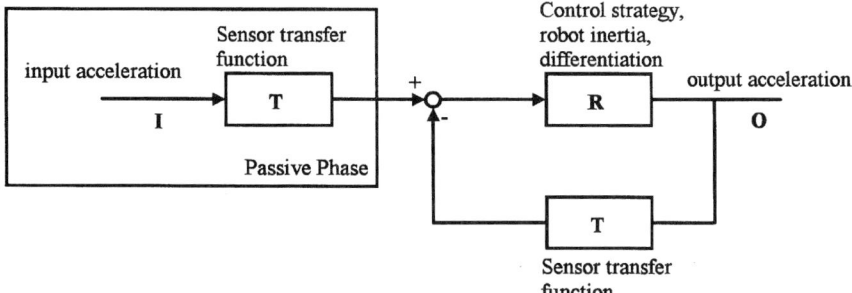

FIGURE 9. Negative feed-back formulation of the replication paradigm. The passive motion provides the input (I) to the inertial sensors (T) whose output feeds the feedback loop. It is assumed that a dynamic memorized representation of passive motion is available. The motor block R covers joystick and robot parameters. The output acceleration (O) feeds the inertial sensors whose output is subtracted from the incoming input.

double integration, if a position profile is stored), and R the robot transfer function (inertia and differentiation to yield robot acceleration) together with the control strategy of the subject. The input/output relationship can be rewritten as

$$\frac{O}{I} = \frac{R \cdot T}{1 + R \cdot T} = \frac{1}{1 + (R \cdot T)^{-1}}. \qquad (2)$$

The subjects controlled the robot by the joystick, whose deflection was proportional to the velocity of the robot. Theoretically, a stable, closed control loop, which has a static gain of unity (i.e., the desired set point is attained at the steady state), must have at least I behavior (where I stands for 'integrative') in open loop. This means that the error signal must be integrated. Otherwise, with a simple proportional gain (P behavior), the overall gain of the loop always will be smaller than unity, implying that the input value always will be greater than the output. If this had been the case in our experiment, subjects would never have covered the imposed distance during replication. The necessary integration of the error yields an overall low-pass behavior of the closed-loop system; thus, RT shows I behavior. To achieve this, the error between desired and actual profile, that is, whether it is stored as acceleration, velocity, or position does not matter, has to be converted into a *position* error.

To state this reasoning in a formal way, let us first look at the signal controlling the joystick angle J. This is given by

$$J = C \cdot (T \cdot I - T \cdot O), \qquad (3)$$

where C is the controller gain. Now we can evaluate this controller transfer function. The overall robot transfer function differentiates, since it consists of joystick (a gain value), robot inertia (a low-pass filter that can be reduced to a gain factor), and one differentiation to yield the robot acceleration from velocity, that is,

$$R = C \cdot G_{ro} \cdot s, \qquad (4)$$

where G_{ro} is the overall robot gain, and s the differentiation in Laplace notation. The otolith transfer function T can also be reduced to a gain factor, while the trajectory is computed either in the acceleration, velocity, or position domain:

$$T = G_{ot} \cdot s^{-x}, \tag{5}$$

where G_{ot} is the otolith transfer function and $x=0,1,2$ (acceleration, velocity, position). Thus,

$$R \cdot T = C \cdot G_{ro} \cdot G_{ot} \cdot s^{1-x} \tag{6}$$

Since RT has to show integrative behavior, C has to meet the following equation:

$$C \cdot s^{1-x} = s^{-1}, \tag{7}$$

that is, $C = s^{x-2}$.

For the input signal to the joystick J, we can now write:

$$J = C \cdot (T \cdot I - T \cdot O) = C \cdot T \cdot (I - O) = s^{x-2} \cdot G_{ot} \cdot s^{-x} \cdot (I - O) = s^{-2} \cdot G_{ot} \cdot (I - O). \tag{8}$$

Since $I - O$ is an acceleration, J is a *position* error no matter what value x has, that is, no matter what specific profile (acceleration, velocity, position) has been stored in memory. Thus, a double integrator (s^{-2}) is a necessary model component if the closed-loop model dynamics has to show I behavior. FIGURE 10A shows the Simulink® circuit developed according to the preceding equations.

Simulations

FIGURE 10B shows the results of the simulations obtained with the model under CONTROL (i.e., RE or CD), SW, and LS conditions. In all these cases the simulated output distance attained the stimulus distance independently of the input velocity profile and of whether the joystick produces a pulselike constant velocity output or the maximal velocity output is lower than under the control condition. If the comparator operated on a velocity estimate rather than on distance, the resulting P-behavior of the control loop would not allow adjustment for the errors made under the LS conditions: the error would suddenly decay to zero at the time instant when memorized velocity becomes equal to the actual velocity during deceleration, and the active displacement would be shorter than the passive one. An I behavior, in contrast, "remembers" the sum of the errors. The accumulated error provided by the double integrator can thus be compensated, and eventually the desired final position is attained after covering exactly the input distance.

The model requires that the instantaneous perceived acceleration, velocity, or position relative to the starting time be directly encoded and the time course stored in spatial memory. If a simple scalar estimate of the passively traveled displacement was fed into the comparator, then a dynamic profile could never be replicated in the active phase. Possibly, this latter interpretation could fit the case of those patients and subjects who did not replicate the dynamic profile of the passive phase. A double integration nevertheless would be

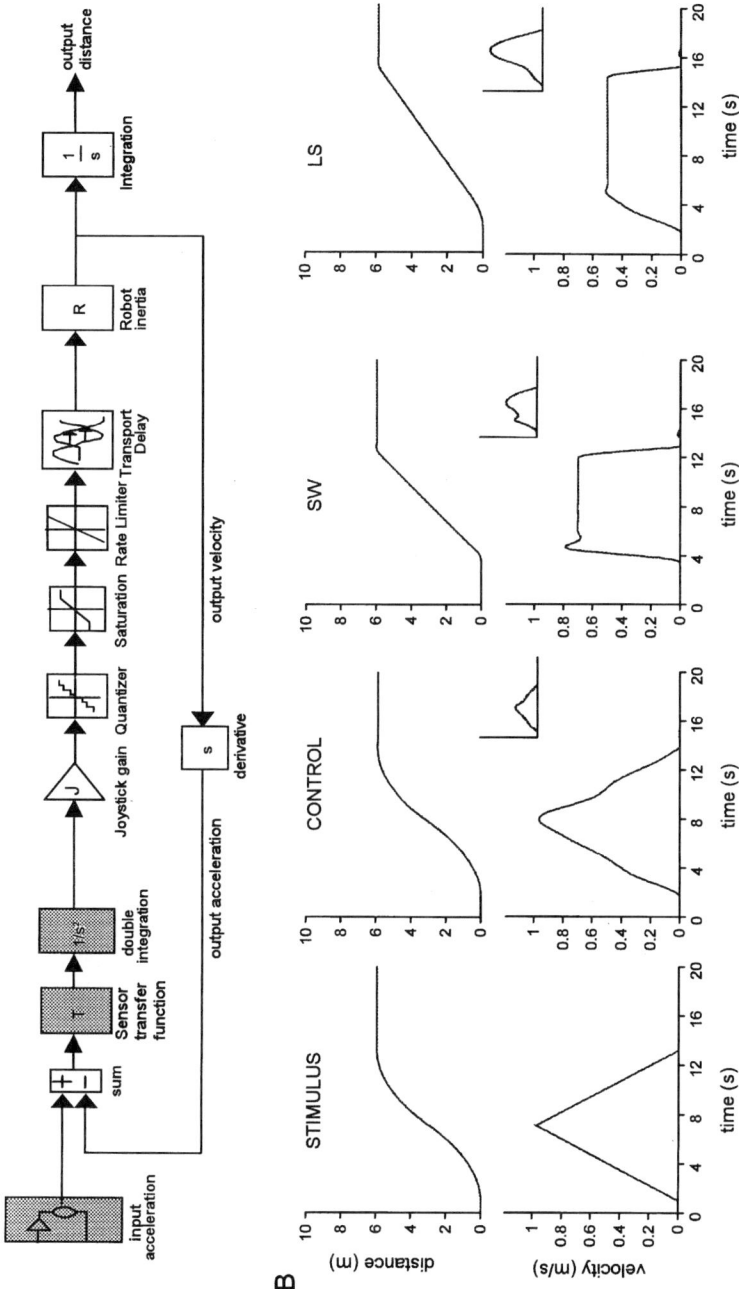

FIGURE 10. Model simulation. (**A**) Simulink® circuit used to produce simulations. The diagram is equivalent to the equations discussed in the text. *Blocks with gray background* refer either to the passive phase or are common to both phases. Note the presence of a double time-integrator ($1/s^2$) of incoming acceleration-related sensory signals. Details are explained in the text. (**B**) *Top diagrams* indicate instantaneous distance. *Bottom diagrams* indicate instantaneous velocity. *Insets* indicate the simulated motor command. From left to right: stimulus, simulation of CONTROL (i.e., RE or CD), SW, and LS experiments. In all cases the stimulus distance (6 m) was attained by the simulation. The stimulus velocity profile was matched only in the control experiment, whereas a pulselike velocity profile was produced in SW. In the LS condition, the stimulus velocity profile was matched only in the rising part of the response until the maximally allowed speed (0.5 m/s) was attained. The simulation results can be compared with actual responses shown in FIGURES 4 and 6.

necessary to work out an estimate of distance from acceleration. The model provides yet another suggestion: details of the sensory transfer function T are not very relevant to the task, since the overall system transfer function results from the closed-loop structure, that is, from the RT product that determines the I behavior. This point may suggest an explanation for why subjects with a partial loss of sensory functions (which imply changes of T) are apparently not impaired in solving the task.

Limitations

The model suggests that the control variable for performing the task is the current position; however, it cannot inform us about the type of dynamic profile stored in memory. In fact, in a purely linear system like the one described here the controller C and the transfer function T cannot be separated. Even though we introduce a nonlinearity by limiting the robot velocity, the preceding considerations still hold, and since, as shown earlier, the error has to be a distance, the simulation correctly replicates the distance. Also, the assumptions about ideal memory must be challenged, and the passive and active phases should have independent sampling rates. New experiments are needed in order to gain a better insight into the memory storage of perceived motion.

DISCUSSION

A passive displacement can be represented in static as well as in dynamic terms. Which is to say that we could refer to an experienced transit either as a 10-m, 10-s linear displacement, or as a motion at a gradually increasing speed, followed by a deceleration to a stop. These two representations might coexist in short-term memory or, alternatively, only one might be stored. In this latter case, if the dynamic representation is the one available, static estimates could be derived by reprocessing it. On the other hand, motion dynamics would be lost if only static parameters are stored.

The results of the present study and the theoretical analysis performed herein suggest that an intermediate stage for processing spatial information from the sensory signals excited by whole-body motion to higher representations may lie in the representation of the movement dynamics. From an acceleration-related sensory (combined vestibular and somatosensory) input, a temporally ordered array of instantaneous perceived accelerations, velocities, or positions appears to be stored (at least temporarily). Thanks to a double integration process, instantaneous positions may either be directly stored into memory or recomputed from the memory record to allow the replication of a given path length. This array may lose its temporal components when spatial invariants need to be extracted, that is, when there is a transition from the egocentric organization of path integration to the allocentric organization of more abstract spatial representations (see reference 24, pp. 206–208).[24] However, the dynamic aspects may not be completely forgotten when an experienced journey is recollected, for instance, when the journey has been performed by several different means of transport or, on the other hand, when the reverse process, from spatial representation to active production of trajectories, has to be put in place (see references 25 and 26).

Neurophysiological Background: Linking Inertial Signals to Spatial Memory

We have suggested that the inertial stimuli imposed in our experiment can be stored to form a spatiotemporal representation of passive travel. Which structures might have been involved in the task? The hippocampal formation has been recently recognized to be a brain structure specifically linked to path integration[27–30] and navigation[31,32] behavior. Recent findings suggest that hippocampal and perihippocampal structures may monitor whole-body self-motion: for angular motion the semicircular canal input can modify the hippocampal EEG theta-rhythm,[33] the activity of complex CA1 spike cells in the rat,[34] hippocampal place cells,[35] and thalamic head direction cells.[36,37] O'Mara et al.[38] reported that neurons in the primate hippocampus discharge upon linear motion; and some rat hippocampal CA1 complex spike cells have been shown to produce a discharge correlated with passive translation in dark arenas.[39] For linear stimuli it also has been shown that peaks in the activity of place fields in the rat hippocampus may be correlated with fixed or proportional distances from environmental landmarks.[40] These findings suggest that an internal coding of traveled distances may be available in the hippocampus (see also reference 9 for a thorough review). Recent data showing that hippocampus-lesioned rats cannot separate locations that have a high degree of spatial similarity but differ in location relative to the starting point[41] seem to confirm this view. Other areas of the human cortex that receive input from the otoliths, as recently uncovered in fMRI studies,[42] may be involved. Finally, the present study suggests that some cells belonging to the neural structures involved in spatial cognition could specifically encode dynamic features of experienced motion.[43,44]

We still need to clarify the location of the double integrator and its computational characteristics, whether it operates directly upon the incoming sensory information or consists of two different integration steps, the first creating an intermediate velocity-related representation and the second a displacement-related representation. Previous studies of human goal-directed locomotion[3] suggested that a leaky double integrator of perceived motion-related acceleration may work during path integration. However, data from motion perception and eye-movement studies in humans revealed that the otolith-induced neural discharge is integrated over time, leading to a close relationship between the perceived and the actual linear velocity.[45] Additional experiments are therefore needed to further examine this question.

Comparison with Locomotion Experiments

There is a close resemblance between the results from the RE and CD tasks and those obtained by Loomis et al.[15] on locomotion. In their experiment, blindfolded subjects were first walked by the experimenter along paths that were from 2 to 10 m long (the same distances as in the present experiment), after which the subjects had to replicate the same distance while blindfolded and without aid. The reported results are strikingly similar to the present ones: the 2-m distance was overshot by 0.26 m (0.31 ± 0.12 m in the present test), and the 10-m distance was undershot by 1.02 m (0.79 ± 0.33 m here). This resemblance suggests that there are important parallels between active locomotion and the self-controlled passive displacement we have used in the present study. It should be noted that the motion parameters (speed and acceleration) selected for passive motion were very much in the physiological range of normal locomotion.

It may thus be suggested that the inertial and proprioceptive signals generated during the passive transport phase of our experimental paradigm are processed in a very similar way as during locomotion, when motion-related information is considered and when motion is self-driven. The double time integration of the acceleration forces from the specific sensors[3,19] might participate in the updating of position during motion.

Contribution of Somatosensory Input

The results obtained in vestibular-defective patients suggest that somatosensory cues may substitute absent otolith-related cues: arthrokinetic information is indeed known to affect linear self-motion perception.[46-48] Tactile cues may complement vestibular information, providing (1) a signal related to the linear body acceleration (pressure on the back, visceral shifts, etc.), and (2) a signal generated by robot vibrations. It is worth remarking that this latter component cannot work without the first one or, at least, outside a "whole-body motion" context. In the present study, all patients could replicate distance irrespective of their specific sensory deficit (vestibular or kinesthetic). Self-driven velocity profiles, however, revealed that patients behaved differently from normals. The difference may be due to the presence of compensatory strategies, developed to overcome the sensory limitation following the neurological deficit (see also references 16 and 49 for the case of locomotion). In fact, Gianna *et al.*[50] showed that a group of patients with bilateral loss of labyrinthine function displayed thresholds for detection of motion direction during passive lateral whole-body acceleration that largely overlapped those of normals. In general, it is hard to separate different sensory contributions in this type of experiment: the issue of the brain processing of inertial signals should be considered one of integration of a multimodal sensory input.[51,52]

ACKNOWLEDGMENTS

This work was supported by the Galileo project, Ministère des affaires étrangères (France-Italy), the GIS "Sciences de la Cognition" (France), and HFSP: RG71/96B. We gratefully acknowledge Dr. P. Tran-Ba-Huy: Hôpital Lariboisière, Dr. E. Vitte: Hôpital de la Salpêtrière, and Dr. A. Semont: Clinique des Augustines, for their assistance during the experiments with the patients; and the Association des Paralysés de France and the Cercle Sportif des Invalides (Paris), for their kind cooperation.

REFERENCES

1. MITTELSTAEDT, H. & M. L. MITTELSTAEDT. 1973. Mechanismen der Orientierung ohne richtende Aubenreize. Fortschr. Zool. **21:** 46–58.
2. MITTELSTAEDT, M. L. & H. MITTELSTAEDT. 1980. Homing by path integration in a mammal. Naturwissenschaften. **67:** 566–567.
3. MITTELSTAEDT, H. & M. L. MITTELSTAEDT. 1982. Homing by path integration. *In* Avian Navigation, F. Papi & H. G. Wallraff, Eds.: 290–297. Springer-Verlag. Berlin.
4. MAURER, R. & V. SÉGUINOT. 1995. What is modelling for? A critical review of the models of path integration. J. Theor. Biol. **175:** 457–475.
5. GALLISTEL, C. R. & A. E. CRAMER. 1996. Computations on metric maps in mammals: Getting oriented and choosing a multi-destination route. J. Exp. Biol. **199:** 211–217.
6. ETIENNE, A. S., E. TERONI, C. HURNI & V. PORTENIER. 1990. The effect of single light cue on homing behaviour of the golden hamster. Anim. Behav. **39:** 17–41.
7. BARLOW, J. S. 1964. Inertial navigation as a basis for animal navigation. Theor. Biol. **6:** 76–117.
8. MITTELSTAEDT, H. 1983. Introduction into cybernetics of orientation behavior. *In* Biophysics, W. Hoppe, W. Lohmann, H. Markl, and H. Ziegler, Eds.: 794–801. Springer-Verlag. Berlin.
9. TRULLIER, O., S. I. WIENER, A. BERTHOZ & J. A. MEYER. 1997. Biologically based artificial navigation systems: Review and prospects. Progr. Neurobiol. **51:** 483–544.
10. THOMSON, J. A. 1983. Is continuous visual monitoring necessary in visually guided locomotion? J. Exp. Psychol. Hum. Percept. Perform. **9**(3)**:** 427–443.
11. ISRAËL, I. & A. BERTHOZ. 1989. Contribution of the otoliths to the calculation of linear displacement. J. Neurophysiol. **62**(1)**:** 247–263.

12. KLATZKY, R. L., J. M. LOOMIS, R. G. GOLLEDGE, J. G. CICINELLI, S. DOHERTY & J. W. PELLEGRINO. 1990. Acquisition of route and survey knowledge in the absence of vision. J. Mot. Behav. **22**(1): 19–43.
13. BLOOMBERG, J., G. MELVILL JONES & B. N. SEGAL. 1991. Adaptive modification of vestibularly perceived rotation. Exp. Brain Res. **84**: 47–56.
14. MITTELSTAEDT, M. L. & S. GLASAUER. 1991. Idiothetic navigation in gerbils and humans. Zool. Jahrb. Physiol. **95**: 427–435.
15. LOOMIS, J. M., R. L. KLATZKY, R. G. GOLLEDGE, J. G. CICINELLI, J. W. PELLEGRINO & P. A. FRY. 1993. Nonvisual navigation by blind and sighted: Assessment of path integration ability. J. Exp. Psychol. Gen. **122**: 73–91.
16. GLASAUER, S., M. A. AMORIM, E. VITTE & A. BERTHOZ. 1994. Goal-directed linear locomotion in normal and labyrinthine-defective subjects. Exp. Brain Res. **98**: 323–335.
17. GUEDRY F. E. & C. S. HARRIS. 1963. Labyrinthine function related to experiments on the parallel swing. NASA Res. Rep., Order R-93. Unpublished.
18. ISRAËL, I. & A. BERTHOZ. 1989. Contribution of the otoliths to the calculation of linear displacement. J. Neurophysiol. **62**: 247–263.
19. ISRAËL, I., N. CHAPUIS, S. GLASAUER, O. CHARADE & A. BERTHOZ. 1993. Estimation of passive horizontal linear whole-body displacement in humans. J. Neurophysiol. **3**: 1270–1273.
20. BERTHOZ, A., I. ISRAËL, P. GEORGES-FRANÇOIS, R. GRASSO & T. TSUZUKU. 1995. Spatial memory of body linear displacement: What is being stored. Science **269**: 95–98.
21. ISRAËL, I., R. GRASSO, P. GEORGES-FRANCOIS, T. TSUZUKU & A. BERTHOZ. 1997. Spatial memory and path integration studied by self-driven passive linear displacement .1. Basic properties. J. Neurophysiol. **77**: 3180–3192.
22. ORMSBY, C. C. & L. R. YOUNG. 1977. Integration of semicircular canal and otolith information for multisensory orientation stimuli. Math. Biosci. **34**: 1–21.
23. YOUNG, L. R. 1984. Perception of the body in space: mechanisms. *In* Handbook of Physiology— The Nervous System III, I. Darian-Smith, Ed.: 1023–1066. American Physiological Society. Bethesda, Md.
24. THINUS-BLANC C. 1996. Animal Spatial Cognition: 1–259. World Scientific. Singapore.
25. TAKEI, Y., R. GRASSO, M. A. AMORIM & A. BERTHOZ. 1997. Circular trajectory formation during blind locomotion: A test for path integration and motor memory. Exp. Brain Res. **115**: 361–368.
26. BERTHOZ, A. 1997. Parietal and hippocampal contribution to topokinetic and topographic memory. Philos. Trans. R. Soc. Lond. B Biol. Sci. **352**: 1437–1448.
27. BENHAMOU, S. & B. POUCET. 1996. A comparative analysis of spatial memory processes. Behav. Processes **35**: 113–126.
28. MCNAUGHTON, B. L., C. A. BARNES, J. L. GERRARD, *et al.* 1996. Deciphering the hippocampal polyglot: The hippocampus as a path integration system. J. Exp. Biol. **199**: 173–185.
29. WHISHAW, I. Q., J. E. MCKENNA & H. MAASWINKEL. 1997. Hippocampal lesions and path integration. Curr. Opin. Neurobiol. **7**: 228–234.
30. MAGUIRE, E. A., N. BURGESS, J. G. DONNETT, R. S. J. FRACKOWIAK, C. D. FRITH & J. O'KEEFE. 1998. Knowing where and getting there: A human navigation network. Science **280**: 921–924.
31. WIENER, S. I., V. A. KORSHUNOV, R. GARCIA & A. BERTHOZ. 1995. Inertial, substratal and landmark cue control of hippocampal CA1 place cell activity. Eur. J. Neurosci. **7**: 2206–2219.
32. WIENER, S. I. 1996. Spatial, behavioral and sensory correlates of hippocampal CA1 complex spike cell activity: Implications for information processing functions. Progr. Neurobiol. **49**: 335–343.
33. GAVRILOV, V. V., S. I. WIENER & A. BERTHOZ. 1995. Enhanced hippocampal theta EEG during whole body rotations in awake restrained rats. Neurosci. Lett. **197**: 239–241.
34. KORSHUNOV, V. A., S. I. WIENER, T. A. KORSHUNOVA & A. BERTHOZ. 1996. Place- and behavior-independent sensory triggered discharges in rat hippocampal CA1 complex spike cells. Exp. Brain Res. **109**: 169–173.
35. SHARP, P. E., H. T. BLAIR, D. ETKIN & D. B. TZANETOS. 1995. Influences of vestibular and visual motion information on the spatial firing patterns of hippocampal place cells. J. Neurosci. **15**: 173–189.

36. BLAIR, H. T. & P. E. SHARP. 1995. Anticipatory head direction signals in anterior thalamus: Evidence for a thalamocortical circuit that integrates angular head motion to compute head direction. J. Neurosci. **15:** 6260–6270.
37. ROBERTSON, R. G., E. T. ROLLS & P. GEORGES-FRANÇOIS. 1998. Spatial view cells in the primate hippocampus: Effects of removal of view details. J. Neurophysiol. **79:** 1145–1156.
38. O'MARA, S. M., E. T. ROLLS, A. BERTHOZ & R. P. KESNER. 1994. Neurons responding to whole-body motion in the primate hippocampus. J. Neurosci. **14:** 6511–6523.
39. GAVRILOV, V. V., S. I. WIENER & A. BERTHOZ. 1998. Discharge correlates of hippocampal complex spike neurons in behaving rats passively displaced on a mobile robot. Hippocampus **8:** 475–490.
40. O'KEEFE, J. & N. BURGESS. 1996. Geometric determinants of the place fields of hippocampal neurons [see comments]. Nature **6581:** 425–428.
41. GILBERT, P. E., R. P. KESNER & W. E. DECOTEAU. 1998. Memory for spatial location: Role of the hippocampus in mediating spatial pattern separation. J. Neurosci. **18:** 804–810.
42. LOBEL, E., J. F. KLEINE, D. L. BIHAN, A. LEROY-WILLIG & A. BERTHOZ. 1998. Functional MRI of galvanic vestibular stimulation. J. Neurophysiol. **80:** 2699–2709.
43. WIENER, S. I., C. A. PAUL & H. EICHENBAUM. 1989. Spatial and behavioral correlates of hippocampal neuronal activity. J. Neurosci. **9(8):** 2737–2763.
44. BURGESS, N., J. G. DONNETT, K. J. JEFFERY & J. O'KEEFE. 1997. Robotic and neuronal simulation of the hippocampus and rat navigation. Philos. Trans. R. Soc. Lond. B Biol. Sci. **352:** 1535–1543.
45. YOUNG, L. R. 1984. Perception of the body in space: mechanisms. *In* Handbook of Physiology—The Nervous System III, I. Darian-Smith. Ed.: 978–1023. American Physiological Society. Bethesda, Md.
46. BLES, W., M. JELMORINI, H. BEKKERING & B. DE GRAAF. 1995. Arthrokinetic information affects linear self-motion perception. J. Vestib. Res. **5:** 109–116.
47. DE GRAAF, B., J. E. BOS, S. WICH & W. BLES. 1994. Arthrokinetic and vestibular information enhance smooth ocular tracking during linear (self-)motion. Exp. Brain Res. **101:** 147–152.
48. HLAVACKA, F., T. MERGNER & G. SCHWEIGART. 1992. Interaction of vestibular and proprioceptive inputs for human self-motion perception. Neurosci. Lett. **138:** 161–164.
49. TAKEI, Y., R. GRASSO & A. BERTHOZ. 1996. Quantitative analysis of human walking trajectory on circular path in darkness. Brain Res. Bull. **40:** 491–496.
50. GIANNA, C., S. HEIMBRAND & M. GRESTY. 1996. Thresholds for detection of motion direction during passive lateral whole-body acceleration in normal subjects and patients with bilateral loss of labyrinthine function. Brain Res. Bull. **40:** 443–437.
51. IVANENKO, Y. P. & R. GRASSO. 1997. Integration of somatosensory and vestibular inputs in perceiving the direction of passive whole-body motion. Cogn. Brain Res. **5:** 323–327.
52. SEIDMAN, S. H., G. A. BUSH, G. D. PAIGE & D. L. TOMKO. 1998. Perception of translational motion in the absence of non-otolith cues. *In* 28th Society Meeting: 416. Society for Neuroscience. Los Angeles.

Artificial Gravity Considerations for a Mars Exploration Mission

LAURENCE R. YOUNG[a]

Massachusetts Institute of Technology, 77 Massachusetts Avenue, Building 37-219, Cambridge, Massachusetts 02139, USA

> ABSTRACT: Artificial gravity (AG), as a means of preventing physiological deconditioning of astronauts during long-duration space flights, presents certain special challenges to the otolith organs and the adaptive capabilities of the CNS. The key issues regarding the choice of AG acceleration, radius, and rotation rate are reviewed from the viewpoints of physiological requirements and human factors disturbances. Head movements and resultant Coriolis forces on the rotating platform may limit the usefulness of economical short centrifuges for other than brief periods of intermittent stimulation.

INTRODUCTION

"Zero-G and I'm feeling fine," said Mercury astronaut John Glenn when he became the first American in orbital flight in 1962. "One G and I'm feeling fine," said Senator John Glenn when he returned from his second space flight, of nine days, in 1998. But after the last reentry, when he tried to walk around and balance himself, things were not so fine, although he eventually recovered. It is well known that about 70% of all space travelers experience a form of motion sickness soon after going weightless, and that many of them are either nauseous or have posture and gait instability after landing. The response of the body to weightlessness can lead to serious problems after return. Of particular concern are the loss of bone and muscle, cardiovascular deconditioning, loss of red blood cells and plasma, possible compromise of the immune system, and finally, an inappropriate interpretation of otolith system signals, which are so necessary to avoid falling over upon return to a gravity field.

Numerous remedies for these effects have been employed, but at this time only a few seem to have any documented beneficial effect. These few are the ingestion of ionically balanced water before reentry, inflation of an anti-g suit during reentry, and the maintenance of a vigorous and time-consuming exercise regimen nearly daily while in orbit.[1] Despite the successes to date of life support for shorter missions, the space traveler coming back to Earth after a year in weightlessness would likely be unable to walk normally for a day or more. Although the returning astronauts apparently recover in time, they are not in shape for safe vigorous or coordinated activity immediately after landing. They would be poor risks for the first interplanetary exploration. Before embarking on the next major step in space exploration, human travel to Mars, it is imperative to develop safe ways of protecting astronauts from these debilitating effects. The issue of appropriate

---

[a]Address for telecommunication: phone: 617/253-7759; fax: 617/258-8111; e-mail: lry@mit.edu

"countermeasures" to overcome the effects of weightlessness on humans ranks as one of the chief safety concerns about a Mars mission. Along with protection against radiation exposure beyond the Earth's magnetic field, these weightlessness countermeasures require urgent development before committing to the Mars exploration.

The current status of countermeasures is confused. In 1988, Burton concluded that "little progress had been made on methods to prevent the physiologic decay that accompanies long-duration stays in such a space environment. . . . Using methods that simply treat the symptoms of physiologic deconditioning is an approach that superficially is attractive, but, because the gravitation effect is ignored, should be expected to meet with some degree of failure."[2]

The Russian view, based upon extensive experience with exercise countermeasures as well as lower-body negative pressure and drugs, is that the in-flight countermeasures on their long missions were insufficient to prevent postflight physiological symptoms such as orthostatic intolerance and musculoskeletal deterioration. The typical Russian exercise protocol for missions lasting more than a month consists of up to 2 h/day of bicycling, resistive exercises using elastic cords, and jogging on the treadmill. This stands in addition to the 8 h/day use of the Penguin Suit, and a 30-min exposure to lower-body negative pressure of 30 mmHg every 4 days.[3] [In concluding that these countermeasures were both insufficient and time-costly, Shipov made a strong case for the creation of artificial gravity (AG) for interplanetary flights.[4]] Furthermore, the endurance exercise protocols used by the Russians, which afford a measure of protection for the cardiovascular system, do little to prevent weight-bearing muscles from losing strength and changing innervation patterns, and have almost no beneficial effect on bone density.[1] These countermeasures were included as part of the International Space Station health maintenance possibilities only because no better techniques have yet been demonstrated. Several other options have been suggested to deal with long-duration weightlessness; however, they remain untested.[5,6]

The armamentarium of operational space medicine offers a variety of countermeasures, but none are yet deemed fully effective against all the physiological problems for all individuals.[7] For example, a bicycle exercise that might provide adequate stress to the heart probably does not provide the impulsive loading needed to promote bone growth. Nor is it clear that the shift in blood volume footward during a long-duration soak in lower-body negative pressure will stimulate the physiological responses required to avoid fainting after or even during reentry. The incremental contribution of the Penguin Suit, which forces the astronaut to exercise the "antigravity" muscles each time a limb is extended, has not been formally assessed. Lack of understanding of the dynamics of drug action in space is but one of the problems with pharmacological approaches to the problem of microgravity. Diet control and dietary supplements have not been effective in countering the calcium loss from bone in space, but there has been relatively little attention to some of the newer drug treatments available for osteoporosis, or of aggressive manipulation of the astronauts' diet. Electrical stimulation of both muscle and bone has been suggested but never evaluated. Finally, various attempts to preadapt astronauts to the unusual vestibular conditions of weightlessness have never been adequately tested. The Soviet/Russian approach to the problem of in-flight sensory conflict was to pre-expose cosmonauts to a variety of extreme motion conditions, including a rotating gondola on a centrifuge and extensive underwater maneuvering. By these means they hoped to develop a reduced sensitivity to sensory conflict and an absence of space motion sickness. The approach has apparently been abandoned. In the U.S. program, a preflight adaptation trainer was built at the Johnson Space Center to expose astronauts to an altered visual-field motion in association with head and body tilts, but it, too, was never adequately assessed and is not in general use. Biofeedback techniques have similarly been tried and not validated for control of space motion sickness.[1]

HISTORY OF ARTIFICIAL GRAVITY

Hardly had the concept of artificial satellites and human space travel been conceived, when the pioneers imagined the creation of artificial gravity by centrifugation. In 1903 Tsiolkovsky, the Russian space visionary, imagined interplanetary travel using AG to counter the effects of weightlessness.[8] His ideas influenced Korolev, the father of the Soviet space program, who had a design team working on a tethered flexible-cable AG system for the Voskhod manned mission in the early 1960s.[9,10] In a book that influenced many of the early space engineers 50 years ago, Willy Ley envisioned a spinning space station erected in orbit by splitting the spaceship into two parts connected by a tether and accelerated to spin about their common center of gravity.[11,12] Von Braun was among the early enthusiasts for human space exploration who stressed the potential need for AG. His 125-foot torus, slowly spinning at 3 rpm (FIG. 1), was the basis for science fiction as well as technology.[13] Arthur Clarke described human space travel even before the first artificial satellite,[14] and in the popular film *2001: A Space Odyssey*, based on Clarke's stories, Stanley Kubrick had his crew passing from AG to weightlessness and back with nary a thought about Coriolis forces. Engineers were working on practical ways of turning these dreams into technical realities long before the first manned orbital flights.[15]

In this paper we presume that artificial gravity will be required for an interplanetary mission and consider ways to assure its effectiveness without unnecessary cost or complexity.

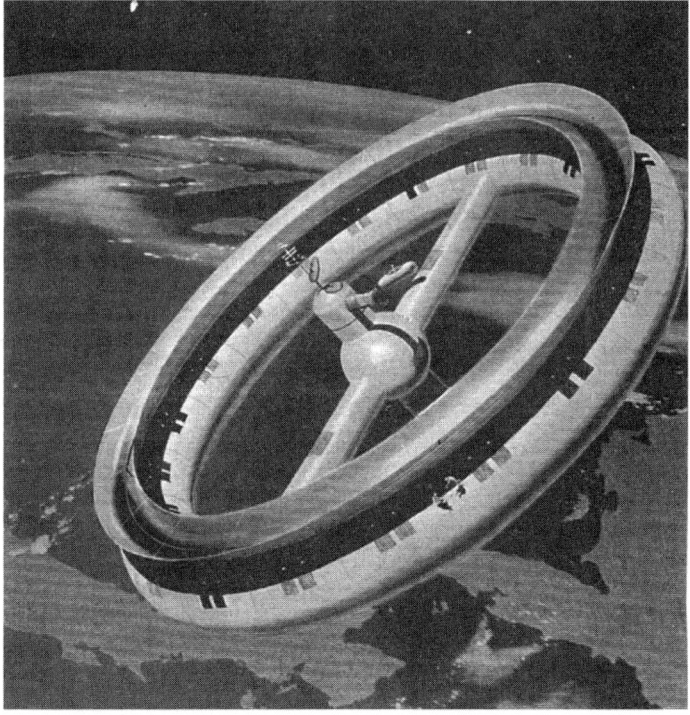

FIGURE 1. Von Braun's rotating space station.

The conservative designs of a large rotating spacecraft or of two spacecraft connected by a tether are undesirably complex and expensive. We will discuss factors influencing the choice of the key parameters that determine the nature of a practical device: radius, angular velocity, g level, and duration. The requirements for artificial gravity for a given mission profile will be considered from three overall perspectives: the physiological requirements, the human factors and performance requirements, and the practicality and cost of construction and operation.

PHYSIOLOGICAL REQUIREMENTS FOR ARTIFICIAL GRAVITY

If we are to commit to a spinning vehicle, the physiological questions differ for the two main types: a full-time rotating habitat, or intermittent stimulation—a kind of gravity gym.

For *sustained continuous rotation*, the first question is what level of acceleration at foot level is the minimum required to maintain normal function? Various means of achieving artificial gravity will be considered later, but we must first concentrate on the size and speed needed to provide sufficient centripetal acceleration. We can be reasonably certain that 1 g will suffice, but do not know if a lesser level will be equally effective.[16] Can we avoid deconditioning by spinning continuously at a level of 0.38 g to match the Martian gravity? (If not, will some sort of artificial gravity or other countermeasure be needed to condition the astronauts during a year's stay on the Martian surface?[3]) Should the appropriate g level for a nine-month trip to Mars consist of eight months at 1 g, and a final month at 0.38 g to prepare for the Martian environment?

Data to answer these questions are not available at this time. To find the answers, research *in orbit* must be conducted, first using animals and then human subjects. (The reason that we cannot conduct all the supporting research on earth centrifuges, of course, is that the constant 1-g bias cannot be eliminated except for short periods during parabolic flight. For example, the Pensacola Slow Rotating Room experiments, discussed below, all suffered from the inherent limitation that every head movement in the rotating room was accompanied by a new orientation relative to Earth's gravity, in addition to the centrifugal force.) The first attempt to spin two space vehicles about their common center was nearly a disaster. On the Gemini 8 mission in 1966, after docking with the Agena target vehicle and beginning a rotation test, one of the thrusters on the Gemini became stuck open and continued to accelerate the pair up to a rotation rate of one revolution per second, which continued even after separation. This serious malfunction was overcome only by the skillful use of the main reentry thruster by Astronaut Neil Armstrong. During Skylab, in the early 1970s, we saw astronauts running around an interior track to generate their angular velocity and achieve some artificial gravity. However, except for brief periods of exposure to linear acceleration on the D-1 Vestibular Sled, and rotation while displaced 1 m from the axis, on IML-1 and on Neurolab, no human AG experiments have yet been performed in space.

On the other hand, our knowledge about animals' responses to centrifugation is far more substantial. We are likely to achieve the earliest answers to the artificial gravity requirements with further experiments on animals. We already know that rats that have been centrifuged during their space flight do not show the major deterioration in bone, muscle, and cardiovascular response that is exhibited by their free-floating counterparts—but that is only at 1 g. The current International Space Station plans include a centrifuge to carry up to eight modules for rodents, fish, and eggs, but it will not accommodate the primates that many feel are needed to adequately model human responses. This variable-gravity animal centrifuge not only serves as a 1-g control for the zero-g experiments but also allows one to explore the entire range from 0 to 2 g's for a variety of species. Furthermore, it can be used to test a zero-g raised animal by measuring its reactions to 1 g at various times during the mission, without the artifacts normally associated with reentry.

In contrast to sustained continuous rotation, *intermittent artificial gravity stimulation* presents a number of potential advantages, particularly in its application to human subjects. Several ground studies using intermittent g exposure to counter the effects of bed rest have been conducted. As part of our normal circadian rhythm, the very g-dependent processes that result in fluid loss and bone deconditioning are probably turned off during normal sleeping hours. Extended periods of bed rest, however, produce effects on the skeleton, muscles, and cardiovascular system that resemble those occurring in space. This simulation of space flight is made more accurate if the bed rest is conducted with a 6 deg head-down tilt to accelerate the shift of fluid toward the head, and if the subject lies partially immersed, although dry, in a hightech water bed. The ready applicability of bed rest as a deconditioning model has made it possible to examine the efficacy of intermittent exposure to a real 1-g field, with or without exercise, to assess the protection afforded by intermittent centrifugation of various duration.[17] Furthermore, with the use of a centrifuge for short periods there is no reason to be restricted to 1 g. Just as for other physical stimuli, there must certainly be some dose–response relationship, which remains undetermined. Preliminary tests were conducted on subjects lying horizontal on the MIT 2-m-radius short-arm centrifuge, with 100% gravity gradient. We determined that at least 1.5 g at foot level was required to elicit the same change in blood pressure as that produced by moving from supine to upright in 1 g.[18]

Some preliminary data concerning intermittent stimulation are encouraging in terms of its use as a space countermeasure. Experiments at Douglas in the 1960s showed that most of the musculoskeletal problems of bed rest, excepting plasma loss, could be prevented by four short (11.2-min) daily exposures on a very short arm (1.1-m-radius) centrifuge.[19] Vernikos and colleagues at NASA Ames recently showed the potential protection afforded by 2–4 hours of standing or walking in preventing orthostatic intolerance, plasma loss, or calcium loss—but not in maintaining aerobic capability.[20,21] A new small centrifuge incorporating a bicycle-drive has been developed at Ames to provide exercise as well as g forces.

In our laboratory Diamandis first investigated the concept of an Artificial Gravity Sleeper and showed the practicality of sleeping in a 2-m-long bed rotating about an axis through the head at 24 rpm to produce a centripetal acceleration at the feet of 1 g.[22] Cardus extended the short arm centrifuge concept and proposed treatment of patients with little mobility using a sleep rotator.[23] However, we can only make educated guesses at this time as to the details of artificial gravity required to maintain fitness of all physiological systems. A combination of short-duration and long-term studies in ground centrifuges and rotating rooms can be useful in answering many of the key questions concerning the application period, frequency, and intensity of centripetal acceleration. They will also be useful in determining the seriousness of the problem of dual adaptation to both rotating and nonrotating environments. (Clearly, if astronauts experience a form of motion sickness each time they change from weightlessness to a rotating centrifuge and back, the treatment will be worse than the disease and will not be accepted. The development of context-specific dual adaptation, on the other hand, would eliminate this problem. It would permit crews to move easily, move repeatedly between gravito-inertial environments, and immediately adopt the appropriate new postural and sensorimotor reactions.) Finally, these studies may shed light on the physiological importance of a gravity gradient across the body if one is to truly proceed with rotators having a radius comparable to the subject's height. These ground gravitational physiology studies are essential for effective use and interpretation of the space variable-gravity centrifugation tests to be mentioned.

HUMAN FACTORS ISSUES

Even if a rotating space vehicle adequately addresses the main physiological issues, important human factors issues remain concerning its practicality for continuous exposure.

The g force at any radius is simply proportional to the radius times the square of the angular velocity. The Apollo lunar locomotion experience provides some support for the human factors adequacy of 1/6 g, and most researchers assume that at least 0.2 g will be needed for comfortable locomotion. As discussed earlier, a large enough rotating station, with diameter of the order of one kilometer and rotating at only 1 rpm, could produce a centripetal acceleration of 1 g at the rim and be tolerated with no difficulty. The construction and operation costs would be prohibitive, however, and therefore emphasis has been placed on reducing the g level, reducing the radius, and increasing the spin rate without destroying the habitability conditions. All these trends introduce potential problems associated with working and living in a rotating environment.[24]

Stone concisely summarized the human factors comfort and performance issues 25 years ago.[25] The *gravity gradient* as one moves from the rim toward the hub is the difference of radii divided by the radius at the rim. As the spacecraft radius shrinks to only double an astronaut's height, for example, the artificial gravity at head level is reduced to only half that at foot level, and the gradient is 50%. This gravity gradient could create awkward materials handling problems. Consider the astronaut in FIGURE 2a and 2b who bends to pick up an experiment module at foot level and place it into a rack at eye level, where it weighs only half as much. Will he learn to reduce his supporting force with height or will the package tend to fly up out of his grip?

A certain problem, at least in the early hours and days, is cross-coupled angular acceleration every time the astronaut makes a head movement about any axis except parallel to the spacecraft spin axis. The magnitude of this acceleration is equal to the product of the spin velocity and the head velocity out of the plane of spacecraft rotation. It is directed orthogonally to those two vectors, and therefore occurs in a direction that depends upon both the direction of head movement and the orientation of the astronaut relative to the spacecraft. The astronaut illustrated in FIGURE 3, who is initially facing in the direction of

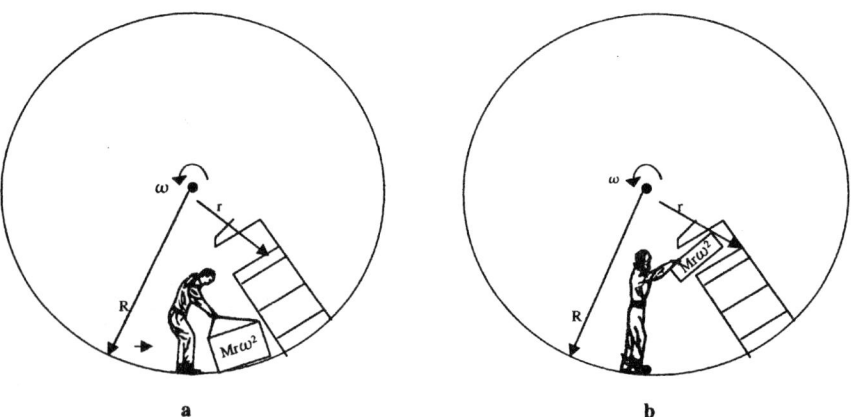

FIGURE 2. Gravity gradient. (**a**) Object of mass M "weighs" $Mr\omega^2$ on the rim. (**b**) Object of mass M "weighs" $Mr\omega^2$ at radius r.

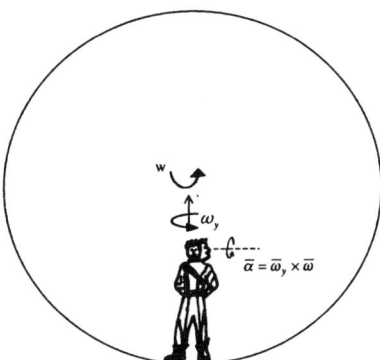

FIGURE 3. Cross-coupled acceleration. Yaw head velocity ω_y perpendicular to the spacecraft spin velocity ω produces an acceleration α about the naso–occipital axis. [Adapted from fig. 8, "Influences of artificial gravity on locomotion." (caption), in R. W. Stone, "An overview of artificial gravity," 5th Symposium on the Role of the Vestibular Organs in Space Exploration, Pensacola, Fla., Aug. 1970 (NASA SP = 314: 28).]

the spacecraft spin axis and turns his head to the left, will experience an unexpected acceleration about the naso–occipital axis, right ear down. The vestibular system will correctly pick up this acceleration and signal the brain about it, but the surrounding visual field will remain fixed and fail to confirm such an illusion. Of course, if the astronaut were initially facing the opposite direction, the same head movement would produce a left-ear-down acceleration. Habituation to such a complex pattern of visual–vestibular conflict is time-consuming and may not prove possible for all subjects. Distinctive interior color schemes to enable the astronaut to immediately identify orientation relative to the spin axis might reduce the severity of this problem. The sensory conflict is sure to produce symptoms of motion sickness, at least at first, unless the rotation rate is as low as 1 rpm or the astronaut is immobile.

Radial Coriolis forces may unexpectedly push the astronaut toward or away from the center of rotation, depending on the direction of locomotion. No Coriolis force occurs when walking parallel to the spin axis. However, an astronaut walking in the direction of the rim velocity will feel heavy, as in proceeding from FIGURE 4a to 4b, and she will feel light walking away from the rim, as in FIGURE 4c. The magnitude of the Coriolis force is the body mass times the product of the spin rate and the astronaut's speed. As her walking speed approaches rim speed, traction is drastically reduced to a point where walking is no longer possible. The ratio of the walking speed to the rim speed is the same as the ratio of the radial Coriolis force to the weight. Stone suggests that this ratio be kept less than 0.25. Finally, the astronaut must deal with the tangential Coriolis force, which acts to push one toward or away from the direction of the spacecraft spin when moving toward or away from the central hub, as shown in FIGURE 5. Ladders to ascend or descend in an AG spacecraft might be designed so that the astronaut is always pressed toward rather than away from the ladder, which implies two-sided ladders for ascending and descending. These forces also mean that an object will not fall directly down toward the floor, but will follow a curved trajectory. In time, astronauts will learn to compensate for this curvature, providing that the spacecraft interior walls are clearly designated, by color or otherwise, to indicate the direction of the spin.

FIGURE 4. Radical Coriolis forces during locomotion. (**a**) Standing. (**b**) Moving in direction of spin. (**c**) Moving in direction opposite of spin. [Adapted from fig. 8, "Influence of artificial gravity on locomotion" (caption), in R. W. Stone, "An overview of artificial gravity," 5th Symposium on the Role of the Vestibular Organs in Space Exploration, Pensacola, Fla., Aug. 1970 (NASA SP = 314: 28).]

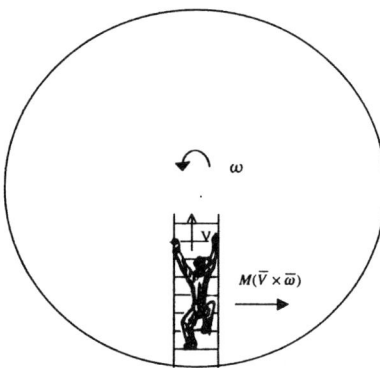

FIGURE 5. Tangential Coriolis forces. Tangential Coriolis forces push sideways on an astronaut moving radically in the spinning spacecraft. [Adapted from fig. 8, "Influence of artifical gravity on locomotion." (caption), in R. W. Stone, "An overview of artificial gravity," 5th Symposium on the Role of the Vestibular Organs in Space Exploration, Pensacola, Fla., Aug. 1970 (NASA SP = 314:28).]

The choice of maximum permissible rotation rate is thus critical for a rotating spacecraft. Pioneering experiments carried out on the Slow Rotating Room in Pensacola showed that most people could adapt to living in these conditions, at least up to 6 rpm on earth. Increasing the spin rate to 10 rpm, however, requires specialized habituation techniques, including programmed head movements over a two-week period, with incremental increases in the rotation rate. After exposure some readaptation time is again needed.[26–28] Current testing on the Slow Rotating Room at Brandeis University by Lackner and DiZio is extending this line of research to investigate strategies for visuomotor adaptation.[29] Graybiel's early studies give reason to believe that early short-term direction-specific adaptation effects are gradually replaced by a more general nondirection-specific long-term adaptation, which would enable astronauts to function easily in a rotating or nonrotating environment during a long space voyage.[30] Stone[25] and Shipov[4] both assumed that cross-coupling limits the spin rate to 6 rpm. Acceleration at the rim is kept to less than 1 g for sufficient conditioning, and the ratio of Coriolis force to weight is limited to 0.25, assuming that walking is at 1 m/s. The rim speed of at least 6 m/s keeps the "vertical" g difference between walking and standing to 10–15%. [The walking speeds that appear best for head and eye stabilization on a treadmill are in the vicinity of 1.4 m/s, somewhat higher than those assumed by early AG researchers. (Raphan, private communication, 1998).]

DESIGN ENVELOPES

Loret summarized the design envelopes for radius, g level and rotation rate used by many of the early proposals for rotating spacecraft.[31] As indicated in FIGURE 6, after Loret most of the earlier designs, whether for a dumbbell or for a torus, are in the region of 3–4 rpm, with a radius of 1–40 m. At the extreme ends of the designs were Schnitzer's 6-m radius torus and Loret's (PGV) expanded dumbbell with a 55-m radius. (Stone's optimistic minimum was a radius of 15 m and Shipov thought that 20 m was the minimum.) A stepwise habituation schedule and appropriate unambiguous visual orientation cues should allow a build-up to 10 rpm, which would drastically reduce the minimum radius. Diamandis's 2-m short-radius Artificial Gravity Sleeper is also illustrated.

The additional design point in FIGURE 6, labeled "MIT," represents a goal for a practical short-radius continuous-rotation spacecraft that might satisfy the need for preventing deconditioning while being acceptable from the human-factors standpoint. With a radius of only 4 m and a g level at the rim of 0.5, it would require a spin rate of 10 rpm. With these parameters the MIT design would be small and relatively inexpensive, and could be spun to 10 rpm slowly, over a period of several weeks during the early phase of a long mission. Two or three astronauts might be spun on it simultaneously, reducing to some extent the need for balancing weights. The gravity gradient might be as high as 50% for a 2-m astronaut, but that would probably be acceptable, based on the preliminary cardiovascular data from Burton's lab and our own. The rim speed would be only 4 m/s, rather than the 6 m/s recommended earlier. This would mean that fast motion in the direction opposite to the spin might have to be accomplished by pushing off the floor and bounding, rather than walking, but normal walking at 1–1.5 m/s could be accomplished easily. The use of a 0.5-g level at the rim is highly speculative, but hardly more so than the 0.6-g level recommended by others. Research will be needed to validate this number for adequate prevention of cardiovascular and muscular deconditioning.

The Coriolis ratio for a 1-m/s walk would be high (0.5) as opposed to the upper limit of 0.25. Once again, it is felt that by properly distinguishing the walls so that the direction of motion is obvious, astronauts will adapt to the direction-specific side force while walking. Finally, the level of cross-coupling for a typical head movement is raised to 190 deg/s^2, from the level of 115 deg/s^2 for the design recommended by Stone. This issue should also be resolved by careful stepwise adaptation protocols, including astronaut voluntary head movements.

SPACECRAFT DESIGN

If it were really easy to investigate artificial gravity in space, we would have done it by now. After the early spinning torus concept, the issue was seriously pursued at NASA's Langley Research Center, and by 1962 a number of spinning spacecraft were proposed, consistent with the Apollo module and Saturn launch vehicles. Several versions of inflatable designs allowed the entire station to be carried aloft on a single booster, but they were abandoned because of their susceptibility to puncture by micrometeorites. (With improved materials the concept of inflatable spacecraft has recently been revived and remains under consideration as a backup plan for the International Space Station.) Other concepts provided for a series of rigid modules to be assembled into a rotating station on orbit. The artificial gravity concepts that have been under discussion in recent times can be divided into three categories: tethered devices, fixed-truss assemblies, and small internal centrifuges.

The use of a tether that could be unreeled on orbit and cut loose if necessary is quite appealing. In the "Mars Direct" concept, the entire long combination is accelerated up to 1 rpm to provide 0.38 g for the trip to Mars, with no g protection offered for the return trip. A very ambitious tethered design for a variable gravity spacecraft[32] provided for a laboratory and habitation module connected to a large counterweight by a cable that could extend to a radius as long as 4700 m and provide variable g loads up to 2 g's.

A rigid-truss design presents certain advantages to counter its complexity, weight, and cost. It makes it easier to provide a zero-gravity hub with intermediate radius stations at different g levels. The International Space University developed their concept for Newton, whose variable radius could be as large as 100 m and could spin at up to 3 rpm. They suggested a unique sliding sleeve assembly to permit the effective radius of the living quarters to be varied.

Finally, let us turn to the potential use of a short-arm centrifuge for intermittent artificial gravity exposure of astronauts. This concept has been validated in numerous ground studies as an effective way to overcome the deconditioning of bed rest, and it could be attached to or even built into a space station module. The early Douglas tests[33] were conducted with the subject crouched in a 1.4-m radius wheel. Burton and Meeker proposed a

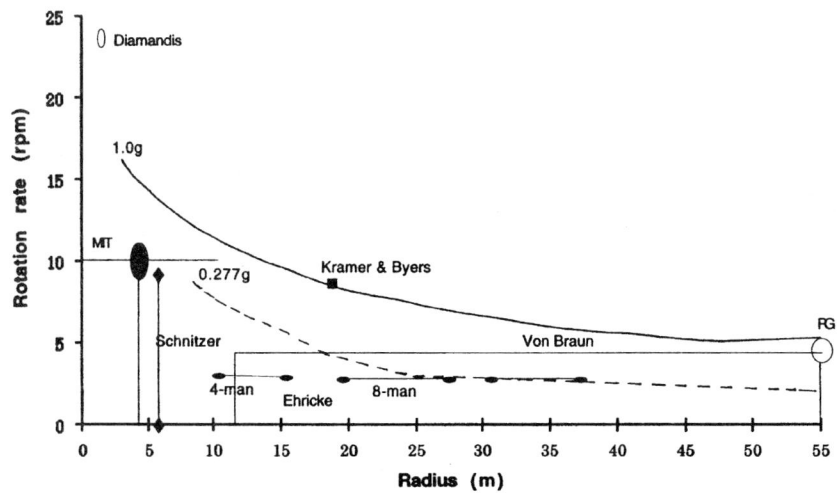

FIGURE 6. Artifical gravity design points.

similar design for a Short Arm Centrifuge that would accommodate several crew members at once for a "spin in the gym."[34–36] Both the Ames Research Center Short Arm Centrifuge[37] and the earlier Artificial Gravity Sleeper could be accommodated on the station or within the Space Shuttle Bay.

CONCLUSION

The requirements for artificial gravity as a countermeasure against the deconditioning of long-duration space flight remain poorly defined. A research program utilizing both ground studies and space experiments is required if the community is to be able to respond to the challenge of assuring human health and well-being for a Mars exploration mission. The ground studies should use rotating rooms for long-duration adaptation research *and* short-arm centrifuges combined with bed rest to assess the human factors and physiological effects. The space studies can start on the Space Station with a human support module to be attached to the animal centrifuge for brief exposures, and then continue with a stand-alone artificial-gravity research facility as described earlier. Within a decade we should have all the answers we need concerning not only the requirements for artificial gravity, but also the magnitude, radius, rotation rate, and frequency of the exposure. Only then will we be able to judge whether artificial gravity is justified as a protective mechanism for human interplanetary travel.

ACKNOWLEDGMENTS

The National Aeronautics and Space Administration supported this work through the NASA Cooperative Agreement NCC 9-58 with the National Space Biomedical Research Institute. This paper was also supported by MIT's Apollo Program Chair in Astronautics, particulaly by the generosity of Dr. and Mrs. Robert Seamans. The development of the author's longtime friend and colleague, the late Volker Henn. Preparation of the manuscript was assisted by Dawn Hastreiter, Sean Tytler, and Anna Tomassini of the Man-Vehicle Laboratory. Marsha Warren edited the manuscript and prepared my adaptions of Stone's drawings.

REFERENCES

1. NATIONAL AERONAUTICS AND SPACE ADMINISTRATION. 1997. NASA Task Force on Countermeasures: Final Report.
2. BURTON, R. 1988. A human-use centrifuge for space stations: proposed ground-based studies. Aviat. Space Environ. Med. **59** (6): 579–582.
3. DIAMANDIS, P. H. 1997. Countermeasure and artificial gravity. *In* Fundamentals of Space Life Sciences, S. E. Churchill, Ed.: 159–175. Krieger. Malabar, Fla.
4. SHIPOV, A. A. 1997. Artificial gravity. *In* Space Biology and Medicine: Humans in Spaceflight, Vol. 3, Book 2; A. E. Nicogossian, S. R. Mohler, O. G. Gazenko and A. I. Gregoriev, Eds.: 349–363. Am. Inst. Aeronaut. Astronaut. Reston, VA.
5. VERNIKOS, J. *et al.* 1996. Effect of standing or walking on physiological changes induced by head-down bed rest: implications for spaceflight. Aviat. Space, Environ. Med. **67**(11): 1069–1079.
6. GRYMES, R. A, C. E. WADE & J. VERNIKOS. 1995. No title. Life Sciences Division, NASA Ames Research Center, Moffett Field, Calif.
7. NICOGOSSIAN, A. E., C. F. SAWIN & A. I. GRIGORIEV. 1994. Countermeasures to space deconditioning. *In* Space Physiology and Medicine, 3rd ed., A. E. Nicogossian, C. L. Huntoon, and S. L. Pool, Eds. Lea & Febiger. Philadelphia.
8. TSIOLKOVSKY, K. E. 1911. Exploration of global space with jets. *In* Collected Works (1954, in Russia), Vol 2: 100–139. Nauka. Moscow.

9. KOROLEV, S. P. 1957. The practical significance of Konstantin Tsiolkovsky's proposals in the fields of rocketry. (English Rep.). *In* History of the USSR: New Research. (Soc. Sci. Today **5**: 59, 1986).
10. HARFORD, J. 1973. Korolev: 7, 13, 186. Wiley. New York.
11. LEY, W. 1944. Rockets: The Future of Travel Beyond the Stratosphere. Viking. New York.
12. LEY, W. 1968. Rockets, Missiles and Men in Space, rev. ed. Viking. New York.
13. VON BRAUN, W., F. L. WHIPPLE & W. LEY. 1953. Conquest of the Moon, C. Bonestell and C. Ryan, Eds.: 11. Viking. New York.
14. CLARKE, A. 1957. The Making of a Moon: The Story of the Earth Satellite Program. Harper. New York.
15. HILL, P. R. & E. SCHNITZER. 1962. Rotating manned space stations. Astronautics **7**: 4–28.
16. SANDLER, H. 1995. Artificial gravity. Acta Astronaut. **35**: 363–372.
17. GUNJI, A., Ed. 1997. Proc. 2nd Symp. on Inactivity and Health: Effects of Bed Rest on Health. J. Gravitat. Physiol. **4**(1): S1–S106.
18. HASTREITER, D. & L. R. YOUNG. 1997. Effects of gravity gradient on human cardiovascular responses. J. Gravitat. Physiol. **4**(2): P23–P26.
19. BURTON, R. 1989. Acceleration simulation in space. Intersociety Conference on Environmental Systems. SAE Technical Paper Series: 24–26.
20. VERNIKOS, J., & D. A. LUDWIG. 1994. Intermittent Gravity: How much, How Often, How Long? NASA Technical Memo 108800, NASA Ames Research Center, Moffett Field, Calif.
21. VERNIKOS, J. 1997. Artificial gravity: intermittent centrifugation as a space flight countermeasure. J. Gravitat. Physiol. **4**(2): P13–P16.
22. DIAMANDIS, P. H. 1988. The artificial gravity sleeper: a deconditioning countermeasure for long duration space habitation. S.M. Aeronaut. Astronauti. MIT. Cambridge.
23. CARDUS, D., W. G. MCTAGGART & S. CAMPBELL. 1991. Progress in the development of an artificial gravity simulator. Physiologist **34**(S1): S224–S225.
24. YOUNG, L. R. 1971. Modelling disorientation in a rotating spacecraft. AIAA/ASMA Weightlessness and Artificial Gravity Meeting, Williamsburg, Va.
25. STONE, R. W., Jr. 1970. An overview of artificial gravity. 5th Symp. on the Role of the Vestibular Organs in Space Exploration, Pensacola, Fla. (NASA SP-314: 23–33).
26. GRAYBIEL, A. 1969. Structural elements in the concept of motion sickness. Aerosp. Med. **40**: 351–367.
27. GRAYBIEL, A., B. CLARK & J. J. ZARRIELLO. 1960. Observations on human subjects living in a "slow rotation" room for periods of two days. Arch. Neurol. **3**: 55–73.
28. REASON, J. T. & A. GRAYBIEL. 1970. Progressive adaptation to Coriolis accelerations associated with one rpm increments of velocity in the slow-rotation room. Aerosp. Med. **41**(1): 73–79.
29. LACKNER, J. R. 1993. Orientation and movement in unusual force environments. Psychol. Sci. **4**: 134–142.
30. GRAYBIEL, A. 1975. Velocities, angular accelerations and Coriolis accelerations. *In* Foundations of Space, Biology and Medicine. Vol. II, Book I, M. Calvin and O. G. Gazenko, Eds.: 247–304. NASA. Washington, D.C.
31. LORET, B. J. 1963. Optimization of space vehicle design with respect to artificial gravity. Aerosp. Med. **34**: 430–441.
32. SMITH, M., P. WERCINSKI, *et al.* 1990. A Conceptual Design Study of a Variable Gravity Spacecraft. Space Exploration Projects Center Rep. NASA/Ames Research Center. Moffett Field, Calif.
33. WHITE, W. J., J. W. NYBERG, P. D. WHITE, R. H. GRIMES, & L. M. FINNEY 1965. Biomedical Potential of a Centrifuge in an Orbiting Laboratory, Douglas Rep. SM-48703 and SSD-TDR-64-209-Supplement (July 1995), Douglas Aircraft, Santa Monica, Calif.
34. BURTON, R. R. & L. J. MEEKER. 1992. Physiologic validation of a short-arm centrifuge for space applications. Aviat. Space Environ. Med. **63**: 476–481.
35. MEEKER, L. J., *et al.* 1996. A human-powered, small-radius centrifuge for space application: a design study. SAFE J. **26**(1): 34–43.
36. BURTON, R. R., L. J. MEEKER & J. H. RADDIN. 1991. Centrifuges for studying the effects of sustained acceleration on human physiology. IEEE Eng. Med. Bio. 56–65.
37. VERNIKOS, J. 1997. Artificial gravity: intermittent centrifugation as a space flight countermeasure. J. Gravitat. Physiol. **4**(2): P13–P16.

The Role of Somatosensory Input for the Perception of Verticality

DIMITRI ANASTASOPOULOS,[a,b,d] ADOLFO BRONSTEIN,[c]
THOMAS HASLWANTER,[b] MICHAEL FETTER,[b] JOHANNES
DICHGANS[b]

[a]*Department of Neurology, University of Ioannina, P.O. Box 1186,
45110 Ioannina, Greece*

[b]*Department of Neurology, Eberhard-Karls-Universität Tübingen,
Hoppe-Seyler-Str. 3, 72076 Tübingen, Germany*

[c]*Medical Research Council Human Movement and Balance Unit,
Section of Neuro-Otology, National Hospital for Neurology and
Neurosurgery, Queen Square, London, WC1N 3BG, UK*

INTRODUCTION

In the absence of a visual frame of reference, a normal subject is able to set a line to the vertical accurately, within 1 or 2 deg, when sitting upright (subjective visual vertical, SVV).[1] He perceives uprightness of his body within an inverted cone with a base of about 5 deg diameter, when passively tilted in the roll and pitch planes at low velocity.[2] When normal subjects sitting in a motor-driven gimbal are asked to actively set themselves to vertical by means of a joystick control, their judgments deviate on average 1.7 deg from the true gravitational vertical.[3]

Aubert was the first to describe an apparent systematic tilt of the SVV depending on the orientation of the head.[4] When the head (or whole body) is tilted to one side, a vertical line appears to the observer tilted to the opposite side, so that, when asked to set it vertical, he rotates its upper edge toward the side he is tilted (A-effect). This effect has been reported to be absent after complete proprioceptive sensory loss below the neck due to polyneuropathy.[5] While patients with acute unilateral vestibular lesions show strong deviations of the SVV toward the lesion side, their postural vertical judgments remain veridical.[2,3] Thus, conflicting orientation responses to the gravity vector can coexist, depending on the multiple sensory inputs and processing structures involved. In fact, the contribution of each system (visual, vestibular, proprioceptive) is not yet fully understood.

In the following we assessed the contribution of the somatosensory system to the perception of verticality by asking patients with various degrees of sensory disturbances to set the SVV in upright (sitting) position and when lying sideways. The perception of the subjective postural vertical (SPV) was also evaluated.

---

[d]To whom correspondence may be addressed. Phone: T/K; fax: +30-651-78265; e-mail: danastas@cc.uoi.gr

METHODS

Apparatus

Subjects sat upright in a chair located inside a spherical cabin, with the head positioned in the center of the sphere and the torso restrained in vacuum cushions. The cabin was supported by the inner ring of a four-axis gimbal system.[6] Appropriate positioning of the middle ring, which could be rotated around an axis perpendicular to that of the outer ring, allowed tilts of the subjects about a naso–occipital axis (i.e., in the roll plane). A 10-cm-long luminous line, generated by a laser, could be projected at eye level on the cabin wall in front of the subject. Both the cabin and the target line could be rotated in the roll plane either by the subject by means of separate joysticks or by the experimenter.

Procedures

The SVV was determined by means of six adjustments from a random angular offset of the target line (variable up to ±40 deg from earth vertical) in darkness. The subject was required to set the line "so that it appears to be vertical." The angular deviation from the true vertical in degrees was measured by a potentiometer and read by the experimenter. SVV estimates were repeated after tilting the chair 80 deg in the roll plane randomly to the right and left. The SPV was determined by 12–16 adjustments of the position of the cabin by the subject so that he "feels his body is vertical" after a random offset in the roll plane of ± 5, 15, or 25 deg. Acceleration and deceleration of the stimuli used to offset the subjects was 10 deg/s^2, thus resulting in triangular velocity profiles with maximal velocity of 7.0, 12.1, and 15.5 deg/s respectively. Subjects responses were self-paced, with an upper velocity limit of either 2 deg/s or 3 deg/s randomly set when subjects were adjusting themselves to earth vertical. This procedure guaranteed the absence of time cues for repositioning to upright. Measurements were determined manually from the computer-stored displacement traces.

Subjects

Eighteen patients with acute and chronic somatosensory disturbances (3 with brain and spinal cord tumors, 2 with cerebral haemorrhage, 6 with cerebral and spinal ischaemia, 4 with a first episode of demyelination, and 3 with polyneuropathy) were compared with 20 age-matched normal subjects (mean age 50.2, SD 10.8 years). Not all patients could be tested in both paradigms. In two patients the sensory loss was extreme, including complete unilateral loss of position sense. The cause was an acute cerebral hemisphere infarction (FIG. 1A and 1B); only SVV estimates were obtained from these two patients. Their tilt estimates were obtained lying in bed and adjusting a luminous line presented on a computer screen by means of a hand-held trackball.

RESULTS

Subjective Visual Vertical

In the upright position, the SVV estimates of the normal and patient groups were similar (1.9 SD 1.4 deg; 1.9 SD 1.7 deg, respectively; unsigned values). Only one patient with

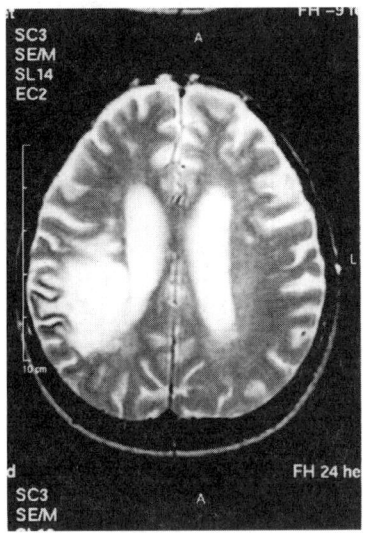
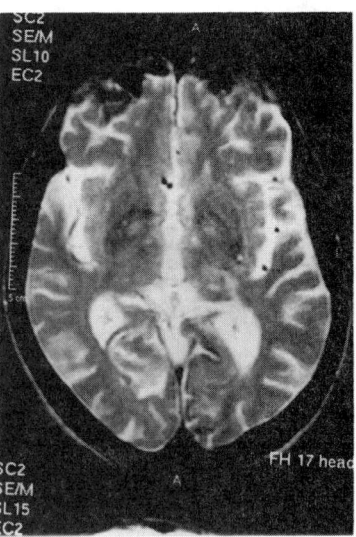

FIGURE 1. MRIs of the two patients lacking A-effect on the SVV when lying on the hypoesthetic side, showing ischemic infarctions in the right parietal lobe (**A**) and left thalamic region (**B**). The visual vertical data of these patients is shown in FIGURE 2, with symbols ▲ and ★, respectively.

complete hemisensory loss due to a parietal infarct (FIG. 1A) showed a marginal ipsiversive tilt of 4.9 deg (FIG. 2A). In the lateral body position, SVV settings of normal subjects deviated in the direction of body tilt (normal values—16.5 SD 10.4 deg and 14.2 SD 7.8 deg for left- and right-tilted position, respectively). In the two patients with almost complete hemisensory loss the A-effect was essentially abolished when lying on the hypoesthetic side (FIG. 2B). The remaining had SVV settings deviated in the same direction as body tilt by a normal amount (i.e., normal A-effect).

Subjective Postural Vertical

Only in 2 patients with moderate hypesthesia was there a bias of the SPV, toward the hypoesthetic side. One of them had an acute infarct of the right internal capsule (SPV–5.3 deg) and the other a right temporoparietal glioblastoma (SPV–4.6 deg) (FIG. 2A, showing unsigned data). They were the two most severely hypoesthetic patients tested in the SPV task. As group data, SPV setting from the normal subjects (1.6 SD 1 deg) and patients (2.1 SD 1.5 deg) were not different (unsigned values).

DISCUSSION

Our patients with sensory loss did not show significant tilts of the SVV in the upright, seated position. In contrast, while lying sideways, unilateral sensory loss completely abolished the apparent displacement of the SVV in the opposite direction of body tilt in our

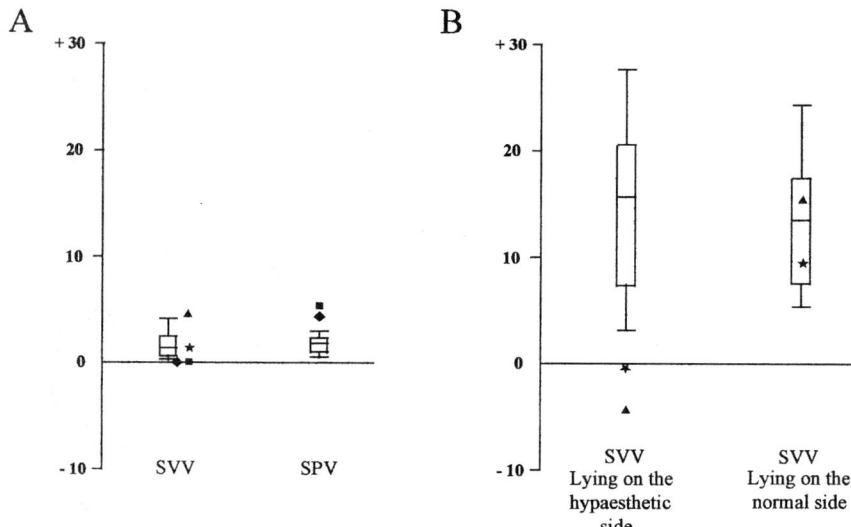

FIGURE 2. (A) SVV in the upright position in 4 patients with moderate to severe hemihypoesthesia, and SPV in 2 of these patients in whom it could be tested. Normal control data shown are medians plus 75th and 90th percentiles. All data points are unsigned. (B) Visual vertical settings in two patients with severe hemianesthesia when lying on the normal and anesthetic side. Median, 75th and 90th percentiles of the normal group are shown, arbitrarily taking left side down for control of the hypoesthetic side. These latter values for normal controls and patients have been inverted to facilitate comparison with the unaffected side. Note absence of the A-effect in the patients when lying on the hemianesthetic side.

two patients with parietal and thalamic infarcts. It appears that the origin of tilt-mediated effects on the visual vertical (A-effect) is primarily somatosensory, not otolithic. Consistent with this hypothesis are the findings of symmetrical A-effect in patients with unilateral peripheral vestibular disease[7] and the abnormally large A-effect in patients with bilateral absence of vestibular function.[8] An additional patient with a long-standing thalamic infarct previously tested also showed an absence of A-effect when lying on the hypoesthetic side (Anastasopoulos and Bronstein, submitted manuscript), as well as a patient with a severe polyneuropathy.[5] The findings in patients with hypoesthesia have the general implication that somatosensory input is capable of influencing the visual perception of verticality. However, this influence seems to come into play only during body tilt, as documented by the finding that even patients with acute hemihypoesthesia do not show significant tilts of the visual vertical while upright. The functional significance, if any, of the displacement of the SVV during lateral tilt (A-effect) is unknown. Mittelstaedt explains its occurrence as secondary to an internal drive that tends to rotate the SVV in the subject's main body axis (the idiotropic vector; Mittelstaedt, this issue). It would be reasonable to assume that such a body-centered vector relies on symmetric somatosensory input and that severe anesthetic lesions could have a profound influence on its magnitude or orientation.

It has been concluded that the SPV, at least while sitting with the body strapped to the seat, is overwhelmingly determined by somatosensory input, so that even acute vestibular imbalance does not bias it.[2,3] Only two patients with moderate to severe hemihypoesthesia showed a bias towards the side of the sensory loss. Our results are also of significance in showing that mild to moderate unilateral sensory losses are not sufficient to produce a bias of the SPV. It is likely that the remaining somatosensory, and possibly otolithic, information is able to counteract any bias induced by unilaterally reduced contact cues. It could be argued that the lesions in these patients may have interfered with central vestibular pathways or cortex leading to abnormal perception of the SPV, as it is known to occur with the SVV.[9,10] Similarly, the loss of the A-effect on the SVV might have arisen from interference with central vestibular processing by the lesions. However, the absence of vestibular symptoms and signs as well as the normal visual vertical settings while upright strongly argue against this possibility. On a neurological basis, the findings reported here for both the SVV and the SPV can only be attributed to severe asymmetry in the somatosensory system.

For a comment on our article, see the paper by Mittelstaedt on page 334 of this volume.

REFERENCES

1. HOWARD, I. P. 1982. Human Visual Orientation: 412. Chichester, Wiley. England.
2. BISDORFF, A. R., C. J. WOLSLEY, D. ANASTASOPOULOS, A. M. BRONSTEIN & M. A. GRESTY. 1996. The perception of body verticality (subjective postural vertical) in peripheral and central vestibular disorders. Brain. **119**:1523–1534.
3. ANASTASOPOULOS, D., T. HASLWANTER, A. BRONSTEIN, M. FETTER & J. DICHGANS. 1997. Dissociation between the perception of body verticality and the visual vertical in acute peripheral vestibular disorders. Neurosci. Lett. **233**: 151–153.
4. AUBERT, H. 1861. Eine scheinbare bedeutende Drehung von Objecten bei Neigung des Kopfes nach rechts oder links. Virchows Arch. **20**: 381–393.
5. YARDLEY, L. 1990. Contribution of somatosensory information to perception of the visual vertical with body tilt and rotating visual field. Percept. & Psychophys. **48** (2):131–134.
6. KOENIG, E., H. WESTERMANN, K. JÄGER, K. BECHERT, M. FETTER & J. DICHGANS. A new multi-axis rotating chair for oculomotol and vestibular functional testing in humans. Neuro-Ophthalmology. **16**: 157–162.
7. BOEHMER, A. & J. RICKENMANN. 1994. The subjective visual vertical as a clinical parameter of vestibular function in peripheral vestibular diseases. J. Vest. Res. **5**: 35–44.
8. BRONSTEIN, A. M., L. YARDLEY, A. P. MOORE & L. CLEEVES. 1996. Visually and posturally mediated tilt illusion in Parkinson's disease and in labyrinthine defective subjects. Neurology. **47**: 651–656.
9. FRIEDMANN, G. 1970. The judgment of the visual vertical and horizontal with peripheral and central vestibular lesions. Brain. **93**: 313–328.
10. BRANDT, TH., M. DIETERICH & A. DANEK. 1994. Vestibular cortex lesions affect the perception of verticality. Ann. Neurol. **35**: 403–412.

Otolith-Vestibular-Evoked Potentials in Humans

Intensity, Direction of Acceleration (Z+, Z−), and BESA Modeling of Generators

P. M. BAUDONNIÈRE,[a] S. BELKHENCHIR, J. C. LEPECQ AND
S. MERTZ

*Neurosciences Cognitives et Imagerie Cérébrale, CNRS-UPR 640,
Universite Paris VI, Hôpital de la Salpétrière, 75651 Paris Cedex 13*

The recording of averaged bioelectrical responses to physiological stimuli provides an important tool for monitoring the function of sensory organs and localizing their pathways in human subjects. The development of evoked-response recording techniques has also helped determine lesion sites in sensorial systems. Indeed, evoked potentials have been used to detect auditory, visual, and somatosensory dysfunctions. Vestibular-evoked potentials (VsEP) have received considerably less attention, essentially because of the difficulty in providing appropriate stimuli.

In order to elicit a vestibular-evoked response it is necessary to deliver intense repetitive, reproducible, identical, and adequately short stimuli. Few studies have attempted to elicit and record VsEP, and most of these were performed with rotational stimuli designed to stimulate the semicircular canals. The aim of this study was to record otolithic-evoked potentials, generated noninvasively in humans.

HYPOTHESIS

The electrophysiological responses can be considered as being of vestibular origin if (1) the same response is obtained for equivalent accelerometric patterns (up acceleration = down deceleration; down acceleration = up deceleration); (2) the amplitude of the response increases as a function of stimulus intensity; (3) and no response can be recorded in patients whose vestibular function is completely lost, bilaterally. This manipulation may also provide information regarding any somatosensory contribution to the VsEP.

METHOD

Fourteen healthy subjects (8 females and 6 males, ranging in age from 25 to 50 years, with a mean of age of 32.3) participated in the experimental session. One subject (female, aged 35) who was suffering from a complete bilateral loss of the vestibular function was also examined for comparison.

---

[a]To whom correspondence may be addressed. Phone: 33 1 44 24 52 92; fax: 33 1 44 24 39 54; e-mail: lenapmb@ext.jussieu.fr

The otolithic stimulation was provided by the seated subject's linear displacement along an earth vertical Z-axis (Z+, Z−) with three intensity values (0.1g ; 0.2g ; 0.4g) of acceleration and deceleration, and with a duration of 30 ms (the order of presentation was counterbalanced). An accelerometer was used to monitor movements during the vertical displacement. Each subject was blindfolded and exposed to a white noise.

Evoked potentials were recorded through 21 cutaneous electrodes, which were proof against movement. The electrode placement was standardized amongst the different subjects using an "electroCap" helmet according to the 10-20 international standard system. Linked earlobe electrodes were used as reference. The vertical EOG was recorded from infraorbital and supraorbital electrodes, so it could be removed during the processing of data.

RESULTS

Comparison Between Patterns Obtained with the Accelerometer

The same response is obtained for equivalent accelerometric patterns. The up-acceleration pattern is similar to the down-deceleration one. This is also verified for the two other patterns (down-acceleration = up-deceleration). However, the difference between accelerometric patterns, the cinematics of which are opposite to each other (up acceleration and down acceleration vs. up-deceleration and down-deceleration), is significant ($p = 0.01$).

Comparison Between VsEP of Healthy Subjects for the Same Patterns of Acceleration

The VsEP is constituted by a biphasic negative wave with a prominent peak on the vertex (Cz) and the frontal scalp locations (Fz) (FIG. 1). Both up-acceleration and down-deceleration produced the same evoked response in terms of amplitude ($Cz = 7.22$ μV ; $Fz = 7.05$ μV) and latency (86 ms).

Variation of VsEP with Stimulus Intensity

The amplitude of the response given by the accelerometer placed on the head increases linearly with stimulus intensity. Furthermore, whatever the intensity of the stimulation, the peak of the acceleration response is always the same (56 ms). In all healthy subjects, the amplitude of averaged evoked responses increases linearly with stimulus intensity ($p = 0.001$). The response latency is constant for the three stimulus intensities (86 ms).

BESA modelization shows three active areas in a time windows 50–300 ms after the stimulation (see FIG. 1): Brodmann's area (Ba), 10 median frontal gyrus, 2 bilateral Ba 4 precentral gyrus (SMA), and 2 bilateral Ba 19 median occipital gyrus.

DISCUSSION

Since the visual and auditory inputs are blocked, the recorded Eps cannot be generated by these modalities. Both recorded topography and chronology are incompatible with a somatosensory origin. The similarity of Eps in both the acceleration and deceleration conditions is incompatible with an exclusively somatosensory hypothesis. The patient's lack of response strengthens the otolithic origin of the Eps recorded in healthy subjects.

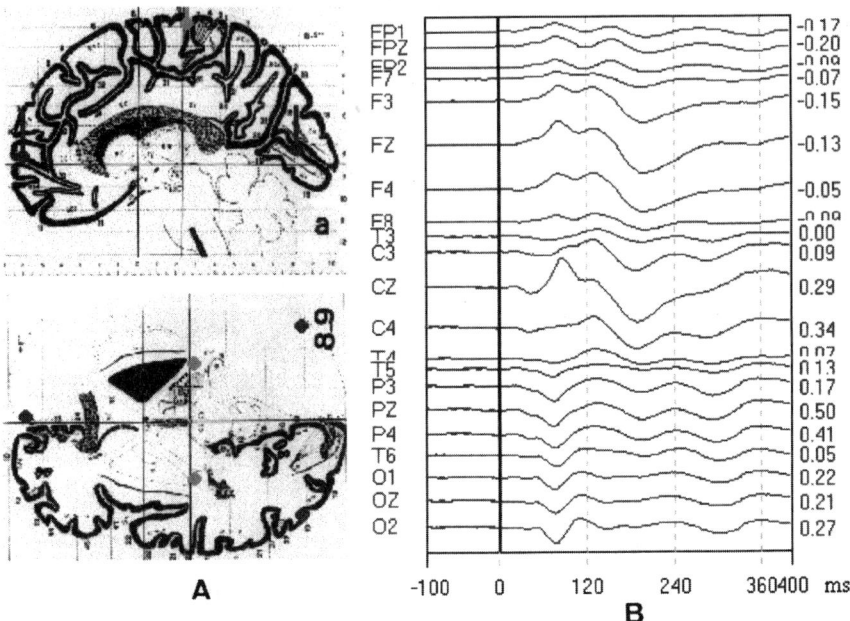

FIGURE 1. (**A**) Ba 10 medium frontal gyrus; Ba precentral gyrus SMA; Ba 19 medium occipital gyrus. (**B**) Evoked potential for UP acceleration.

This set of results strongly suggests that we have shown evoked potentials of seemingly genuine otolithic origin. This noninvasive technique can be of major interest for diagnosis in clinical applications.

Measuring the Otolith–Ocular Response by Means of Unilateral Radial Acceleration

A. H. CLARKE,[a] A. ENGELHORN, CH. HAMANN AND U. SCHÖNFELD

Vestibular Research Laboratory, HNO-Klinik, Universitätsklinikum Benjamin Franklin, Freie Universität, Berlin

During constant-velocity rotation about the earth-vertical axis, eccentric displacement of the head can be used to generate adequate stimulation of the otolith organs. In general, this is exploited to generate bilateral stimuli with radii ranging between 50 cm and 10 m. More recently, various authors have reported the use of a variable radius rotatory chair, which permits a controlled modulation of the radial, or centripetal acceleration, to achieve linear acceleratory frequencies much lower than with a conventional linear sled.[1] A further refinement is achieved by displacing the subject minimally from the rotatory axis, so that one labyrinth becomes aligned on-axis, while the second labyrinth alone is exposed to the radial acceleration. This unilateral stimulus technique has previously been employed in studies of subjective vertical[2] and otolith–ocular response.[3,4] As will be demonstrated here, the technique can also be used to examine the gain/phase relationships and threshold of the otolith–ocular response (OOR) at extremely low frequencies.

METHOD

Assuming an interlabyrinth distance of 7 cm, a lateral displacement of 3.5 cm would align one labyrinth with the rotatory axis, with the second labyrinth set at a radius of 7 cm. The procedure commenced with an earth-vertical z-axis velocity ramp (5 deg/s^2) up to a constant angular rate of 300 deg/s. After cessation of perrotatory nystagmus (approximately 2 min), the subject was oscillated along the interaural axis (trapezoidal or sinusoidal ± 3.5 cm). This eccentric displacement profile (EDP) yields a unilateral radial acceleration component of maximum 1.96 m/s^2, equivalent to a gravito-inertial acceleration (GIA) vector tilt of 11.3 deg at the stimulated labyrinth. At the two highest test frequencies (0.5 and 1.0 Hz) the lateral acceleration of the chair movement (and the Coriolis acceleration) is of comparable magnitude to the centripetal acceleration, and was included in the calculation of OOR gain.

Three-dimensional eye movements were recorded with binocular video oculography.[5] Images of the left and right eyes were captured with two miniature video cameras mounted in a head-mounted mask assembly. Automatic image-processing algorithms yield a discretely sampled measure (25/s, resolution < 0.1 deg) of the horizontal, vertical, and torsional eye position in Fick coordinates. Numerical routines for correction of projection

---

[a]To whom correspondence may be addressed. Phone: +49 30 8445 2434; fax: +49 30 834 2116; e-mail: clarke@zedat.fu-berlin.de

error[6] were included in the algorithms employed. To facilitate torsional measurement, high-contrast artificial tincture landmarks were applied to the limbus. This guarantees a near ideal contrast profile for the polar correlation algorithm employed.

Examination of unilateral function was carried out with the trapezoidal function illustrated in FIGURE 1A. Frequency response and threshold testing were performed using a sinusoidal stimulus profile. For threshold estimation, the radial acceleration level was adjusted by modifying angular rate. During all tests the subject was in the dark and instructed to look straight ahead. The torsional component of the OOR was averaged over ten cycles of lateral translation and individually normalized using the root-mean-square value of the averaged response as factor.

FINDINGS AND DISCUSSION

Unilateral Function

Despite the small amplitude of the elicited unilateral OOR, in normal subjects ($N=11$) it proves to be highly conjugate, and symmetrical for stimulation of the left and right utricles. Unilateral OOR gain was found to have a median value of 0.09, compared to 0.16 for an equivalent bilateral, head-tilt (11.3 deg) stimulus.[4] In the tested patients ($N=5$) the pathological nature of the OOR is reflected clearly in the individual responses (FIG. 1B). These findings demonstrate the unilateral specificity of EDP testing of peripheral utricular function. Moreover, given that the patients were tested up to 15 months after discharge, the present test procedure facilitates recognition of the unilateral deficit, regardless of any CNS compensation.

The torsional component of the OOR observed during unilateral stimulation accords with current understanding of utricular function. During the plateau intervals of the stimulus cycle employed, the eccentric labyrinth is subjected to the radial acceleration, while the on-center labyrinth is free of any radial acceleratory stimulus. In the eccentrically displaced labyrinth, this would elicit an increase in discharge rate in those units spread across the midmedial area of the utricular macula of the eccentric labyrinth, where the hair cells respond maximally to ipsilaterally directed forces.[7] Concomitantly, the discharge rate of the cells on the on-center macula would merely return to their resting frequency, that is, rather than being sheared in the opposite (inhibitory) direction, as is the case during physiological bilateral stimulation.

In comparison to head-tilt or standard centrifugation approaches, the radial acceleration technique is unique in that it facilitates unilateral linear acceleratory stimulation to each of the utricles. This specificity offers considerable advantage in the examination of unilateral otolith (dys)function in both health and disease.

Frequency Response

The gain and phase relationships of the unilateral OOR in normal subjects ($N=10$) are presented in FIGURE 2A together with the results from the bilateral OOR study by Lichtenberg et al.[8] The apparent low-pass characteristic of the OOR over the measured range of 0.03–1.0 Hz may be compared to the neurophysiological recordings from the otolith afferent fibers reported by Fernández and Goldberg,[9] which show a near flat response over the frequency range of dc to 2.0 Hz. On the basis of this comparison, the existence of a leaky integrator mechanism for the OOR, analogous to the velocity storage mechanism for the canal-mediated angular VOR, as suggested by Anderson and Precht[10]

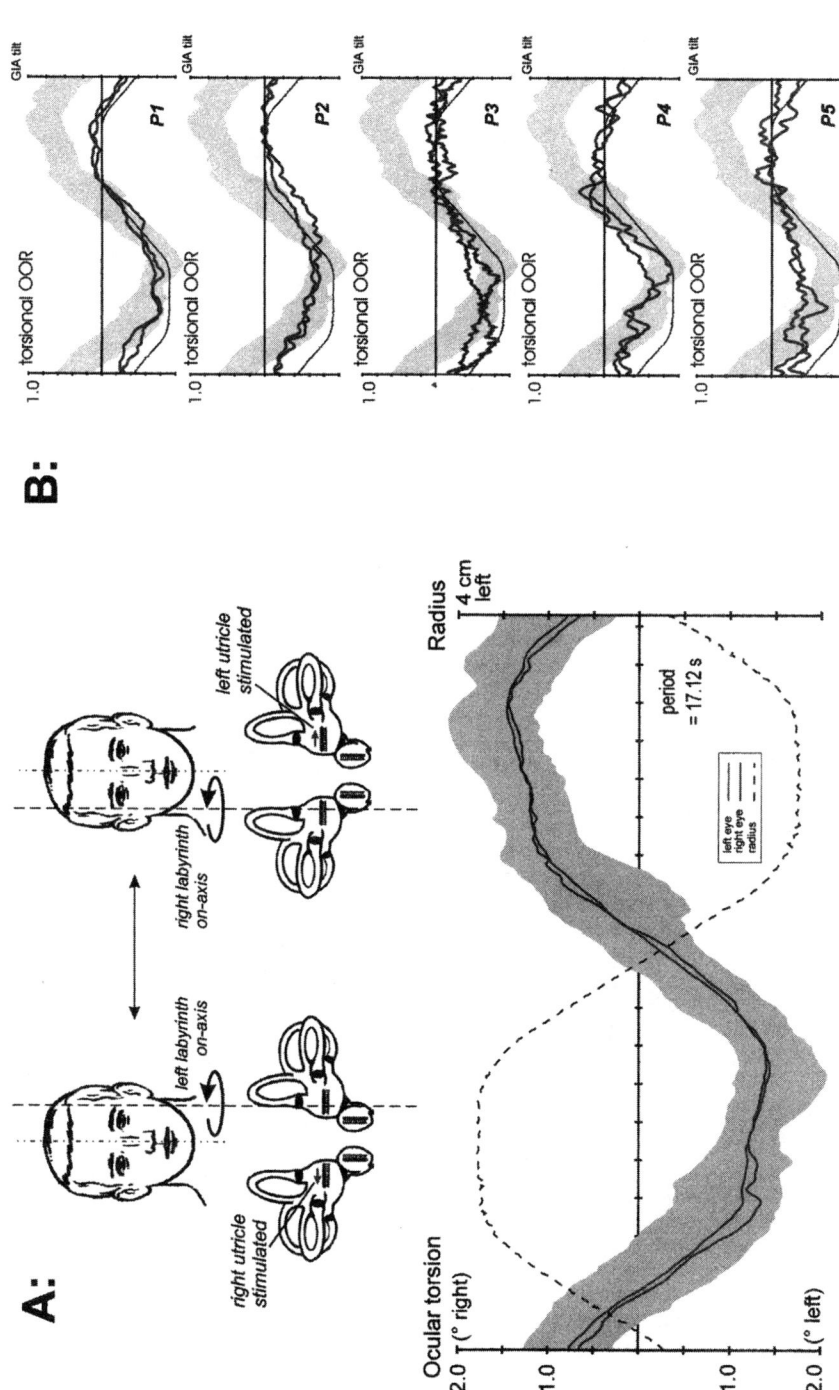

FIGURE 1. (A) Illustration of EDP stimulation, and the OOR waveform, represented by the median and 100% range for right and left eyes, over twelve normal subjects. Note that the Lichtenberg study involved bilateral stimulation (B) Torsional OOR response to EDP stimulation (individually normalized), as measured in five patients after unilateral deafferentation.

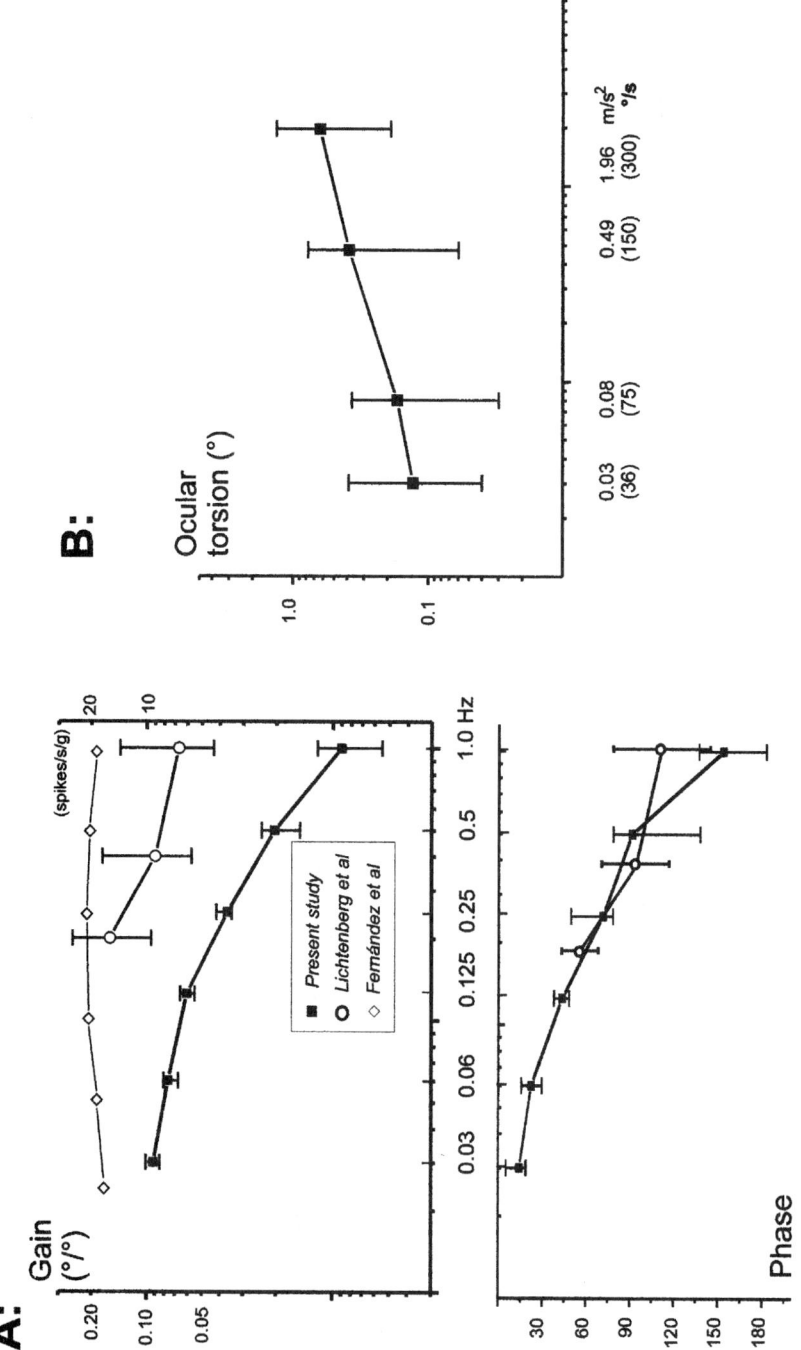

FIGURE 2 (**A**) Frequency/phase response for the human torsional OOR to oscillatory linear acceleration. The measured data (median and interquartile range) for response gain (*left axis*: deg OOR /deg GIA tilt) are presented together with those of the earlier human studies by Lichtenberg et al.[8] and of Fernández and Goldberg[9] on the peripheral neurons in the squirrel monkey (*right axis*: units in spikes/s/g). Note that the Lichtenberg study involved bilateral stimulation. (**B**) Data for threshold estimation; torsional OOR amplitude (median and interquartile range) is plotted against acceleration level, as generated at different angular rates.

may be considered. The low-pass characteristic of the OOR may also be related to the proposed low-pass filtering required by recent models for the discrimination of tilt and translation. It is also worthy of mention, however, that in comparison to tilt with respect to gravity where both utricular and saccular hair cells are stimulated, horizontal plane translation stimulates only the utricular cells.

Threshold

The subjective threshold for the perception of lateral acceleration has been estimated by Benson *et al.*[11] to be of the order of 0.06 m/s^2. For the 3.5-cm radius, as employed in the EDP stimulus, this would be generated by an angular velocity of 53 deg/s. The preliminary data on threshold estimation during sinusoidal stimulation at 0.125 Hz are shown in FIGURE 2B. The findings ($N = 5$) demonstrate a detectable OOR at acceleration levels as low as 0.03 m/s^2. Measurement at lower levels was restricted by the resolution of the video oculography system.

REFERENCES

1. SEIDMAN, S. H. & G. D. PAIGE. 1996. Perception and eye movements during low frequency centripetal acceleration. Ann. N.Y. Acad. Sci. **781**: 693–695.
2. WETZIG, J., M. REISER, E. MARTIN, N. BREGENZER & R. J. BAUMGARTEN. 1990. Unilateral centrifugation of the otoliths as a new method to determine bilateral asymmetries of the otolith apparatus in man. Acta Astronaut. **21**: 519–525.
3. CLARKE, A. H., A. ENGELHORN & H. SCHERER. 1996. Ocular counter-rolling in response to asymmetric radial acceleration. Acta Otolaryngol. (Stockh.) **116**: 652–656.
4. CLARKE, A. H. & A. ENGELHORN. 1998. Unilateral testing of utricular function. Exp. Brain Res. **121**: 457–464.
5. CLARKE, A. H., W. TEIWES & H. SCHERER. 1991. Video-oculography—An alternative method for measurement of three-dimensional eye movements. *In* Oculomotor Control and Cognitive Processes, D. Schmidt and D. Zambarbien, Eds.: 432–443. Elsevier. Amsterdam.
6. MOORE, S. T., T. HASLWANTER, I. S. CURTHOYS & S. T. SMITH. 1996. A geometric basis for measurement of three-dimensional eye position using polar cross-correlation. Vision Res. **36**: 445–459.
7. ROSENHALL, U. 1972. Vestibular macular mapping in man. Ann. Oto. Rhinol. Laryngol. **81**: 339–351.
8. LICHTENBERG, B. K., L. R. YOUNG & A. P. ARROTT. 1982. Human ocular counterrolling induced by varying linear acceleration. Exp. Brain Res. **48**: 127–136.
9. FERNÁNDEZ, C. & J. M. GOLDBERG. 1976. Physiology of peripheral neurons innervating otolith organs of the squirrel monkey. III Response dynamics. J. Neurophysiol. **39**(5): 996–1008.
10. ANDERSON, J. H. & W. PRECHT. 1979. Otolith responses of extraocular muscles during sinusoidal roll rotations. Brain Res. **160**: 150–154.
11. BENSON, A. J., B. A. SPENCER & J. R. R. SCOTT. 1986. Thresholds for the detection of the direction of whole body, linear movement in the horizontal plane. Aviat. Space Environ. Med. **11**: 1088–1096.

Saccular Dysfunction in Ménière's Patients

A Vestibular-Evoked Myogenic Potential Study

CATHERINE DE WAELE,[a,d] PATRICE TRAN BA HUY,[b]
JEAN-PIERRE DIARD,[c] GEORGES FREYSS,[b]
AND PIERRE-PAUL VIDAL[a]

[a]*Laboratoire de Neurobiologie des Réseaux Sensorimoteurs, ESA 7060, Paris, France*

[b]*Service ORL, Hôpital Lariboisière, Paris, France*

[c]*CPEMPN, Percy, France*

INTRODUCTION

Ménière's disease is a frequent inner ear pathology that induced episodic rotatory vertigo, low-frequency hearing loss, tinnitus, and aural fullness. It probably results from a dysfunction of the endolymphatic sac. The auditory function progressively deteriorates with time to become permanent in a few years. In contrast, the vestibular function often appears less impaired when tested between the vertigo attacks. However, this could be due to the fact vestibular function is usually assessed by means of caloric bithermal and rotatory tests that mainly explore the horizontal canalar function. In this study, we have investigated the sacculospinal functionality in Ménière's patients by monitoring the vestibular myogenic potentials (VEMPs) evoked by high-level clicks in the sternomastoid muscles (SCM). The VEMPs are now recognized as a reliable test to explore the saccular function: guinea-pigs' saccular afferents and vestibulospinal neurons of the lateral and of the descending vestibular nuclei were shown to respond to loud clicks.[5-7] This response induces the initial excitatory–inhibitory potentials (P13–N23) observed in the tonically contracted ipsilateral SCM when loud clicks are delivered unilaterally in the ear of normal subjects.[2-4] We have also tried to correlate the saccular deficit with the degree of hearing loss and canal paresis.

METHODS

Patients

Fifty-nine patients (34 males and 25 females) suffering from a unilateral Ménière's disease (26 left and 33 right) were included in this study. They ranged in age from 18 to 74

[d]Address for communication: Catherine de Waele, MD PHD, Laboratoire de Neurobiologie des réseaux sensori-moteurs, 15 rue de l'Ecole de Médecine, 75 270, Paris Cedex 06. Phone: 01 43 29 54 43; fax: 01 44 07 36 81; e-mail: cdw@CCR.Jussieu.fr

years, with an average age of 51.7 ± 13.5 years. All subjects had a typical history of Ménière's disease with unilateral hearing loss, tinnitus, and vertigo. They were all examined in the ENT department of the Lariboisiere Hospital by means of the classic vestibular clinical tests, the bithermal caloric test, a pure-tone audiometric test, and by VEMP testing. Patients presenting a spontaneous nystagmus were excluded from the study. In seven patients, the caloric testing was interrupted because of nausea and emesis. The endocochlear nature of the hearing loss was always confirmed on the BERA test. In particular, we checked that the latency of the fifth wave never exceeded 6 ms. Two variables were used as indicators of the degree of hearing loss: low-frequency (250–1000 Hz) and high-frequency (4–8k Hz) pure tone averages. Finally, patients were not selected if their pure tone audiometric test showed a significant conductive component associated with their perceived hearing loss. Indeed, such a conductive hearing loss caused an abolition of the VEMP, because clicks were not able to mechanically activate the saccule in these patients.

VEMPs Recordings

Each ear was stimulated twice in a row, that is, four trials per patient (two trials on the left ear and two trials on the right ear). The test was performed as follows: stimulation of the left ear twice and then stimulation of the right ear twice. For each of the four trials, the EMG responses were averaged over a series of 512 click stimuli. The clicks consisted of 0.1-ms rarefactive square waves of 100 dB HL, delivered by calibrated TDH 39 headphones. The clicks were delivered at a frequency of 6 Hz, and their amplitude was 145 dB SPL, since most of the subjects could not hear clicks softer than 45 dB. Surface EMG activity was recorded using skin electrodes located symmetrically on the upper half of each SCM. The reference surface and the ground electrode were located over the upper sternum and the central forehead, respectively. VEMP recordings were performed with a Nicolet Viking 4 with a 4-channel averaging capacity. The EMG from each side was amplified, bandpass filtered (10 Hz–1.6 kHz), and averaged using a 2.5-kHz sampling rate for each channel. Patients lay supine on a bed and were asked to raise their head straight ahead off the bed to activate their SCMs bilaterally and symmetrically. Simultaneous average EMG of both SCMs were collected starting at 20 ms before the clicks and ending 80 ms afterwards.

Data Analysis

The mean peak latency (in ms) and peak-to-peak amplitude (in µV) of the two early potentials (P13 and N23) of the VEMP were measured alone because they were previously shown to be of saccular origin.[1] The late VEMPs (N34 and P44 waves) were not studied, since they have been shown to be of cochlear origin.[1] The SCM EMG activity (RMS) was measured before the first trial (first left ear stimulation) and after the last trial (last right ear stimulation). Indeed, the P13–N23 peak-to-peak amplitude was previously shown to fluctuate with the SCM electromyographic amplitude.[8]

RESULTS

Only the latencies and amplitudes of the early VEMPs in the SCM ipsilateral to the stimulated ear (intact and affected one) were analyzed. The VEMPs evoked in the contralateral SCM were too variable in the control groups to be taken into consideration.[9]

The uncrossed saccular response evoked by the stimulation of the affected ear was abolished in 32 of the 59 patients (54%). In 27 (46%) of the 59 Ménière's patients, VEMPs were detected ipsilaterally following stimulation of the affected ear. The mean latency of the P13 potentials evoked by the stimulation of the affected ear in the ipsilateral SCM muscle was 11.3 ± 1.3 ms (min 8.9–max 14.5), and that of the N23 potential was 18.8 ± 2.0 ms (min 14–max 23.1), respectively. These values were not significantly different from those measured in the control subjects.[9] The mean P13–N23 peak-to-peak amplitude (left and right uncrossed SCM VEMP amplitude pulled together) in these 27 patients amounted to 63.8 ± 62.6 µV (min 11 µV–max 349 µV), which was not significantly different from the value recorded in the control group.[9] In 25 patients (42%), stimulation of the affected ear evoked no P13–N23 response in the ipsilateral SCM, whereas stimulation of the intact ear resulted in a normal ipsilateral P13–N23 response. In 7 patients (11%), there was no VEMP in the ipsilateral SCM following stimulation of either the affected or intact ears. Finally, in 2 patients (3%), there was no VEMP in response to the stimulation of the intact ear, whereas stimulation of the affected ear resulted in a detectable P13–N23 biphasic wave. The contralateral or bilateral absence of the VEMP in some Ménière's disease patients was confirmed by testing the subjects on another day.

Absence of Relationship Between VEMP and Canal Paresis

Canal paresis was only investigated in 52 of the 59 Ménière's patients because caloric testing had to be interrupted in 7 cases due to major neurovegetative signs. In the 30 (57.7%) patients with no VEMP following stimulation of the affected ear, the mean canal paresis was 32.8 ± 32.7%. The mean value was 17.0 ± 24.1% in the 22 patients (42.3%) who had intact VEMP following stimulation of the affected side. The difference was not significantly different ($p = 0.07$). Some patients with canal paresis equal to or above 60% displayed intact saccular responses.

Relationship Between VEMP and Hearing Loss

The low-frequency hearing loss of the affected ear was significantly greater in patients who did not exhibit ipsilateral VEMPs than in patients who had intact VEMPs ($p=0.02$). Following the stimulation of the affected ear, the 32 patients without VEMPs had a mean

---

FIGURE 1. (**Upper part**) Vestibular-evoked myogenic potentials recorded from the SCM of a subject suffering from a right unilateral Ménière's disease. *Traces 1 and 3* correspond to the evoked potentials obtained on the left (1) and on the right (3) SCM when 100-dB clicks were delivered to the left ear (A). *Traces 5 and 7* illustrate the evoked potentials obtained on the left (5) and right (7) SCM when 100-dB clicks were delivered to the right ear (B). Notice that these loud monaural clicks failed to evoke any early P13–N23 potentials on the right SCM when delivered on the right affected ear. In contrast, normal VEMP were observed in the left SCM muscle when loud clicks were delivered to the left intact ear. *Horizontal scale*: interval between two points: 10 ms. *Vertical scale:* interval between two points: 20 µV. (**Middle part**) Histogram representing the number of the Ménière's patients presenting (*black squares*) or not presenting VEMP (*gray stripes*) in the SCM ipsilateral to the stimulated ear according to the degree of canal paresis as assessed by bithermal caloric testing. (**Lower part**) Histogram representing the number of Ménière's patients presenting (*black squares*: VEMP+) or not presenting (*gray stripes*: VEMP-) ipsilateral VEMP in the SCM ipsilateral to the stimulated ear according to the mean (250–1000 Hz) low-frequency hearing loss. Note that VEMPs were absent in patients in whom low-frequency hearing loss exceeded 60 dB.

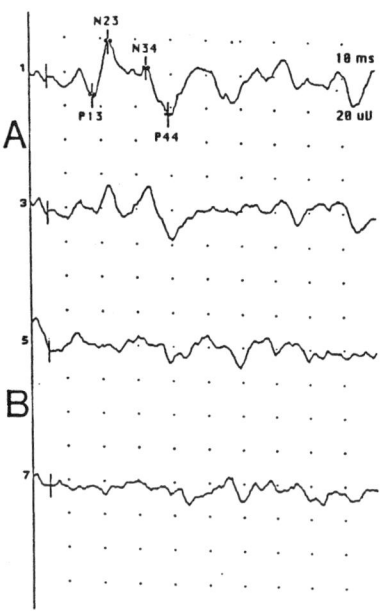
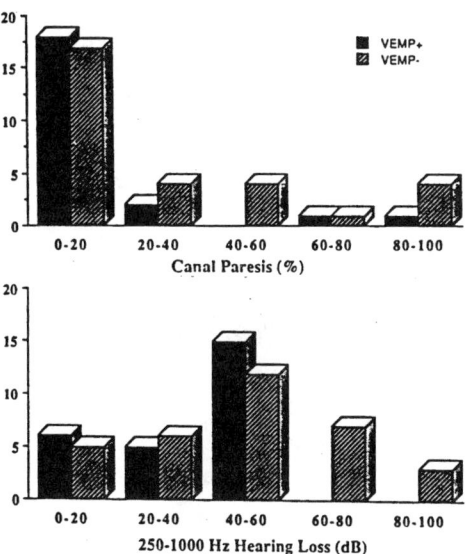

low-frequency hearing loss of 51.1 ± 20 dB, whereas the mean value was 39.4 ± 16.5 dB in the 27 patients who had normal VEMPs. Of the 49 patients with low-frequency hearing impairment ranging between 0 and 60 dB, the VEMP was intact in 26 (53%) and absent in 23 (47%). It was absent from the affected side in all patients with low-frequency hearing loss of more than 60 dB.

The mean high-frequency hearing loss of the 50 patients who presented normal, uncrossed saccular responses following the stimulation of the nonhydropic ear was 29 ± 20.9 dB. In these patients, early VEMP was detected on the intact ear even when the 4–8-kHz hearing loss was greater than 60 dB. The mean 4–8-kHz hearing impairment of the 27 patients with normal, uncrossed VEMP following the stimulation of the affected ear was 48.3 ± 18.7 dB in the affected ear. Four of these patients had a normal saccular response and a 4–8-kHz hearing loss of more than 60 dB on the affected side. In summary, low-frequency (250–1000 Hz), but not high-frequency (4–8 kHz) hearing loss correlated with the loss of VEMP. When the dynamic control of posture was tested by means of dynamic compturerized posturography in Ménière's patients that exhibited a saccular dysfunction, the tests showed that these latter patients were more prone to falls.[9]

CONCLUSION

This study showed that the sacculus may be dysfunctional in almost half of Ménière's patients. Therefore, the view that Ménière's disease poorly affects the vestibular system should be reconsidered. A recent 3-D analysis of the spontaneous nystagmus in Ménière's patients demonstrating an abnormal function of the semicircular canal supports this hypothesis.[10] In addition, endolymphatic hydrops caused more severe lesions to cochlear and saccular cells than to horizontal canalar cells, which appeared more resistant.

Important clinical implications can be derived from this study: (1) the saccular function of Ménière's patients should be systematically investigated using the VEMP; (2) if VEMPs are abolished, the dynamic equilibrium of these patients should be tested; (3) when they exhibit equilibrium loss during that examination, these patients should be considered at risk and sent for vestibular rehabilitation;[9] and (4) equilibrium loss is particularly important in elderly patients, because of the detrimental consequences of falls in that population.

ACKNOWLEDGMENTS

The authors thank Franck Zamith, Nelly Bellalimat, and Thérèse Dabbadie for their excellent technical assistance.

REFERENCES

1. COLEBATCH, J. G., G. M. HALMAGYI & N. F. SKUSE. 1994. Myogenic potentials generated by a click-evoked vestibulocollic reflex. J. Neurol. Neurosurg. Psychiatry **57**: 190–197.
2. DIDIER, A. & Y. CAZALS. 1989. Acoustic responses recorded from the saccular bundle on the eight nerve of the guinea pigs. Hear. Res. **37**: 123–128.
3. MCCUE, M. P. & J. J. GUINAN. 1994. Acoustically responsive fibers in the vestibular nerve of the cat. J. Neurosci. **14**: 6058–6070.
4. MCCUE, M. P. & J. J. GUINAN. 1995. Spontaneous activity and frequency selectivity of acoustically responsive vestibular afferents in the cat. J. Neurophysiol. **74**: 1563–1572.
5. MCCUE, M. P. & J. J. GUINAN. 1997. Sound-evoked activity in primary afferent neurons of a mammalian vestibular system. Am. J. Otol. **18**: 355–360.

6. MUROFUSHI, T., I. S. CURTHOYS & D. P. Gilchrist. 1995. Responses of guinea pig vestibular neurons to clicks. Exp. Brain Res. **111**: 149–152.
7. MUROFUSHI, T. & I. S. CURTHOYS. 1997. Physiological and anatomical study of click-sensitive primary vestibular afferents in the guinea pig. Acta Otolaryngol. (Stockh.) **117**: 66–72.
8. LIM, C. L., P. CLOUSTAN, G. SHEEAN, *et al*. 1995. The influence of voluntary EMG activity and click intensity on the vestibular click evoked myogenic potential. Muscle Nerve **18**: 1210–1213.
9. DE WAELE, C., P. TRAN BA HUY, J. P. DIARD, G. FREYSS & P. P. VIDAL. 1999. Saccular dysfunction in Meniereís disease. Am. J. Otol. **20**: 223–232.
10. TOSHIAKI, Y., O. YOSHIO, S. KAYO, K. ERIKO, *et al*. 1997. 3-D analysis of nyskagmus in peripherical vertigo. Act. Otolaryngol. (Stockh.) **117**: 135–138.

VOR Gain Modulation in the Monkey Due to Convergence of Otolith and Semicircular Canal Afferences During Eccentric Sinusoidal Rotation

L. FUHRY,[a,c] J. NEDVIDEK,[b] C. HABURCAKOVA,[a] S. GLASAUER,[a] G. BROZEK,[b] AND U. BÜTTNER[a]

[a]*Department of Neurology, Klinikum Großhadern, Ludwig-Maximilians-Universität München, Marchioninistr. 15, D-81377 München*

[b]*Department of Physiology, Czech Academy of Sciences, Prague, Czechoslovakia*

INTRODUCTION

Under natural conditions the semicircular canals, which are responsible for the detection of angular acceleration, and the otoliths, which detect linear acceleration, have to work together to achieve stabilization of gaze and stance. One can test the convergence of canal and otolith inputs by applying eccentric rotation to a subject. With this type of stimulation the angular acceleration remains stable independently of the distance of the subject from the center of rotation, thus providing a constant stimulus to the semicircular canals. The effect of the stimulation on the otoliths, however, depends on the distance and two types of linear acceleration: *centripetal* acceleration acts along the radius and depends on angular velocity; *tangential* acceleration is dependent on angular acceleration. To achieve variable tangential acceleration, one has to apply sinusoidal stimuli at different frequencies and amplitudes.

It has been shown that the gain of the vestibulo-ocular reflex (VOR) is significantly enhanced during eccentric sinusoidal rotation when compared to centric rotation.[1-4] This can be explained by the additional tangential acceleration acting along the interaural axis. VOR gain, however, also depends on target distance to be fully compensatory during eccentric rotation.[5]

In the present study eccentricity, frequency, and target distance were systematically varied, while the monkey was placed in two different orientations (nose-in and nose-out) with respect to the rotation axis to test whether the VOR gain is fully compensatory under all conditions.

[c]To whom correspondence may be addressed. Phone: +49/89/7095-3690; fax: +49/89/7095-3677; e-mail: fuhry@lrz.uni-muenchen.de

METHODS

Eye Movement Recordings

Eye movements in the rhesus monkey (*Macaca mulatta*) were recorded with scleral search coils, which were implanted into the left eye under sterile conditions. Surgical procedures have been described in detail elsewhere.[6]

Stimulation

The monkey was sinusoidally rotated centrally and eccentrically at different distances from the center of rotation (0–50 cm) with the nose facing inward or outward. Rotational frequencies ranged from 0.25 Hz to 1.43 Hz and amplitudes were 5–20 deg. Eye movements were recorded either during fixation of earth-stationary lit LED targets, which were located 12–180 cm in front of the monkey's eye in otherwise complete darkness, or in the light when the monkey looked at a cylindrical projection wall with a diameter of 283 cm.

RESULTS

In the light, using a cylindrical projection-wall VOR gain modulation derived from different experimental sessions and different stimulus paradigms, the data almost perfectly aligns with the theoretically required values for target fixation, indicating complete compensation (FIG. 1). Note that the ideal line is curved since target distance depends on eccentricity in this paradigm.

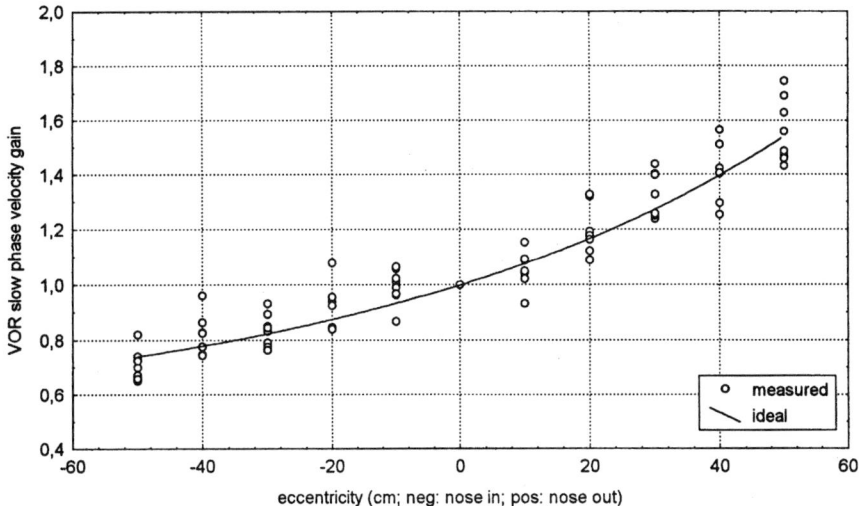

FIGURE 1. VOR slow phase-velocity gain modulation derived from different experimental sessions ($n = 8$) and different stimulus paradigms (0.25–1.43 Hz) in the light using a cylindrical projection wall with a diameter of 283 cm (*circles*). VOR gain during centric rotation was taken as reference value. These data almost perfectly align with the calculated values (*line*), indicating nearly complete compensation. Note that the ideal line is curved, as target distance depends on eccentricity in this paradigm.

The effect of target distance on the VOR gain modulation was systematically investigated with stationary lit LED targets in otherwise complete darkness (FIG. 2). Based on *geometric considerations*, only small VOR gain modulation is expected during centric rotation. The small expected deviation from gain = 1 for near targets is due to the laterality of the eyes, which is 1.6 cm. This predicted pattern of almost absent gain modulation could be confirmed *experimentally* (middle row in FIG. 2). During eccentric sinusoidal rotation VOR gain enhancement strongly depends on target distance. In the nose-out condition target distance and gain enhancement are inversely related (back row in FIG. 2). When the nose is facing inward, VOR gain decrement is expected from *simulation*, while

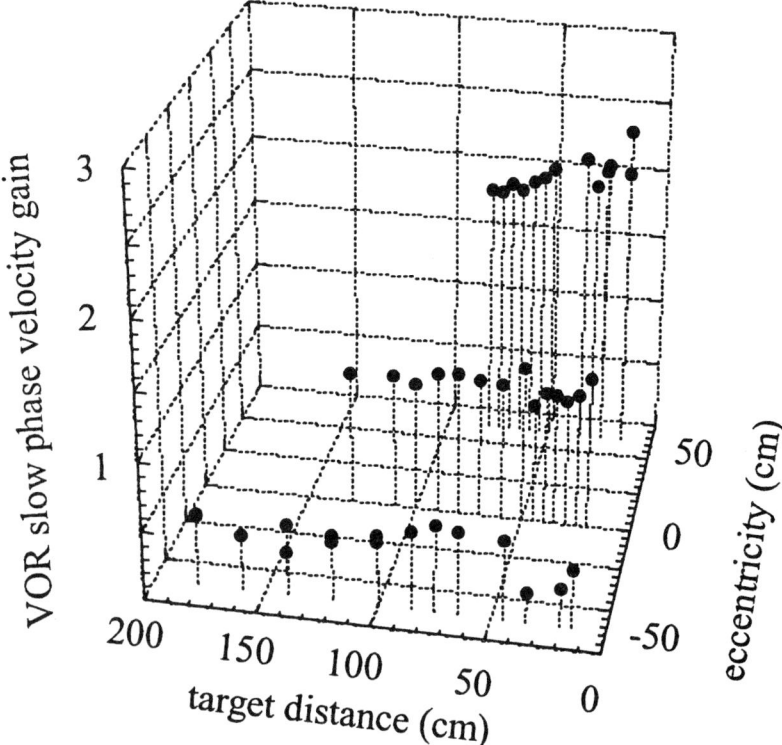

FIGURE 2. Gain (independent of sign) of VOR slow phase velocity for 3 positions of the monkey with respect to the center of rotation (centric, 50-cm eccentric nose facing inward and outward). Eye movements were recorded during fixation of earth-stationary lit LED targets that were located 12–180 cm in front of the monkey's eye in otherwise complete darkness while the monkey was sinusoidally rotated at 1.43 Hz with an amplitude of 5 deg. As expected, there was hardly any gain modulation during centric rotation (*middle row*). During eccentric sinusoidal rotation target distance and gain enhancement are inversely related in the nose-out condition (*back row*). When the nose is facing inward, VOR gain decrement is obtained as predicted from simulation (*front row*). With the target on the axis of rotation the VOR has to be suppressed to be fully compensatory ("inverse VOR suppression"). Strongest gain decrement is found for targets around the axis of rotation, though VOR gain is not zero (*front row, third data point from the right*; VOR gain: 0.22). For targets between the monkey and axis of rotation, the VOR gain increases when compared to centric targets (*front row, first two data points from the right*). Phases of VOR, however, are shifted by approximately 180 deg in this situation (not displayed here).

the target is behind the axis of rotation. With the target on the axis of rotation the VOR has to be suppressed to be fully compensatory ("inverse VOR suppression"). When the target is between the monkey and the axis of rotation, VOR should be inverted, that is, the VOR gain rises again and the phase of the VOR response is inverted. Our *experimental data* demonstrate the expected VOR gain decrement for targets that are behind the axis of rotation (front row in FIG. 2). This decrement is strongest for targets around axis of rotation, though VOR gain is not zero as expected (front row in FIG. 2, third data point from the right). The VOR gain increases for targets between the monkey and axis of rotation when compared to centric targets (front row in FIG. 2, first two data points from the right). Phases of VOR responses are not displayed here.

DISCUSSION

Our preliminary data demonstrate that VOR gain modulation during eccentric rotation is almost perfectly compensatory for many paradigms, even at frequencies above 1.0 Hz when smooth-pursuit mechanisms do not play a major role. This suggests that otolith contribution is a major factor for many VOR conditions. There are some paradigms that demonstrate this in particular: (1) VOR gain enhancement for near targets during eccentric rotation with the nose facing outward; (2) zero VOR for targets on the center of rotation ("inverse VOR suppression"); and (3) VOR reversal for targets between the monkey and the center of rotation when the center of rotation is in front of the monkey. Up to now we have not been able to find zero VOR gain during "inverse VOR suppression," although there was a clear tendency to obtain the smallest gain in this situation. This may at least in part be due to the fact that zero gain is hard to distinguish from episodes in which the monkey was drowsy, and so were removed from analysis. Further experiments are needed to solve this problem. As predicted from simulation, VOR gain obtained for targets between the monkey and the center of rotation was increased when compared to the "inverse VOR suppression" paradigm.

Taken together, our data demonstrate that the predicted behavior of VOR gain modulation can be shown for all tested eccentricities and target distances except for centric targets over a wide range of frequencies.

ACKNOWLEDGMENT

This work was supported by DFG (Fu 291/2-1) and GACR (309/97/0556).

REFERENCES

1. GRESTY, M. A., A. M. BRONSTEIN & H. J. BARRATT. 1987. Eye movement responses to combined linear and angular head movement. Exp. Brain Res. **65**: 377–384.
2. SNYDER, L. H. & W. M. KING. 1992. Effect of viewing distance and location of the axis of head rotation on the monkey's vestibuloocular reflex. I. Eye movement responses. J. Neurophysiol. **67**: 861–874.
3. MCCONVILLE, K. M., R. D. TOMLINSON & E.-Q. NA. 1996. Behavior of eye-movement-related cells in the vestibular nuclei during combined rotational and translational stimuli. J. Neurophysiol. **76**: 3136–3148.
4. TELFORD, L., S. H. SEIDMAN & G. D. PAIGE. 1998. Canal-otolith interactions in the squirrel monkey vestibulo-ocular reflex and the influence of fixation distance. Exp. Brain Res. **118**: 115–125.
5. VIIRRE, E., D. TWEED, K. MILNER & T. VILIS. 1986. A reexamination of the gain of the vestibuloocular reflex. J. Neurophysiol. **56**: 439–450.
6. BOYLE, R., U. BÜTTNER & G. MARKERT. 1985. Vestibular nuclei acivity and eye movements in the alert monkey during sinusoidal optokinetic stimulation. Exp. Brain Res. **57**: 362–369.

An Alternative Approach to the Central Processing of Canal and Otolith Signals

ANDREA M. GREEN[a] AND HENRIETTA L. GALIANA

Department of Biomedical Engineering, McGill University, 3775 University Street, Montreal, Quebec, Canada H3A 2B4

INTRODUCTION

The translational and rotational vestibulo-ocular reflexes (TVOR and RVOR) share a common goal to stabilize visual targets on the retinas during head movement. These reflexes differ significantly in their dynamic characteristics at both the sensory and motor levels, however, implying a requirement for different central processing of canal and otolith signals. In the frequency range of natural movements, the activity of semicircular-canal afferents encodes angular head velocity. In contrast, primary otolith afferents modulate their activities in phase with linear head acceleration, suggesting that an additional "integration" of otolith signals is required for the TVOR. Behaviorally, however, the RVOR is compensatory down to about 0.01 Hz, while the TVOR only exhibits a robust response above 0.5 Hz.[1,2] Hence, two simple integrations of otolith signals are not sufficient to describe behavioral observations. All current hypotheses for the central processing of the TVOR suggest that otolith signals undergo a preliminary filtering process, followed by a neural integration that is shared with the RVOR.[2,3,4] We propose an alternative scheme for the processing of vestibular signals that does not require additional *central* low-pass filters for the TVOR.[5]

DYNAMIC PROCESSING STRATEGY

The proposed strategy is illustrated using the simple unilateral structure shown in FIGURE 1. Reciprocal connections between the vestibular nucleus (VN) and the prepositus hypoglossi (PH) form a positive feedback loop that acts as a distributed oculomotor integrator. Notice that premotor vestibular neurons (PVN) receive head rotation signals ($\dot{H}_{ang}$) directly from the canal afferents, while linear acceleration information ($\ddot{H}_{lin}$) is conveyed from the otolith organs only indirectly via the PH and a nonpremotor VN cell. The implication of differential projection sites for the sensory signals is illustrated by considering the following expression relating conjugate ocular deviations to vestibular stimuli in the dark (see FIG. 1)

$$E(s) = -a\text{PVN}(s)P(s) = -\frac{paGK_p}{(T_i s + 1)}\left[\frac{(T_f s + 1)}{(T_p s + 1)}\right]\frac{T_e s}{T_c s + 1}\dot{H}_{ang}(s) - \frac{gabGK_f K_p}{(T_i s + 1)(T_p s + 1)}\ddot{H}_{lin}(s),$$

[a] To whom correspondence may be addressed. Phone: 514/398-6734; fax: 514-398-7461; e-mail: andrea@eyebeam.biomed.mcgill.ca

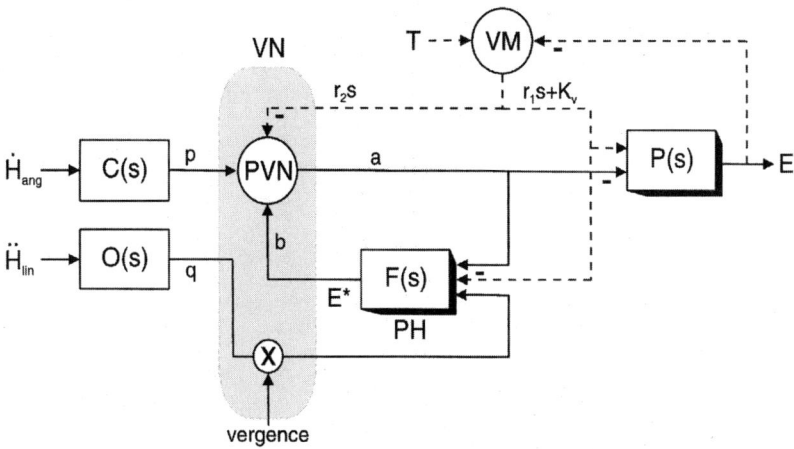

FIGURE 1: Schematic of model used to explore shared central processing of canal and otolith signals for movements in the horizontal plane. *Circular nodes* represent summing junctions, while *boxes* represent dynamic elements or filters. Parameter labels on connection lines indicate the strength of a projection. *Dashed pathways* are activated by a visual or cognitive goal. Areas labeled VN (*shaded*), PH, and VM represent the vestibular nuclei, prepositus hypoglossi, and visuomotor areas, respectively. PVN: a lumped premotor cell in the VN representing the combined responses of position–vestibular–pause (PVP), eye–head–velocity (EHV), and burst–tonic (BT) neurons.[6] The neural filter in PH, $F(s) = K_f/(T_f s+1)$, is proposed to represent a scaled internal model of the eye plant, $P(s) = K_p/(T_p s+1)$, (i.e., $T_f = T_p$). PH output neurons provide an internal estimate, E^*, of eye position, E, under most conditions. Stimuli are angular head velocity ($\dot{H}_{ang}$) sensed by the semicircular canals, $C(s)$, and linear head acceleration ($\ddot{H}_{lin}$) sensed by the otolith organs, $O(s)$. T describes conjugate target position relative to the head. At frequencies below ≈ 10 Hz, semicircular canals are high-pass filters of angular head velocity ($C(s) = T_c s/(T_c s+1)$), while otolith afferents carry signals approximately in phase with linear head acceleration ($O(s) = 1$). The strength of otolith projections (gain q) is presumed modulated by vergence state to simulate changes in TVOR with target distance. Model parameters used in simulations: $a = 0.244$; $b = 1.68$; $p = 1$; $q = 2.5$; $K_f = 2.4$; $K_p = 1$; $K_v = 2.51$; $r_1 = -0.1$; $r_2 = 0.51$; $T_c = 5$; $T_f = 0.28$; $T_p = 0.28$.

where $G = 1/(1-abK_f)$. The time constant $T_i = T_f/(1-abK_f) \gg T_p$ and provides a central integration of sensory signals. When the PH filter, $F(s)$, is an internal model of the eye plant, $P(s)$, (i.e., $T_f = T_p$), the zero associated with the angular component of the ocular response cancels the eye-plant pole (term in square brackets). However, no such cancellation occurs in the translational component. Thus, compensation for eye-plant dynamics is only provided for the RVOR. As a result, canal afferent signals are integrated only once by the central neural integrator, while otolith afferent signals are filtered a second time by the eye plant.

RESULTS

Behavioral Responses

In FIGURE 2, the simulated RVOR exhibits a gain of 0.87 in the dark with compensatory phase at both frequencies. In contrast, the TVOR response is low at 0.2 Hz with a large phase lead (≈70 deg) relative to the ideal compensatory response, but rises significantly to a gain of 0.31 deg/cm/MA at 4 Hz with modulation in phase with head velocity.[1,2] In the

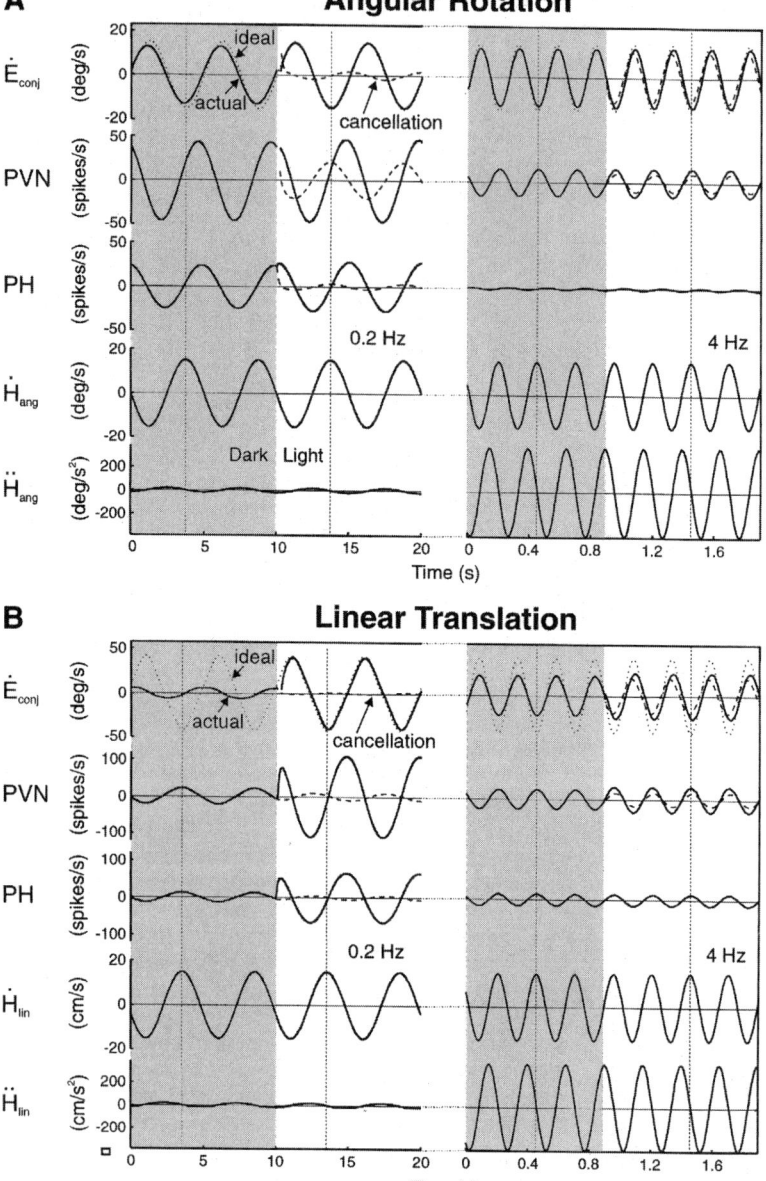

FIGURE 2: Model simulations of behavioral and central responses during (**A**) pure angular rotation while viewing a far target, and (**B**) interaural translation while viewing a target 20 cm away (17 deg vergence for an interocular distance of 6 cm). Stimuli are provided at 0.2 Hz and 4 Hz in both dark (*shaded*) and light conditions. Responses while viewing an earth-fixed target (*solid lines*) are plotted superimposed on those during head-fixed target viewing (*dashed lines*). *Dotted curves* represent ideal compensatory responses. Simulations were performed with SIMULINK (Mathworks, MA) at 100 Hz, using the parameters given in the FIGURE 1 caption.

presence of visual feedback, RVOR and TVOR responses are nearly ideal during both earth-fixed and head-fixed target viewing at the lower frequency, but performance deteriorates beyond the pursuit bandwidth (≈ 1 Hz) as observed.

Central Responses

During pure head rotation, the PVN cell lags ipsilaterally directed head velocity at lower frequencies reflecting its lumped premotor nature, but modulates closely in phase with head velocity at high frequency. PVN sensitivity is reduced but still significant during RVOR cancellation. PH neurons at the output of $F(s)$ code for contralaterally directed eye position under all conditions during a pure head rotation. During the TVOR, the PVN cells modulate in phase with translational head *velocity* across all frequencies in the dark and at high frequency in the light. Hence, otolith signals appear to have been integrated relative to the sensory acceleration signal by the time they reach the PVN.[7,8] In contrast to the robust activity observed during RVOR cancellation, PVN cell responses during TVOR cancellation are very small at low frequency where cancellation performance is good. The PH neuron provides an accurate estimate of eye position at low frequency in keeping with its postulated role as an efference copy signal. However, during translation at 4 Hz the PH cell modulates more closely in phase with eye *velocity*.

CONCLUSIONS

The general behavioral characteristics of the TVOR and RVOR may be simulated using a single shared central processor without requiring additional *central* low-pass filters for the processing of otolith signals. The proposed approach offers several testable predictions: (1) the frequency associated with the second filtering of the otolith signal should correspond well with the dominant pole of the eye plant, and TVOR behavior is expected to reflect higher-order plant dynamics; (2) during TVOR cancellation PVN neurons should exhibit little activity at frequencies where cancellation performance is robust; (3) position-sensitive PH cells are predicted to exhibit increasing phase leads at high frequencies during translation such that modulation is more closely in phase with eye *velocity* .

REFERENCES

1. PAIGE, G. D., G. R. BARNES, L. TELFORD & S. H. SEIDMAN. 1996. Ann. N.Y. Acad. Sci. **781**: 322–331.
2. TELFORD, L., S. H. SEIDMAN & G. D. PAIGE. 1997. J. Neurophysiol. **78**: 1775–1790.
3. ANGELAKI, D. E. & B. J. M. HESS. 1996. Ann. N.Y. Acad. Sci. **781**: 332–347.
4. RAPHAN, T., S. WEARNE & B. COHEN. 1996. Ann. N.Y. Acad. Sci. **781**: 348–363.
5. GREEN, A. M. & H. L. GALIANA. 1998. J. Neurophysiol. **80**: 2222–2228.
6. SCUDDER, C. A. & A. F. FUCHS. 1992. J. Neurophysiol. **68**: 244–264.
7. MCCONVILLE, K. M. V., R. D. TOMLINSON & E.-Q. NA. 1996. J. Neurophysiol. **76**: 3136–3148.
8. CHEN-HUANG, C. & R. A. MCCREA. 1998. J. Vest. Res. **8**: 175–184.

Otolith Signal Processing and Motion Sickness

ERIC GROEN,[a] JELTE BOS, BERND DE GRAAF, AND
WILLEM BLES

*TNO Human Factors Research Institute, Kampweg 5, 3769 DE
Soesterberg, The Netherlands*

The otolith organs register the resultant gravito–inertial force (GIF), consisting of the vector sum of all linear accelerations acting on the head, including gravity. In many cases, a low-pass filter is appropriate to characterize the neural processing by which the subjective vertical (SV) is constructed from the GIF. The time constant of this low-pass filter is generally assumed to be about 20 s, based on measurements of the oculogravic illusion during centrifugation.[1] In contrast with this, we recently observed a much shorter time constant when we measured the somatogravic illusion in a centrifuge.[2] Five subjects were asked to continuously align a tactile rod with the direction of the perceived vertical, while the centrifuge (with an arm of 5.6 m) was accelerated with 10 deg/s^2 up to a constant level of 0.5 G. The group's mean time constant by which the rod was lined up with the GIF amounted only to 5 s. What then is the best value to represent the time constant of the neural filtering of the otolith signals? Here, we will provide an argument in favor of the shorter time constant, based on model simulations predicting motion sickness during linear oscillations along the vertical longitudinal axis (heave motion).

The model is based on the idea that motion sickness generally occurs when a correct sense of gravity is compromised by the motion stimulus: subjects only develop motion sickness when their SV is at stake.[3] Taking this role of the SV into account, the sensory rearrangement theory of Reason and Brand[4] can be refined to: "Motion sickness arises when the sensed vertical as determined on the basis of integrated information from the the eyes, the vestibular system, and the non-vestibular proprioceptors is at variance with the expected vertical as predicted on the basis of previous experience."

To operationalize this so-called "Subjective Vertical conflict theory" we extended the mathematical model of Oman[5] on motion sickness. The essence of the extended model is shown in FIGURE 1. The gravity vector (or sensed vertical, $\mathbf{v}_{sens}$) is explicitly determined from the total set of sensory information, including linear and angular motion information from the otoliths and the semicircular canals, as well as orientation information from the visual system. In the simple case of one-dimensional heave motion, the model only requires low-pass filtering of the otolith signals. Analogous to Oman's original model, an internal model has been implemented to obtain an optimal estimate of the gravity vector, or expected vertical ($\mathbf{v}_{exp}$). Because the internal model uses the same low-pass filter as the sensory path, $\mathbf{v}_{exp}$ will generally be the same as $\mathbf{v}_{sens}$ during normal locomotion. During external motion, however, the two vectors may differ because the motion will be input to the sensory path, but not to the internal model.[3] According to the SV-conflict theory, it is the difference, or conflict, between $\mathbf{v}_{exp}$ and $\mathbf{v}_{sens}$ that correlates with motion sickness. It

---

[a]To whom correspondence may be addressed. Phone: +31 346 356371; fax: +31 346 353977; e-mail: groen@tm.tno.nl

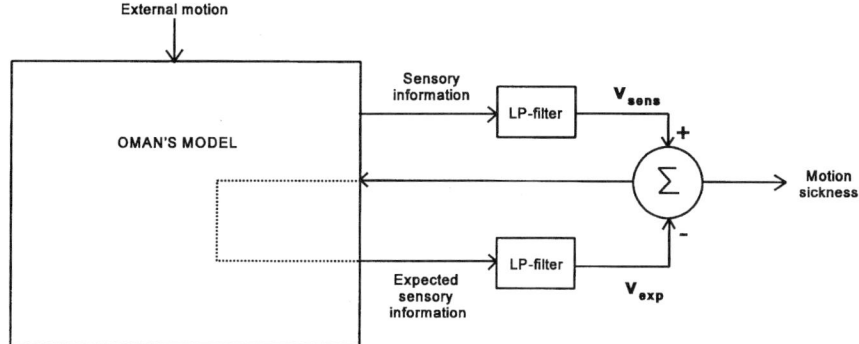

FIGURE 1. Schematic representation of the subjective vertical conflict model. The model of Oman[5] has been extended by modules to calculate the sensed vertical (v_{sens}) from the sensory information (*upper path*). In the case of linear motion (*upright*), the neural processing of the otolith signals can be appropriately simulated by a low-pass filter. In parallel, the same calculations are performed in an internal model (*lower path*) to derive the *expected* vertical (v_{exp}) from the *expected* sensory information. The difference (or conflict vector) between v_{sens} and v_{exp} is assumed to correlate with motion sickness. This conflict vector is also used for feedback into the internal model in order to update v_{exp} (indicated by the *dotted line*).

is interesting to note that v_{exp} can be updated by feedback of the conflict vector into the internal model, which, in the case of linear motion, may induce an extra time lag, since it places the low-pass filter of the internal model in series with the low-pass filter of the sensory path. As a result, the conflict vector can be expected to reach a maximum at a certain stimulus frequency, which is primarily determined by the time constant of the low-pass filter, and to a lesser extent by the weighting of the internal feedback.

We used the time constant of 5 s to simulate the conflict vector for oscillatory heave motion for frequencies up to 1 Hz. The results were compared with a study of O'Hanlon and McCauley,[6] who investigated the motion sickness incidence (MSI) in more than 500 subjects during pure heave motion. Their data showed a maximum of the MSI around a frequency of 0.16 Hz. To make the model's output directly comparable with MSI, the conflict vector was scaled into a predicted MSI value by means of a nonlinear normalization Hill-function, followed by a second-order leaky integrator accounting for accumulation in time. The upper limit of MSI was set to 80% in accordance to the results of O'Hanlon and McCauley. FIGURE 2 shows that the predicted MSI curve closely resembles the experimental data, and most importantly, the predicted maximum of MSI coincides with the observed maximum. It should be emphasized again that the location of this maximum is not affected by the amplitude scaling afterwards. It depends predominantly on the difference between v_{exp} and v_{sens}, and to a smaller degree on the internal feedback weighting.

In conclusion, using a time constant of 5 s the maximum MSI predicted by the model is in agreement with data from the literature. Thus it seems that the time constant for the determination of the SV from the otolith signals is shorter than suggested by the time course of the oculogravic illusion. The time constant of 5 s should not be considered too precise, since the model's output can be fine-tuned by manipulating the feedback weighting. It is our impression, however, that the data of O'Hanlon and McCauley can be simulated even better when using a time constant of *less* than 5 s. A value of 20 s is clearly too high for the SV-conflict model. It remains unclear why the time constant differs so much between the oculogravic and the somatogravic illusion, especially because the measurements discussed here were performed under approximately the same conditions.[1,2] Presumably, the difference is inherent in the different sensorimotor functions tested.

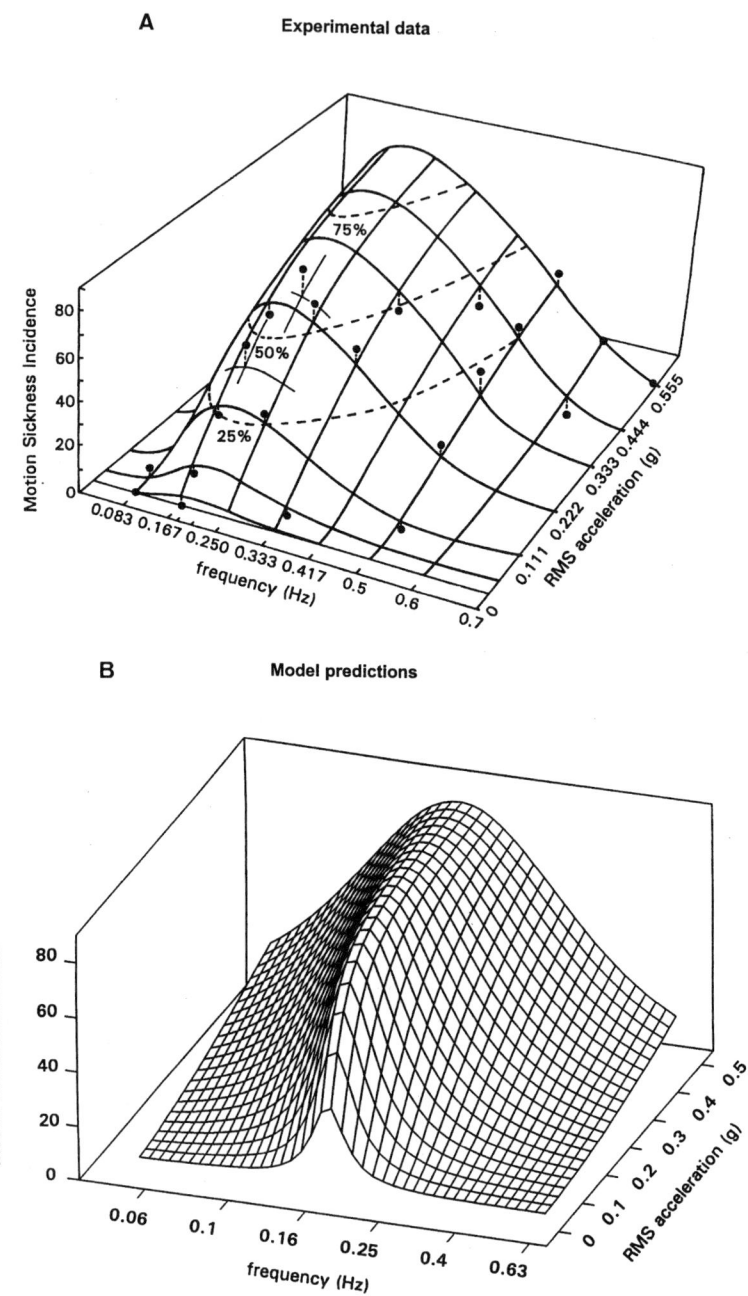

FIGURE 2. Comparison between experimental and predicted data on the motion sickness incidence (MSI) for heave motion. (**A**) Curve-fitted experimental data (adapted from O'Hanlon and McCauley[6]); (**B**) the model predictions. MSI is plotted as a function of the frequency and the amplitude of vertical oscillation. Both curves show a maximum at 0.16 Hz.

REFERENCES

1. STOCKWELL, C. W. & F. E. GUEDRY. 1970. The effect of semicircular canal stimulation during tilting on the subsequent perception of the visual vertical. Acta Otolaryngol. (Stockh.) **70**: 170–175.
2. DE GRAAF, B., J. E. BOS, W. TIELEMANS, F. RAMECKERS, A. H. RUPERT & F. E. GUEDRY. 1996. Otolith contribution to ocular torsion and spatial orientation during acceleration. Naval Aerospace Medical Research Laboratory, Tech. Memo. 96-3, Pensacola, Fla.
3. BLES, W., J. E. BOS, B. DE GRAAF, E. GROEN & A. H. WERTHEIM. 1998. Motion sickness: only one provocative conflict? TNO Rep. TM-98-A033. TNO Human Factors Research Institute, Soesterberg, The Netherlands.
4. BOS, J. E. & W. BLES. 1998. Modelling motion sickness and subjective vertical mismatch detailed for vertical motions. Brain Res. Bull. **47**: 537–542.
5. OMAN, C. M. 1982. A heuristic mathematical model for the dynamics of sensory conflict and motion sickness. Acta Otolaryngol. (Stockh.) Suppl.: **392**.
6. O'HANLON, J. F. & M. E. MCCAULEY. 1974. Motion sickness incidence as a function of the frequency and acceleration of vertical sinusoidal motion. Aerospace Med. **45**: 366–369.

Otolith–Canal Interaction During Pitch While Rotating

T. HASLWANTER,[a] R, JAEGER, AND M. FETTER

Department of Neurology, University Hospital Tübingen, Hoppe-Seyler-Str. 3, D-72076, Tübingen, Germany

INTRODUCTION

Natural movements of the head in space consist of combined rotations and translations, even during such simple tasks as walking around a corner. Investigations of the respective contributions of the vestibular sensors involved, the otoliths and the semicircular canals, rely on paradigms that allow the simultaneous stimulation of both sensory systems. While most paradigms only induce a static or temporary stimulus to the otoliths or the canals, pitch while rotating (PWR) leads to a continuous stimulation of both systems. It is well known that in monkeys such a stimulation pattern leads to eye movements that are compensatory to the continuous rotation of the body in space.[1,2] But to date no experiments have been conducted to investigate human responses to such complex canal–otoliths stimulation. We have therefore recorded 3-dimensional eye movements in human subjects who actively oscillated their head in the pitch plane during a continuous rotation about an earth-vertical axis.

METHODS

Subjects were seated on a turntable and were firmly secured to the chair. At the beginning of the experiment the head was oriented such that Reid's line was oriented about 10 deg nose up. Three-dimensional eye movements were recorded with the dual search coil technique, and were sampled at 100 Hz. We tested 11 healthy subjects (age 24 ± 2.4a) during active head pitching while rotating about an earth-vertical axis. Before the beginning of the PWR experiments the pitching movement of the head was practiced at 1/3 Hz and 2/3 Hz, with an amplitude of ± 20 deg, while the subject was looking straight ahead at an earth-fixed target. The head movement was restricted by a helmet, which was mounted such that the subject could move the head around a horizontal axis located about 2 cm behind and above the intervestibular line. The speed of the sinusoidal movement was indicated by a metronome. Then the sphere was closed, all lights were turned off, and the subject was accelerated in complete darkness to a constant velocity of 100 deg/s. After 2 min of constant-velocity rotation the subject was instructed to pitch the head sinusoidally and keep on looking straight ahead. No visual target was presented. During this head pitching the metronome was adjusted to 1/3 Hz, and the clicks of the metronome were

---

[a]To whom correspondence may be addressed. Phone: +41-1-255-5550; fax: +41-1-255-4507; e-mail: haslwant@neurol.unizh.ch

transmitted to the subject through an intercom system. After 40 s of head pitching the subject kept the head stationary for 1 min while the turntable continued to rotate at 100 deg/s. Then the head pitching was repeated for 40 s with a frequency of 2/3 Hz. After 2 more minutes of constant velocity rotation with the head stationary the subject was decelerated. When the postrotatory nystagmus had died down the whole experiment was repeated in the opposite direction.

From the recorded data the 3-dimensional angular eye velocity was calculated,[3] and the fast phases of the nystagmus were removed.[4] The data analysis was restricted to the horizontal and torsional velocity offsets, as well as the corresponding modulation amplitudes. These were determined by fitting the function

$$\text{eye\_vel} = \text{offset} + \text{amplitude} * \sin(2\pi\nu t + \Delta\phi)$$

to a hand-selected data interval of the eye velocity traces (ν is the frequency of the head pitching). For each subject the values for rotations to the right and the left were averaged such that the offset values correspond to a rotation to the left. The amplitudes of the active, approximately sinusoidal vertical head movements were highly variable, and the execution of the paradigm also showed substantial intersubject variability: while 8 of the 10 subjects kept the eyes approximately stable in space (modulation 1–7 deg), two subjects made eye-in-space movements that were larger than the head movements. Thus the vertical data traces were not analyzed, and no phase relationships were determined for the horizontal and torsional data.

RESULTS

A set of experimental data is shown in FIGURE 1. Over all subjects, the head pitching had an amplitude of 13 deg ± 3 deg. While there was no offset in the torsional component, the horizontal eye velocities showed for both frequencies of head pitching a small but significant offset (1.7 ± 1.3 deg/s at 1/3 Hz, and 1.0 ± 1.1 deg/s at 2/3 Hz). The offset was directed such that it was opposite to the sustained movement of the body in space.

The eye-velocity modulation had the same frequency as the head pitching. Its amplitude increased for the torsional component significantly with the frequency (from 7.9 ± 2.3 deg/s at 1/3 Hz, to 10.8 ± 4.5 deg/s at 2/3 Hz), but stayed approximately constant for the horizontal component (from 4.3 ± 1.6 deg/s to 4.4 ± 2.1 deg/s). There was a significant correlation between the amplitude of the head movements and the magnitude of the torsional eye-velocity modulation ($r = 0.78$).

MODELS

To better understand the task faced by the central nervous system in determining the movement in space from the dynamic canal and otoliths signals, we simulated the stimulation of the canals during PWR. Computational details of the calculation, which consider the mechanical properties of the canals, have been described elsewhere.[5]

FIGURE 2 indicates the complexity of the task: simply changing the frequency of the head pitching in FIGURE 2 by a factor of 10 changes not only the relative magnitude of the stimulation of the canals but also dramatically affects their phase relationships.

We also tested to what extent a model of canal–otolith interaction, which was based on ideas from Merfeld,[6] and which we developed to predict three-dimensional eye movements during OVAR, would be able to reproduce the observed eye velocities. This model reproduced the main features seen in the recorded data: modulations in all three velocity

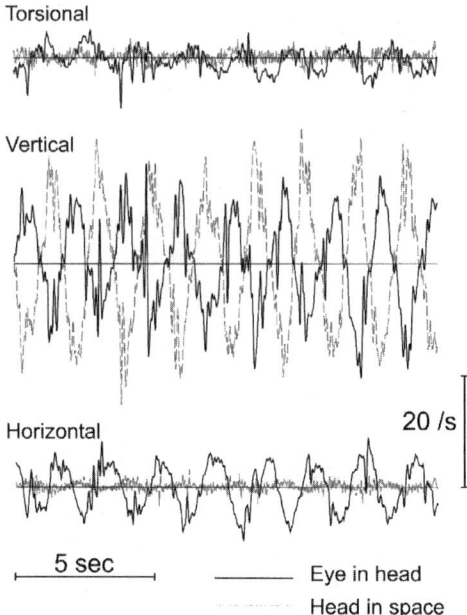

FIGURE 1. Torsional, vertical, and horizontal eye- and head-velocity components during pitch while rotating (PWR). The rotational velocity was 100 deg/s, and the pitching was executed with 2/3 Hz. The *thick solid lines* indicate the velocity of the eye-in-the-head, and the *thin dashed lines* the velocity of the head with respect to the rotating turntable.

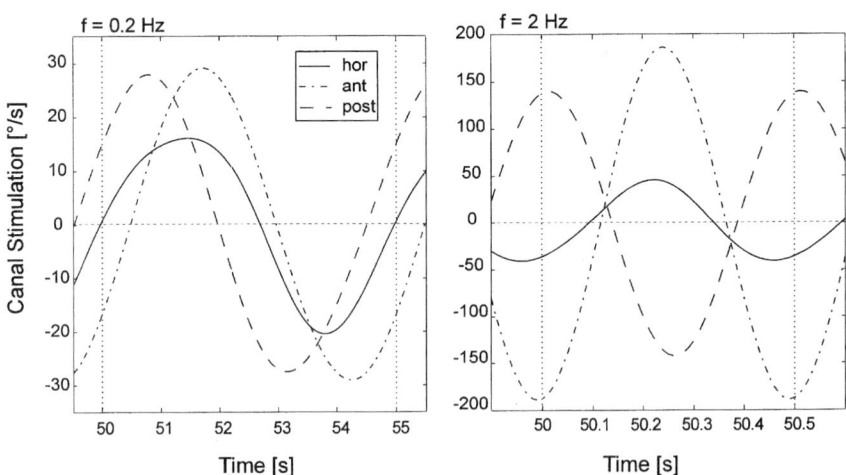

FIGURE 2. Stimulation of the horizontal, anterior, and posterior canal on the right side, during PWR with 0.2 Hz (**left**) and 2 Hz (**right**). The data correspond to a rotation about an earth-vertical axis with 100 deg/s, and a sinusoidal oscillation of the head in the pitch plane by ± 20 deg. In the zero position the head was oriented such that Reid's line was 10 deg nose up with respect to the earth vertical. These positions are marked by the *dotted vertical lines*. The units indicate the velocity "sensed" by each canal during this movement.

components, but no clear offset. When tested with the parameters that simulate the behavior of rhesus monkeys during OVAR, the model also correctly reproduced a compensatory horizontal velocity offset, as found in the experiments.

DISCUSSION

In this study, the first ever of PWR in humans, we found a horizontal velocity offset that is compensatory for the movement in space, but that is—unlike in monkeys[1]—extremely small. This confirms earlier findings in our laboratory, indicating drastically reduced otolith–canal interaction in humans compared to monkeys.[7] A surprising finding has been the increase in modulation of the torsional velocity component with increasing pitch-frequency. The models do not show this increase. They show, however, that any explanation of the origin of the compensatory eye-movement response in monkeys must consider the mechanical properties of the semicircular canals, since they lead to a frequency-dependent phase shift of the stimulation patterns in the canals.

In our experiment the subjects started with the head in a comfortable upright position, which resulted in a nose-up tilt of the lateral semicircular canals by about 30 deg. As a result the frequency of the modulation of the horizontal and torsional velocity was the same as the frequency of the head pitching. In contrast, in experiments with monkeys the head is usually initially oriented such that the lateral canals are parallel to the earth-horizontal axis. In that case, the frequency of the modulation of the horizontal eye-velocity is twice the frequency of the head pitching.[2]

ACKNOWLEDGMENTS

This study was supported by the Deutsche Forschungsgemeinschaft, SFB 307-A10.

REFERENCES

1. RAPHAN, T., B. COHEN, J.-I. SUZUKI. & V. HENN. 1983. Nystagmus generated by sinusoidal pitch while rotating. Brain Res. **276**: 165–172.
2. HESS, B. J. M. & D. E. ANGELAKI. 1993. Angular velocity detection by head movements orthogonal to the plane of rotation Exp. Brain Res. **95**: 77–83.
3. ANASTASOPOULOS, D., T. HASLWANTER, M. FETTER & J. DICHGANS. 1998. Smooth pursuit eye movements and otolith-ocular responses are differently impaired in cerebellar ataxia. Brain. **121**(8): 1497–1505.
4. HOLDEN, R. H., S. L. WEARNE & I. S. CURTHOYS. 1992. A fast, portable desaccading program J. Vestib. Res. **2**: 175–179.
5. HASLWANTER, T. & L. B. MINOR. 1999. Nystagmus induced by circular head shaking in normal human subjects. Exp. Brain Res. **124**(1): 25–32.
6. MERFELD, D. 1995. Modeling human vestibular responses during eccentric rotation and off vertical axis rotation. Acta Oto-Laryngol. (Stockh.) **520**: 354–359.
7. FETTER, M., J. HEIMBERGER, R. BLACK, W. HERMANN, F. SIEVERING & J. DICHGANS. 1996. Otolith-semicircular canal interaction during postrotatory nystagmus in humans. Exp. Brain Res. **108**: 463–472.

Phase Adaptation of the Linear Vestibulo-Ocular Reflex

S. HEGEMANN,[a,b,c] M. J. SHELHAMER,[b] AND D. S. ZEE[b]

[a]*Department of Neurology, Friedrich-Schiller-University, Philosophenweg 3, 07740 Jena, Germany*

[b]*Department of Neurology, The Johns Hopkins University, Pathology 2-210, 600 Wolfe Street, Baltimore, Maryland 21287, USA*

INTRODUCTION

It was recently demonstrated that the brainstem velocity to position neural integrator (NI) can be adaptively modified, changing the phase of the angular vestibulo-ocular reflex (AVOR) in response to sinusoidal rotation.[1] Since otolith receptors do not work as mechanical integrators in the same way as the cupula, a twofold integration is expected for translational stimuli, whereas ocular counterroll with head tilt might not need integration at all. Whether otolithic signals are at least partly processed by the same or a similar NI as semicircular canal signals is unknown.

MATERIAL AND METHOD

We induced adaptation of the phase of the linear VOR (LVOR) in nine healthy volunteers. Subjects were transported along the interaural axis for 20 min using a sinusoidal stimulus (0.5 Hz, 0.3-g peak acceleration). Subjects viewed a computer-driven stereoscopic visual display, which presented a wall at a virtual distance of 1 m. This visual image was moved 53 deg out of phase with respect to sled (head) position during adaptation, so as to require an LVOR phase lead or lag for compensatory tracking of the display. Before and after adaptation, the gain and phase of the LVOR were measured in darkness using scleral search coils: a red-light-emitting diode (LED) at 1-m distance was lit for 5 s in complete darkness. Immediately after it was extinguished, the sled began to move and subjects tried to fixate the remembered target. Cycles highly deformed by saccadic intrusions were omitted. After saccades had been removed from the recorded data, phase (sled position vs. eye position) was calculated. *T*-test was performed to check for statistical significance of the phase changes.

[c]To whom correspondence may be addressed. Phone: *49- (0)3641-35005; fax: *49- (0)3641-35399; e-mail: hegemann@landgraf.med.uni-jena.de

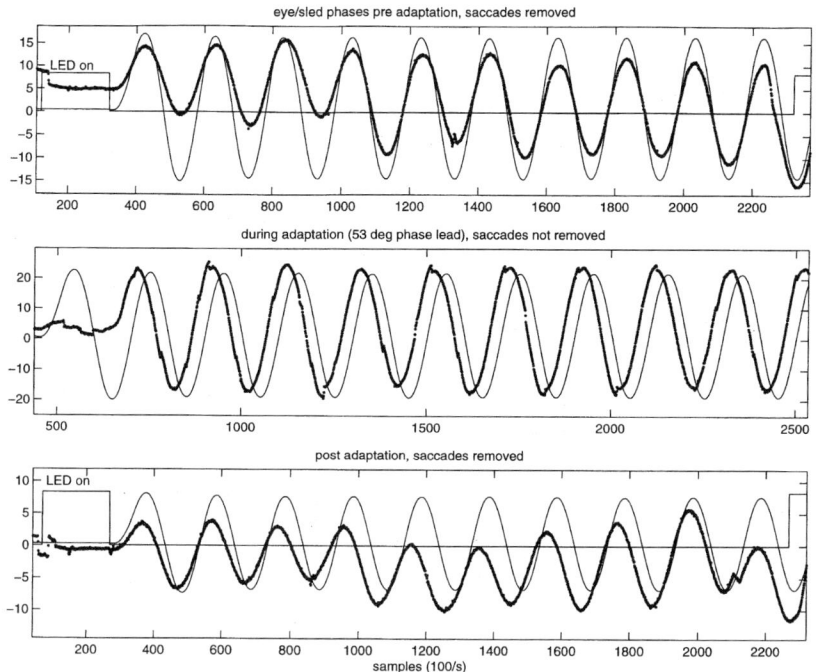

FIGURE 1. LVOR phase before, during and after 20-min adaptation with a 53 deg phase-lag paradigm. *Solid line*: sled position; *dotted line*: right-eye position. Y-axis values indicate horizontal eye position in degrees. Amplitude of sled position has been adapted for better visualization of phase shift.

RESULTS

An impressive example of phase adaptation is shown in FIGURE 1. In this subject, a phase change of 53 deg was asked for, and the measured phase change was 45 deg. Seven of the subjects showed similar significant phase adaptation according to the respective lead or lag paradigm. One subject (no. 5) showed no significant change after lag adaptation, and one (no. 3) showed an adaptive change in the lead direction (reduction in phase lag) after lag adaptation. These data are summarized in TABLE 1.

TABLE 1: Results of LVOR Phase-Adaptation Experiments

Subject No	Adaptation Paradigm	Phase (deg) Preadaptation	Phase (deg) Postadaptation	Phase Change After Adaptation	T-test p-value
1	53° lag	4.2 (8.5)	−13.7 (6.4)	−17.9	0.1279
2	53° lag	−16.9 (9.8)	−0.8 (12.4)	+16.2	0.0038
3	53° lag	11.4 (6.9)	−8.0 (7.6)	−19.4	0.0001
4	53° lag	34.2 (16.6)	−11.2 (16.0)	−45.4	0.0000
5	53° lag	0.9 (8.4)	−2.8 (13.9)	−3.7	0.2514
6	53° lead	−0.8 (9.1)	41.6 (15.7)	+42.4	0.0001
7	53° lead	5.2 (10.1)	29.1 (11.6)	+23.8	0.0001
8	53° lead	−16.2 (6.7)	8.5 (11.8)	+24.7	0.0138
9	53° lead	−13.7 (3.8)	5.1 (9.4)	+18.8	0.0001

During adaptation, all subjects showed good tracking of the visual display, at a phase shift close to the desired value and with little variability. Subject 5, who had little or no adaptation had lightly phase-lagged tracking with a high standard deviation during the adaptation period.

DISCUSSION

The results of the majority of subjects show clearly that the phase of the LVOR can be adaptively modified. This might be achieved by adapting the central structures that process otolithic signals, such as the neural integrator. The weak adaptation in subject 5 might be due to bad following during adaptation. The behavior of subject 3, who showed a lead adaptation after a lag paradigm, is still unexplained.

CONCLUSION

Our results provide evidence that otolithic signals evoked by linear acceleration might be processed by a neural integrator (under adaptive control) similar or identical to that used by the AVOR.

REFERENCES

1. KRAMER P. D., M. J. SHELHAMER & D. S. ZEE. 1995. Short term adaptation of the phase of the vestibulo-ocular reflex (VOR) in normal human subjects. Exp. Brain Res. **106**: 318–326.

Separation Between On- and Off-Center Passive Motion in Darkness

I. ISRAËL[a,c] & S. GLASAUER[b]

[a]*LPPA, Collège de France ~ CNRS, Paris*
[b]*Center for Sensorimotor Research, Dept. of Neurology, LMU Munich*

INTRODUCTION

It is well known that perception of passive linear motion of the whole body is governed by a linear velocity threshold (ca. 0.22 m/s).[1] Similarly, time to detect an angular acceleration step is reached when a critical velocity is attained (1–3°/s).[2-4] However, nothing is known about how to detect the linear component of a circular trajectory, that is, the time and the criterion to perceive the presence of linear velocity and/or centrifugal acceleration in circular motion, and the time to decide about the absence of a linear component in a purely rotational (on-center) motion. We then devised an experiment we hoped would help us answer these questions.

METHODS

Healthy subjects were submitted to a passive motion in complete darkness, seated on the Robuter™ (Robosoft, France). They were instructed to push an identified button as soon as they were sure about the type of ongoing motion: off-center (CIR) or on-center (DOT) rotation. The button signal was recorded, from which the decision time was measured, and all the contingent values of linear and angular velocity and acceleration were obtained. All stimuli had a triangular velocity profile (angular acceleration was constant) with an overall duration of approximately 6 s.

Condition A (Minimal Radius)

All stimuli had 180° total angle, the radius for the CIR stimuli was 0.4 m (the smallest radius usable with the Robuter, which was turning on one wheel). Three different accelerations were used (range 15–20°/s^2). Six trials were applied in random order: 3 CIR (off-center) and 3 DOT (on-center), with slow (S), medium (M), and fast (F) motion. Nineteen healthy subjects took part in the experiment, during which no feedback of the responses was given.

---

[c]To whom correspondence may be addressed. Phone: 33-(0)1-44271288; fax: 33-(0)1-44271382; e-mail: isi@ccr.jussieu.fr

Condition B (Variable Radii)

All stimuli had 6 s total duration, yielding angles of 85° or 170°. The radius was 0.5 m (CIRC), 1 m (CIRA), or 2 m (CIRB); peak angular velocity was either 0.5 rad/s (CIRA-B) or 1 rad/s (CIRC), and angular acceleration was 0.166 rad/s^2 (CIRA-B) or 0.33 rad/s^2 (CIRC). Again, six trials were applied in random order, with the same angular characteristics for DOT and CIR. Ten healthy subjects took part in this experiment.

RESULTS

Condition A

The first result concerns the errors (false detection of DOT or CIR): in the 116 trials performed, 18 errors were reported (four subjects erred twice, one subject four times). Most errors were on CIR—only two errors were committed on DOT—at the first trial. Altogether, 6 errors were made on the first trial, and 6 on the second trial. This means that the subjects made mistakes at the beginning of the session, and apparently understood that they were wrong when they experienced the second stimulus type. They probably needed to experience each of the two stimulus-types before giving correct answers. Trials with false answers were omitted from the subsequent analyses.

The subjects needed, on average, 2.72±1.08 s (mean ± SD) to decide (correctly) for CIR, and 2.54±1.21 s for DOT. No consistent change in reaction time was observed during the experiment.

The 11 subjects who had made mistakes in CIR–DOT selection (M-subjects) were separated from the 8 subjects who had immediately correctly perceived the type of motion they were submitted to (C-subjects). We supposed that the C-subjects probably had used the correct criterion to decide and therefore made no errors. When comparing the delay for all 6 types of trials, we found a significant ($p < 0.002$) difference between the two groups of subjects (C-subjects 2.6±0.3 s; M-subjects 3.4±0.4 s). This supports our hypothesis that the C-subjects used a different criterion to decide than did the M-subjects.

Two parameters are considered to be decision criteria: the tangential velocity Vx (0.26 ± 0.07 m/s, n = 24, when C-subjects pushed the button, FIG. 1A), and the centrifugal acceleration Ay (0.19 ± 0.09 m/s^2, FIG. 1B). In order to determine which of the parameters is used by the subjects, we computed the absolute difference between each Vx (at the time when the button was pressed) and the average Vx of the 25 trials (all CIR trials of the C-subjects), divided by this average; we then did the same for Ay. The result of this computation was 0.24 ± 0.15 (no units) for Vx and 0.41 ± 0.26 for Ay, a very significant difference ($p = 0.008$). This shows that tangential velocity (and not radial acceleration) was indeed the most constant and therefore the most relevant parameter used by C-subjects. The results for M-subjects ($p = 0.012$; Vx at pushing was 0.28 ± 0.07 m/s, and Ay was 0.20 ± 0.08 m/s^2, FIG. 1) suggest that they may have used, in contrast to the C-subjects, either the Ay criterion in some trials or a value composed of both tangential and centrifugal acceleration.

Condition B

In this condition, two stimuli had an identical Vx profile (CIRA-C, FIG. 2A) and two had an identical Ay profile (CIRB-C, FIG. 2B). We therefore computed Vx and Ay at the time when subjects pushed the button. Only two false detection cases are reported, therefore the subjects could not be separated into different groups according to errors.

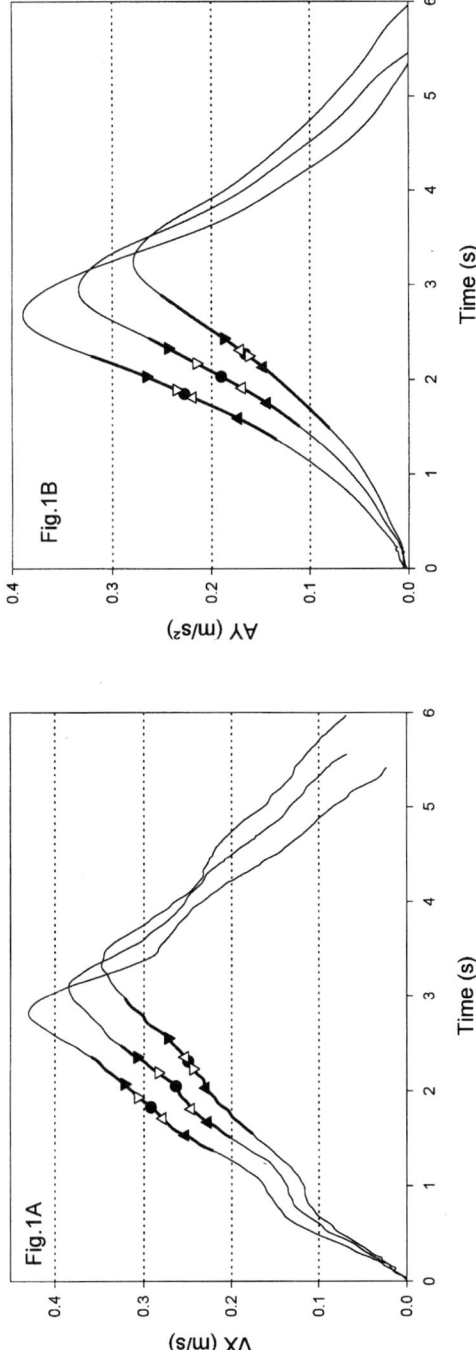

FIGURE 1 (**A**) Tangential velocity profile in condition A, for the three CIR trials (slow, medium, fast). ●: Mean Vx when the button is pushed (correctly) for all subjects; △: C-subjects; ▲: M-subjects; ▽: quick group (see condition B); ▼: slow group. *Heavy line*: mean ± SD of all subjects. (**B**) Centrifugal acceleration in condition A; same notation as in FIGURE 1A.

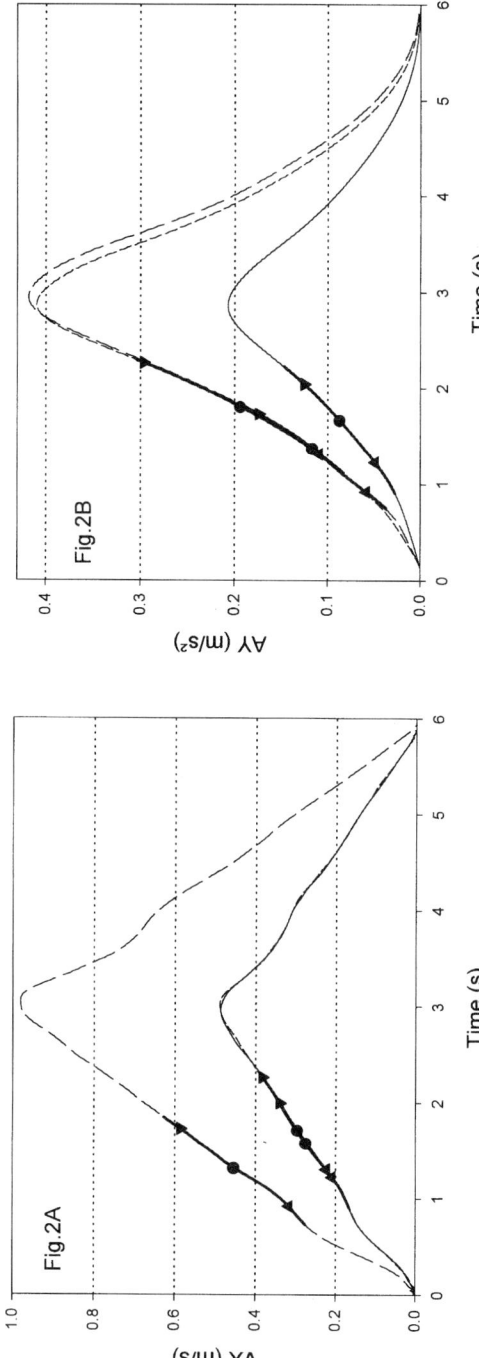

FIGURE 2 (A) Tangential velocity profile in condition B, for the three CIR trials (CIRA: *solid*; CIRB: *long-dash*; CIRC: *short-dash*). Same notation as in FIGURE 1 (without *C*- and *M*-subjects). (B) Centrifugal acceleration in condition B; same notation as in FIGURE 2A.

As expected from the results of condition A, tangential velocity Vx in CIRA (0.27 ± 0.11 m/s) and CIRC (0.29 ± 0.09 m/s) was not significantly different, while Vx for CIRB (0.45 ± 0.18 m/s) was different from both other CIR stimuli ($p < 0.05$), thereby confirming the "velocity hypothesis" (FIG. 2A).

Centrifugal acceleration Ay (CIRA: 0.09 ± 0.06 m/s≈; CIRB: 0.12 ± 0.08 m/s≈; CIRC: 0.19 ± 0.11 m/s≈) was only different between CIRA and CIRC ($p < 0.05$), but not between CIRA and CIRB, so the radial acceleration could not be a criterion (FIG. 2B).

The subjects needed more time to decide that rotation was on-center (DOT) than off-center (CIR) for all stimuli but CIRC–DOTC. Since we found a large variability in response delays, we separated the subjects into a Quick group ($n=5$, 1.17 ± 0.31 s) and a Slow group ($n=5$, 2.23 ± 0.42 s). A comparison of the different profiles was more significant for these groups. For the Quick group, a Vx of 0.21 ± 0.01 m/s for CIRA and 0.22 ± 0.03 m/s for CIRC were found (FIG. 2A). When we applied this method to the data of condition A (FIG. 1), we found a Vx of 0.24 ± 0.06 m/s for the Quick group, and a highly significant difference ($p = 0.002$) between the "normalized" Vx and Ay.

CONCLUSION

Both experiments show that when subjects decided upon off-center rotation (CIR), the tangential velocity Vx of the motion—but not centrifugal acceleration or other parameters, such as angular velocity—was constant regardless of the motion profile imposed. Subjects with short response times decided on Vx values very close to the 0.22-m/s threshold proposed for linear motion detection. This suggests that the decision between in-place rotation (DOT) and circular motion (CIR) is not made by detecting centrifugal acceleration, but by detecting linear motion.

ACKNOWLEDGMENTS

This work was supported by the GIS (Sciences de la Cognition, France) and HFSP: RG71/96B; S. Glasauer was supported by the NEB (France).

REFERENCES

1. YOUNG, L. R. 1984. Perception of the body in space: mechanisms. In Handbook of Physiology— The Nervous System III, I. Darian-Smith, Ed.: 1023–1066. American Physiological Society. Bethesda, Md.
2. MERGNER, T., C. SIEBOLD, G. SCHWEIGART & W. BECKER. 1991. Human perception of horizontal trunk and head rotation in space during vestibular and neck stimulation. Exp. Brain Res. **85**: 389–404.
3. BENSON, A. J., E. C. HUTT & S. F. BROWN. 1989. Thresholds for the perception of whole body angular movement about a vertical axis. Aviat. Space Environ. Med. **60**: 205–213.
4. HUANG, J. & L. R. YOUNG. 1981. Sensation of rotation about a vertical axis with a fixed visual field in different illuminations and in the dark. Exp. Brain Res. **41**: 172–183.

Development of Synaptic Innervation in the Rodent Utricle

ANNA LYSAKOWSKI[a]

Department of Anatomy and Cell Biology, University of Illinois at Chicago, Chicago, Illinois 60612, USA

We decided to reexamine early postnatal synaptic innervation in relation to ongoing studies of the development of physiological, morphological, and molecular features of hair cells in the mouse utricular macula.[1,2] Physiological studies[3,4] demonstrated that mammalian type I and type II hair cells have distinctively different potassium conductances. Type I hair cells have a conductance termed g_{KL} for low-voltage-activated K⁺ conductance because it turns on at about −90 mV. This conductance is absent at birth, but begins to be expressed at PD4 in normal, excised utricles. Coincident with this event, calyces begin to form. The two events are not inextricably linked because g_{KL} is expressed at PD4 even in maculae that have been cultured from PD1. Thus, *ongoing* synaptic innervation does not appear to be necessary for the expression of this conductance. Another motivation for this study was to provide baseline data for spaceflight experiments in which mammals are born or develop in microgravity.

An early study by Favre and Sans[5] was the only quantitative developmental data available for vestibular hair cells. Their data indicated that the number of synaptic ribbons in type I hair cells decreased in cat crista by 93% between birth and adulthood. The neonatal cat appears to be equivalent to our PD4 stage in that about 50% of the type I hair cells possess calyces. The 93% reduction led these authors to conclude that synaptic ribbons in type I hair cells "regressed and disappeared in the adult because they were no longer needed," since the calyx could provide a form of electrotonic transmission. In a previous study,[6] we showed that there were about 10–20 synaptic ribbons per type I hair cell regardless of region; thus, we were curious about whether the type I hair cell started out with 100–200 ribbons and lost 93% of them by adulthood.

One problem with examining synaptic innervation in relation to hair-cell type at early postnatal stages, is that morphologically it has proven difficult to distinguish type I hair cells without their primary characteristic, the calyx. The calyx has been the classic definition of a type I hair cell, but there are secondary attributes that are also present. For example, patch clamp experiments frequently use the shape of the hair cell to distinguish hair-cell types. Several studies have shown that stereociliar diameter and bundle size,[7,8] as well as the apical surface convexity of a hair cell[8] can also be used to distinguish the two types. We have previously measured some of these features in adult material, and then worked in reverse chronological order through PD28 and PD7 utricular hair cells whose calyces were present.[1,9] In so doing, we found that these morphometric differences between hair-cell types were statistically significant. Thus, we felt comfortable extending the use of these "secondary attributes" to stages before the presence of a full calyx.

---

[a]Address for telecommunications: Phone: 312/996-5990; fax: 312/413-0354; e-mail: aLysakow@uic.edu

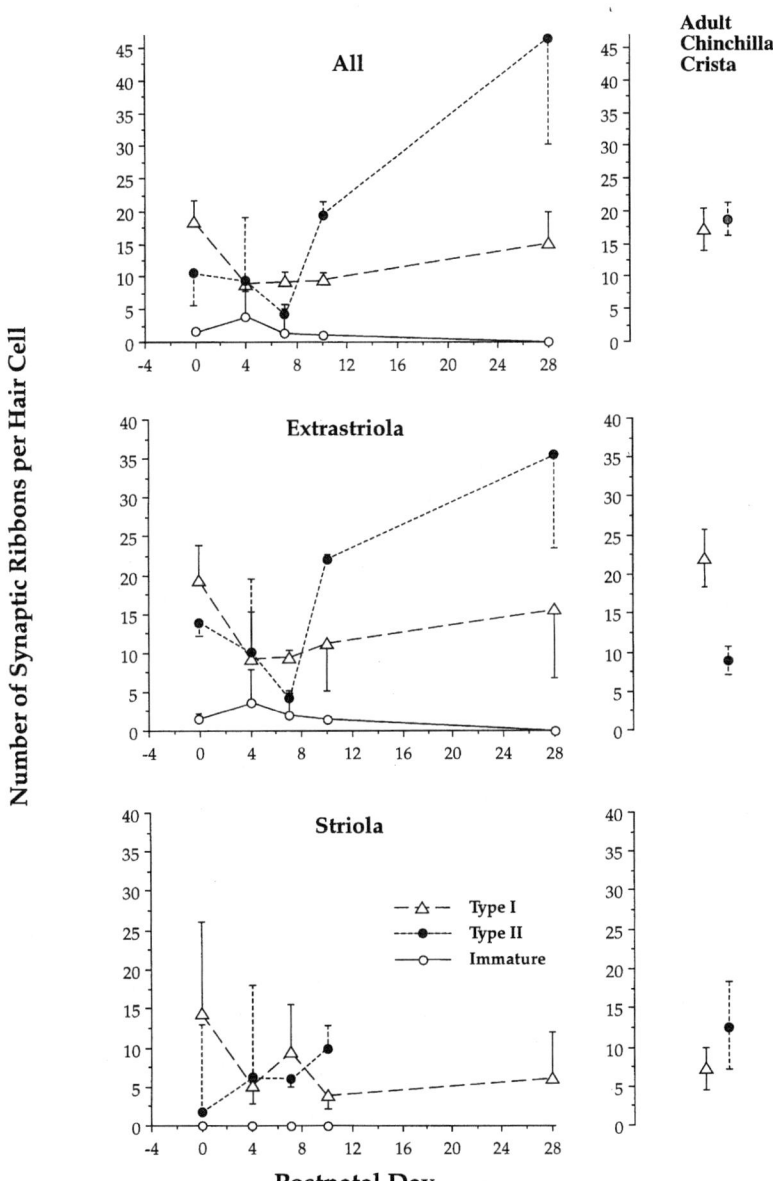

FIGURE 1. Development of synaptic innervation over the first postnatal month in mouse utricle. The key for the entire figure is given in the *bottom* (striola) *panel*. Values for adult chinchilla crista are shown at right; they are taken from reference 6 and represent all (*top*), the peripheral zone, equivalent to the extrastriola (*center*); and the central zone, equivalent to the striola (*bottom*).

In the present study, samples were taken from Harlan ICR mice at postnatal days 0 (birth), 4, 7, 10, and 28. Three samples of 30 serial sections 70 nm thick were taken at each stage. Each series of sections was cut perpendicular to the striola and contained a cross section of all three regions of utriculus (medial extrastriola, striola, and lateral extrastriola). Photomontages at 5000 × magnification were assembled and used to collect data about the numbers, positions, and shapes of synaptic ribbons. Medial and lateral extrastriola were combined. Synaptic ribbon counts were done using the dissector method as described previously.[6] So far, two samples from each stage have been examined.

Results are shown in FIGURE 1 and are compared to counts taken from adult chinchilla crista.[6] Sample sizes are smaller in the striola, which constitutes only 10% of the epithelium in a transverse section, and indeed, we have no type II hair-cell equivalents yet at PD28. Nonetheless, we have so far examined the equivalent of 26 striolar and 97.5 extrastriolar hair-cells for a total of 123.5 fully reconstructed hair cells across all stages. While the results are still preliminary, they indicate that after an initial decrease between PD0 and PD4, there is an orderly increase in numbers of ribbons over all stages, with PD28 mouse utricular hair cells being roughly comparable in numbers to the adult chinchilla crista. In the extrastriola, type II hair cells in particular tend toward increased numbers at older stages.

At younger stages, we noticed greater numbers of ribbon clusters and pairs (not shown). Hollow and spherical ribbons were also more common in type II hair cells. In addition, we noticed more "free ribbons"[10] at PD4 and PD7, that is, ribbons floating free in the cytoplasm. Thus, we would agree with the results in the cochlea of Sobkowicz and colleagues,[10] that these free ribbons probably represent misplaced and free ribbons liberated after the initial afferent denervation. Even with these "floating ribbons" included ($\approx 5\%$ of the total), numbers of ribbons tended to be less at PD4 than at later stages, indicating that synaptogenesis continues after this initial "pruning back" of innervation.

In conclusion: (1) approximately 14,500 hair-cell profiles have been examined, equivalent to about 125 complete hair cells; (2) synaptic innervation in the postnatal utricular macula appears to proceed at an orderly rate, after a decrease from neonatal stages; however, sample sizes are still small, and more samples need to be taken through larger portions of the striola; (3) hollow ribbons are more common in type II hair cells, as in the chinchilla crista, but not significantly more, suggesting that they are not *per se* an immature form; and (4) synaptic innervation is another marker of hair-cell development that may be specific for one hair-cell type versus another.

ACKNOWLEDGMENTS

This work was supported by PHS Grant R01 DC-02290 and NASA Grant NAG 5-4593. We gratefully acknowledge the excellent technical assistance of Mr. Steve Price. Ms. Parveen Ahmed assisted in the early stages of this work.

REFERENCES

1. RÜSCH, A., A. LYSAKOWSKI & R. A. EATOCK. 1998. Postnatal development of type I and type II hair cells in the mouse utricle: acquisition of voltage-gated conductances and differentiated morphology. J. Neurosci. **18**: 7487–7501.
2. CARRANZA, A., A. LYSAKOWSKI, L. C. BARRITT, M. A. VOLLRATH, D. L. HIMES, K. W. BEISEL & R. A. EATOCK. 1998. Expression of Kv subunits in rodent vestibular hair cells. ARO 21st Midwinter Meet. Abstr.: p. 76. ARO. St. Petersburg Beach, Fla.

3. RENNIE, K. J. & M. J. CORREIA. 1994. Potassium currents in mammalian and avian isolated type I semicircular canal hair cells. J. Neurophysiol. **71**: 317–329.
4. RÜSCH, A. & R. A. EATOCK. 1996. A delayed rectifier conductance in type I hair cells of the mouse utricle. J. Neurophysiol. **76**: 995–1004.
5. FAVRE, D. & A. SANS. 1979. Morphological changes in afferent vestibular hair cell synapses during the postnatal development of the cat. J. Neurocytol. **8**: 765–775.
6. LYSAKOWSKI, A. & J. M. GOLDBERG. 1997. A regional ultrastructural analysis of the cellular and synaptic architecture in the chinchilla cristae ampullares. J. Comp. Neurol. **389**: 419–443.
7. PETERSON, E. H., J. R. COTTON & J. W. GRANT. 1996 Structural variations in ciliary bundles of the posterior semicircular canal: quantitative anatomy and computational analysis. Ann. N.Y. Acad. Sci. **781**: 85–102.
8. LAPEYRE, P. N. M., A. GUILHAUME & Y. CAZALS. 1992. Differences in hair bundles associated with type I and type II vestibular hair cells of the guinea pig saccle. Acta Otolaryngol. (Stockh.) **112**: 635–642.
9. LYSAKOWSKI, A. 1996. Morphometric criteria for hair cell types. J. Vestibular Res. **S4**: S86.
10. SOBKOWICZ, H. M., J. E. ROSE, G. L. SCOTT & C. V. LEVENICK. 1986. Distribution of synaptic ribbons in the developing organ of Corti. J. Neurocytol. **15**: 693–714.

Human Gaze Stabilization for Voluntary Off-Centric Head Rotations

W. P. MEDENDORP,[a] B. J. BAKKER, J. A. M. VAN GISBERGEN, AND C. C. A. M. GIELEN

Department of Medical Physics and Biophysics, Nijmegen University, Nijmegen, The Netherlands

Natural head movements often have both rotation and translation components due to an off-centric location of the rotation axis of the head (see Fig. 1A). Therefore, a complete description of the head movement should specify either the rotation and translation of the head or the orientation and location of the head rotation axis. Both descriptions are equivalent. Recently, we have shown that the axes for the repertoire of natural head movements occupy consistently different spatial locations[1] (see FIG. 1B). For purely horizontal movements the location of the rotation axis remains fixed at a point midway between the two ear canals. For vertical head movements, however, the rotation axis does not stay fixed in space but shifts downwards with respect to the body for larger movement amplitudes. Consequently, since the rotation axes are not fixed in space, stabilization of gaze in near vision during natural head movements will require complex eye movements.

The purpose of the vestibulo-ocular reflex (VOR) is to stabilize retinal images during head movements. As can be seen in FIGURE 1A, to correctly stabilize gaze, the VOR must take into account not only head rotation but also eye translation and the target distance. This requires the gain of the VOR (eye velocity/head velocity) to exceed 1.0 for normal head movements. Movements of the head are detected by the semicircular canals, responding to angular acceleration, and by the otoliths, which measure linear acceleration. Passive-rotation experiments suggest that the compensatory eye movements are controlled by combined canal and otolith responses, which also depend on target distance.[2] From this perspective, the location of the rotation axis, and its shift in vertical head movements, is expected to modulate the gain of the VOR on a moment-to-moment basis.

In this report we present some preliminary results of gaze stabilization in human subjects making voluntary horizontal or vertical sinusoidal head movements (0.25–2 Hz) at two different amplitudes (15 and 30 deg) in a darkened room. Meanwhile, subjects were instructed to maintain gaze on a target LED at straight-ahead (distance: 0.5 or 1.5 m) that disappeared after a fixed time interval. Head movements were recorded with an OPTOTRAK 3020 system. Two-dimensional eye position was measured using the scleral search-coil technique in a very large magnetic field system (3.3-m sidelength).

FIGURE 2 presents the results of a subject performing a roughly sinusoidal vertical head movement with an amplitude of about 30 deg while fixating the near target (0.5 m) that disappeared after 9 s. The upper two panels depict the rotation of the head and the eye, respectively. Note that the eye excursions are larger. The third panel illustrates the actual gaze (thin) together with the optimal (i.e., no retinal slip) gaze signal (bold). While a unity-gain rotational VOR, with no compensation for translation,

---

[a]Address for communication: W. P. Medendorp, Dept. Medical Physics and Biophysics, University of Nijmegen, Geert Grooteplein 21, NL-6525 EZ Nijmegen, The Netherlands. Phone: 31 24 3614237; fax: 31 24 3541435; e-mail: pieter@mbfys.kun.nl

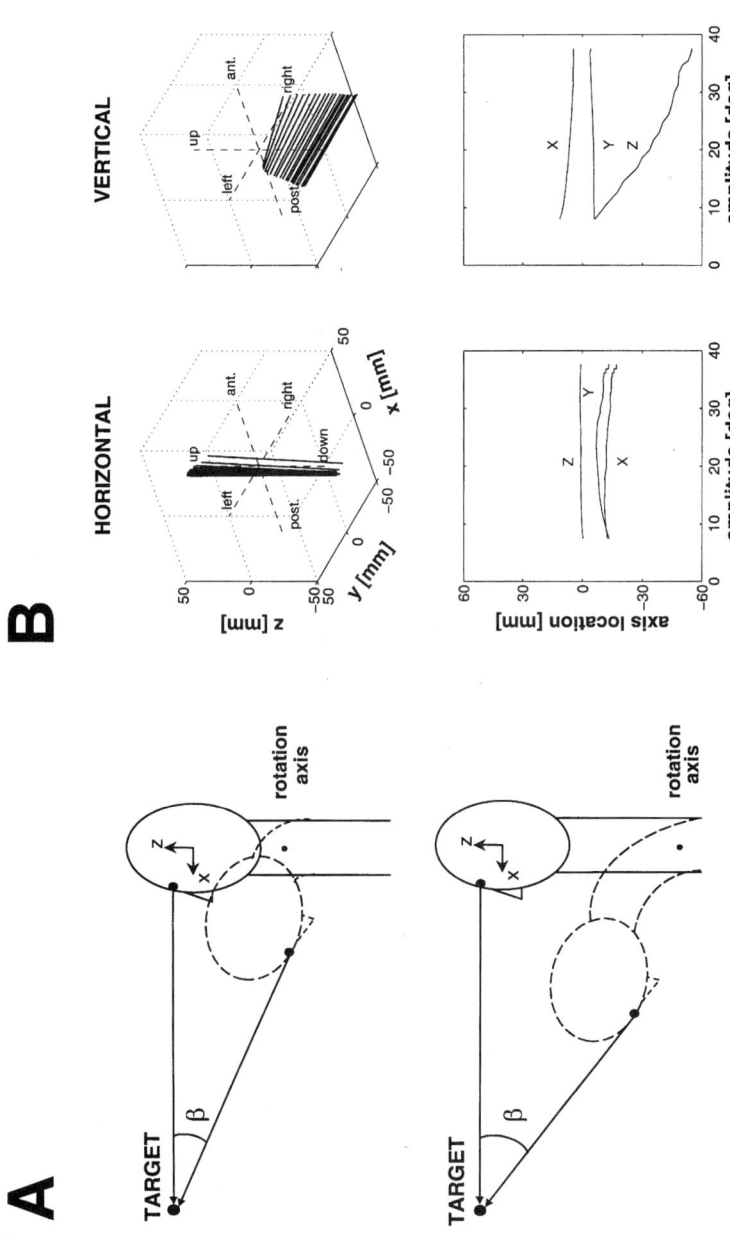

FIGURE 1. (**A**) The translation of the head is determined by the eccentricity of the rotation axis. A more eccentric rotation axis requires a larger compensatory eye movement for fixating a near target. (**B**) For horizontal head movements, the rotation axis remains at a fixed location throughout the movement. For vertical movements, however, the rotation axis shifts to lower locations with respect to the skull for larger movement amplitudes. (Adapted from Medendorp et al.[1])

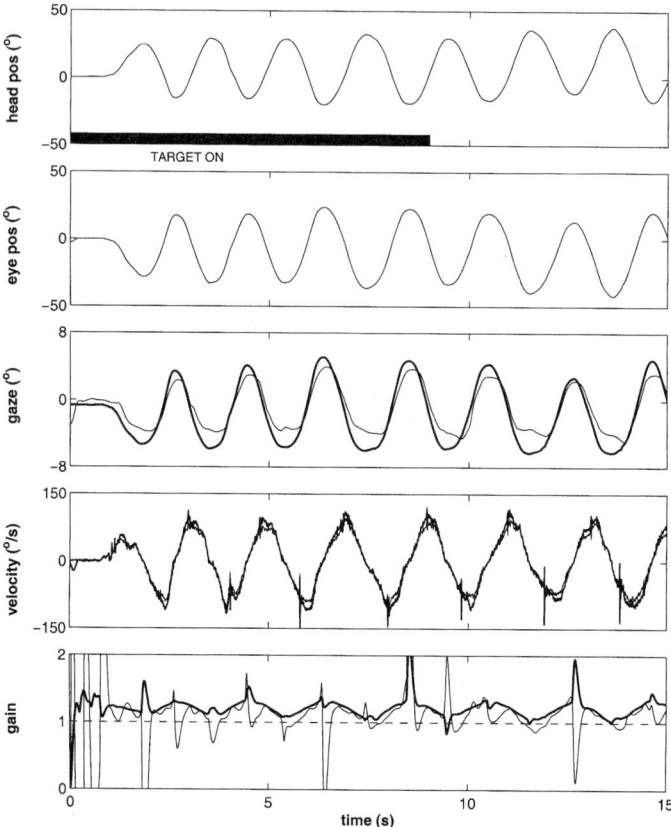

FIGURE 2. Gaze stabilization during an active sinusoidal vertical head movement (0.5 Hz). The *upper two panels* show head position and eye position, respectively. The *third panel* shows gaze (*thin*: measured gaze; *bold*: optimal gaze). The *fourth panel* shows eye velocity (inverted) exceeding head velocity, thereby partially compensating for translation of the eyes by the head movement. The *bottom panel* shows the instantaneous VOR gains and their modulation with head position (*bold*: required gain; *thin*: measured gain).

would have yielded a gaze-signal constant at zero, the actual gaze excursions clearly realize a considerable percentage of what is required to keep the eye on the near target. How is this accomplished? The fourth panel illustrates head velocity (thin) and inverted eye velocity (bold) during the head movement. Eye velocity exceeds head velocity to compensate for eye translation relative to the target. The bottom panel shows that, as needed, the instantaneous above-unity VOR gains (bold: optimal gain) are modulated with head position. The VOR gain increased with target proximity for both vertical and horizontal head movements. Cross-correlation of inverted eye velocity and head velocity revealed virtually no phase difference in either case. Yet, we observed a time delay between measured gaze and required gaze. Our explanation for this apparent discrepancy is that the inverted eye-velocity signal consists mostly of a signal in phase with head velocity, along with a small out-of-phase signal. The latter, representing eye-translation compensation, shows up in panel 3 in a pure form, which allows the phase difference to be revealed.

A noteworthy observation in the vertical task was a tendency for a higher VOR gain during large as compared to small amplitudes. We did not observe this for horizontal head movements. Cautious interpretation of this result suggests a possible difference in movement pattern of the subject in executing either small or large vertical head movements. In the latter case, our earlier work suggests that the head rotation axis will reach a more eccentric (lower) position, thereby requiring additional compensation. By contrast, in horizontal movements the rotation axis stays at a fixed location, and therefore requires no extra compensation for larger amplitudes. Much additional work is required to more rigorously assess the effect of the head rotation axis in human gaze stabilization.

REFERENCES

1. MEDENDORP, W. P., B. J. M. MELIS, C. C. A. M. GIELEN & J. A. M. VAN GISBERGEN. 1998. Off-centric rotation axes in natural head movements: implications for vestibular reafference and kinematic redundancy. J. Neurophysiol. **79**: 2025–2039.
2. SNYDER, L. H. & W. M. KING. 1992. Effects of viewing distance and location of the axis of head rotation on the monkey's vestibuloocular reflex. I. Eye movement responses. J. Neurophysiol. **67**: 861–874.

A Simple Model of Vestibular Canal–Otolith Signal Fusion

THOMAS MERGNER[a,c] AND STEFAN GLASAUER[b]

[a]*Neurologische Klinik, Neurozentrum, Universitätsklinikum, Breisacher Strasse 64, D-79106 Freiburg, Germany*

[b]*Klinikum Grosshadern, Zentrum für Sensomotorik, Marchioninistrasse 23 81377 München, Germany*

INTRODUCTION

When being moved passively with no visual or auditory space or motion cues available, we still can perceive self-motion with respect to the external space. Information about angular and linear acceleration is transmitted to the CNS with the help of the vestibular organ, a biological 6-DoF inertial measuring device. It consists of two subsystems, the otolith system and the canal system. The *otolith system*, which functions like a differential densitometer,[1] measures both orientation of the head relative to the gravitational force (**g**) vector and inertial forces arising during translatory head accelerations. The *canal system* represents an angular speedometer; inertia of the endolymph-cupula mass upon angular acceleration is partially counteracted by viscous forces, with the result that the neural afferent signal that is evoked is proportional to angular head velocity. However, this acceleration-to-velocity integration is imperfect, showing a time constant of approximately 5 s.[2]

The nonideal properties of the two vestibular subsystems pose a serious problem to the brain when it tries to obtain a veridical estimate of body motion and orientation in space. To this end, it has to distinguish between gravitational and inertial forces in the otolith afferent signal [the "gravito-inertial force (GIF) resolution problem"],[3] on the one hand, and, on the other hand, it has to improve the signal of angular velocity received from the canals. Nowadays, it is generally assumed that these two problems are solved mainly by a central *canal–otolith signal fusion*. By this we mean that canal information is used for the GIF resolution, and that information on angular position changes contained in the otolith signal is used to improve the canal signal for earth-vertical head rotations. Psychophysical work suggests that the brain's solution of these problems is not ideal, because a number of typical errors are found with human self-motion perception. For instance, pure linear body translations may lead to an erroneous perception of body tilt (Hill Top illusion).[4,5] We hold that these errors can be used to infer, from the outside of the brain, basic aspects of the computation that the brain performs for the canal–otolith signal fusion.

In the following we consider what the main problems with the fusion might be and how these problems might relate to the illusions that are found in vestibular psychophysics. We found that the GIF resolution may easily be solved. This applies if the canal signal of angular head velocity were ideal, which is not the case, however. We

---

[c]To whom correspondence may be addressed. Phone: 49 761 270 5313; fax: 49 761 270 5310; e-mail: mergner@sun1.ruf.uni-freiburg.de

therefore hold that the main problem of the fusion arises from the time constant in the canal signal. This time constant is prolonged, behaviorally, only from 5 to approximately 20 s (e.g., human self-motion perception for earth-horizontal rotations)[6] by central mechanisms. It has been speculated that the prolongation is limited by the biological noise in the system.[6] A further improvement by otolith-derived information is assumed for earth-vertical rotations (see earlier).

Our approach was first to design a model for the GIF resolution (the most parsimonious one we could think of), then to add the canal time constant and, finally, to search for a simple solution that, at least partially, allows for the time constant error.

CONVENTIONS AND ASSUMPTIONS

For the purpose of modeling and simulation we used the following conventions and assumptions: (a) the vestibular signals are represented in head coordinates, with the sensors being located, without spatial dimension, in the center of the head; the center is at the intersection of the three axes of head rotation (x, y, z). (b) Signal transfer through the otolith peripheral system is considered to be ideal, with the otolith input signal representing a three-dimensional GIF vector $\mathbf{a}_{GI}$ that is, the sum of the inertial acceleration vector $\mathbf{a}$ and the gravitational vector $\mathbf{g}$, $\mathbf{a}_{GI} = \mathbf{a} + \mathbf{g}$. This signal may be composed of inputs from other linear accelerometers, for example, in the trunk, as well. (c) The otolith system is considered to provide *jerk*, the first derivative of $\mathbf{a}_{GI}$ ($\dot{\mathbf{a}}_{GI}$); this assumption relates to the finding of a jerk-like signal in the irregular fibers of the otolithic part of vestibular nerve.[7] (d) The canal signal gives head angular velocity (ω_{CAN}) in the same three orthogonal planes of space as the GIF vector, showing a decay time constant of 5 s, which is prolonged centrally to 20 s. (e) The canal–otolith interaction transforms the two input signals ($\mathbf{a}_{GI}$, ω_{CAN}) into three output signals, that is, estimates of head angular velocity ($\hat{\omega}$), the direction of $\mathbf{g}$ ($\hat{\mathbf{g}}$), and head translatory acceleration ($\hat{\mathbf{a}}$), which represent intermediate steps of the central processings for the stabilization of gaze [vestibulo-ocular reflex (VOR)] and body equilibrium as well as for spatial orientation and self-motion perception. (f) Integration in the model is performed with respect to a known initial state.

The model was implemented in Simulink/Matlab. For "stimulation" of the model we used rotations and translatory accelerations in space coordinates. Transformation of these signals from space into head coordinates was performed with the help of quaternion integrators.

RESULTS

The three steps performed are shown in FIGURE 1A, which gives the signal fusion part of the model in head coordinates. Note that in the first two steps signal transfer in the canal system is taken to be ideal (model without dotted parts).

1. *Estimate of g.* The canal input ω_{CAN} is used to yield an estimate of how the direction of gravity changes in relation to a defined starting condition (e.g., initially upright head, no translatory acceleration). To this end, the signal is transformed back into the dimensions of a three-dimensional acceleration vector by means of a vector product with an estimate of $\mathbf{g}$ ($\hat{\mathbf{g}}$), which is provided by a local feedback and an integration over time (initial state defined by starting condition). In mathematical terms,

$$\dot{\hat{\mathbf{g}}} = \hat{\omega} \times \hat{\mathbf{g}} \tag{1}$$

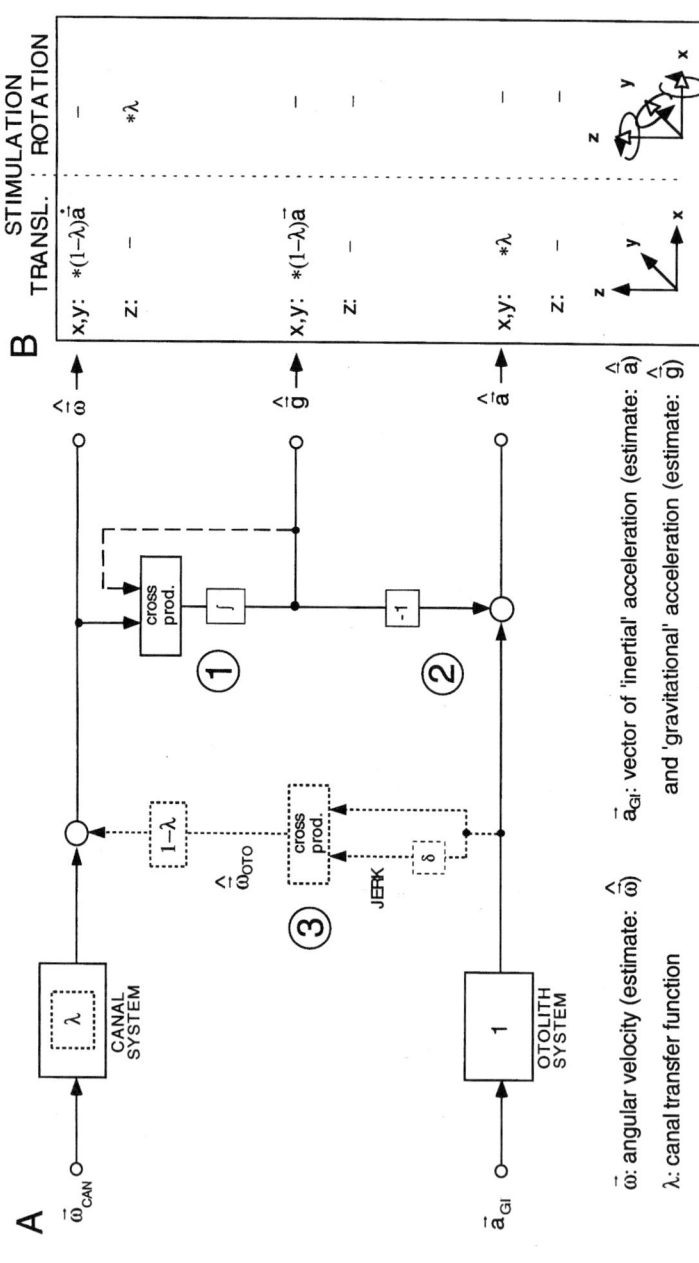

FIGURE 1. Model of otolith–canal interaction. (**A**) Signal fusion part of the model in head coordinates. *Full lines* give the steps that lead to the internal estimates of the gravitational vector ($\hat{g}$; 1) and of the translatory acceleration (**a**; 2). *Dotted lines* show the addition of the canal time constant (λ) and the fusion of canal and otolith-derived information on head-in-space rotation (3). (**B**) The model's predictions of estimation errors during head rotations and translations in space. These errors are related to the canal time constant λ. Further details in text.

This differential equation simply mimics the physical rotation of the **g**-vector by an angular velocity vector ω with respect to the head.

2. *Estimate of translatory acceleration.* Knowing $\hat{\mathbf{g}}$ (integral of $\dot{\hat{\mathbf{g}}}$), the estimate of translatory acceleration $\hat{\mathbf{a}}$ is obtained by simply subtracting $\hat{\mathbf{g}}$ from the otolith input signal, that is, the GIF vector $\mathbf{a}_{GI} = \mathbf{a} + \mathbf{g}$. In mathematical terms,

$$\hat{\mathbf{a}} = \mathbf{a}_{GI} - \hat{\mathbf{g}} \qquad (2)$$

Note that all simulation that we undertook with this model yielded veridical results if **g** was held constant; errors did not occur until the canal time constant was introduced.

3. *Sensor fusion of canal and otolith-derived information on head angular velocity.* The two signals are $\lambda\omega_{CAN}$, the input signal from the canals, with λ representing its deficient transfer function, and $\hat{\omega}_{OTO}$, an estimate of head angular velocity derived from otolith input. In the model the latter signal is obtained by a vector product of the normalized otolith input $[\mathbf{a}_{GI}/|\mathbf{a}_{GI}|^2]$ with the first derivative of this input ($\dot{\mathbf{a}}_{GI}$; "jerk" in FIG. 1A). Fusion is obtained in the form of $\hat{\omega} = \lambda\omega_{CAN} + (1-\lambda)\hat{\omega}_{OTO}$. Note that $\hat{\omega}_{OTO}$ is veridical with head rotations in earth-vertical planes, and thus yields a veridical $\hat{\omega}$ ($\hat{\omega} = \hat{\omega}_{OTO}$), whereas it does not contribute to $\hat{\omega}$ with rotations in the earth-horizontal plane ($\hat{\omega} = \lambda\omega_{CAN}$). A mathematical description would be

$$\hat{\omega} = \lambda\omega_{CAN} + (1-\lambda)*(\dot{\mathbf{a}}_{GI} \times \mathbf{a}_{GI})/|\mathbf{a}_{GI}|^2 \qquad (3)$$

with × denoting the vector cross product. The improved $\hat{\omega}$ then is fed into Eq. 2 and used for vestibular functions, for example, the perception of self-motion during rotations.

Note that this vector product yields erroneous results for off-vertical axis rotation (OVAR), because $\hat{\omega}_{OTO}$ deviates from ω_{CAN} by $\mathbf{a}_{GI}*(\mathbf{a}_{GI}*\omega)/|\mathbf{a}_{GI}|^2$. The error is not corrected for at present; OVAR rotation does evoke an erroneous self-motion perception,[8,9] which qualitatively resembles the simulation results obtained without the correction.

FIGURE 1B gives basic predictions of our model in terms of errors that result from the canal time constant (stimuli in space coordinates). The prediction for rotations (right column) is that, obviously, the $\hat{\omega}$ response to horizontal rotation carries λ (high-pass characteristics of the canal signal), while $\hat{\mathbf{g}}$ and $\hat{\mathbf{a}}$ do not show this error. Translations (left column) in the *z*-direction are not affected, but translations in the *x*- and *y*-directions are affected in that $\hat{\mathbf{a}}$ carries λ, while $\hat{\mathbf{g}}$ is affected by $\mathbf{a}*(1-\lambda)$ ($1-\lambda$, low-pass behavior) and $\hat{\omega}$ by the first derivative of this error. In other words, the prediction is that the perception evoked during translation shows high-pass characteristics that are related to the canal time constant and, in addition, includes a tilt illusion with low-pass characteristics (compare above, Hill Top illusion).

The basic aspects of the signal processing described by Eqs. 1 and 2 have been used before in other models of otolith–canal interaction.[10] However, our approach of building the sensor fusion around the transfer function of the canals (Eq. 3) is novel, and, as a comparison of our simulations with psychophysical data shows, is promising.

REFERENCES

1. ROBERTS, T. D. M. 1979. Otoliths and uprightness. Progr. Brain Res. **50**: 493–499.
2. FERNANDEZ, C. & J. M. GOLDBERG. 1971. Physiology of peripheral neurons innervating semi-circular canals of the squirrel monkey. II. Response to sinusoidal stimulation and dynamics of peripheral vestibular system. J. Neurophysiol. **34**: 661–675.

3. MERFELD, D. M. 1995. Modeling the vestibulo-ocular reflex of the squirrel monkey during eccentric rotation and roll tilt. Exp. Brain Res. **106**: 123–134.
4. COHEN, M. M., R. I. CROSBIE & L. H. BLACKBURN. 1973. Disorienting effects of aircraft catapult launchings. Aerospace Med. **44**: 37–39.
5. GLASAUER, S. 1995 Linear acceleration perception: frequency dependence of the hill-top illusion. Acta Otolaryngol. Suppl **520**: 37–40.
6. MERGNER, T., C. SIEBOLD, G. SCHWEIGART & W. BECKER. 1991. Human perception of horizontal head and trunk rotation in space during vestibular and neck stimulation. Exp. Brain Res. **85**: 389–404.
7. FERNANDEZ, C. & J. M. GOLDBERG. 1976. Physiology of peripheral neurons innervating otolith organs of the squirrel monkey. III. Response dynamics. J. Neurophysiol. **39**: 996–1008.
8. BENSON, A. J. 1974. Modification of the the response to angular acceleration by linear acceleration. *In* Handbook of Sensory Physiology, Vol. VI/2, H. H. Kornhuber, Ed.: 281–320. Springer-Verlag. Berlin.
9. GUEDRY, F. E. 1974. Psychophysics of the vestibular sensation. *In* Handbook of Sensory Physiology, Vol. VI/2, H. H. Kornhuber, Ed.: 3–154. Springer-Verlag. Berlin.
10. GLASAUER, S. & D. MERFELD. 1997. Modelling three dimensional vestibular responses during complex motion stimulation. *In* Three-Dimensional Kinematic Principles of Eye-, Head-, and Limb Movements in Health and Disease, M. Fetter, D. Tweed, and H. Misslisch, Eds.: 337–345. Harwood. Amsterdam.

Centrifugal Force Affects Perception but not Nystagmus in Passive Rotation

MARIE-LUISE MITTELSTAEDT[a] AND WILLI JENSEN

Max-Planck-Institut fuer Verhaltensphysiologie, D-82319 Seewiesen, Germany

INTRODUCTION

During eccentric rotation in the dark the vestibular-ocular reflex (VOR) is influenced by a combination of linear and angular acceleration. Perceived changes in attitude and perceived angular velocity during and after acceleration and deceleration phases do not reflect the dynamics of the VOR.[1-3] In the following study, in which we look for the causes of this difference, the perceived angular velocity and the VOR during constant rotation are measured *simultaneously* under varied combinations of linear and angular inputs.

METHODS

Subjects were payed or employed volunteers of both sexes aged 24 to 54 who had given informed consent. They were positioned (blindfolded and earphoned with pink noise) face forward on a rotating platform at a radial distance of 0.15 to 1.6 m. Subjects were standing and gripped a radial bar with both hands. They were then accelerated within 1 second to a constant velocity of $\omega = 0.52$ or 0.87 rad/s. The centripetal acceleration ranged from 0.02 to 0.122 g. The latter causes a 7.08-deg (roll) tilt of the resulting vector. Maximal tangential acceleration at $r = 1.6$ m was 2.79 m/s^2. After about 160 s the subjects were decelerated within 1 or 2 s to a full stop. They estimated their angular velocity after acceleration [rotation phase (RP)] and after deceleration to the full stop [postrotation phase (PRP)]. Subjects indicated whenever they either felt rotated through 180 degrees, or when they felt no further rotation. It has been shown that subjective angular velocity can be deduced from perceived angular displacement.[4] During the whole experiment the horizontal and vertical VOR was measured. (2D VOG-video-Oculography, SensoMotoric Instruments GmbH, Teltow, Germany). Only the horizontal slow phase velocity was evaluated.

RESULTS

Subjective velocity and slow phase velocity of the VOR declined exponentially with time (FIG 1). The amount of time constants showed large interindividual differences. But in all cases the time constant and the duration of perceived rotation in RP depended

---

[a]To whom correspondence may be addressed. Phone: 49 (0) 8157 932-354; fax: 49 (0) 8157 932-209; e-mail: m.l.mittelstaedt@mpi-seewiesen.mpg.de

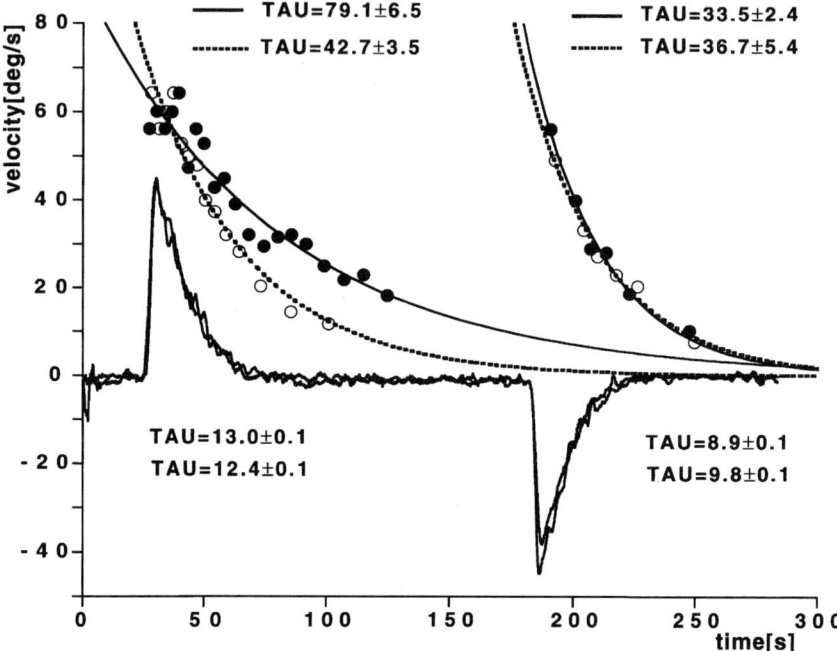

FIGURE 1. Angular velocity of perception and horizontal VOR during RP and PRP for one subject at two radial positions. The subject was accelerated within 1 s to a constant velocity of $\omega = 0.87$ rad/s (50 deg/s) and after about 160 s decelerated to a full stop. *Dots*: subjective velocity (means over every successive indicated 180 deg) at radius $r = 1.6$ m (*filled dots*) and $r = 0.15$ m (*open dots*). Points are fitted with $m_1 * \exp(-t/TAU)$. *Lower curves*: smoothed velocity of the horizontal VOR (50-Hz measurement). Points are fitted with $(k_1 * \exp(-(t-t_0)/TAU_{long})) - (k_2 * \exp(-(t-t_0)/TAU_{short})) + k_3$. The fitting curves are only drawn for data on perception.

linearly on the amount of the centrifugal force (FIG 2): The centrifugal force has an enhancing effect on perceived angular velocity and the duration of perceived rotation. The time constants during PRP were independent of the radius and disk velocity of the preceding rotation, as would be expected if the added orthogonal force caused the enlarged time constants in RP. The subjective velocity of the first 180 deg was overestimated by most of the 15 subjects.

By contrast, the time constants of the VOR did not depend on the amount of centrifugal force during RP (FIG 2). On average these constants were slightly higher than during PRP. The maximal slow phase velocity of the VOR was only dependent on disk velocity, not on radius, and showed virtually no difference between RP and PRP. It was always lower than disk velocity.

Time constants and maximal-velocity perception estimates did not correlate with the respective parameters of the VOR for all the subjects. Some of our subjects showed asymmetries of rotatory sign. But in line with the preceding results, the idiosyncratic asymmetries of perception did not correlate with the asymmetries of the VOR.

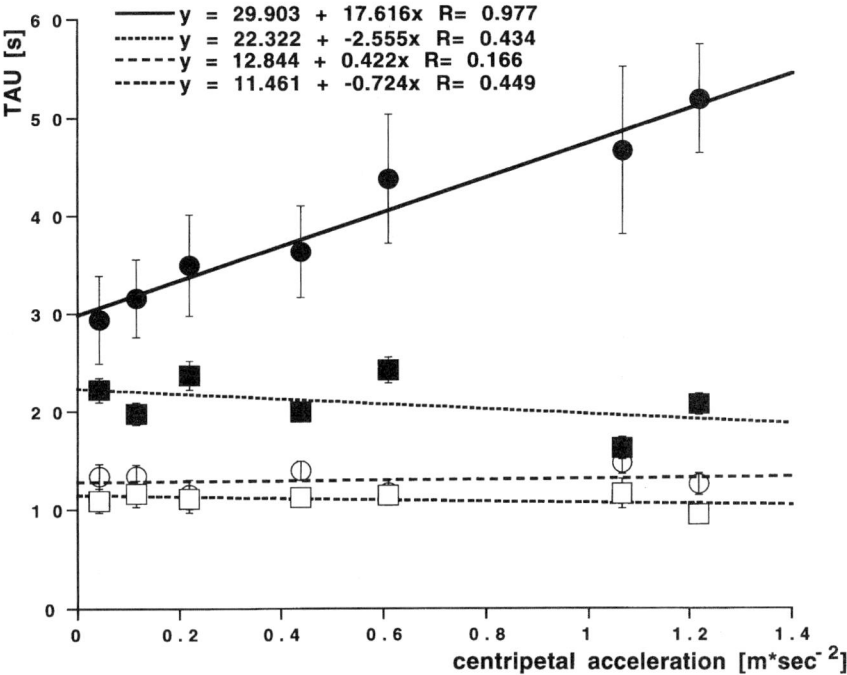

FIGURE 2. Time constants *TAU* as a function of centripetal acceleration. Means of 7 subjects. *Error bars*: standard error; *filled symbols*: *TAU* of perception; *open symbols*: *TAU* of horizontal VOR; *circles:* during RP; *squares*: during PRP. Rotation velocities of the disk: ω = 0.52 and 0.87 rad/s (30 and 50 deg/s); radii: r = 0.15, 0.8, 1.4, and 1.6m.

DISCUSSION

The results are another example of the differences of the dynamics between VOR and perception. Extant centrifuge experiments revealed large differences in the amount and time course of tilt sensation between RP and PRP.[1-3,5-7] In simultanous situations, however, the VOR did not. On the other hand, as shown in references 1 and 8, the VOR was dependent on the direction of the tangential and centripetal acceleration during RP, while the perception of rotation was not.[9]

During VOR the eye tries to track earth-fixed, *exocentric*, targets based on the output of the semicircular canals and otoliths as well as on the seen or imagined distance of the target. But if the subjects are asked to estimate their movement in the horizontal plane, the same inertial information is part of the *egocentric* path integration system. Therefore the central integrative processes are expected to be different for VOR and perception. The extant results show that mutual conclusions from eye movements to sensations cannot be drawn.

ACKNOWLEDGMENTS

We thank R. Ströbele and K. Fisher for the construction of the rotating disk. Many thanks are also due to the subjects for their cooperation.

REFERENCES

1. GUEDRY, F. E., A. H. RUPERT, B. J. MCGRATH & C. M. OMAN. 1992. The dynamics of spatial orientation during complex and changing linear and angular acceleration. J. Vestib. Res. **2**: 259–283.
2. MCGRATH B. J., F. E. GUEDRY, C. M. OMAN & A. H. RUPERT. 1995. Vestibulo-ocular response of human subjects seated in a pivoting support system during 3 Gz centrifuge stimulation. J. Vestib. Res. **5**: 331–347.
3. GUEDRY, F. 1996. Spatial orientation perception and reflexive eye movements—A perspective, an overview, and some clinical implications. Brain Res. Bull. **40**: 505–512.
4. MERGNER, T., A. RUMBERGER & W. BECKER. 1996. Is perceived angular displacement the time integral of perceived angular velocity? Brain Res. Bull. **40**: 467–471.
5. HOLLY, J. E. 1997. Three-dimensional baselines for perceived self-motion during acceleration and deceleration in a centrifuge Vestib. Res. **7**: 45–61.
6. GLASAUER, S. 1992. Human spatial orientation during centrifuge experiments: nonlinear interaction of semicircular canals and otoliths. *In*: Proc. 17th Barany Society Meeting, H. Krejcova and J. Jerabek, Eds.: 102–105. Slovakian Academy of Sciences. Bratislava, Slovakia.
7. CURTHOYS, I. S. 1996. The delay of the oculogravic illusion. Brain Res. Bull. **40**: 407–412.
8. MERFELD, D. M. 1996. Vestibulo-ocular reflex of the squirrel monkey during eccentric rotation with centripetal acceleration along the naso-occipital axis. Brain Res. Bull. **40**: 303–309.
9. MITTELSTAEDT, M. & H. MITTELSTAEDT. 1995. The influence of otoliths and somatic graviceptors on angular velocity estimation. J. Vestib. Res. **6**: 355–366.

Oculomotor, Postural, and Perceptual Asymmetries Associated with a Common Cause

Craniofacial Asymmetries and Asymmetries in Vestibular Organ Anatomy

D. ROUSIE,[a] J. C. HACHE,[a] P. PELLERIN,[a] J. P. DEROUBAIX,[a] P. VAN TICHELEN,[a] AND A. BERTHOZ[b,c]

[a]*Centre Hospitalo-Universitaire, Lille, France*

[b]*Laboratoire de Physiologie de la Perception et de l'Action CNRS, Collège de France, Paris, France.*

INTRODUCTION

Cerebral asymmetries in form and volume, associated with cranial asymmetries, are a common feature of the human race and are often associated with facial asymmetries.[1-3] This asymmetry is, in many cases, related to asymmetric cerebral growth, which is mosty accomplished *in utero* when the future bony envelopes of the cephalic pole are still modifiable,[4,5,6] although they may also have a local origin, for instance, in the case of mandibular asymmetry. The main parts of these envelopes are the vault, the face, and the base of the skull. The base itself can be divided into anterior base with a portion of the orbits containing the eye globes; the median base; a true hinge; and the posterior base containing the vestibular organs.

In a previons study we described the existence of vestibular organ asymmetries associated with these craniofacial asymmetries (CFA).[7] In the present work we report some functional tests that reveal functional deficits associated with these asymmetries. In addition we report the partial compensation for these deficits by using an appropriate treatment with prisms. This study was prompted by observations of systematic complaints of patients suffering from these craniofacial asymmetries and from associated postural anomalies, deviations of the eyes, deficits in the perception of the vertical and horizontal, plus occasional reports of spatial disorientation during navigation in outdoor environments.

---

[c]Address for communication: Pr. A. Berthoz, Collège de France, 11 Place Marcelin Berthelot, 75005 Paris, France. Phone: 331-4427-1299; fax: 331 4427 1425; e-mail: aber@ccr.jussieu.fr

METHODS

We used different populations of patients selected at random (in alphabetical order), from a larger population. Several systematic investigations were conducted with these patients with their informed consent for their participation, including a routine medical examination in the clinic. Each subject was free of neurological or psychiatric disease. Magnetic resonance imaging (MRI) investigations were performed in order to localize the vestibular and cranial asymmetries that were often reflected by asymmetries in the faces of the patients (FIG. 1). Two groups of patients were studied. The first population consisted of 90 patients (Groupe G90) with visible facial asymmetries, and was composed of 34 men and 56 women ranging in age from 13 to 72 years. They were all subjected to MRI of the head using the protocol described by Baudrillard and Rousie.[8] A second population of 250 patients (Groupe G250), was composed of 195 women and 55 men ranging in age from 13 to 72 years. They all had identifiable CFAs, and were subjected to a systematic indentification of clinical signs. In this preliminary communication we shall only report part of the results of the tests that were performed :

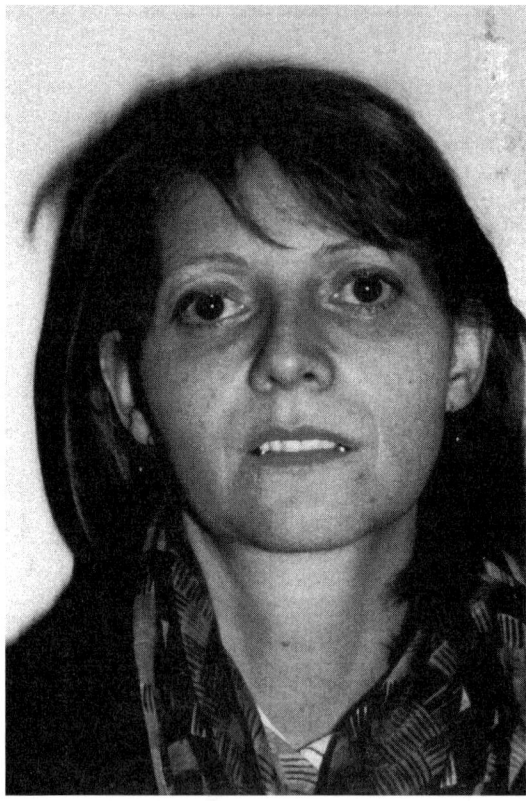

FIGURE 1. Craniofacial asymmetry. Note the position of ears, eyes, and orbits in this important asymmetry. This woman presents a rightward head tilt in agreement with the left torsional eye.

1. The central and peripheral visual acuity of each subject was measured. Also, classic tests were used to explore oculomotor function and eye deviation. The positions of the fovea were measured with a scanning laser ophthalmoscope SLO,[9] which was modified in order to measure very precisely the angular position of the head during the test, to obtain a precisely calibrated measure of ocular tilt or skew deviation. A clinometer was fixed by clips to the head of the partients (acuracy 1°). The foveal positions were measured as follows: the SLO prints a copy of the fundus on which the foveal area is seen as a little cross; we draw a line *OA* from the center of the retina to the middle of the cross, and a second horizontal line *BO* from the middle of the cross to the center of the retina (in fact, it is the extension of the horizontal line of the printed cross), as shown on FIGURE 2. The value of the torsion is given by the angle *AOB*. This construction is drawn for each eye and allows us to compare the two angular values. By convention, we chose the negative values for the angles situated under the horizontal line that crosses the center of the retina.

2. Deviations of the rachis and postural deficits were assessed by several techniques: (a) X rays of the rachis; (b) the Fukuda test, which consists of asking the subjects to walk in place with their eyes closed for several minutes, and then measuring the deviation in the heading of the body; (c) measurement of the deviations of the head, trunk, and limb posture using a video-computerized technique (VICON) during free upright static posture (we measured the motion of a set of 3 head markers, on the front, the top, and the back of the subject's head); (d) otolith deficits in perception of the nonvisual subjective horizontal were also assessed by a classic test: subjects were asked to stand upright, eyes closed, and to point with their index fingers to the horizontal plane in front of them. It has been suggested that otolith asymmetries will be revealed in this test by a vertical asymmetry in the horizontal position of the two fingers. We measured the total vertical deviation of the finger positions with respect to the horizontal.

3. Vestibulo-ocular reflexes were tested by the video–nystagmo–oculography (VEONIS) system. We used classic vestibular tests: (a) recording spontaneous nystagmus during fixation and in darkness with the eyes open; (b) sometimes a head-shaking test (for 10 s); (c) horizontal rotation in darkness according to the protocol of Vitte and Sémont,[10] which consists of the following manoeuvers: rotation of the chair by 180 degrees at a speed of 30°/s, followed by measuring the postrotary nystagmus; this test is performed first with the head upright, then in the roll plane; (d) cervico-ocular reflex test: the head is immobilized, then the trunk is rotated 20 deg rightward and leftward.

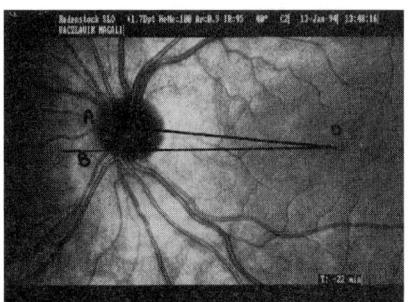 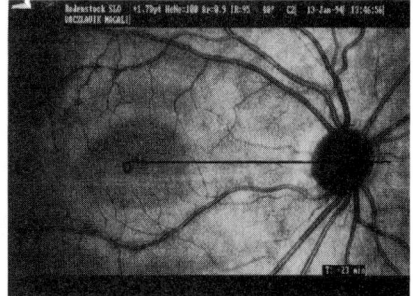

AOB = 5° left eye   AOB = 0° right eye

FIGURE 2. SLO print. Note that the cross establishes the center of the fovea (this cross is fixed by the patient's eye while the photo is taken). (**Left**) Left eye: AOB = 5 deg. (**Right**) Right eye AOB = 0 deg. The center of the fovea is on the line of the center of the retina.

RESULTS

In the 90 patients with visible facial asymmetries, explored by an MRI (G90), we found several types of asymmetries: 24 had a facial asymmetry, 3 an asymmetry of the vault, and 63 a CFA. The most common CFAs are characterized by an asymmetry of the whole cephalic pole, including orbits and labyrinths (FIG. 3A). In all these CFA subjects, the MRI scan has revealed, an asymmetry of the anterior base of the skull, with asymmetries of the orbits associated with an asymmetry of two posterior hemibases, which contains the vestibular organs. The semicircular canals and otoliths had *positional and/or orientational* asymmetries. To measure their spatial position, we located the right and the left horizontal canals on the MRI slices successively in the coronal, saggital, and axial planes: for 80% of the subjects, the right labyrinth was found to be lower than the left one with respect to the neural sagittal median plane,[5] without any sign of orientation anomaly. For the last 20% of the cases, we found asymmetries in the orientation of the plane of the two horizontal canals.

In the second groupe of 250 patients with craniofacial asymmetries (G250), we performed an oral review of their clinical symptoms, which led to the following observations: 97% of the patients experienced lateralized cervical pain, most of the time on the right side; 90% experienced occipito suborbitary headache; 30% suffered from right lateralized migraines; 78% experienced ocular fatigue, difficulty in fixation and diplopy with fatigue; 68% had pseudovertigo and deviation during navigation (walking during stress or in a dark environment); and 76% had orofacial pain (temporomandibular joint lesions). Remarkably, 100% of the patients exhibited lateral head tilt in the roll plane, most of the time to the right.

    1. Oculomotor tests: Several oculomotor asymmetries were found: (a) deficiency in convergence; (b) SLO tests performed on a group of 30 subjects selected from the 250 patients mentioned in the Methods section showed 22 exocyclotorsions on the left eye and 8 exocyclotorsions on the right eye. *The results suggest that, in humans with common CFA, the positon of the fovea is lower on the left eye most of the time.* The difference in torsion between the two eyes fluctuates from 2 to 16 degrees, the mean value being 6 degrees. We found a positive correlation between the degree of exocyclotorsion and the importance of the CFA. But the anatomical CFA measurement is still quantitative, so we now need to quantify the structural asymmetry of the different parts of the head using 3D-reconstruction software.

    2. Visual Acuity: In 75% of cases the visual acuity test revealed astigmatism in the same eye as the one exhibiting exocyclotorsion. In 15% of the cases, the astigmatism was in the eye opposite the torsional eye, and in 10% of the cases, we found no astigmatism.

    3. Rachis and Postural Test: In a population of 120 patients (taken from Groupe G250), ranging in age from 13 to 55 years, including 82 women and 38 men, rachis and postural tests revealed that:

    (a) In 95% of cases the X-ray examination showed the presence of a *lumbar scoliotic postural bias* or of a true lumbar scoliosis associated with the head tilt in roll: the head was tilted to the opposite side to the lumbar convexity. In 75% of cases the lumbar convexity was found to the left and the head tilt to the right, in accordance with a right-side muscular hypertonia.

    (b) The Fukuda test showed a deviation in the direction of the body during blind walking in place that was opposite to the head tilt (the measurements of body deviations were found in the range from 30 to 120 degrees, with a mean value of 50 degrees) (a value smaller than 30 degrees is considered to be normal). We found a majority of leftward deviations (75%), which could be due to a decreased left-side vestibulospinal tonus or a deviation to the left of the subjective straight ahead.

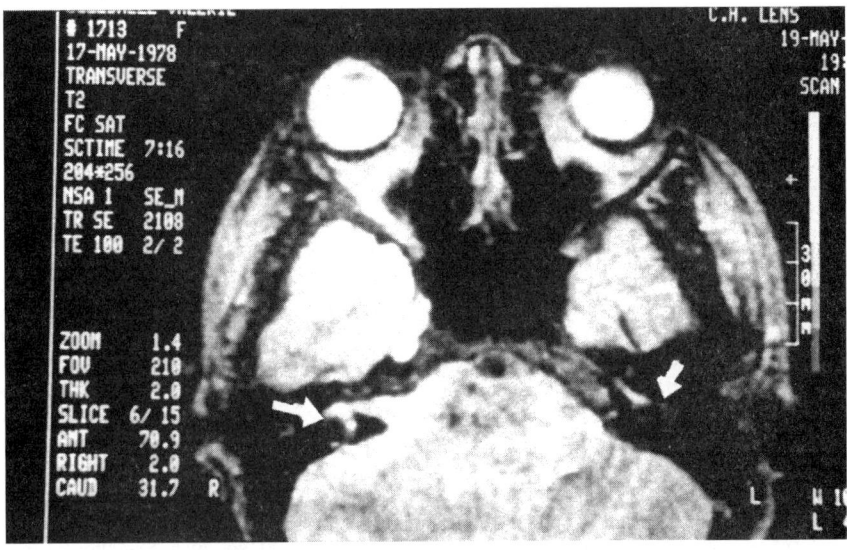

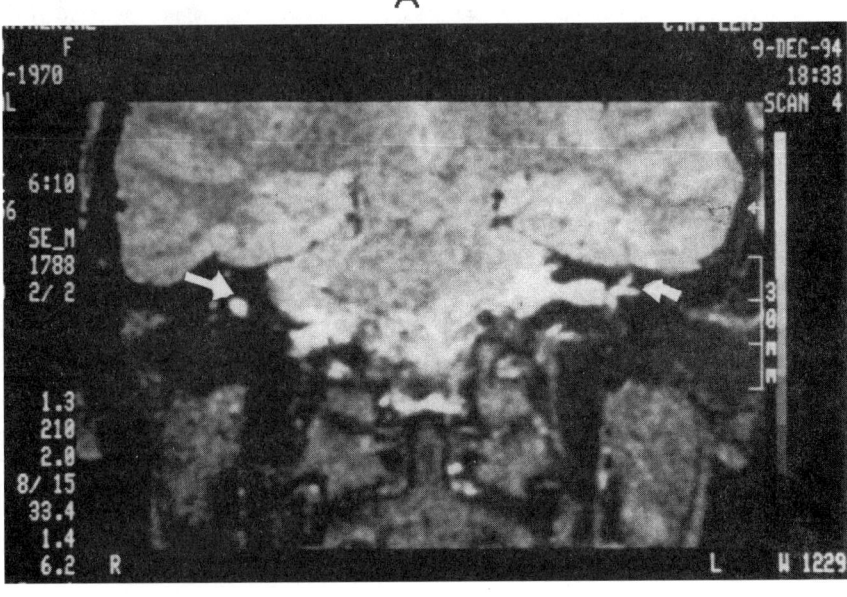

FIGURE 3. MRI images of CFA obtained by the intracranial reference plane. (**A**) Transverse (horizontal) slice. Note the asymmetry on the anterior part of the basicranium with asymmetry of the orbits and the divergence of eye's axis. Note the asymmetry of the posterior part of the basicranium with positional asymmetry of the labyrinths. The right horizontal canal and the right aqueduct appear clearly. Nothing is visible on the left side. (**B**) Coronal (frontal) slice. Note the asymmetry between the two temporal inferior lobes. The left horizontal canal and the left vestibular aqueduct appear clearly, while on the right side, a part of the cochlea is visible.

(c) The test of the nonvisual subjective horizontal with the index fingers pointing task was measured as following: 30 patients out of this subgroup of 120 of G250 were photographed by a simple camera placed 3 meters in front of them. On the fixed picture we measured the slope of the line given by the two index fingers. In 30 patients, on whom these measures were taken, we observed 20 rightward inclinations of the nonvisual subjective horizontal for 10 leftward inclinations. For the rightward inclinations, the range was 5 to 20 degrees, and the mean value was 10.75 degrees. For the leftward inclinations, the range was from 9 to 17 degrees, and the mean value was 13.6. This preponderance of rightward tilt of the nonvisual subjective body horizontal agrees with the greater number of right head tilts previously found.

4. Vestibulo Ocular Tests: The tests were applied on 19 patients (from Groupe G250), ranging in age from 22 to 72 years, including 1 man and 18 women. With eyes open and fixed, we never observed any nystagmus. With eyes open (in darkness), but not fixed, there was a small dominance of leftward nystagmus. Note that a vestibular spontaneous nystagmus reflects an imbalance between the two labyrinths, and generally occurs on the opposite side of the labyrinthine deficit, but is rarely on the same side (irritative lesion). A spontaneous nystagmus can highlight a peripheral or central vestibular dysfunction. We found 3 rightward bilateral and 5 leftward nystagmus in the 19 subjects. We can conclude from this first test that the majority of 19 CFA subjects (11/19) have a central and/or peripheral pathology, or at least a functional unbalance between the two sides. These deficits agree with their clinical symptoms. Vestibulo-ocular tests during horizontal rotation for the same 19 subjects showed 11 rightward peripheral deficits, 6 leftward peripheral deficits, and 2 symmetrical peripheral results. In addition, we found 3 central deficits. In 4 patients the deficits were thought to have cervical origin, which may have been related to the importance of the associated head tilt.

5. Compensation of the Deficits by Prisms: To try to decrease the symptoms of the patients, who gave their informed consent, as did the University of Lille, we attempted to correct the motor and perceptual biases by prisms placed equally on the two eyes. Two prisms of 2 dioptries (" Press on " from 3M) were placed on standard glasses according to a precise method: for a left-eye torsion, the edges of each prism were placed vertically toward the right. The ray of light is therefore deviated to the base and to the left on the retina (the eye is therefore induced to make a recentering saccade to the opposite side, in this case, to the right).

This prismation was performed on a population of 100 volunteer subjects, ranging in age from 13 to 58 years, including 40 males and 60 females: they had to wear the prims 4 to 6 hours per day, and they were recontrolled 2 months later: after prismation, we observed the following results:

(a) Subjective clinical pains: 78 subjects exhibited real improvement (we used the international pain scale of 1 to 10 grades), 15 subjects did not improve, and 7 exhibited an increase in some of their symptoms, chiefly of their cervical pair: however, these last 7 subjects had important lesions of the cervical rachis.

(b) Ocular torsions: for 57 subjects, a small decrease in the torsion (1 to 4 degrees) was observed in the eye that exhibited exocyclotorsion before prismation, and a weak exocyclotorsion on the opposite eye (1 to 4 degrees) was observed, which induced a symmetrization of the two foveal positions. We did not find any change in the other 43 subjects.

(c) Posture: We used a video-computerized technique (VICON) to control the action of the prisms on the head tilt in roll. Prismation caused a correction of the head tilt to a more upright position.

(d) Subjective symptoms: the 100 patients were asked to report any change in their subjective symptoms: 74 patients found a real improvement of their cervical pain and pseudovertigo; 20 patients found a weak improvement in their cervical pains, but a real improvement in their pseudovertigo; and 6 patients were unchanged.

DISCUSSION

The MRI measurements of the basicranium asymmetries are in agreement with left hemispheric preponderance. The asymmetrical growth of the brain, which develops before the bone of the cranial structures stabilizes, could be responsible for the spatial asymmetry of the orbits and of the two posterior hemibases containing the labyrinths. This could explain the anatomically higher left labyrinth and the frequent asymmetrical orientation berween left and right vestibular organs. This study has shown a correlation between these anatomical asymmetries and the functional ones exhibited at differents levels. The preponderance of the right head tilt and a hypoactivy of the right labyrinth in CFA patients agrees with the hypothesis of Diamond and Markham,[11,12] who suggested the possibility of a congenital labyrinthine asymmetry in humans. The hypoactivity of the right labyrinth, associated with a left lumbar convexity, is consistent with a possible vestibulospinal right hypertonia. The tonic lateralized hypertonia of the spinal muscles (paravertebral) could progressively create the frontal convex curvature on the opposite side of the spine to the hypertonia. This is observed in the idiopathic scoliosis, where the deformation is always on the weak side. This lateralization of the anomalies of the rachis is always noted in the statistical reports concerning idiopathic scoliosis.[13-15]

The localization of exocyclotorsion in the left eye in 73.3% of cases was unexpected. It differs from the torsion found in the ocular tilt reaction described by Dietrich and Brandt,[16] who studied the eye positions on patients with destroyed labyrinths. In our study, the two labyrinths are functional, with each labyrinth acting on both eyes, but they present a functional asymmetry. This lateralization of the torsion could also be due to a bias in otolithic control of asymmetry. This lateralization of the torsion bias is suggested by a tonic deviation of the subjective horizontal indicated by the finger test. In 26% of the cases, however, cyclotorsion was found in the eye ipsilateral to the head tilt: these subjects had a large CFA associated with an important vestibular deficit revealed by the vestibulo-ocular tests.

Further studies will be required to define the precise role of the asymmetry of the orbits, because some of these torsions could be due to the asymmetrical insertions of the extraocular muscles. These observations show that the clinical signs are related to the amplitude of the asymmetries. More work is also needed to define precisely the relation between the various types of craniofacial asymmetries and the observed vestibular, postural, and oculomotor deficits.

REFERENCES

1. GESCHWING, N. & W. LEVITSKY. 1968. Human brain: left-right asymmeties in temporal speech region. Science **161**: 186–187.
2. GALABURDA, A. M. & M. LE MAY. 1978. Right left asymmetries in the brain. Science 852–856.
3. LE MAY, M. 1976. Morphological cerebral asymmetries of modern man, fossil man and nonhuman promate. Ann. N.Y. Acad. Sci. **280**: 349–366.
4. BLEDSCHMIDT, M. 1976. Principles of biodynamic differentiation in human. In Development of the Basicranium, Vol. 4: 54–80. Nat. Inst. Health. Bethesda, Md.
5. ENLOW, D. H. 1976. The prenatal and postnatal growth of the basicranium. In Development of the Basicranium, Vol. 12: 192–204. Nat. Inst. Health, Bethesda, Md.
6. DELAIRE, J. 1965. Malformations faciales et asymétries de la base du crâne. Rev. Stomatol. **66**: 379–396.
7. ROUSIÉ, D. & J. C. BAUDRILLARD. 1997. Apport du plan neurosagittal médian dans l'étude des asymétries craniofaciales. Biom. Hum. Anthropol. 1–2: 55–64.
8. BAUDRILLARD, J. C. & D. ROUSIÉ. 1995. I.R.M. des canaux semi-circulaires dans les asymétries craniofaciales. J. Radiol. (Paris) **76**: 579–585.
9. WEBB, R. H. 1987. Confocal scanning laser ophtalmoscope. Appl. Opt. **26**: 8.

10. VITTE, E. & A. SEMONT. 1995. Assesement of vestibular function by video-nystagmography. J. Vest. Res. **95**: 8–9.
11. DIAMOND, S. G. & C. H. MARKHAM. 1987. Binocular counterrolling in human with labyrinthectomy and in normal controls. Ann. N.Y. Acad. Sci. **374**: 69–79.
12. DIAMOND, S. G & C. H. MARKHAM. 1991. Prediction of space motion sickness susceptibility by disconjugate eye torsion in parabolic flight. Aviat. Space Environ. Med. **62**: 1158–1162.
13. GEISSELE, A. 1990. Magnetic resonnance imaging the brain stem in adolescent idiopathic scoliosis. Spine **16**: 7.
14. GEISSELE, A. 1950. The tonic neck reflex and symmetric behaviour. J. Pediatr. **36**: 165–176.
15. BARRACK, R. 1990. Proprioception in idiopathic scoliosis. Spine **9**: 7.
16. DIETRICH, M. & TH. BRANDT. 1992. Subjective visual vertical and eye-head coordination (roll) with brain stem lesions. *In* The Head-Neck Sensory Motor System, A. Berthoz, W. Graf, and P. P. Vidal, Eds.: 640–643. Oxford Univ. Press. Oxford.
17. MARKHAM, C. H. & S. G. DIAMOND. 1992. Further evidence to support disconjugate eye torsion as a predictor of space motion sickness. Aviat. Space Envir. Med. **63**: 118–21.

Role of Otoliths in Spatial Orientation During Passive Travel in a Curve

I. SIEGLER,[a,c] G. REYMOND,[b] AND P. LEBOUCHER[a]

[a]*LPPA, Collège de France, 11 place Marcelin Berthelot, 75005 Paris, France*

[b]*Renault, Research Department, 1, av. du Golf, 78288 Guyancourt Cedex, France*

INTRODUCTION

Individuals can estimate their self-displacement in space in a fairly accurate manner even in the absence of visual cues, during simple movements such as straight lines[1] and rotations in place.[2] In the case of passive bidimensional displacement in the ground plane, the transformation of linear and angular acceleration stimuli into position and orientation estimates is *a priori* more complex. Theoretically, linear acceleration must be combined with direction angle before being integrated twice to yield a displacement value in an earth-fixed reference frame.

Experiments involving the interaction of both stimuli have shown, for instance, that the centripetal acceleration has an enhancing effect on the estimation of self-rotation in a centrifuge,[3] whereas transient estimation of rotation may be unaffected by concurrent inertial cues when subjects are required to indicate only their perceived orientation during complex two-dimensional (2D) trajectories.[4]

We designed a new experiment, using a mobile robot, in order to identify whether the performance in a spatial-orientation task, which necessarily required the use of both linear and angular vestibular inputs, could be affected significantly by the characteristics of the motion stimuli. We used several cornerlike trajectories and the manual task of pointing toward a close object, as being representative of a common experience (driving or walking around an obstacle).

METHODS

Eleven subjects (Ss) aged between 20 and 38 participated in the experiment after giving their informed consent. Ss were passively carried along a corner trajectory on a remote-controlled mobile robot (Robuter®, Robosoft), on which they were seated with head restrained on trunk. In order to mask auditory spatial cues, Ss were blindfolded and wore headphones delivering wideband noise. The trajectory consisted of 3 phases: a linear path (acceleration phase of length 1.5 m), a corner (two arcs of clothoid traveled at

[c]To whom correspondence may be addressed. Phone: (33) 1. 44. 27. 14. 07; fax: (33) 1. 44. 27. 13. 82; e-mail: siegler@cdf-lppa.in2p3.fr

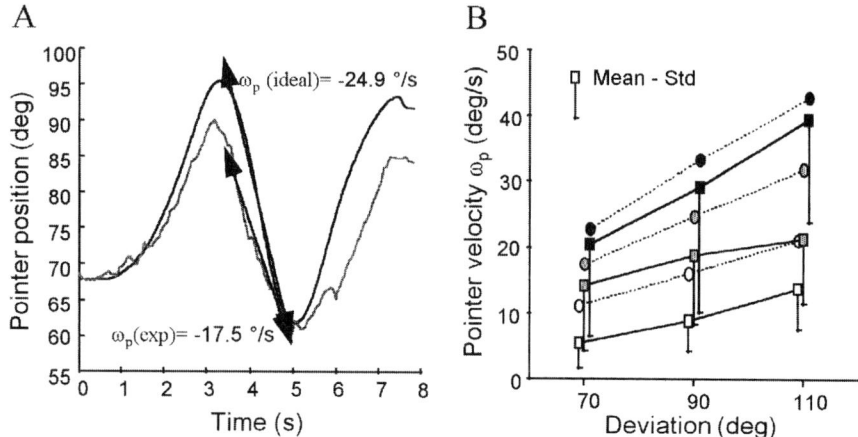

FIGURE 1. (**A**) Example of a subject's pointing response (*gray line*) and the corresponding ideal response (*black line*). The angular pointer velocity ω_p is averaged during the turning phase of the trajectory. (**B**) Ideal pointer velocity ω_{id} (*circles*) and mean "experimental" pointer velocity (*squares*) at different angular deviations and tangential velocities (0.5 m/s: *white*, 0.75 m/s: *gray*, 1 m/s: *black*).

constant tangential velocity), and again a linear path (deceleration phase). The length of the trajectories was constant (7 m), but different deviations to the left (Dev = 70, 90, 110°) and tangential velocities (V = 0.5, 0.75, 1.0 m/s) were applied. In the curved section, the resulting angular velocity (ω_r) and centripetal acceleration ($\gamma_c = \omega \cdot V$) of the robot had triangular profiles, with peak values ranging from 26°/s to 72°/s, and from 0.05g to 0.13g, respectively.

Subjects were asked to continuously indicate by means of a manual angular pointer the position of a previously seen and memorized target, close to the starting position (3.2 m left, 1.3 m ahead). Both the angular position of the pointer, measured by a potentiometer, and the odometry of the robot were recorded at a sampling rate of 100 Hz.

Ss performed each of the nine different runs twice. We concentrated our investigations on the second phase of the trajectory, where both angular and linear accelerations were involved. We computed the mean angular velocity of the pointer ω_p, which reflected the dynamic response to transient motion stimuli (FIG. 1A), and did not carry the cumulative error of the pointer angle itself. We also analyzed the velocity ratio (VR) of ω_p to the mean velocity ω_{id} of an ideal response, as an indicator of the Ss' performance (FIG. 2).

A control experiment was carried out on six of the eleven subjects. They were submitted to nine rotations in place, during which they had to perform the same pointing task. The angular velocity profiles of those rotations were the same as those submitted during the curves in the main experiment.

RESULTS

The mean values of ω_p and ω_{id} are presented in FIGURE 1B as functions of deviation and tangential velocity, respectively. In order to quantify whether ω_p was closer to ω_{id} than to ω_r (i.e., to check whether subjects did not simply compensate for the robot angular velocity), a regression slope was calculated between ω_p and ω_{id} (respectively, ω_p and ω_r) for each subject. All the regressions were significant (p<0.01), with the following mean slopes and

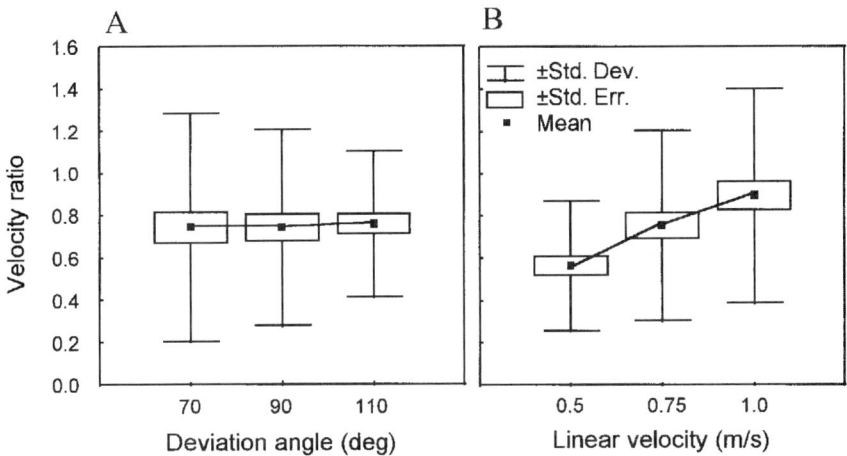

FIGURE 2. Subjects' velocity ratio distributions as a function of deviation angle (**A**) and tangential velocity (**B**).

correlation coefficients: 0.81 ± 0.30 (respectively, 0.64 ± 0.24) and 0.77 ± 0.09 (respectively, 0.74 ± 0.09).

In addition, the respective influence of Dev and V on the subjects' responses, ω_p, and ideal responses, ω_{id}, was compared. Ideal responses, ω_{id}, increased with Dev, V, and with an interaction between the two parameters (FIG. 1B). A similar influence of Dev and V on ω_p was shown by an ANOVA test: there was indeed a significant effect of Dev [$F(2,20) = 40$, $p<0.001$] and V [$F(2,20) = 47$, $p<0.001$] on ω_p. A significant interaction between those two effects was also observed [$F(4,40) = 8.1$, $p<0.001$].

In order to compare the performance of subjects across the different trajectories, we analyzed the velocity ratio VR = ω_p/ω_{id}. When all tangential velocities were pooled, the mean value of VR remained relatively constant, and variability was large, but diminished with increasing deviations (FIG. 2A). With all deviations pooled, the mean, VR, and standard deviation of VR increased significantly [$F(2,20) = 20.8$, $p<0.001$] with tangential velocity (FIG. 2B).

Control Experiment

For each of the six subjects, the linear regression between ω_p and ω_{id} was significant. The mean slope and coefficient correlation were 1.03 ± 0.26 and $r = 0.90 \pm 0.04$. No significant influence of maximum (or mean) angular robot velocity on VR (= $\omega_p / \omega_{id} = \omega_p / \omega_r$) was observed.

DISCUSSION

A comparison of regression slopes (ω_p with respect to ω_{id} and ω_r, respectively) shows us that subjects took their linear displacement relative to the target into account. Indeed, if the subjects simply compensated for the angular velocity of the robot, the regression slope between ω_p and ω_r would be close to unity, which was not the case; ω_p was in fact closer to the ideal response ω_{id}. Moreover, ω_p was shown to bear the same dependence on the motion parameters V and Dev, as the ideal response. Those two results indicate that on average subjects could use the combined otolith and canal stimuli to perform the task.

The analysis of the velocity ratio VR (FIG. 2A) showed that performance did not vary across the trajectories. Therefore, the chosen pointing task was little affected by the geometry of the traveled path under our conditions, although distance and heading relative to the target varied across the different trajectories.

A noticeable result was a significant increase in the mean VR, that is, performance, with tangential velocity. In the second phase of the trajectory, inertial stimuli were only the combined angular and lateral accelerations, and this modification of the subjects' responses could be predominantly attributed to one of those two stimuli. Based on the matching performances of the subjects in estimating self-rotations at various speeds, as shown in our control experiment and by Ivanenko et al.[4] under similar conditions, we propose that the performance of the central integration process responsible for combining the vestibular signals of linear and angular accelerations into an estimation of self-displacement depends to great extent on the nature of the lateral acceleration stimuli.

ACKNOWLEDGMENTS

This work was supported by Human Frontiers (HFSP: RG71/96B); also Mr. Reymond benefits from a CIFRE grant.

REFERENCES

1. BERTHOZ, A., I. ISRAËL, P. GEORGES-FRANÇOIS, R. GRASSO & T. TSUZUKU. 1995. Spatial memory of body linear displacement: what is being stored. Science **269:** 95–98.
2. GUEDRY, F. E. 1974. Psychophysics of vestibular sensation. *In* Handbook of Sensory Physiology, Vol. 6, No. 2, H.H. Kornhuber, Ed.: 3–154. Springer-Verlag. New-York.
3. MITTELSTAEDT, M. L. 1995. Influence of centrifugal force on angular velocity estimation. Acta Otolaryngol. (Stockh.) **115** (Suppl.): 307–309.
4. IVANENKO, Y., R. GRASSO, I. ISRAËL & A. BERTHOZ. The contribution of otoliths and semicircular canals to the perception of two-dimensional passive whole-body motion in humans. J. Physiol. (Lond.) **502:** 223–233.

Comparison of Tilt Estimates Based on Line Settings, Saccadic Pointing, and Verbal Reports

A. D. VAN BEUZEKOM[a] AND J. A. M. VAN GISBERGEN

Department of Medical Physics and Biophysics, University of Nijmegen, Geert Grooteplein 21, NL-6525 EZ Nijmegen, The Netherlands

INTRODUCTION

Numerous experiments relying on the luminous-line method have established that roll-tilted subjects make considerable systematic errors in estimating the subjective visual vertical (SVV). It has been suggested that these errors are not simply a direct consequence of an erroneous internal representation of body orientation, but reflect tilt-dependent properties of the spatial-vision system. There is evidence that subjects perform better in estimating their body orientation than in the SVV task, but this issue has received less attention than the SVV paradigm. In the present work we have investigated whether the general picture obtained with the luminous-line method is also valid when the assessment paradigm relies on a different system. In this vein, we have used the oculomotor system as an alternative pointer. Apart from the question of whether this would yield different results, the experiments are also of interest for the study of oculomotor control as such. Studies demonstrating that the slow phase of optokinetic after-nystagmus in tilted animals may reorient its rotation axis to the spatial vertical[1] have raised the question of how body tilt affects other oculomotor subsystems, the saccadic system in particular.

METHODS AND RESULTS

In our experiments, we measured the subject's ($n = 6$) ability to estimate his or her roll tilt in three different ways. First, the SVV was measured using the classic luminous-line method. The luminous line, adjusted by remote control, subtended a visual angle of 17 deg. Second, subjects were instructed to make eye movements aligned with their subjective oculomotor horizontal (SOH) and subjective oculomotor vertical (SOV). Two-dimensional eye position was measured using the scleral search-coil technique. Third, subjects were asked to estimate their tilt angle, expressed in a clock scale as minutes past/to the hour.

The three paradigms were tested for the complete range of roll angles (ρ from –180 to 180 deg, in steps of 10 deg). In each trial the subject was rotated in the dark from upright

---

[a]Address for telecommunications: Phone: ++31 24 3614237; fax: ++31 24 3541435; e-mail: anton@mbfys.kun.nl

to a random orientation with a constant velocity of 15 deg/s. To allow signals from the semicircular canals to wear off, tasks started 24 s after completion of the rotation. Since the subject had 12 s to set the luminous line, the final adjustment was made 36 s after completion of the chair rotation. Before the oculomotor response was made or the line was set, the subject reported the perceived roll angle verbally and was turned back to the initial upright position at the end of the trial. Room lights were switched on for at least 10 s to give the subject a chance to reorient to vertical again.

The results of the three paradigms are shown in FIGURE 1. At small roll angles, most subjects accurately aligned the visual line with the gravitational vertical, and only one of our six subjects showed a minor E-effect. However, at large tilt angles, all subjects showed an A-effect. In the oculomotor paradigm, subjects again showed strong A-effects. Note that clear differences exist between the horizontal and vertical response. Whereas the SOH shows only an A-effect, the SOV also shows a clear E-effect in some subjects. Note the similarity between the SVV and the SOH. So, subjects made large errors in estimating the direction of gravity when using the luminous line or the oculomotor system as a pointer. However, they were quite good in estimating their body position in space using verbal reports (FIG. 1). Although the scatter in the verbal reports was larger than in the other paradigms, mean errors were much smaller.

In order to describe intersubject differences, we applied a principal-components analysis. Sets of eigenvectors were calculated separately for the SVV, SOV, and SOH responses of all subjects. The corresponding eigenvalues denote the ability of the eigenvector to describe variability within each set of responses, yielding the eigenvector with the largest eigenvalue as the most valuable descriptor. The shape of these first eigenvectors, for each task, closely resembled the average response. To describe the response of any subject completely, one would need the contribution of all eigenvectors. However, as illustrated in FIGURE 2, we found that the SVV response of each subject, $R(\rho)$, could be described already quite well as a combination of a weighted version of this first principal component, $P(\rho)$, and a noise term (ε):

$$R(\rho) = a \cdot P(\rho) + \varepsilon(\rho).$$

Note that $P(\rho)$ is common for all subjects and that $R(\rho)$, a and $\varepsilon(\rho)$ are subject specific.

DISCUSSION

The present study shows that subjects are quite good in estimating their body position in space, yet make large errors in verticality and horizontality judgments. Intersubject variability, which is quite considerable, can be described in terms of a general basis function (the first principal component), representing the pattern of systematic errors, and random noise. The weighting factor of the basis function varies from subject to subject, but its shape is invariant. The idiotropic vector, proposed by Mittelstaedt,[2] can fit the responses from subjects with a large A-effect quite well. For subjects with a weak A-effect, the scheme also predicts an E-effect so that the concept of a common basis function would not be expected. A remarkable result is that the oculomotor horizontal and the visual vertical are so similar. In each of these tasks, the A-effect was the most striking and the most robust phenomenon, while the E-effect was weak or absent. We have confirmed the earlier finding of Pettorossi et al.[3] that the oculomotor vertical and horizontal show different characteristics. The presence of the E-effect in the oculomotor vertical, but not in the oculomotor

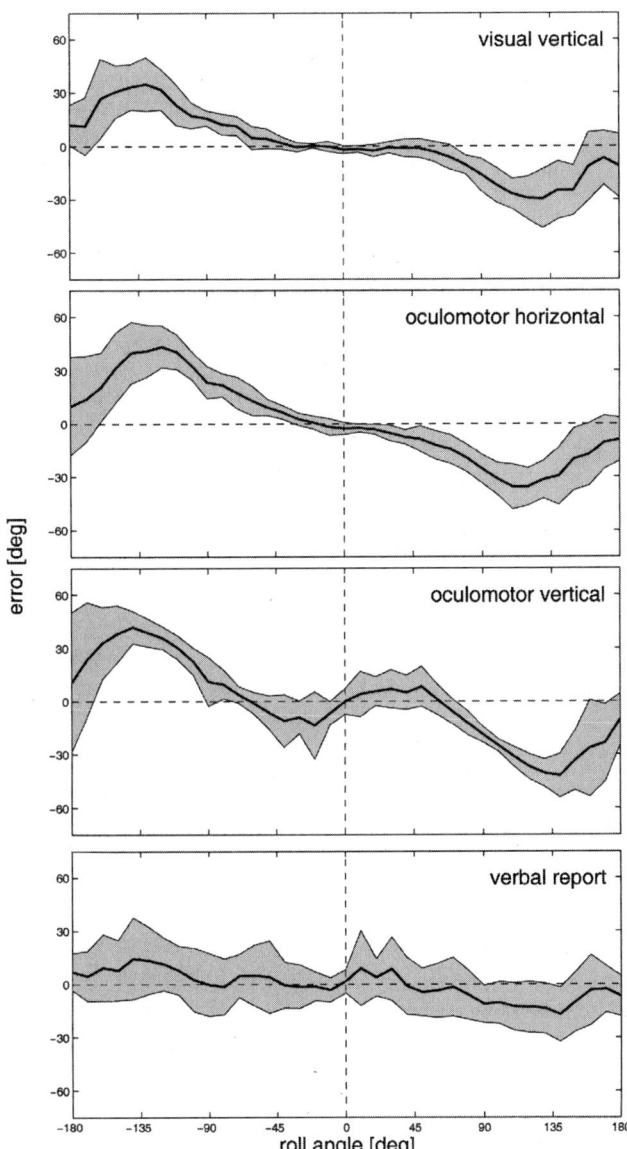

FIGURE 1. Error functions of the four measures. The average responses (—) and the standard deviations (*shaded*) are shown. Note the similarity between the visual vertical and the oculomotor horizontal, which both show an A-effect. The oculomotor vertical, however, was different in showing an E-effect along with an A-effect. The verbal-report data demonstrate only a small A-effect.

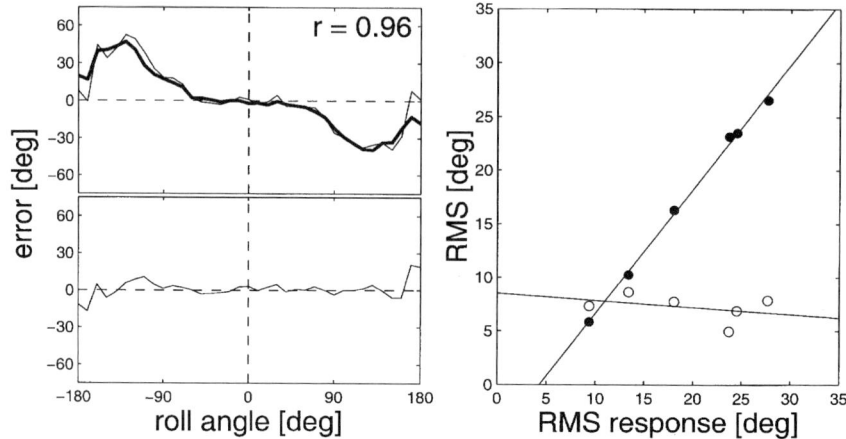

FIGURE 2. Decomposition of the total errors in the SVV task in the first principal component contribution and a random noise term. In the **left-hand panels**, an example of one subject is shown. The *upper panel* depicts the error in line settings (—) and the scaled first principal component (━). The residue (*lower panel*) scatters around zero and therefore represents noise. The **right-hand panel** shows the results of all subjects. The rms values of both the scaled first principal component (•) and the residue (○) are shown as a function of the rms value of the total response error. Note that subjects vary widely in the weight (a) of the first principal component, but that the random noise (ε) is virtually constant.

horizontal, may be a peculiarity of the oculomotor system. The alternative explanation, that external perceptual space is nonorthogonal, would be supported if the visual vertical and the visual horizontal are also different, a possibility that we are presently testing.

REFERENCES

1. DAI, M. J., T. RAPHAN & B. COHEN. 1991. Spatial orientation of the vestibular system: dependence of optokinetic after-nystagmus on gravity. J. Neurophysiol. **66**: 1422–1439.
2. MITTELSTAEDT, H. 1983. A new solution to the problem of the subjective vertical. Naturwissenschaften **70**: 272–281.
3. PETTOROSSI ,V. E., D. DAMBAGIONI, A. M. BRONSTEIN & M. A. GRESTY. 1998. Assessment of the perception of verticality and horizontality with self-paced saccades. Exp. Brain Res. **121**: 46–50.

Vestibular Projections in the Human Cortex

P. P. VIDAL,[a] C. DE WAELE,[a] P. M. BAUDONNIÈRE,[b,d] J. C. LEPECQ,[b] AND P. TRAN BA HUY[c]

[a]*Laboratoire de Neurobiologie des Réseaux Sensorimoteurs, ESA 7060, CNRS-Paris 6-Paris 7, 15 rue de l'Ecole de Médecine, 75270 Paris CEDEX 06, France*

[b]*Neurosciences Cognitives et Imagerie Cérébrale, UPR 640 CNRS, Groupe Hospitalier La Pitié-Salpétrière, 47 Boulevard de l'Hôpital, 75651 Paris Cedex 13 Paris, France*

[c]*Service ORL, Hôpital Lariboisière, 3 rue Ambroise Paré, 75010, Paris, France*

There is convincing evidence from animal studies that several cortical areas such as area 2v at the tip of the intraparietal sulcus, area 3v in the central sulcus, the parieto-insular vestibular cortex (PIVC) adjacent to the posterior insula, and area 7 in the inferior parietal lobule are involved in the processing of vestibular information. In humans, recent PET-scan and f-MRI imaging studies tend to confirm these data. However, because of the poor temporal resolution of these two methods and the mainly canalar vestibular stimulation used in humans, two questions remained unanswered: first, are the vestibulocortical pathways in humans mainly trisynaptic as previously shown in monkeys using electrophysiological recordings? Second, are the vestibulocortical pathways segregated in otolith and canalar cortical neuronal networks, depending on their canalar or otolith origin?

In this study, we try to answer these questions by using the evoked-potentials method and two different vestibular stimuli, a per-operative electrical stimulation of the vestibular nerve in anesthetized patients and high-level clicks (100 dB) in normal awake subjects. Due to its excellent temporal resolution, the evoked-potential method is ideally suited to investigate the short-latency cortical vestibular potentials (SLVP). The evoked potentials were recorded through subcutaneous active electrodes on the scalp, and dipole sources were calculated using the Besa program (version 2.0, 1993).[1]

First, the vestibular nerve was stimulated per-operatively using bipolar electrodes in seven anesthetized patients operated from an acoustic neurinoma resection or for a vestibular neurotomy because of an intractable Ménière's disease. Five distinct cortical zones were activated with a latency to onset of responses of 6 ms (see FIG. 1, left side): the ipsilateral superior and medial frontal gyrus (Brodman area—Ba 10), the controlateral supplementary motor area (Ba 6), the precentral gyrus (Ba 4), and the superior parietal sulcus (Ba 19). The 6-ms latency recorded for each of these areas tends to indicate that several trisynaptic pathways link the primary vestibular afferents to these cortical areas via the vestibular nuclei and thalamic neurones.

---

[d]To whom correspondence may be addressed. Phone: 33 1 44 24 52 92; fax: 33 1 44 24 39 54; e-mail: lenapmb@ext.jussieu.fr

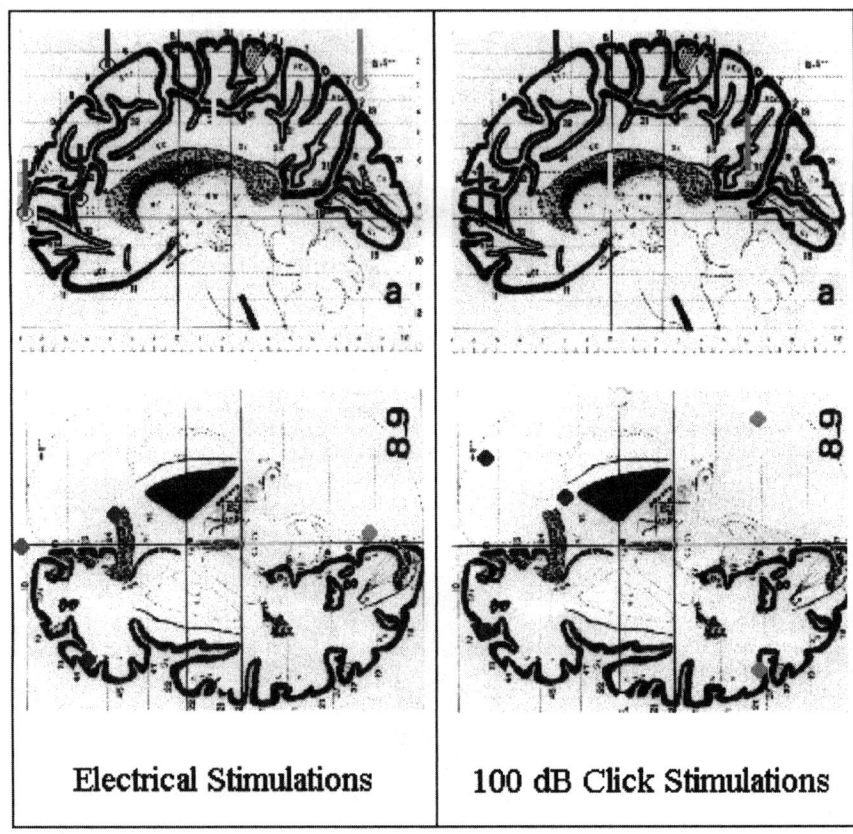

FIGURE 1.

Second, clicks were delivered in three alert subjects either at 100 dB to stimulate both the saccular and the cochlear hair-cells, or at 60 dB to activate the acoustic system only. Results show (see FIG. 1, right side): dipoles 1 and 2 are symmetric and localized on the ipsilateral temporoparietal area (Ba 22); dipoles 3 and 4 are symmetric and localized at the limit of the ipsilateral frontal and prefrontal lobe on the superior frontal gyrus (Ba 10); dipoles 5 and 6 are symmetric and localized on the angular gyrus (Ba 39); and dipole 7 is localized on the contralateral anterior portion of the supplementary motor (Ba 6 around the supplementary eye field).

Comparison of these two sets of data suggested that the saccular afferents projected bilaterally to the superior frontal gyrus (Ba 10) and to the angular gyrus (Ba 39), to the controlateral supplementary motor area (Ba 6), and most probably to the posterior part of the superior frontal gyrus (Ba 22), although this projection could not be differentiated from the parallel activation of the auditory cortex (Ba 22).

Two zones, the fronto marginal gyrus (Ba 10) and the contralateral superior parietal sulcus (Ba 19), may receive an utricular and/or canalar input but not a saccular input. On the other hand, the fact that the prefrontal, the SMA, and the temporoparietal areas were activated by the two types of stimulations indicate that saccular information is processed in these areas. Does canalar and otolithic information converge in some of these areas? Our paradigms cannot respond to that question, because an area input solely by the saccular information could be activated by both types of stimulations. Nevertheless, because none of these areas where selected by the electrical stimulations, canalar and otolithical information have to converge in some of these areas. This should not come as a surprise, because, first, convergence of the otolithic and canalar inputs are common at the second-order vestibular neuron level, as documented in animal studies. Second, selective electrical stimulation of the branches of the VIIIth nerve have led to the conclusion that both the otolithic and the canalar afferences project to areas SII, 2V, and 3a in the cat.[2,3] Third, in humans, we and others authors have recently published in abstract form that the superior temporal gyrus and the SMA were activated following otolothic stimulation.[4,5]

REFERENCES

1. SCHERG, M. & J. S. EBERSOLE. 1993. Models of brain sources. Brain Topogr. **5**: 419–423.
2. ANDERSON, S. & B. E. GERNANDT. 1954. Cortical projection of vestibular nerve in cat. Acta Otolaryngol. Suppl. 10–18.
3. JIJIWA, H., T. KAWAGUCHI, S. WATANABE & H. MIYATA. 1991. Cortical projections of otolith organs in the cat. Acta Otolaryngol. Suppl. **481**: 69–72.
4. BRANDT, T. & M. A. DIETERICH. 1998. Cortical areas associated with vestibular input and perception of verticality. *In* Abstract of Otolith Function in Spatial Orientation and Movement. Zurich
5. DE WAELE, C., P. M. BAUDONNIÈRE, M. BALLESTER, P. BELIN, Y. SAMSON, P. TRAN BA HUY & P. P. VIDAL. 1998. Localization of cortical otolithic vestibular projections in humans: an evoked potentials and PET-scan study. *In* Abstract Book of the Forum Meeting of European Neuroscience. Berlin.

Spatial Properties of Otolith Units Recorded in the Vestibular Nuclei

SERGEI B. YAKUSHIN,[a] THEODORE RAPHAN,[b] AND
BERNARD COHEN[a,c,d]

[a]*Department of Neurology, Mount Sinai School of Medicine, 1 East 100th Street, New York, New York 10029, USA*

[b]*Department of Computer and Information Science, Brooklyn College of the City of New York, Brooklyn, New York, 11210, USA*

[c]*Department of Physiology and Biophysics, Mount Sinai School of Medicine, 1 East 100th Street, New York, New York 10029, USA*

INTRODUCTION

Central otolith-recipient units, which are found largely in the lateral and descending vestibular nuclei (LVN, DVN), have a wide range of polarization vectors, amplitudes, and phase characteristics.[1-8] Some central otolith neurons with canal-convergent inputs have spatial characteristics that are invariant. Other central otolith units respond to stimulation in all planes during sinusoidal oscillation around a spatial horizontal axis.[1,2,6-8] The temporal phase of the response of these units varies as a function of head orientation. These units have been called spatiotemporal convergence (STC) cells.[2] It has been suggested that STC cells have convergent inputs from canal and otolith afferents with different spatial and temporal properties, but the specific canal input to these otolith units has not been demonstrated. Using a technique for identifying specific canal input,[9] we analyzed the spatial characteristics central otolith neurons with and without STC characteristics in the alert monkey and identified the specific semicircular canals that provide input to these neurons.

METHODS

Conclusions about canal and otolith input were inferred from the following tests: animals were sinusoidally oscillated about a spatial vertical axis while statically pitched forward and backward to demonstrate the relation of unit activity to the lateral and/or vertical canals. (In this study, we considered that all convergent inputs are excitatory.) If a unit has afferent input only from the lateral semicircular canals, then it will be maximally modulated when an animal is pitched forward about 30°.[3,9] If the unit has afferent input only from the vertical canals, it will be maximally modulated when the animal is tilted about 60° backward. If the neuron has convergent input from horizontal and vertical

---

[d]To whom correspondence may be addressed. Phone: 212/241-7068; fax: 212/831-1610; e-mail: syakush@smtplink.mssm.edu

canals, then maximal unit modulation will occur somewhere between these two planes, resulting in complex phase relations. Otolith input, related to dorsoventral linear acceleration, is reflected as a bias frequency, when the head is tilted forward and backward during sinusoidal oscillation about a spatial vertical axis.

To determine if a unit has vertical-canal input, the animal is sinusoidally oscillated about a spatial horizontal axis with the head in different orientations to the plane of oscillation. Modulation of unit activity is plotted as a function of head orientation and fit with a sine function to obtain gain and phase of the spatial response. Assuming that an increase of unit activity will occur only with ipsilateral canal activation, then excitatory input can be related to a particular vertical canal if the spatial phase of the response lies close (<22.5°) to the canal plane. During sinusoidal oscillation around a spatial horizontal axis, otolith input can mimic a vertical-canal response, but otolith sensitivity will be detected when the unit is tested with static head tilt at the same orientation.

RESULTS

Sixteen neurons, predominantly sensitive to head orientation relative to gravity, were recorded in the vicinity of the LVN and DVN in three cynomolgus monkeys (*Macaca fascicularis*). Neck proprioceptive input was tested by pressing on the neck muscles. Only units that did not have clear neck proprioceptive input and no relation to velocity storage were analyzed in this study. The resting discharge of these units varied from 11.8 to 60.6 imp/s (average 31.0 ± 15.8 imp*s^{-1}) (TABLE 1). The average coefficient of variation (SD/average frequency) ranged from 0.2 to 0.55 (average 0.36 ± 0.07). Therefore, the units were classified as irregular.

Units were tested with static tilts while the head was differently oriented relative to the spatial horizontal axis. To determine the orientation of the response vector, neuronal activity was plotted as a function of head orientation in tilt and fitted with a sinusoid (TABLE 1).

TABLE 1: Summary Table of Central Otolith Neurons Characteristics

Convergent Inputs	Response-Vector Orientation			Resting Discharge	Sensitivity (imp*s^{-1}g^{-1})		Monkey ID
	Static Stim.	Sine 0.25 Hz	Ovar	Ave+SD	Static Stim.	Sine 0.25 Hz	Unit Id
None	—	97°	≈64°	14.9 ± 1.8	—	47.3	M9357(07)
None	—	282°	296°	60.3 ± 9.1	—	33.3	M9304(68)
None	315°	321°	314°	32.1 ± 2.9	35.6	38.0	M9358(07)
c-PC	126°	127°	113°	80.5 ± 4.1	43.3	89.5	M9357(38)
c-PC	288°	261°	258°	53.8 ± 4.6	22.1	38.1	M9304(69)
i-AC	347°	358°	332°	36.3 ± 1.2	25.0	73.1	M9358(09)
i-PC	—	—	123°	34.6 ±8.8	—	—	M9304(22)
i-AC & *I*-PC	125°	89°	105°	17.7 ± 2.6	14.8	50.1	M9357(46)
c-AC & *c*-PC	—	283°	278°	45.3 ± 2.8	—	51.4	M9304(70)
Body muscles	—	—	51°	66.4 ± 11.4	—	—	M9304(33)
Spatiotemporal Convergence (STC) Cell							
i-AC	67°	80°	50°	39.3 ± 3.4	18.4	31.3	M9357(51)
i-AC & *i*-PC	82°	89°	79°	40.4 ± 6.6	17.3	27.8	M9358(11)
i-AC & *i*-PC, *c*-LC	97°	112°	89°	24.5 ± 1.2	17.0	40.5	M9358(10)
bi-PC	—	181°	138°	77.2 ± 7.7	—	31.0	M9357(17)
bi-PC	—	186°	178°	16.2 ± 2.9	—	34.0	M9357(40)
bi-AC	346°	10°	348°	37.7 ± 1.5	17.8	37.4	M9357(19)

During off-vertical-axis yaw rotation (OVAR), the unit response vectors led head orientation for both directions of rotation. The direction of the response vector was defined as the phase of the maximal response averaged for clockwise and counterclockwise rotation (TABLE 1). The orientation of the response vector based on static or sinusoidal stimulation around a spatial horizontal axis or based on OVAR, was about the same.

Ten otolith-related neurons were activated during sinusoidal rotation about a spatial vertical axis while the head was pitched forward or back from 0° to 90°. In nine units, it was due to convergent vertical canal, and in one to lateral-canal input.

When tested with sinusoidal rotation around a spatially fixed horizontal axis, the average unit sensitivity to pitching was 45.9 ± 17.3 imp*s^{-1}*g^{-1}. Ten units had stable temporal phase leads of about 30° (varying from 0° to 50°) during pitching in various head orientations (FIG. 1A). In contrast, temporal phases varied as a function of head orientation in six units (FIG. 1B). These units had STC characteristics. That is, their activity was in phase with head position in one plane and in phase with head velocity in the orthogonal plane. Four STC units had vertical-canal inputs. Activity of another two units did not modulate during sinusoidal oscillation around a spatial vertical axis, but modulated in phase with head velocity during sinusoidal pitch forward/backward. They probably had equal excitatory inputs from both posterior canals (TABLE 1).

Four STC neurons that were tested during sinusoidal modulation about a spatial vertical axis had a bias change as a function of head tilt. In one, the bias activity was increased with tilt forward and decreased with tilt backward. In another three, biases were minimal with the animal upright, but increased when the animal was tilted forward and backward (FIG. 1D). These units are similar to otolith afferents described as +Z.[10,11] One of the units had an activity increase when it was tilted laterally to the left or right in the roll plane (FIG. 1E). This unit may have had convergent input from two +Z afferents, which lay in two orthogonal planes. Thus, activity of at least four of six STC units in our study had saccular-type activation.

During sinusoidal oscillation around a spatial horizontal axis, the head passed through the upright position twice during each cycle. Therefore, Z units should modulate at double the frequency of oscillation. This was observed only in a few cases (FIG. 1C). In Baker et al.[2] several STC units indeed had modulation at a double frequency. They noted that "a few cells responded to both direction of rotation in one or more planes. . . ." In the other neurons in our study, it is possible that another modality, like vertical-canal input had modified the unit response, so that double peak did not appear.

DISCUSSION

This study shows that central otolith neurons located in LVN and DVN receive a wide range of convergent inputs from the semicircular canals. Our sample of otolith-related neurons, in the alert monkey had similar distribution of response vectors as shown for decerebrate cats.[1-5] Using our identification technique, it was possible to identify the specific canal input to the spatially identified otolith neurons. There were units that received input from a single canal (5/16). In other units (7/16), the input came from two canals, located on one or both sides. Of the six STC cells, four that were adequately tested had saccular input. Thus, we have demonstrated that the STC property of some central otolith neurons could be due to convergent input from the saccular maculae and specific vertical canals. Although preliminary, this study shows that there are sufficient types of otolith/canal-convergent neurons to implement the behavioral responses required for orientation and compensation during movement in three dimensions.

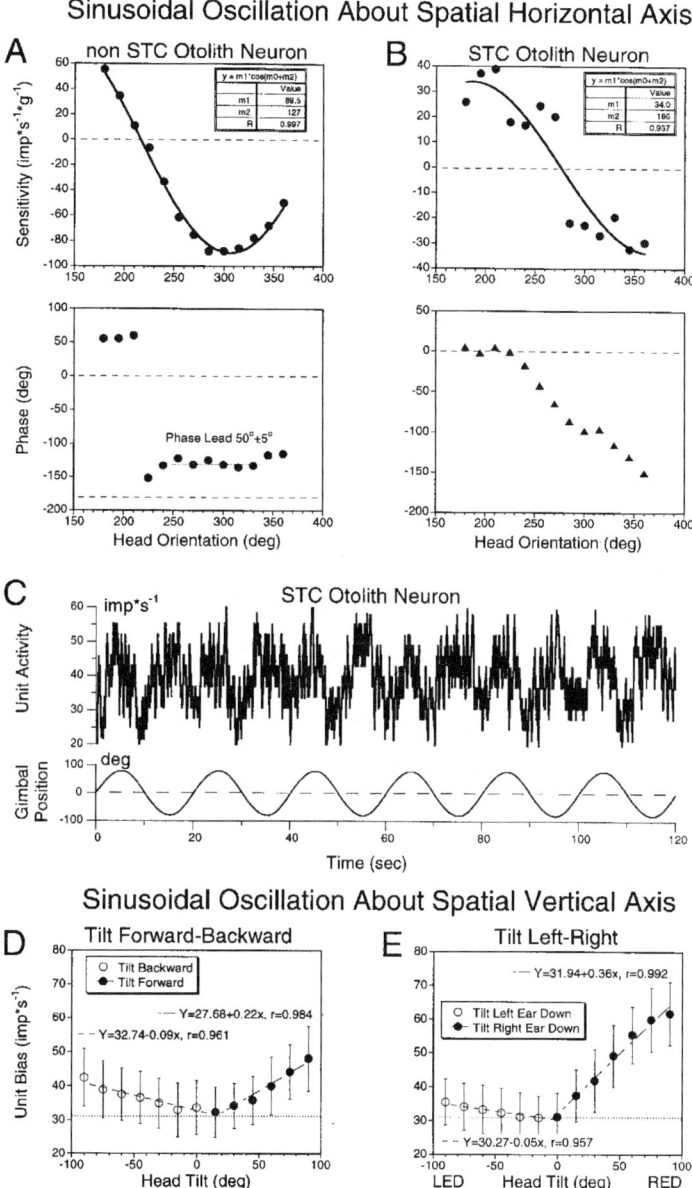

FIGURE 1. Otolith neurons modulated during sinusoidal oscillation around spatially fixed horizontal axis. Gain was defined as positive (A and B, top) if the phase of unit modulation was less then 90° re stimulus position. Head orientation in tilt was defined as: 0°—nose down; 90°—right ear down; 180°—nose up; 270°—left ear down. **(A)** Unit with stable temporal phase (A, bottom). **(B)** Phase of modulation of STC unit varied as a function of head orientation (B, bottom). **(C)** Sinusoidal oscillation at some head orientations had clear modulation at double frequency, indicating saccular afferent convergent input to this neuron. Unit activation by forward–backward **(D)** and left–right **(E)** tilts. Increase in activity for each direction was linear. The slope of the regression lines represents sensitivity for that direction of tilt.

ACKNOWLEDGMENTS

This work was supported by grants from the National Institutes of Health EY02296, EY11812, EY04148, DC03787, DC03284, and EY01867.

REFERENCES

1. SCHOR, R. H., A. D. MILLER & D. L. TOMKO. 1984. Responses to head tilt in cat central vestibular neurons. I. Direction of maximum sensitivity. J. Neurophysiol. **51**(1): 136–146.
2. BAKER, J., J. GOLDBERG, G. HERMANN & B. PETERSON. 1984. Spatial and temporal response properties of secondary neurons that receive convergent input in vestibular nuclei of alert cats. Brain Res. **294**(1): 138–143.
3. BAKER, J., J. GOLDBERG, G. HERMANN & B. PETERSON. 1984. Optimal response planes and canal convergence in secondary neurons in vestibular nuclei of alert cats. Brain Res. **294**: 133–137.
4. SCHOR, R. H. 1974. Responses of cat vestibular neurons to sinusoidal roll tilt. Exp. Brain Res. **20**: 347–362.
5. SCHOR, R. H., A. D. MILLER, J. B. STEPHEN, J. B. TIMERICK & D. L. TOMKO. 1985. Responses to head tilt in cat central vestibular neurons. II. Frequency dependence of neuronal response vectors. J Neurophysiol. **53**(6): 1444–1452.
6. ANGELAKI, D. E. 1992. Two-dimensional coding of linear acceleration and the angular velocity sensitivity of the otolith system. Biol. Cybern. **67**(6): 511–521.
7. ANGELAKI, D. E., G. A. BUSH & A. A. PERACHIO. 1992. A model for the characterization of the spatial properties in vestibular neurons. Biol. Cybern. **66**(3): 231–240.
8. BUSH, G. A., A. A. PERACHIO & D. E. ANGELAKI. 1993. Encoding of head acceleration in vestibular neurons. I. Spatiotemporal response properties to linear acceleration. J. Neurophysiol. **69**(6): 2039–2055.
9. YAKUSHIN, S. B., M. DAI, J.-I. SUZUKI, T. RAPHAN & B. COHEN. 1995. Semicircular canal contributions to the three-dimensional vestibuloocular reflex: a model-based approach. J. Neurophysiol. **74**(6): 2722–2738.
10. FERNANDEZ, C. & J. M. GOLDBERG. 1976. Physiology of peripheral neurons innervating otolith organs of the squirrel monkey. I. Response to static tilts and to long duration centrfugal force. J. Neurophysiol. **39**: 970–984.
11. TOMKO, D. L., R. J. PETERKA & G. D. PAIGE. 1981. Responses to head tilt in cat eight nerve afferents. Exp. Brain Res. **41**: 216–221.

Index of Contributors

Anastasopoulos, D., 379–383
Andersen, R. A., 282–292
Angelaki, D. E., 136–147, 148–161

Baker, R., 1–14
Bakker, B. J., 426–429
Baudonnière, P. M., 384–386, 455–457
Belkhenchir, S., 384–386
Berthoz, A., 313–323, 439–446
Betts, G. A., 27–34, 173–180
Bles, W., 406–409
Böhmer, A., 221–231
Bos, J., 406–409
Bradley, D. C., 282–292
Brandt, T., 293–312
Bremmer, F., 272–281
Bronstein, A. M., 232–247, 324–333, 379–383
Brozek, G., 398–401
Burgess, A. M., 27–34, 173–180
Büttner, U., 81–93, 398–401
Büttner-Ennever, J. A., 51–64

Cartwright, A. D., 27–34
Clarke, A. H., 387–391
Chen-Huang, C., 65–80
Cohen, B., xi, xiii–xiv, 94–122, 181–194, 458–462
Crowell, J. A., 282–292
Curthoys, I. S., 27–34, 173–180, 195–204

Dai, M., 181–194
De Graaf, B., 406–409
Deroubaix, J. P., 439–446
De Waele, C., 392–397, 455–457
Diard, J.-P., 392–397
Dichgans, J., 379–383

Dickman, J. D., 136–147
Dieterich, M., 293–312

Eatock, R. A., 15–26
Engelhorn, A., 387–391

Fetter, M., 379–383, 410–413
Freyss, G., 392–397
Fuhry, L., 398–401
Fukushima, J., 248–259
Fukushima, K., 248–259

Galiana, H. L., 402–405
Georges-François, P., 345–366
Gielen, C. C. A. M., 426–429
Gilland, E., 1–14
Glasauer, S., 81–93, 345–366, 398–401, 417–421, 430–434
Glonti, L., 81–93
Grasso, R., 345–366
Green, A. M., 402–405
Gresty, M. A., 232–247
Groen, E., 406–409
Grüsser, O.-J., 313–323

Haburcakova, C., 398–401
Hache, J. C., 439–446
Halmagyi, G. M., 27–34, 173–180, 195–204
Hamann, Ch., 387–391
Haslwanter, T., 379–383, 410–413
Hegemann, S., 414–416
Henn, V., 181–194
Hess, B. J. M., xi, 148–161
Highstein, S., 35–50
Hoffmann, K.-P., 272–281
Holt, J. R., 15–26

Imagawa, M., 162–172
Israël, I., 345–366, 417–421
Isu, N., 162–172

Jaeger, R., 410–413
Jensen, W., 435–438

Katsuta, M., 162–172
Kleine, J. F., 81–93, 313–323
Kubischik, M., 272–281
Kushiro, K., 162–172

Lappe, M., 272–281
Le Bihan, D., 313–323
Leboucher, P., 447–450
Lempert, T., 232–247
Lepecq, J. C., 384–386, 455–457
Leroy-Willig, A., 313–323
Lobel, E., 313–323
Locke, R., 35–50
Lysakowski, A., 422–425

MacDougall, H. G., 27–34, 173–180
Maruta, J., 181–194
Mast, F., 221–231
McCrea, R. A., 65–80
McHenry, M. Q., 136–147
Medendorp, W. P., 426–429
Mergner, T., 430–434
Mertz, S., 384–386
Miles, F. A., 260–271
Mittelstaedt, H., 334–344
Mittelstaedt, M.-L., 435–438

Nedvidek, J., 398–401
Newlands, S. D., 136–147

Ogawa, Y., 162–172

Paige, G. D., 123–135
Pekel, M., 272–281

Pellerin, P., 439–446

Raphan, T., 94–122, 181–194, 458–462
Reymond, G., 447–450
Rousie, D., 439–446

Sato, H., 162–172
Sato, T., 248–259
Schönfeld, U., 387–391
Seidman, S. H., 123–135
Shelhamer, M. J., 414–416
Sheliga, B. M., 94–122
Shenoy, K. V., 282–292
Siebold, C., 81–93
Siegler, I., 447–450
Silvers, A., 94–122
Snyder, L. H., 282–292
Suwa, H., 1–14
Suzuki, J.-I., 181–194

Tran Ba Huy, P., 392–397, 455–457

Uchino, Y., 162–172

Van Beuzekom, A. D., 451–454
Van de Moortele, P-F., 313–323
Van Gisbergen, J. A. M., 426–429, 451–454
Van Tichelen, P., 439–446
Vautrin, J., 35–50
Vidal, P.-P., 392–397, 455–457
Vollrath, M. A., 15–26

Waespe, W., 181–194
Walker, M. F., 205–220

Yakushin, S. B., 94–122, 458–462
Young, L. R., 367–378

Zakir, M., 162–172
Zee, D. S., 205–220, 414–416